Flavour Science
Recent Developments

Flavour Science

Recent Developments

Edited by

A. J. Taylor
University of Nottingham

D. S. Mottram
University of Reading

THE ROYAL
SOCIETY OF
CHEMISTRY
Information
Services

The proceedings of the Eighth Weurman Flavour Research Symposium, held in Reading, UK on 23–26 July 1996.

Special Publication No. 197

ISBN 0-85404-702-6

A catalogue record for this book is available from the British Library.

Published by The Royal Society of Chemistry,
Thomas Graham House, Science Park, Milton Road,
Cambridge CB4 4WF, UK

Printed by Great Britain by Hartnolls Ltd, Bodmin, UK

Preface

Flavour Science is a multidisciplinary subject encompassing biochemical, chemical and physical aspects of food science, the organic chemistry of natural products as well as the physiology and psychology of sensory perception. Over the past 25 years the science has developed from a systematic study of organic compounds found in the volatiles of food into a science which aims to provide an understanding for all aspects of flavour from its generation in the food to its perception during eating. These advances have been possible through the advancement in analytical methodology and have been driven by the desire of scientists, both in academia and industry, to gain a better understanding of flavour quality in foods and beverages.

A number of important conferences have been organised over the past three decades to review different aspects of flavour science, and among these the Weurman Flavour Research Symposia have become established as one of the premier meetings for flavour scientists. They were the idea of Cornelius Weurman, one of the founders of flavour research in Europe, who sadly died shortly before the first meeting in 1975. Since then the Symposia have been held every three years in different European countries and have been named in his memory.

The 8th Weurman Flavour Research Symposium, held from 23rd–26th July 1996 in Reading, England, was attended by 105 invited participants from 17 countries in Europe America, Asia and Australia. The organisers tried to adhere to Cornelius Weurman's guiding principles for the meeting which are to encourage discussion around the latest research in flavour science through a combination of lectures, posters and workshops. Unlike most scientific conferences, the Weurman Symposia are limited to invited participants all of whom make active contributions.

This book contains, the 86 lectures, posters and workshop presentations which were made at the 8th Symposium. These are mainly research papers and are divided into seven subject areas reflecting the major divisions of flavour science. The section on Flavours of Biological Origin covers flavour formed from both plant and animal sources, while those primarily resulting from Thermal Generation are included in a separate section covering both food products and model systems. Biotechnological Production of Flavour is an area of increasing interest to the food and flavouring industries and topics in this section include plant cell culture and yeast and microbial fermentations. An important development in flavour science in recent years has been the ability to separate enantiomers, and the section on Chirality in Flavour contains five reports on aspects of the enantiomeric composition of natural flavours. Advances in flavour science have always been dependent on developments in analytical methodology and the section on Novel Methods in Flavour provides accounts of recent progress in both the isolation and the analysis of aroma. Sensory Evaluation of Flavour is central to flavour science and many of the contributions in this book reflect this; however, a section devoted to this topic contains those papers which primarily deal with the sensory analysis of flavour. In recent years the most significant development in flavour science has been the growing interest in measuring Flavour Binding and Flavour Release. It is appropriate, therefore, that one of the largest sections is devoted to this topic. The subject includes contributions on the effect of food components on flavour retention and the measurement of flavour release in the mouth. Part of the discussions which took place during the Symposium were centred around three Workshops on selected topics. Reports on these three workshops are also included.

Manuscripts for this book were submitted on diskette which permitted editing to provide a uniform style throughout the book. All the papers were refereed for their scientific content and we thank our referees for their considerable assistance in this task. We are also grateful to all authors whose prompt submission of manuscripts and co-operation in editing of the texts has enabled rapid publication.

As Organisers of the Symposium we are most grateful for generous sponsorship by the following companies: Dragoco, Givaudan-Roure, Firmenich, Haarman and Reimer, Quest International, Reading Scientific Services Ltd., Tastemaker, R.C. Treat, Unilever and Weetabix. This enabled us to offer places at the Symposium to research students at nominal fees, thus fulfilling another Weurman ideal that young scientists at the start of their careers should have the opportunity to present their work alongside more established workers.

We would also like to acknowledge the many people who helped in the organisation of the meeting and the preparation of these proceedings. In particular to the Members of the Scientific Committee (J.M. Ames, B.D. Baigrie, L.J. Farmer, J.G. Gramshaw, J. Knights, J.A. Mlotkiewicz, H.E. Nursten and J.R. Piggott) who, in addition to planning and running the scientific programme, were responsible for most of the refereeing of the manuscripts. We gratefully acknowledge the contribution of the Symposium Secretary, Mrs. Maria Itta, whose hard work ensured the smooth-running of the conference. The mammoth task of formatting and copy-editing of all the manuscripts fell on John Horton and we are most grateful for all his efforts in the preparation of this book. Lastly we would like to record our thanks to our wives, Angela Mottram and Helen Taylor, for their help, patience and encouragement during the organisation of the Symposium and the preparation of this book.

D.S. Mottram
The University of Reading

A.J. Taylor
The University of Nottingham

September 1996

Contents

Section 2 – **Biotechnological Production of Flavour**

Section 3 – **Chirality and Flavour**

Section 4 – **Thermally Generated Flavour**

SECTION 1

FLAVOUR OF BIOLOGICAL ORIGIN

THE ROLE OF DIET AND ENVIRONMENT IN THE NATURAL FLAVOURS OF SEAFOODS

F.B. Whitfield, F. Helidoniotis and M. Drew

CSIRO, Division of Food Science and Technology, Sydney Laboratory, P.O. Box 52, North Ryde, New South Wales 2113, Australia

1 INTRODUCTION

The flavour of seafoods is determined by a variety of pre- and post-harvest effects, including diet, environment, handling, processing, storage and transportation. Over the past 40 years, post-harvest effects on flavour have been extensively studied[1] but at present little information is available on the roles played by either the animal's diet or the natural environment in which it lived. It was demonstrated in the early 1960s that diet can affect the flavour of seafoods, when the ingestion of certain species of invertebrates by salmon[2] or cod[3] was shown to produce recognisable flavour defects in the processed fish. Although there has been a number of reports in recent years[4,5,6] that certain other natural seafood flavours are also related to the animal's diet, little effort has been made to demonstrate this relationship. Similarly, the impact of the natural environment on flavour has continued to receive little attention, even though some studies have shown that in lakes and ponds (both fresh and saline) environmental effects can be responsible for significant changes of flavour in cultivated fish[7] and crustaceans.[8] By comparison, the effect of industrial pollution on seafood flavour has been widely investigated.[9]

Simple bromophenols, such as 2- and 4-bromophenol, 2,4- and 2,6-dibromophenol, and 2,4,6-tribromophenol, are natural components of the marine environment and are also key flavour components of certain species of fish[6,10] and crustaceans.[4,5,10] Of these compounds, the monobromophenols enhance sweetness and overall seafood flavour, while 2,6-dibromophenol and 2,4,6-tribromophenol provide iodoform-, prawn- and ocean-like flavour characteristics. However, the sources of these compounds in fish and crustaceans has been principally a matter for speculation. For example, Australian workers have suggested that small benthic animals, such as polychaetes, are the source of bromophenols in fish and crustaceans,[6,11] while similar benthic animals and some species of marine algae are the source of these compounds in fish.[6] Danish workers have also indicated that marine worms (annelids) are the likely source of bromophenols in North Atlantic shrimp.[5] Of these, *Polychaeta* are one of the most populous benthic organisms in the marine environment, while algae are ubiquitous throughout the world's oceans. To provide additional information on the source of bromophenols in Australian seafoods, we have recently surveyed a broad range of ocean fish and crustaceans, together with their principal dietary components. A total of 33 species of fish and 12 species of prawn, taken from a number of different locations, were analysed for the target compounds. Furthermore, based on published information of the dietary intake of these animals, 50 species of marine algae, 16 species of polychaetes and a smaller number of bryozoans, sponges and molluscs were similarly surveyed for bromophenols. This paper presents the results from these studies and uses the information obtained to attribute the origin of bromophenols in certain species of fish and crustaceans to their diet and environment.

2 COLLECTION OF SPECIMENS AND ANALYSIS FOR BROMOPHENOLS

Samples of fish and some species of prawns were caught off the coast of New South Wales by the fisheries research vessel M.V. Kapala. Additional samples of prawns were obtained from the Sydney Fish Marketing Authority, the CSIRO Division of Fisheries, and commercial interests in New South Wales, Western Australia, Queensland and the Northern Territory. Samples of marine algae and polychaetes were collected from intertidal zones along the New South Wales and southern Queensland coastline, bryozoans from New South Wales, Tasmania and Western Australia, sponges from Western Australia, and molluscs from southern Queensland. All samples were stored in ice until delivered to the laboratory, where they were snap-frozen and stored at −20 °C until required for analysis. All samples of fish and prawns were identified to species by the staff of the New South Wales Department of Fisheries, marine algae by the Sydney Botanical Gardens, polychaetes by the Australian Museum, bryozoans by the University of Tasmania and sponges by the Museum of Natural History, Queensland.

Volatile extracts of the different samples were obtained by simultaneous steam distillation–solvent extraction (SDE), using methods previously described.[4,6] In all cases, the samples were homogenised in purified water, and acidified to pH 1 before distillation. Samples of fish were extracted as separated gut and flesh and prawns as separated head and tail portions. All other materials were extracted whole. Multiple ion detection gas chromatography–mass spectrometry was used for quantitative analysis of the target bromophenols in the extracts, with 3,5-dimethyl-2,4,6-trichloroanisole[4] and 2,6-dibromophenol-d_3[6] as internal standards. These analyses were performed on either a Varian MAT 311A MS coupled to a Varian 1440 GC or an HP 5971 MSD coupled to an HP 5890 GC. Non-polar GC columns coated with either BP1 or HP5 phases were used for these analyses.[4,6] Individual bromophenols were identified by their GC retention time, by the presence of three characteristic ions and by the ratio of the abundances of these ions to one another. Extraction efficiencies were determined for the recovery of bromophenols from fish flesh, prawn meat and algae, and were used to correct the raw data obtained from these analyses. Recoveries of bromophenols from polychaetes, bryozoans and sponges were assumed to be the same as those obtained for marine algae.

3 BROMOPHENOL CONTENT OF OCEAN FISH

Of the 33 species of fish analysed, 21 were categorised as benthic carnivores, six as diverse omnivores, four as piscivorous carnivores and two as restricted omnivores. Bromophenols were found in the gut and carcass of all species with the exception of the four piscivorous carnivores. However, the concentrations of these compounds varied greatly between species of different categories, the highest concentrations being found in the benthic carnivores, followed by the diverse omnivores and restricted omnivores. To illustrate the effect of different feeding habits, eight species, typifying the four categories, have been selected for comparison, and the concentrations of the five target compounds in the gut and carcass are recorded in Table 1.

Table 1   *Concentration of Bromophenols in Ocean Fish from the East Coast of Australia*

Zoological Name Common Name	Sample	Bromophenols (ng g^{-1})[a]					
		2-BP	4-BP	2,4-DBP	2,6-DBP	2,4,6-TBP	Total
Benthic Carnivores							
Nemadactylus douglasii	gut	2.6	100	80	5.8	170	358
Rubber lipped morwong	carcass	trace[b]	0.1	0.1	0.1	3.4	3.7
Branchiostegus wardi	gut	5.2	22	78	5.7	100	211
Pink tilefish	carcass	0.1	0.2	0.4	0.1	0.4	1.2
Piscivorous Carnivores							
Zeus faber	gut	ND[c]	ND	ND	ND	ND	ND
John Dory	carcass	ND	ND	ND	ND	ND	ND
Pseudorhombus arsius	gut	ND	ND	ND	ND	ND	ND
Large toothed flounder	carcass	ND	ND	ND	ND	ND	ND
Diverse Omnivores							
Pseudocaranx dentex	gut	0.2	56	7.3	1.2	26	91
Silver trevally	carcass	trace	1.2	0.4	0.1	1.3	3.0
Meuschenia freycineti	gut	0.7	2.2	2.4	1.3	17	24
Six spined leatherjacket	carcass	0.4	trace	0.3	0.3	4.3	5.3
Restricted Omnivores							
Girella tricuspidata	gut	0.7	ND	19	5.3	32	57
Luderick	carcass	trace	ND	0.2	0.2	0.8	1.2
Kyphosus sydneyanus	gut	0.2	0.8	2.5	2.2	7.0	13
Black drummer	carcass	trace	0.1	0.2	0.2	0.1	0.6

[a] 2-BP = 2-Bromophenol; 4-BP = 4-Bromophenol; 2,4-DBP = 2,4-Dibromophenol; 2,6-DBP = 2,6-Dibromophenol and 2,4,6-TBP = 2,4,6-Tribromophenol;

[b] trace = less than 0.05 ng g^{-1};

[c] ND = Not Detected at a detection limit 0.01 ng g^{-1};
Flavour threshold concentrations in water, 2-bromophenol, 3×10^{-2} ng g^{-1}; 4-bromophenol, 23 ng g^{-1}; 2,4-dibromophenol, 4 ng g^{-1}; 2,6-dibromophenol 5×10^{-4} ng g^{-1}; 2,4,6-tribromophenol, 6×10^{-1} ng g^{-1};[4]
Flavour threshold concentrations in prawn meat, 2-bromophenol, 2 ng g^{-1}; 2,6-dibromophenol, 6×10^{-2} ng g^{-1};[4] 2,4,6-tribromophenol, 50 ng g^{-1}.

All five bromophenols were found in the benthic carnivores, the diverse omnivores and one restricted omnivore; four bromophenols were found in the remaining restricted omnivore, *Girella tricuspidata*. In all species, the total bromophenol content in the gut was greater than that found in the carcass by factors ranging from 5 in the diverse omnivore, *Meuschenia freycineti*, to 200 in the benthic carnivore, *Branchiostegus wardi*. Individual bromophenols were found in the carcass only when they were present in the gut. In the majority of species, if a bromophenol exceeded 5 ng g^{-1} in the gut, then it would be found in the carcass. 2,4,6-Tribromophenol was the major bromophenol in five of the six chosen species and 4-bromophenol in the other, the diverse omnivore, *Pseudocaranx dentex*. 2,4-Dibromophenol was present in significant concentrations in three species, the two benthic carnivores and the restricted omnivore, *G. tricuspidata*. Of special interest are the bromophenols with the lowest flavour threshold concentrations, 2,6-dibromophenol and 2-bromophenol (see Table 1), which were present in relatively low concentrations. This sample group was typical of the bromophenol distribution found across the 33 species covered by the survey.

The influence of the bromophenols on the flavour of these species can be assessed from the flavour threshold concentrations of individual compounds (Table 1) and their concentration in the carcass. 2,6-Dibromophenol, with a threshold concentration in prawn meat of 6×10^{-2} ng g^{-1}, would have an impact on the flavour of all species containing this compound. At threshold levels, this compound would be expected to produce iodoform-, iodine- and sea salt-like flavours in the flesh.[4,12] In addition, it has been claimed that, even at sub-threshold concentrations, bromophenols can have an impact on the general flavour acceptability of the fish.[12] In this regard, 2-bromophenol will add to its sweetness and 2,4,6-tribromophenol to its ocean flavour. By comparison, the flavour of those species that do not contain bromophenols would be expected to be bland and lack such ocean- or sea-like flavours.

4 BROMOPHENOL CONTENT OF OCEAN PRAWNS

A total of 200 samples, representing 12 species of prawn, has been analysed for bromophenol content. Of these, eleven of the species were caught off the eastern, northern and western coasts of Australia, and two species of cultivated prawns were obtained from farms in eastern Australia and the island of New Caledonia. Bromophenols were found in the head and tails of all wild species, including two deep-sea prawns, *Haliporoides sibogae* and *Plesionika martia*.[4] As with fish, the concentrations of these compounds varied greatly between species; however, unlike fish, all mature prawns are benthic carnivores and differences in diet are dictated by the habitat in which the prawns feed. Importantly, the concentrations of bromophenols in cultivated prawns were very low in comparison with those of wild species. To demonstrate the effect of different habitat and consequently diet, six species of wild and two species of cultivated prawns have been selected for comparison. The concentrations of the five bromophenols in the head and tail are recorded in Table 2.

Table 2   *Concentration of Bromophenols in Ocean Prawns from Australian Coastal Waters*

Zoological Name Common Name	Sample	Bromophenols (ng/g)[a]					
		2-BP	4-BP	2,4-DBP	2,6-DBP	2,4,6-TBP	Total
Wild							
Penaeus merguiensis	head	0.9	508	630	1.4	103	1243
Banana prawn	tail	0.1	20	104	trace[a]	4.3	128
Metapenaeus endeavouri	head	32	170	180	280	120	782
Endeavour prawn	tail	0.9	4.6	1.1	5.7	3.0	15
Penaeus esculentus	head	0.7	60	290	1.2	9.1	361
Brown tiger prawn	tail	trace	14	32	0.6	trace	47
Penaeus plebejus	head	0.3	56	10	9.0	2.0	77
Eastern king prawn	tail	0.3	15	2.8	0.1	0.7	19
Haliporoides sibogae	head	1.8	1.2	9.3	30	7.3	50
Royal red prawn	tail	0.1	trace	0.4	1.5	0.3	2.3
Penaeus monodon	head	trace	3.0	5.0	6.0	460	474
Black tiger prawn	tail	trace	0.3	0.6	0.2	12	13
Cultivated							
Penaeus stylirostris	head	trace	0.6	trace	trace	trace	0.6
	tail	0.6	0.7	trace	trace	trace	1.3
Penaeus monodon	head	ND[a]	ND	ND	0.8	0.7	1.5
Black tiger prawn	tail	ND	ND	ND	ND	ND	ND

[a]   For abbreviations see Table 1.

All five bromophenols were found in the six wild species and in the cultivated species, *Penaeus stylirostris*. However, in the second cultivated species, *Penaeus monodon*, only two bromophenols were found and then only in the head in low concentrations. As previously observed with fish, the total bromophenol content in prawn heads was greater than that found in the tails by factors ranging from 4:1 in *Penaeus plebejus* to 50:1 in *Metapenaeus endeavouri*. The exception to this relationship was the cultivated prawn, *P. stylirostris*, where bromophenols were slightly more abundant in the tails than in the heads. This apparent anomaly can be explained by the practice of placing cultivated prawns in brackish water to empty their guts before harvesting. Of the six wild species reported in Table 2, 2,4-dibromophenol was the major bromophenol present in *Penaeus merguiensis* and *Penaeus esculentus*, 2,6-dibromophenol in *M. endeavouri* and *H. sibogae*, and 4-bromophenol in *P. plebejus*. 2,4,6-Tribromophenol, the major bromophenol found in most species of fish, was only identified as the major component in the sample of *P. monodon* caught in the wild. However, it was found in significant concentrations in both *P. merguiensis* and *M. endeavouri* (Table 2). In prawns, the bromophenols with low flavour threshold concentrations, 2,6-dibromophenol and 2-bromophenol, were usually present in greater concentrations than in fish (see Table 1). This result could account, at least in part, for the stronger flavour perceived in prawns compared with that of most fish. As with the results presented in Table 1, the bromophenol concentrations recorded for the seven selected species of prawns are typical of the 12 species and 200 samples covered by this survey, with respect to variation and distribution.

Table 3   *Relationship between Total Bromophenol Content (TBC) and Diet of Individual Species of Fish*

Zoological Name	TBC[a] whole fish (ng g^{-1})	Diet[b]
Benthic Carnivores		
Nemadactylus douglasii	40	(polychaetes, invertebrates and crustaceans)[16]
Branchiostegus wardi	24	(polychaetes, gastropods, molluscs and amphipods)[17]
Piscivorous Carnivores		
Zeus faber	ND	(fish)[18]
Pseudorhombus arsius	ND	(crustaceans and teleosts)[18]
Diverse Omnivores		
Pseudocaranx dentex	17	(crustaceans, polychaetes, molluscs, algae and detritus)[18]
Meuschenia freycineti	10	(crustaceans, algae, bryozoans and polychaetes)[18]
Restricted Omnivores		
Girella tricuspidata	11	(algae, Zostera, crustaceans and detritus)[18]
Kyphosus sydneyanus	5.3	(algae, Zostera, crustaceans and detritus)[17]

[a]   Calculated;
[b]   Listed in order of importance.

As with fish, the influence of the bromophenols on the flavour of prawns can be assessed from the flavour threshold concentration of individual compounds (Table 1) and their concentration in the tails. 2,6-Dibromophenol would have an impact on all species containing this compound, while 2-bromophenol could well influence the flavour observed in *M. endeavouri* and *P. stylirostris*. The high concentration of 2,6-dibromophenol in *M. endeavouri* (5.7 ng g^{-1}) was shown to produce an intense iodoform-like flavour.[4] By comparison, sensory assessments of *P. stylirostris*, which had only trace concentrations of this compound, indicated that the meat was sweet, but generally bland. Sub-threshold concentrations of the bromophenols could also have an impact on the general flavour acceptability of the prawns.[12] The flavour of prawn meat is further complicated by the practice of cooking prawns whole. Studies have shown that this procedure favours the diffusion of flavour compounds from the gut into the meat, thus intensifying the perceived flavour. In general, consumers consider that the flavour of cultivated prawns is bland when compared with that of prawns caught in the wild. The absence, or near absence, of bromophenols from the cultivated material no doubt contributes greatly to this perception.

5 NATURAL SOURCES OF BROMOPHENOLS IN THE ENVIRONMENT

The simple bromophenols have been identified in a variety of soft bottom benthic organisms, in marine algae, sponges and bryozoans.[14] That these compounds are synthesised by some marine worms and algae has been demonstrated by the isolation of specific haloperoxidases from these organisms.[15] Such animals and plants are therefore possible sources of the bromophenols in fish and prawns. The principal dietary components of eight species of fish and their total bromophenol content (TBC) are reported in Table 3, and similar information for the six species of wild and two species of cultivated prawns is reported in Table 4. The calculated TBC for fish varies between not detected in the carnivorous species, *Z. faber* and *Pseudorhombus arsius*, to 40 ng g^{-1} in the benthic carnivore, *Nemadactylus douglasii*, and, with the exception of the restricted omnivore, *Kyphosus sydneyanus*, all other species have values of 10 ng g^{-1} or greater. By comparison, the calculated TBC for prawns varies between 0.5 ng g^{-1} in the cultivated *P. monodon* to 530 ng g^{-1} in *P. merguensis*. With the exception of the two cultivated animals, all other species have values of 20 ng g^{-1} or greater. The explanation for the observed difference in TBC between fish and prawns is the difference in weight ratio of gut to flesh in the two types of animals. For fish, the ratio of gut to carcass is about 1:6 whereas for prawns the ratio of head to tail is only 1:2.

Table 4 *Relationship between Total Bromophenol Content (TBC), Habitat and Diet of Individual Species of Prawns*

Zoological Name	TBC^a whole fish (ng g^{-1})	Habitat^b/Diet^c
Wild		
Penaeus merguiensis	530	mud[19]/crustaceans, polychaetes and molluscs[20]
Metapenaeus endeavouri	290	mud[19]/crustaceans, polychaetes and detritus[20]
Penaeus monodon	180	mud-sand[19]/polychaetes, molluscs and crustaceans[20]
Penaeus esculentus	160	mud-sand[19]/molluscs, crustaceans and polychaetes[20]
Penaeus plebejus	40	sand[19]/crustaceans, molluscs and polychaetes[21]
Haliporoides sibogae	21	mud[19]/not identified
Cultivated		
Penaeus stylirostris	1	mud[19]/fish meal, plant material, prawn and squid meal[22]
Penaeus monodon	0.5	mud[19]/fish meal, plant material, prawn and squid meal[22]

[a] Calculated;
[b] Composition of ocean floor;
[c] Listed in order of importance

Comparison of the diets of the selected fish and prawns (Tables 3 and 4) shows that a number of similarities exists between components eaten by benthic carnivores, diverse omnivores and wild prawns. The common components are crustaceans, molluscs, polychaetes and detritus; omnivorous species also eat marine algae. Of these dietary components, only polychaetes and algae are known producers of bromophenols. The bromophenol contents of eight species of polychaetes and eight species of algae from the east coast of Australia are recorded in Tables 5 and 6. Table 6 also contains data on the concentration of these compounds in two species of bryozoans and two species of sponges. In these tables, the polychaetes are divided according to the type of habitat from which they are collected and the algae according to their colour classification.

Table 5 *Concentration of Bromophenols in Marine Polychaetes from the East Coast of Australia*

Zoological Name	Bromophenols $(ng\ g^{-1})^a$					
	2-BP	4-BP	2,4-DBP	2,6-DBP	2,4,6-TBP	Total
Muddy Habitat						
Barantolla lepte	61	100	11000	32000	8300000	8340000
Nephtys australiensis	11	28	460	2300	880000	883000
Marphysa sanguinea	61	30	500	1500	250000	252000
Australonereis ehlersi	33	69	770	2400	96000	99000
Ceratonereis aequisetis	12	23	340	420	34000	35000
Glycera americana	320	4200	4600	72	660	9900
Rocky Habitat						
Diopatra dentata	10	100	170	80	760	1100
Sandy Habitat						
Australonuphis teres	270	1	51	31	144	500

ª For abbreviations see Table 1

Table 6 *Concentration of Bromophenols in Marine Algae, Bryozoans and Sponges from Australian Coastal Waters*

Botanical/Zoological Name	Bromophenols $(ng\ g^{-1})^a$					
	2-BP	4-BP	2,4-DBP	2,6-DBP	2,4,6-TBP	Total
Red Algae						
Pterocladia capillacea	trace	30	320	440	1800	2590
Amphiroa anceps	1.6	5.1	26	6.4	1300	1339
Haliptilon roseum	7.2	35	110	26	190	368
Plocamium angustatum	ND	2.9	61	23	250	337
Brown Algae						
Phyllospora comosa	ND	98	55	0.2	280	433
Homoeostrichus sinclairii	16	9	35	29	63	152
Green Algae						
Enteromorpha intestinalis	18	260	640	75	1400	2393
Ulva lactuca	0.1	2.2	25	1.2	1200	1229
Bryozoans						
Bugulara dissimilis	ND	53	330	130	610	1123
Orthoscuticella ventricosa	1.5	ND	24	47	540	613
Sponges						
Ascidiacea	3.2	62	110	10	240	425
Niphates sp.	1.0	3.2	7.6	9.0	97	118

ª For abbreviations see Table 1.

The highest TBCs found in the current study were in those polychaetes that were obtained from muddy habitats. In these animals, the TBC varied between 9900 ng g^{-1} in

Glycera americana and 8340000 ng g^{-1} in *Barantolla lepte*. In those species from rocky or sandy habitats, the concentrations were significantly lower, for example, *Diopatra dentata* (1100 ng g^{-1}) and *Australonuphis teres* (500 ng g^{-1}). 2,4,6-Tribromophenol was the major bromophenol in all species with the exception of *G. americana*, in which both 2,4-dibromophenol and 4-bromophenol were present in larger quantities. 2,6-Dibromophenol was present in significant, although minor, concentrations in *Nephtys australiensis*, *Marphysa sanguinea* and *Australonereis ehlersi*. As previously observed, 2,4,6-tribromophenol was the major bromophenol found in most species of fish, while 2,4-, 2,6-dibromophenol and 4-bromophenol were found in greatest concentrations in prawns. It is of interest that these three compounds have been identified as the major bromophenols in other species of marine worms.[14,23] It is therefore likely that the samples of *P. merguiensis* and *M. endeavouri*, with high concentrations of 4-bromophenol, 2,4- and 2,6-dibromophenol, had fed on polychaetes with high concentrations of these compounds. As yet, the only Australian worm studied that is known to contain these compounds as major components is *G. americana* (Table 5). However, it is likely that other Australian polychaetes with similar bromophenol compositions will be found in further study.

It is of interest that those prawns and species of fish with high concentrations of bromophenols are all bottom feeders and those with the highest bromophenol content are usually caught over muddy sections of the ocean floor. By comparison, omnivorous species with low bromophenol content are usually caught over sandy bottoms or close to rocky outcrops, where polychaetes with low concentrations of bromophenols are found. However, at present, it is not possible to prove such relationships as the only part of the polychaete that survives the digestive system of fish and prawns is the jaws, and these are insufficient for identification of species, although in some cases the genus has been identified.

The reason for the difference in bromophenol content in polychaetes from different habitats is currently in dispute. Bromophenols are known to be powerful bactericides, and it has been suggested that these compounds are synthesised by polychaetes for this purpose.[14] Consequently, animals from a muddy environment could require more protection and hence higher concentrations of bromophenols than those from a sandy or rocky habitat. However, recent studies have shown that some species of bacteria thrive in the vicinity of polychaetes and, as a consequence, it has been suggested that these compounds might alternatively act as deterrents for predators and competitors.[23] The fact that polychaetes are major dietary components of Australian prawns and some species of fish would tend to lessen the likelihood that these compounds are antipredator factors.

The concentrations of bromophenols in marine algae are all appreciably lower than those found in polychaetes from muddy environments, but are comparable with levels found in worms from sand or rocky habitats (Table 6). For the eight selected species of algae, the TBC varied between 152 ng g^{-1} in the brown alga, *Homoeostrichus sinclairii*, and 2100 ng g^{-1} in the red alga, *Pterocladia capillacea*. 2,4,6-Tribromophenol was the major bromophenol found in all of these species of algae. 2,6-Dibromophenol was found in significant concentrations in *P. capillacea*, while 4-bromophenol and 2,4-dibromophenol were equally significant in *Haliptilon roseum*, *Phyllospora comosa* and *Enteromorpha intestinalis*. Of the species of algae recorded in Table 6, the green algae, *E. intestinalis* and *Ulva lactuca*, are major dietary components of the restricted omnivores, *G. tricuspidata* and *K. sydneyanus*. *E. intestinalis* and *U. lactuca* are used extensively by the recreational fisherman to catch both of these species of fish. In addition, most diverse omnivores also include algae among their normal dietary intake (Table 3).[18] They also feed upon *E. intestinalis* and *U. lactuca*, as well as a range of other algae, including *P. capillacea*,

Amphiroa anceps and *H. sinclairii*. Furthermore, marine biologists frequently report fish grazing on a variety of red and green algae on both submerged reefs and the ocean floor.[24]

Bryozoans and sponges are also dietary components of diverse omnivores,[18] of which *M. freycineti* is a typical example (Table 3). Data on these colonising animals show that the TBC of bryozoans ranged from 610 ng g^{-1} for *Orthoscuticella ventricosa* and 1100 ng g^{-1} for *Bugulara dissimilis*, while for sponges this content varied between 120 ng g^{-1} for a *Niphates* species and 430 ng g^{-1} for an *Ascidiacea*. These values cover a similar range of concentrations as previously observed for marine algae. However, unlike algae and polychaetes, it is not known whether bryozoans or sponges are capable of synthesising bromophenols or whether they extract these compounds from the environment.

Based on the above evidence, marine algae are a major dietary source of bromophenols in restricted omnivores, while algae, bryozoans and sponges contribute to the bromophenol content of diverse omnivores. The latter animals also obtain these compounds by eating polychaetes (Table 3). Some authors have indicated that bromometabolites, including bromophenols, possess an anti-herbivore function.[25] The current work would suggest the opposite, with certain species of restricted omnivores, such as *G. tricuspidata*, attracted to those algae with high levels of these compounds.

6 CONCLUSION

The current investigation has shown that the simple bromophenols are present in ng g^{-1} concentrations in most species of ocean fish and prawns caught in Australian coastal waters and that the source of these compounds is the animal's diet. However, the quantity of bromophenols present is dependent on the environment in which the animal feeds. Thus benthic carnivores that feed over muddy bottoms are more likely to eat polychaetes that are high in bromophenols than those that feed over sandy bottoms. Similarly, with restricted omnivores, the floral environment in which they feed is the determining factor and the bromophenol content of these animals tends to depend on the algae growing in that region. The presence of bryozoans and sponges would add to the bromophenol intake of diverse omnivores. Outbreaks of iodine- or iodoform-like flavours in fish and prawns can now be attributed to their dietary intake.

In this study, bromophenols were not detected in the four species of piscivorous carnivores examined. However, it is likely that some piscivorous carnivores that feed on benthic carnivores or omnivores would receive a bromophenol input from such sources. This could account for the relatively high concentrations of bromophenols found in some North American species, such as the pink and chinook salmon (*Oncorhynchus gorbuscha* and *O. tshawytscha*) of Alaska.[10]

ACKNOWLEDGEMENT

This work was supported, in part, by a research grant from the Fisheries Research and Development Corporation. The authors wish to thank K. Graham for the supply of samples of fish and prawns, and K.J. Shaw, D. Svoronos and G.L. Ford for GC–MS analyses.

REFERENCES

1. D.B. Josephson, in 'Volatile Compounds in Foods and Beverages', ed. H. Maarse, Marcel Dekker, New York, 1991, p. 103.
2. T. Motohiro, *Mem. Fac. Fish Hokkaido Uni.*, 1962, **10**, 1.

3. J.C. Sipos and R.G. Ackman, *J. Fish Res. Board Can.*, 1964, **21**, 423.
4. F.B. Whitfield, J.H. Last, K.J. Shaw and C.R. Tindale, *J. Sci. Food Agric.*, 1988, **46**, 29.
5. U. Anthoni, C. Larsen, P.H. Nielsen and C. Christophersen, *Biochem. System. Ecol.*, 1990, **18**, 377.
6. F.B. Whitfield, F. Helidoniotis, D. Svoronos, K.J. Shaw and G.L. Ford, *Water Sci. Technol.*, 1995, **31**, 113.
7. M. Yurkowski and J.A.L. Tabachek, *J. Fish Res. Board Can.*, 1974, **31**, 1851.
8. R.T. Lovell and D. Broce, *Aquaculture*, 1985, **50**, 169.
9. F.B. Whitfield, *Water Sci. Technol.*, 1988, **20**, 63.
10. J.L. Boyle, R.C. Lindsay and D.A. Stuiber, *J. Food Sci.*, 1992, **57**, 918.
11. F.B. Whitfield, K.J. Shaw and D.I. Walker, *Water Sci. Technol.*, 1992, **25**, 131.
12. J.L. Boyle, R.C. Lindsay and D.A. Stuiber, *J. Aquatic Food Prod. Technol.*, 1992, **1**, 43.
13. F.B. Whitfield and D.J. Freeman, *Water Sci. Technol.*, 1983, **15**, 85.
14. J.L. Boyle, R.C. Lindsay and D.A. Stuiber, *J. Aquatic Food Prod. Technol.*, 1993, **2**, 75.
15. S.L. Neidleman and J. Geigert, 'Biohalogenation – Principles, Basic Roles and Applications', Ellis Horwood, Chichester, 1986, p. 49.
16. M. Goman, Museum of Victoria, personal communication, 1994.
17. J. Paxton, Australian Museum, personal communication, 1994.
18. Anon, 'The Ecology of Fish in Botany Bay – Biology of Commercially and Recreationally Valuable Species', State Pollution Control Commission, Sydney, 1981.
19. D.L. Grey, W. Dall and A. Baker, 'A Guide to the Australian Penaeid Prawns', Northern Territory Government Printing Office, Darwin, 1983.
20. D.J.W. Moriarty and M.C. Barclay, *Aust. J. Freshwater Res.*, 1981, **32**, 245.
21. T.J. Wassenberg, CSIRO Division of Fisheries, personal communication, 1996.
22. D. Smith, CSIRO Division of Fisheries, personal communication, 1995.
23. C.C. Steward, J. Pinckney, Y. Piceno and C.R. Lovell, *Mar. Ecol. Prog. Ser.*, 1992, **90**, 61.
24. A.K. Millar, *Aust. Syst. Bot.*, 1990, **3**, 293.
25. J. Lubchenco and D.J. Carlson, *Bull. Ecol. Soc. Am. Abstracts*, 1988, **69**, 212.

VOLATILE ACIDS AND NITROGEN COMPOUNDS IN PRAWN POWDER

K.B. de Roos and J.A. Sarelse

Tastemaker B.V., P.O. Box 414, 3770 AK Barneveld, The Netherlands

1 INTRODUCTION

In spite of the many studies on the volatile flavour of shrimp and prawn,[1-3] it is still difficult to duplicate the flavour of this seafood by reconstitution from volatile compounds. This suggests that part of the volatile flavour is still unknown.

In almost all studies done so far, some form of steam distillation was used to isolate the flavour compounds. In our studies on shrimp and prawn flavour, we observed that even after exhaustive vacuum steam distillation, the residue left after the distillation still possessed a strong characteristic smell. The investigation of this poorly steam-volatile fraction of prawn flavour is the subject of the present paper.

2 FLAVOUR ISOLATION

Prawn powder 1505 (Nikken Foods, Japan) was selected as the starting material for this analysis because of its strong, dried, fermented-seafood character. Preliminary experiments had demonstrated that the poorly steam-volatile fraction of prawn flavour is very water soluble and, therefore, the flavour isolation was begun with extraction using methanol–water, 1:1. The resulting extract contained the characteristic prawn flavour. The solid residue possessed only a weak fishy smell.

Investigation of the nature of the water-soluble flavour compounds present in the aqueous extract revealed that a major part consisted of nitrogen compounds and organic acids. Salt formation between amines and acids prevented their isolation by solvent extraction or steam distillation at neutral pH. Therefore, the isolation of the flavour compounds was carried out in two steps. First, the neutral and nitrogen compounds were isolated by solvent extraction at pH 10. Subsequent extraction at pH 3 yielded a fraction containing the organic acids. To allow high recovery of water-soluble flavour compounds, a continuous version of the modified Folch extraction method was developed for this purpose.[4] The Folch extraction allows isolation of natural products without problems due to emulsion formation.

The 'Folch extracts' were further fractionated and analysed as outlined in Scheme 1. All fractions were evaluated organoleptically and only the most interesting fractions were analysed.

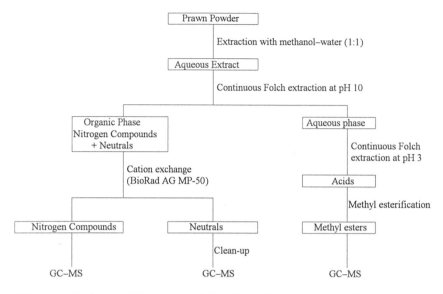

Scheme 1 *Isolation of the water soluble flavour from prawn powder*

3 DISCUSSION OF RESULTS

High concentrations of fatty acids (C_3–C_{18}) and amino-acid derived acids, such as 3-methylbutanoic acid, 2-hydroxy-3-methylbutanoic acid and 2-oxo-3-methylbutanoic acid (all derived from leucine), were found. Further, the occurrence of the homologous series ω2-oxo-alkanoic acids (C_5–C_{16}), alkanedioc acids (C_2–C_{12}) and 4-oxo-alkanoic acids (**1**) (C_5, C_9 and C_{10}) is worth mentioning. It is possible that (part of) the 4-oxo-alkanoic acids were originally present in the form of β,γ-unsaturated γ-lactones (**2**), but were hydrolyzed during the continuous Folch extraction at pH 10. During the present analysis only α-angelicalactone (from 4-oxopentanoic acid) was found.

Among more than 70 nitrogen compounds found, amides, pyrazines, amines and pyridines were most abundant. The pyrazines and amines are organoleptically by far the most interesting components. The major nitrogen compound found in the prawn powder was 2-piperidone (δ-valerolactam (**3**)) which constituted more than 10% of total volatile fraction. This compound has previously been identified in fish sauce.[5] Other related cyclic amides (**4–8**) were also present as well as a wide range of aliphatic (N,N-dimethyl)amides. The aroma properties of the amides are not very interesting. In general, the amides have a dry seafood, fishy odour. More interesting are the taste properties of some of the amides. For example, 2-piperidone has a bitter taste and N,N-dimethylbenzamide has a cooling effect in the mouth like menthol. Interesting, also, is the occurrence of fatty acid primary amides which have recently been identified in the cerebrospinal fluid of sleep-deprived cats.[6] These compounds induce sleep when injected into rats. One might speculate about the role that these amides play in determining the preference of cats for seafood. It is possible the cats like seafood not only for its flavour but also because of the presence of the amide narcotics in the food.

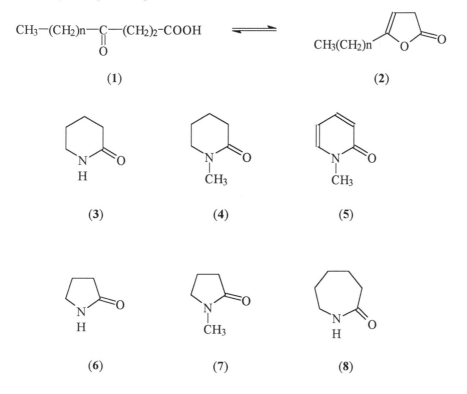

Figure 1 *Structures of flavour compounds identified in prawn powder*

The contribution of the amines and acids to the total flavour impact was found to depend heavily on the pH and the fat content of the application medium. For example, at the relatively high pH of prawn powder (pH of an aqueous suspension is 8.5), the contribution of the amines to the total flavour is high and that of the acids low. At normal food pH, however, the reverse holds. In order to better understand the effect of pH and fat content on the perception of acids and amines, the effect of these variables on the volatility (headspace concentration) of these compounds was investigated in more detail.

3.1 Effect of the Medium on the Volatility of Amines and Acids

It is the concentration of the flavour compound in the nasal cavity that determines at what strength an aroma compound is perceived. Therefore, the effect of pH and fat content on flavour impact can be estimated from the effect of these variables on the headspace concentrations. Due to the extremely low concentrations of amines and acids in the headspace over foods, it is very difficult to measure these concentrations accurately. Therefore, the equilibrium headspace concentrations were calculated with the aid of the relationships described below.

Under equilibrium conditions, the amount of a flavour compound released from a product into the headspace is given by:

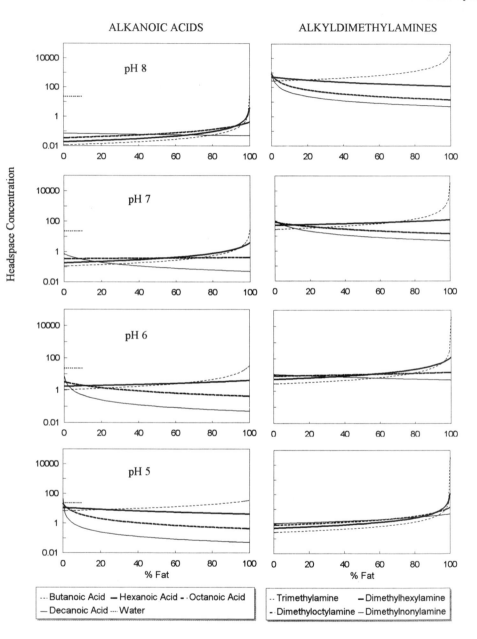

Figure 2   *Relative equilibrium headspace concentrations of amines and acids as a function of pH and fat content of the product assuming that $V_a = V_p$.*

$$\frac{X_a}{X_t} = \frac{V_a/V_p}{P_{pa} + V_a/V_p} \tag{1}$$

where X_t is the total amount (g) of the flavour compound present in the system and X_a is the amount released into the headspace. V_a and V_p are the total volumes (l) of air and product phase, respectively. P_{pa} is the product-to-air partition coefficient which is defined as $P_{pa} = C_p/C_a$, where C_p and C_a are the concentrations (g l^{-1}) of the flavour compound in product and air phase. P_{pa} can be calculated from the available oil-to-air and water-to-oil partition coefficients P_{oa} and P_{wo} as follows:

$$P_{pa} = \frac{C_p}{C_a} = \frac{f_o C_o + f_w C_w}{C_a} = P_{oa}(f_o + f_w P_{wo}) \tag{2}$$

where f_o and f_w are the fractions of oil/fat and water in the product phase. The partitioning of amines and acids between air and product phases is complicated by the fact that in the aqueous phase these compounds can occur in both ionic and neutral form. For example, the partitioning of the amines in the product phase is governed by the following equilibria:

$$N\ (fat) \ \overset{P_{wo}}{\rightleftharpoons} \ N\ (water) + H^+ \ \overset{pK_a}{\rightleftharpoons} \ NH^+\ (water) \tag{3}$$

K_a is here the ionization constant. Since only the neutral species is able to partition between all phases, the following relationship exists between the water-to-oil partition coefficient P_{wo} and the total amount C_w^{Nt} of amine present in the aqueous phase:

$$P_{wo} = \frac{C_w^N}{C_o^N} = \frac{C_w^{Nt}\left(1 + \dfrac{C_w^{H+}}{K_a}\right)}{C_o^N} \tag{4}$$

where C_w^{NH+} and C_w^N are the concentrations of the protonated and neutral amine in the aqueous phase. Combination of equations 1, 2 and 4 gives then the relationship between the volatility of an amine and its pK_a, P_{wo} and P_{oa} as a function of the pH and hydrophobic-hydrophilic balance in the product:

$$\frac{X_a}{X_t} = \frac{V_a/V_p}{P_{oa}\left(f_o + f_w P_{wo}\left(1 + \dfrac{C_w^{H+}}{K_a}\right)\right) + \dfrac{V_a}{V_p}} \tag{5}$$

A similar relationship can be derived for the volatility of acids in foods:

$$\frac{X_a}{X_t} = \frac{V_a/V_p}{P_{oa}\left(f_o + f_w P_{wo}\left(1 + \dfrac{K_a}{C_w^{H+}}\right)\right) + \dfrac{V_a}{V_p}} \tag{6}$$

The headspace concentrations over different products of different pH and fat content can now be calculated with the aid of the data of Table 1. The results are shown in Figure 2. It appears that at pH values higher than 5, the organic acids are always less volatile than water. The same holds for amines at pH lower than 6.5. In general, amines and organic acids are less volatile than water at normal food pH (pH 6 or less). This explains why steam distillation is usually not a suitable method for the isolation of acids and amines. It is noteworthy that ion pair formation (not taken into account here) will result in still further reduction of the headspace concentrations.

Table 1   *Partition Coefficients and pK$_a$ Values of Organic Acids and Amines*

Compound Name	Partition Coefficients			pK$_a$
	$\log P_{aw}$	$\log P_{ao}$	$\log P_{ow}$	
Butanoic acid	−4.78	−4.47	−0.31	4.84
Hexanoic acid	−4.56	−5.39	0.83	4.83
Octanoic acid	−4.29	−6.37	2.08	4.83
Decanoic acid	−3.97	−7.29	3.32	4.83
Trimethylamine	−1.79	−1.40	−0.39	9.76[7]
Dimethylhexylamine	−1.26	−3.91	2.65	10.02[7]
Dimethyloctylamine	−1.05	−4.84	3.79	10.02[7]
Dimethylnonylamine	−0.94	−5.30	4.36	10.02[7]

Figure 3 shows that at low and intermediate pH and fat content, the lower and higher homologues are often less volatile than the compounds of intermediate chain length. The reason is that the lower homologues are mainly present in the non-volatile ionic form in the aqueous phase, whereas the higher homologues are mainly present in neutral form in the fat phase in which they are less volatile than the lower homologues.

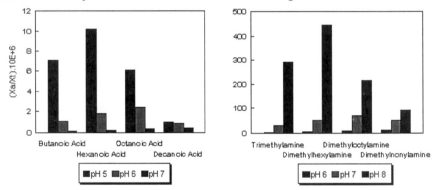

Figure 3   *Effect of pH on volatility in aqueous products containing 5% fat*

REFERENCES

1. K. Kubota, C. Uchida, K. Kurosawa, A. Komuro and A. Kobayashi, in 'Thermal Generation of Aromas,' eds. T.H. Parliment, R.J. McGorrin and C.-T. Ho, A.C.S. Symposium Series 409, American Chemical Society, Washington, 1989, p. 376.
2. S. Mandeville, V. Yaylayan and B. Simpson, *J. Agric. Food Chem.*, 1992, **40**, 1275.
3. Y.J. Cha and K.R. Cadwallader, *J. Food Sci.*, 1995, **60**, 19.
4. I.S. Chen, C.-S.J. Shen and A.J. Shephard, *J. Am. Oil Chem. Soc.*, 1981, **58**, 559.
5. R.C. McIver, R.I. Brooks and G.A. Reineccius, *J. Agric. Food Chem.*, 1982, **30**, 1017.
6. B.J. Cravett, O. Prospero-Garcia, G. Siuzak, N.B. Gilula, S.J. Henriksen, D.L. Boger and R.A. Lerner, *Science*, 1995, **268**, 1508.
7. H.K. Hall, *J. Am. Chem. Soc.*, 1957, **79**, 5441.

EFFECT OF CASTRATION AND SLAUGHTER AGE ON FATTY ACIDS AND PHENOLS IN SHEEP ADIPOSE TISSUE

Michelle M. Sutherland and Jennifer M. Ames

Department of Food Science and Technology, The University of Reading, Whiteknights, Reading, RG6 6AP, U.K.

1 INTRODUCTION

Various factors are known to influence sheepmeat flavour, including breed, feed, weight, slaughter age and sex and castration.[1] A range of chemical compounds has been associated with the species odour of cooked sheepmeat, including certain branched chain fatty acids (BCFA), *viz* 4-methyloctanoic, 4-methylnonanoic and 4-ethyloctanoic acids[2–4] and phenols.[5] Certain BCFA have significantly lower odour threshold values (OTVs), compared to their straight chain counterparts,[6] and those associated with sheepmeat have 'muttony', 'goaty', 'fatty' and 'sweaty' odours.[7] However, on their own, fatty acids do not account for the species flavour of sheepmeat and certain alkylphenols and thiophenol are reported to make a contribution.[5]

Previous studies on lamb flavour have involved either sensory analysis or chemical and/or instrumental investigation. In recent years, we have conducted the first interactive (sensory, chemical and instrumental) study of the effects of castration (within 24 hours of birth) and age at slaughter (12 or 30 weeks) on lamb flavour. We reported that cooked lean plus adipose tissue from 12 week old entires scored more highly for some terms, including 'farmyard', than equivalent samples from castrates.[8] When sensory analysis was extended to cooked tissue from 30 week old animals, scores for 'farmyard' were even higher for entires but about the same for castrates, compared to tissue from the younger animals.[9] Headspace analysis showed that although there were some differences between meat from rams and wethers slaughtered at the same age, they were not statistically significant.[1,9] The effect of castration and slaughter age on levels of BCFA and phenols in sheep adipose tissue are reported here.

2 EXPERIMENTAL

All animals were twin male lambs of the Suffolk breed. One twin of each pair was left entire and the other was castrated within 24 hours of birth. Ten pairs of twins were reared by the Agricultural College, Edinburgh and were slaughtered at 12 weeks of age. A further ten pairs of twins were reared by the University of Reading and were slaughtered at 30 weeks of age. The rearing conditions at both sites were strictly controlled and were exactly the same. Further details of the animals used have been reported previously.[1,10] Adipose tissue (*ca.* 450 g) from the leg region of each animal was recovered for analysis. Tissue from each animal of each castration–age combination was ground, thoroughly mixed and three representative samples were taken for free fatty acids analysis. A further three representative samples were taken for the analysis of phenols. Fatty acids were isolated from uncooked tissue using a modified Bligh and Dyer method and analysed as their methyl esters according to Sutherland and Ames.[10] For the isolation of phenols, adipose tissue was cooked,[1] prior to steam distillation-solvent extraction, using a procedure based on that reported by Heil and Lindsay.[11] Concentrated isopropyl ether:pentane (9:1) extracts were analysed by GC-MS, in the selected ion monitoring (SIM) mode, using a CP WAX column (Chrompack U.K. Ltd., London, U.K.).[9] Statistical analysis of all data was performed by analysis of variance (ANOVA) using Microsoft Excel (version 4.0) software.

3 RESULTS AND DISCUSSION

3.1 Fatty Acids

A total of 15 free fatty acids were identified in this study. Levels of total fatty acids were almost the same for rams and wethers slaughtered at 12 weeks of age. At 30 weeks of age, total levels were 1.5-fold greater in rams compared to wethers; also when compared to levels at 12 weeks of age, levels at 30 weeks were 4-fold greater in wethers and 6-fold greater in rams, respectively. The BCFA, 4-methyloctanoic and 4-methylnonanoic acids, were among the 15 acids identified. At 12 weeks of age, their levels in rams and wethers were not significantly different. However, significant differences were apparent when either rams and wethers of 30 weeks of age or 12 and 30 week old rams were compared ($p<0.001$ and $p<0.05$ for 4-methyloctanoic and 4-methylnonanoic acids, respectively, for both comparisons). The F-value for 4-methyloctanoic acid was particularly high at 6.46×10^7 when 30 week old rams and wethers were compared. The data are shown in Table 1.

In order to assess the contribution of these BCFA to the total aroma of the adipose tissue, their OTVs were considered. Since it is only the protonated form of these acids that is volatile, the pH of the medium must be considered when calculating their odour threshold values. The adipose tissue used in this study had a pH of 6.5. The theoretical odour thresholds of 4-methyloctanoic and 4-methylnonanoic acids at pH 6.0 (representative of the pH of adipose tissue) were calculated as 2.85 and 92.7 mg kg^{-1}, respectively,[10] and compare with experimental values reported in the literature of 0.02 and 0.65 mg kg^{-1}, respectively, at pH 2, when the compounds are fully protonated.[7] Thus, the theoretical log odour unit values (TOV) for both acids in all four adipose tissue samples were obtained by dividing the amount of acid present in each adipose tissue sample by the theoretical odour threshold value at pH 6. Although TOVs provide a guide to the contribution of BCFA to the odour of the adipose samples, lamb adipose tissue comprises about 72% lipid. The BCFA will partition between the aqueous and lipid phases, affecting the odour threshold values. Therefore, although the TOVs take into account some of the parameters of the system and are more valid for these components than values obtained without considering the pH of the system, they still need to be interpreted with caution.

The TOVs were negative for 4-methylnonanoic acid but zero or positive for 4-methyloctanoic acid for all the samples (Table 1). The TOV for 4-methyloctanoic acid increased with both entirety and age. It was very similar for both 30 week castrate and 12 week entire samples but was more than 10-fold greater for 30 week entire samples, compared to rams slaughtered at 12 weeks of age.

Table 1 *4-Methyl Branched Chain Fatty Acids Identified in the Adipose Tissue of Entire and Castrated Lambs Slaughtered and 12 and 30 Weeks of Age*

Fatty acid	Concentration mg kg⁻¹ (standard deviation)			
	Castrate 12 weeks	*Castrate 30 weeks*	*Entire 12 weeks*	*Entire 30 weeks*
4-Methyloctanoic acid	2.85 (0.70) $TOV^a = 0.00$	3.89 (1.00) $TOV = 0.14$	3.79 (1.40) $TOV = 0.12$	50 (1.01) $TOV = 1.25$
4-Methylnonanoic acid	8.72×10^{-2} (0.97) $TOV = -3.03$	3.49×10^{-1} (1.20) $TOV = -2.42$	3.48×10^{-1} (0.65) $TOV = -2.43$	1.40 (0.90) $TOV = -1.82$

[a] TOV = Theoretical log odour unit value.

Both 4-methyloctanoic and 4-methylnonanoic acids have been associated with the 'sweaty' odour of cooked mutton which is often described as 'soo' by the Chinese,[2] and Ha and Lindsay[4] have demonstrated that 4-methyloctanoic acid provides a highly characteristic odour to sheep perinephric adipose tissue. Brennand *et al.*[7] described the odours of both these acids at two dilutions in water. 4-Methyloctanoic acid was 'waxy' and 'goaty' at 1 mg kg⁻¹ and 'goaty-muttony' at 10 mg kg⁻¹. 4-Methylnonanoic acid was 'waxy-sweet', 'soapy', 'fatty' and 'acid-like' at 1 mg kg⁻¹ but 'muttony', 'wet wood' and 'fatty' at 25 mg kg⁻¹. Bearing in mind the sensory work on cooked tissue from the animals used for this study,[8,9] the data presented here provide evidence that 4-methyloctanoic acid contributes to the odour associated with adipose tissue of both wethers and rams, even for animals as young as 12 weeks at slaughter. In contrast, 4-methylnonanoic acid was present below its TOV, even in rams slaughtered at 30 weeks of age. Brennand[3] and Brennand and Lindsay[12] concluded from their studies that 4-methyloctanoic acid provides the backbone to the characteristic flavour of sheepmeat, due to its low OTV, while 4-methylnonanoic acid, with a higher OTV, only becomes important in strong mutton odours. Another branched chain acid implicated in the species note of sheepmeat, specifically mutton, is 4-ethyloctanoic acid.[3] It has been isolated in much higher levels from mature sheep (rams and ewes) than lambs and it is more usually associated with older animals.[4] It has a low odour threshold value in water of 1.8×10^{-3} mg kg⁻¹ [12] and is implicated in the 'goaty' odour associated with mature male goats.[13,14]

This study has demonstrated that there are significant differences in the levels of both 4-methyloctanoic and 4-methylnonanoic acids when rams of 12 and 30 weeks of age and wethers and rams of 30 weeks of age are compared. Levels of both acids were highest in mature rams where it is possible that they act as pheromones. 4-Ethyloctanoic acid has been shown to exhibit pheromone activity in goats.[15]

In the current study, some straight-chain acids were identified at significantly higher levels in adipose tissue of 30 week old rams compared to wethers of the same age and to 12 week old rams.[10] Although these acids were usually present below their TOVs, they may act synergistically and thus contribute to the typical flavour of sheepmeat. For example, octanoic acid, at a level of 50 mg kg⁻¹, has been described as 'goaty'.[7]

3.2 Phenolics

A total of nine phenols were identified (Table 2), comprising six alkylphenols and three sulfur-containing derivatives. All the alkylphenols, with the exception of 2-*iso*propyl-5-methylphenol (thymol), were present in every sample. Thiophenols were only identified in the ram samples and all nine phenols are reported only in tissue from 30 week old rams.

3.2.1 Alkylphenols. While no general trend was apparent for the alkylphenols on comparing samples from entire and castrated animals of the same age, levels of individual compounds nearly always increased with slaughter age. For rams, all the alkylphenols, apart from 2-methylphenol, were present at higher levels at 30 weeks and, for 4-methyl- and 4-ethylphenol, $p<0.001$.

All the alkylphenols identified have been described as 'phenolic'.[5] Other terms often used include 'indole-like' and 'wood preservative-like'. Descriptors applied to the 4-methyl and 4-ethyl compounds include 'animal-like' and 'sheepyard-like', respectively. OTVs in water for the alkylphenols range from 0.6 mg kg^{-1} for 4-ethylphenol, to 1.4×10^{-5} mg kg^{-1} for thymol.[5]

Where the OTV data were available,[5] the calculated log odour unit values (LOVs) were all positive, with the exception of 4-ethylphenol in the wethers slaughtered at 30 weeks (Table 2). The values for 4-methylphenol and thymol were particularly high. In addition, with the exception of 4-ethylphenol in the wethers, the LOVs for the alkylphenols were higher at 30 weeks than at 12 weeks for both rams and wethers. At a slaughter age of 30 weeks, with the exception of 2-methylphenol, all the alkylphenols possessed higher LOVs in the rams than the wethers. Alkylphenols contribute to the aroma of sheepmeat, and the notes provided are not attractive. The data support the sensory analysis carried out on meat from the same animals.[8,9] In general, the importance of these compounds increases with slaughter age, with levels usually being higher in 30 week old rams than in wethers of the same age. Ha and Lindsay[5] studied phenols present in the perinephric tissue of a market-size lamb and a mature ram. They identified a range of alkylphenols at much higher levels in the ram, but concluded that they contributed species notes to both ram and lamb tissue since they were identified well above their OTVs in both. The results reported here support those data.

Ha and Lindsay[5] blended in water authentic alkylphenols at the levels at which they were identified in ram adipose tissue and the mixture resulted in a sheepyard-like odour. By adding 4-methyloctanoic and 4-ethyloctanoic acids to the alkylphenol mixture at levels at which they were identified in ram adipose tissue,[4] a very pronounced mutton-like note was produced. Thus, it appears that certain branched chain fatty acids and selected alkylphenols both contribute to the undesirable notes associated with cooked sheepmeat. It seems that alkylphenols may intensify the species note of sheepmeat, especially in older animals, particularly rams. It is noteworthy that both 4-methyl- and 4-ethylphenol have been implicated in the body odour of live entire boars.[16]

3.2.2 Thiophenols. Thiophenol has been reported previously in lamb and ram adipose tissue[5] and Garbusov *et al.*[17] have identified 2-methylthiophenol in beef. This study is the first report of 3-methylthiophenol in any meat. Thiophenols were only identified in samples from entires and the 3-methyl derivative is only reported in the older rams. Levels of thiophenol and the 2-methyl derivative increased significantly ($p<0.01$) on increasing the slaughter age to 30 weeks. Both these compounds have been described as 'meaty' and 'broth-like' at the 10^{-2} mg kg^{-1} level,[5,17] but terms used to describe the odour of neat thiophenol include 'stench', 'burnt' and 'rubbery'.[5] The OTV in water of thiophenol is 0.05

mg kg^{-1} [3] while the OTV in air of 2-methylthiophenol is 1×10^{-4} mg l^{-1}.[18] Therefore, these compounds may contribute desirable meaty notes at the levels at which they were identified in this study, rather than to undesirable or species odours.

Thiophenol was identified in lamb and ram perinephric tissue by Ha and Lindsay[5] at levels of 8×10^{-3} and 1.8 mg kg^{-1}, respectively, and, in the older animals, they considered that it contributed to the unpleasant mutton odour.

Table 2 *Alkyl phenols and Thiophenols Identified in the Adipose Tissue of Entire and Castrated Lambs Slaughtered and 12 and 30 Weeks of Age*

Phenol	Concentration mg kg^{-1} (standard deviation)			
(Selected ions)	Castrate 12 weeks	Castrate 30 weeks	Entire 12 weeks	Entire 30 weeks
2-Methylphenol (108/107)	4.8 (2.1) LOV^a = 1.7	9.6 (3.8) LOV = 2.0	1.4×10^{-1} (9.0×10^{-2}) LOV = 0.2	4.0×10^{-1} (3.6×10^{-1}) LOV = 0.7
3-Methylphenol (108/107)	9.0×10^{-2} (6.0×10^{-2}) LOV = 0.1	1.7×10^{-1} (5.0×10^{-2}) LOV = 0.4	7.0×10^{-2} (2.0×10^{-2}) LOV = 0.0	2.4×10^{-1} (4.0×10^{-2}) LOV = 0.6
4-Methylphenol (108/107)	4.1×10^{-1} (2.1×10^{-1}) LOV = 2.3	1.3 (9.5×10^{-1}) LOV = 2.8	11 (3.9) LOV = 3.8	29 (9.1) LOV = 4.2
4-Ethylphenol (107/122)	2.2 (1.3×10^{-1}) LOV $-$ 0.6	2.0×10^{-1} (1.1×10^{-1}) LOV $-$ -0.5	1.0 (9.5×10^{-1}) LOV = 0.2	10 (2.7) LOV = 1.2
2,4-Dimethylphenol^b (122/107)	2.4 (7.6×10^{-1})	3.6 (9.8×10^{-1})	5.0×10^{-2} (1.0×10^{-2})	2.5×10^{-1} (4.0×10^{-2})
2-*iso*Propyl-5-methylphenol (thymol) (135,150)	nd	nd	5.2×10^{-1} (1.2×10^{-1}) LOV = 4.6	1.9 (5.0×10^{-2}) LOV = 5.1
Thiophenol (109/110)	nd	nd	4.0×10^{-2} (6.0×10^{-2}) LOV = -0.1	1.1×10^{-1} (9.0×10^{-2}) LOV = 0.3
2-Methylthiophenol^{b,c} (91/124)	nd	nd	6.0×10^{-2} (5.0×10^{-2}) LOV = 2.8	17 (4.4) LOV = 6.2
3-Methylthiophenol^b (91/124)	nd	nd	nd	4.0×10^{-2} (3.0×10^{-2})

^{a} LOV = Log odour unit value. Calculated using literature OTV data.[5,18]
^{b} OTV in water not reported in the literature.
^{c} LOV calculated based on OTV in air.

4 CONCLUSION

It is established that certain branched chain fatty acids and phenols both contribute to the characteristic flavour of sheepmeat,[2-5] and we have shown by sensory analysis that scores for the term 'farmyard' increase with both slaughter age and entirety.[8,9] In this study, levels of 4-methyloctanoic and 4-methylnonanoic acids increase with both age and entirety and are significantly greater in rams slaughtered at 30 weeks of age compared to both 30 week old wethers and 12 week old rams. 4-Methyloctanoic acid makes an increasingly important contribution to sheepmeat aroma with both entirety and slaughter age, but 4-methylnonanoic acid was identified below its TOV in all the samples examined, including tissue from 30 week old rams. Alkylphenols contribute to the odour of sheepmeat in all four samples but their odour impact is greater for the 30 week old than for the 12 week old animals and they make the greatest contribution to the rams slaughtered at 30 weeks. Thiophenols were only identified in ram tissue where they may contribute desirable meaty notes. Their levels also increased with slaughter age.

ACKNOWLEDGEMENTS

The Meat and Livestock Commission, U.K., is thanked for a studentship (to M.M.S.). The Agricultural College, Edinburgh and the University of Reading reared the 12 week old and the 30 week old animals, respectively. Samples of 4-methyloctanoic and 4-methylnonanoic acids were provided by Tastemaker, Milton Keynes, U.K.

REFERENCES

1.   M.M. Sutherland and J.M. Ames, *J. Sci. Food Agric.*, 1995, **69**, 403.
2.   E. Wong, L.N. Nixon and C.B. Johnson, *J. Agric. Food Chem.*, 1975, **23**, 495.
3.   C.P. Brennand, Ph.D. Dissertation, University of Wisconsin-Madison, 1989.
4.   J.K. Ha and R.C. Lindsay, *Lebensm.-Wiss. u. -Technol.*, 1990, **23**, 433.
5.   J.K. Ha and R.C. Lindsay, *J. Food Sci.*, 1991, **56**, 1197.
6.   J.K. Ha and R.C. Lindsay, *J. Dairy Sci.*, 1990, **73**, 1988.
7.   C.P. Brennand, J.K. Ha and R.C. Lindsay, *J. Sensory Studies*, 1989, **4**, 105.
8.   M.M. Sutherland, D.B. MacDougall and J.M. Ames, in 'Trends in Flavour Research', eds. H. Maarse and D.G. van der Heij, Elsevier Science B.V., Amsterdam, 1994, p. 157.
9.   M.M. Sutherland, Ph.D. Dissertation, University of Reading, 1996.
10. M.M. Sutherland and J.M. Ames, *J. Agric. Food Chem.*, submitted.
11. T.P. Heil and R.C. Lindsay, *J. Environ. Sci. Health*, 1988, **B23**, 475.
12. C.P. Brennand and R.C. Lindsay, *Lebensm.-Wiss. u. -Technol.*, 1992, **25**, 357.
13. H. Boelens, H.G. Haring and D. de Reijke, *Perfumer and Flavorist*, 1983, **8**, 71.
14. T. Sugiyama, H. Sasada, J. Masaki and K. Yamashita, *Agric. Biol. Chem.*, 1981, **45**, 2655.
15. T. Sugiyama, H. Matsuura, H. Sasada and J. Masaki, *Agric. Biol. Chem.*, 1986, **50**, 3049.
16. R.L.S. Patterson, *J. Sci. Food Agric.*, 1967, **18**, 8.
17. V.G. Garbusov, W.G. Rehefeld, R.V. Golovnia and M. Rothe, *Die Nahrung*, 1976, **20**, 235.
18. F.A. Fazzalari, 'Compilation of Odor and Taste Threshold Values Data', American Society for Testing and Materials, Philadelphia, 1978.

SCREENING FOR AND CONTROL OF DEBITTERING PROPERTIES OF CHEESE CULTURES

G. Smit,* Z. Kruyswijk,* A.H. Weerkamp,* C. de Jong† and R. Neeter†

*Department of Microbiology and †Department of Analytical Chemistry, Netherlands Institute for Dairy Research (NIZO), P.O. Box 20, 6710 BA Ede, The Netherlands

1 INTRODUCTION

Bitterness is one of the most common off-flavours in cheese.[5,11] We have focused on the possibilities of predicting and controlling the debittering abilities of cheese (adjunct) cultures, which are used for accelerated ripening or flavour modification of Gouda cheese. The bitter-tasting C-terminal part of β-casein, the so-called C-peptide (a.a. 193–209), formed by the action of rennet and starter organisms, is a major cause of bitterness in Gouda cheese,[8,11] as well as in cheddar cheese.[5] It has been found that the use of certain starter cultures and/or adjunct cultures especially give rise to the development of a bitter taste in cheese and therefore these strains are characterized as 'bitter' or 'non-bitter' strains.[11,12] Based on this, an HPLC method was developed, which allows the quick monitoring of the formation and degradation of the C-peptide. Several strains of lactic acid bacteria have been tested for their ability to degrade this peptide in relation to their growth conditions. Moreover, cheese production experiments were undertaken to confirm the results from the laboratory experiments.

2 MATERIAL AND METHODS

2.1 Bacterial Strains, Culture Conditions and Treatments

Strains were obtained from the NIZO culture collection. They were routinely stored at −135 °C and grown in low-fat milk at 20 °C (the Bos starter culture), 30 °C (mesophilic lactococci) or 37 °C (thermophilic strains). Cultures were cultivated in sterilized low-fat milk under acidifying or pH-controlled conditions. In the latter case, the cultures were maintained at a pH of 5.7 using 6.5 M NaOH. Lysed cell extracts were obtained by sonication of a stationary phase culture, using a Heat Systems Sonicator XL four times for 30 s each at 0 °C.

2.2 Purification of C-Peptide and HPLC Analyses

C-peptide was purified according to Vreeman et al.[13] Degradation of C-peptide upon incubation with bacterial cells was monitored by HPLC. Samples were pre-purified (see below) and analysed on a C_{18}-reverse phase column (PLPS 300 Å, 150 × 4.6 mm). A 10%–90% (v/v) acetonitrile gradient was used as the mobile phase, with of 0.1% TFA in both solvents. Eluting compounds were monitored at 260 nm.

2.3 Cheese Trials and Analyses

Gouda cheese was made from 200 l portions of pasteurized (10 s, 74 °C) milk, as described previously.[9] Cheese milk was inoculated with 0.7%–1.0% starter culture, grown for 16 h in low-fat milk, and, if appropriate, 2.5% adjunct culture was added. The adjunct cultures were cultivated for 40 h to minimize acidifying activity (see below). Cheeses were ripened for up to six months at 13°C and analysed at various intervals. Organoleptic analyses were performed at 6, 13 and 26 weeks with a trained panel. Particular attention was focused on the bitterness of the cheeses, which was scored on a scale from 0 (no bitterness) up to 4 (extremely bitter).

2.4 Analysis of Volatile Compounds

Volatile components in cheese were identified using purge-and-trap thermal desorption cold-trap gas chromatography mass spectrometry (PTTDCT–GC–MS), as described previously.[1,7] Briefly, 20 ml of a cheese slurry, obtained by homogenization of a mixture of cheese and double-distilled water (1:2 w/w), were prepared and used immediately after preparation. The samples were purged with 150 ml min^{-1} helium for 30 min at 40 °C and volatile components were trapped on a borosilicate glass trap containing Carbotrap (80 mg, 20–40 mesh, Supelco) and Carbosieve SIII (10 mg, 60–80 mesh, Supelco). The trapped compounds were transferred onto a capillary column of a gas chromatograph using the Chrompack PTI injector (Chrompack, Middelburg, The Netherlands) in the TDCT mode, by heating the trap for 10 min at 250 °C. A narrow injection band was achieved by cryofocusing at −100°C. The conditions for chromatographic separation and mass spectrometry have been described earlier.[14] Structures were assigned by spectrum interpretation, comparison of the spectra with bibliographic data, and comparison of retention times with those of reference compounds.

3 RESULTS AND DISCUSSION

3.1 Development of the Bitter Assay and Screening of Cultures

A laboratory assay was developed in which C-peptide was incubated with bacterial cultures. Three ml of the cultures were harvested at the stationary phase of growth, milk components were clarified by raising the pH to 6.8 (NaOH) and addition of sodium citrate to a final concentration of 1%, and centrifuged in an Eppendorf centrifuge (14000g for 15 s). Subsequently, the cell pellets were resuspended in 50 mM citrate, 500 mM NaCl buffer (pH 5.4) to an absorbance ($A_{578\ nm}$) of 2.5 and incubated with 100 μg C-peptide at 30 °C. With respect to the salt concentration and pH, the conditions in the assay resemble those in the cheese. At regular intervals, 250 μl samples were taken. Enzyme activity in these samples was destroyed by heat treatment (20 min at 66 °C) and, subsequently, the samples were clarified by centrifugation and the amount of C-peptide determined by HPLC. As shown in Figure 1, the amount of C-peptide decreased during the incubation.

Several lactic acid bacteria, grown under acidifying conditions in milk, were screened by means of the assay. For each strain tested, the debittering activity in the assay was expressed as the decrease in C-peptide per hour and per amount of cells (A_{578} units). The strains were found to differ significantly in their ability to degrade the bitter-tasting C-peptide (Table 1). Some strains were hardly able to degrade the C-peptide, while others were capable of degrading the peptide very rapidly. Moreover, the peptide profiles which were formed during the degradation of the C-peptide were different for a number of strains.

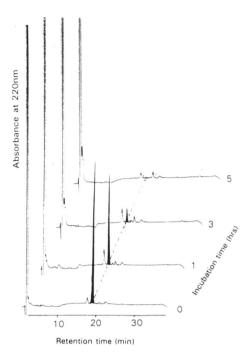

Figure 1 *Degradation of the bitter-tasting C-peptide during incubation with* L. acidophilus *I233 cells in the assay*

Table 1 *Debittering Activity of Several Lactic Acid Bacteria[a]*

Strain or Culture	Debittering Activity (μg C-pep h^{-1} A^{-1})
L. helveticus T75	0.0
Mixed strain culture T57	0.2
L. helveticus T19	1.8
Lactococcus lactis B48	4.6
L. helveticus T96	5.6
Mixed strain culture TM5	6.5
L. helveticus B5	6.5
Lactococcus lactis B65	7.3
L. helveticus T172	18.5
L. helveticus T100	19.0
L. acidophilus T74	20.0
L. helveticus T18	34.0
L. acidophilus I233	38.0

[a] Debittering activity is expressed as the decrease of C-peptide per hour.

3.2 Effect of Growth Conditions on Debittering Activity

Figure 2 shows that growth conditions of the culture significantly affect the debittering activity. In general, cells grown under pH-controlled conditions have a stronger debittering ability than cells grown under acidifying conditions. Therefore, strains cannot simply be marked as 'bitter' or 'non-bitter'.[12] This result might open the possibility of using cultures for cheese-making which had previously been disregarded because they had been marked as 'bitter'.

3.3 Possible Role of Cell Lysis in Debittering Activity

Since the C-peptide is too large to cross the cell membrane of the lactic acid bacteria, such peptides are not degraded by the cell envelope proteinase of the lactococci[3,4] and since Tan *et al.*[10] proposed lysis of starter cultures to be involved in debittering activity, we hypothesised that lysis of the bacteria might play an important role in the ability of the cells to degrade the C-peptide. To test whether the different cultures have the enzyme potential to do this, lysed cell extracts were tested in the assay. All strains tested were found to be able to degrade the C-peptide upon lysis (Figure 2); therefore, differences in the release of intracellularly-located enzymes (*e.g.*, peptidases such as PepN[10]) play a major role in debittering activity.

Preliminary results indicate that different growth conditions (see above) of the bacteria result in different sensitivity of the cells to lysis; *e.g.*, cultures grown under pH-controlled conditions appear to be more sensitive to lysis than cells grown under acidifying conditions. This difference might explain the differences in debittering activity (Figure 2). Moreover, it cannot be excluded that different levels of enzyme activity, as described for some peptidases by Meijer *et al.*,[6] may also contribute to this difference.

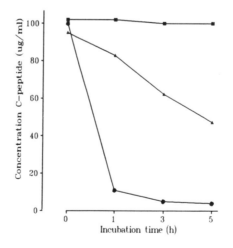

Figure 2 *Influence of growth conditions on debittering activity of culture T57 – square acidifying; triangle pH-controlled; •: acidifying/lysed*

3.4 Cheese Trials

In order to test whether results from the debittering assay can be used to predict bitterness development in cheese, cheeses were made with a number of adjunct cultures in conjunction with the mesophilic starter Bos as the acidifying culture. As shown in Table 2, results from the assay can indeed be used to predict the debittering activity of cultures in real cheese.

It is important to note that the effect of culture conditions on the debittering activity of adjunct cultures was not only found in the laboratory assay, but also in the cheese trials. For instance, culture TM5 was found to have an increased debittering activity when grown under pH-controlled conditions. The same culture resulted also in non-bitter cheeses, whereas the reference cheese, prepared with the addition of TM5 grown under acidifying conditions, *did* result in bitterness (Table 2).

Table 2   *Debittering Activity of Thermophilic Adjunct Cultures in the Assay and in Cheese after Three Months of Ripening*

Starter culture	*Adjunct culture*	*Growth conditions[a]*	*Debittering activity[b]*	*Bitter score in cheese[c]*
Bos	T19	acidifying	1.8	1.7
Bos	T172	acidifying	18.5	<0.1
Bos	I233	acidifying	38.0	<0.1
Bos	TM5	acidifying	6.5	1.9
Bos	TM5	pH-controlled	10.1	0.1

[a]   Cultures were grown for 40 h under acidifying or pH-controlled conditions at pH 5.7;
[b]   Debittering activity of adjunct cultures determined in the bitter assay (see Table 1);
[c]   Bitter scored on a scale of 0 (absent) to 4 (very strong).

Several strains with the highest debittering activity were tested for their ability to debitter cheese. A starter which is known to develop a strong bitter taste in cheese was used. As shown in Table 3, selected strains with high debittering activity were indeed able to significantly reduce bitterness in such cheeses. These results show the usefulness of the laboratory assay in the selection of debittering strains and in growth conditions of adjunct cultures by which bitterness in cheese can be controlled.

Table 3   *Influence of* L. acidophilus *I233 on the Development of Bitterness in Cheese after Three months of Ripening*

Starter Culture	*Adjunct Culture*	*Bitter Score in Cheese[a]*
Bos/T72	None	2.7
Bos/T72	I233	0.2
13M/C17	None	1.5
13M/C17	I233	0.1

[a]   Bitter scored on a scale of 0 (absent) to 4 (very strong).

It is noteworthy, that debittering activity in the assay should not be used as the absolute or sole criterion in the characterization of strains with respect to their debittering capacity, since it cannot be excluded that peptides other than the C-peptide contribute to the bitterness of cheeses.[2] Moreover, compounds other than peptides might contribute to or enhance the bitterness.[14] This might also explain the strain-dependency which was observed for some of the strains in the cheese trials (data not shown). In such cases, bitter-tasting peptides other than the C-peptide might have been released in the cheese or degradation products of the C-peptide might still be large enough to result in a bitter taste.

3.5 Effect of Growth Conditions on Production of Volatile Flavour Components

Volatile flavour components in cheese were determined in order to establish whether growth conditions of the adjunct culture might affect the formation of flavour components other than bitter ones. For this purpose, cheeses made with culture TM5, grown under different growth conditions, were examined since there was a clear difference with respect to bitterness in these cheeses (Table 2). However, when the same cheeses were analysed for volatile flavour components, no significant difference could be detected (Figure 3). Organoleptic evaluation of the cheeses corroborated the latter results, since, apart from the bitter score, the cheeses were judged similar. Therefore, for this culture, the effect of growth conditions apparently does not strongly affect the formation of volatile flavour components.

In conclusion, the debittering assay allows rapid screening for and prediction of the debittering activity of lactic acid bacteria. The mechanism which accounts for the differences found among strains might be a result of differences in (sensitivity to) lysis under cheese conditions. Bitterness in cheese can be controlled by adaptation of the growth condition and/or by the use of highly debittering strains as adjunct culture.

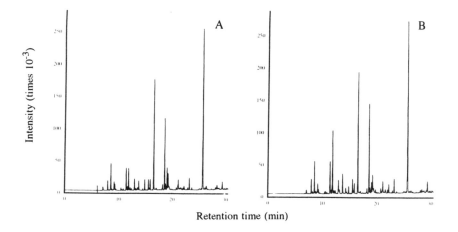

Figure 3   *GC-aromagrams of three-month-old cheeses prepared with culture TM5 grown under acidifying (A) and pH-controlled (B) conditions*

ACKNOWLEDGEMENTS

This work was financially supported by EC grant AIR$_2$-CT93-1531. The authors thank W. Meijer for discussions and critical reading of the manuscript.

REFERENCES

1. H.T. Badings, C. de Jong and R.P.M. Dooper, *J. High Res. Chrom. Comm.*, 1985, **8**, 755.
2. E. Bumberger and H.-D. Belitz, *Z. Lebensm. Unters. Forsch.*, 1993, **197**, 14.
3. F.A. Exterkate, A.C. Alting and C.J. Slangen, *System. Appl. Microbiol.*, 1995, **18**, 7.
4. E.R.S. Kunji, A. Hagting, C.J. de Vries, V. Juillard, A.J. Haandrikman, B. Poolman and W.N. Konings, *J. Biol. Chem.*, 1995, 1569.
5. L. Lemieux, R. Puchades and R.E. Simard, *J. Food Sci.*, 1989, **54**, 1234.
6. W.C. Meijer, J.D. Marugg and J. Hugenholtz, *Appl. Environ. Microbiol.*, 1996, **62**, 156.
7. R. Neeter and C. de Jong, *Voedingsmiddelentechnologie*, 1992, **25**, 11:9.
8. J. Stadhouders, G. Hup, F.A. Exterkate and S. Visser, *Neth. Milk Dairy J.*, 1983, **37**, 157.
9. J. Stadhouders, L. Toepoel and J.T.M. Wouters, *Neth. Milk Dairy J.*, 1988, **42**, 183.
10. P.S.T. Tan, T.A.J.M. van Kessel, F.L.M. van de Veerdonk, P.F. Zuurendonk, A.P. Bruins and W.N. Konings, *Appl. Environ. Microbiol.*, 1993, **59**, 1430.
11. S. Visser, C.J. Slangen, G. Hup and J. Stadhouders, *Neth. Milk Dairy J.*, 1983, **37**, 181.
12. S. Visser, F.A. Exterkate, C.J. Slangen and G.J.C.M. de Veer, *Appl. Environ. Microbiol.*, 1986, **52**, 247.
13. W.J. Vreeman, P. Both and C.J. Slangen, *Neth. Milk Dairy J.*, 1994, **48**, 63.
14. R. Warmke and H.-D. Belitz, *Z. Lebensm. Unters. Forsch.*, 1993, **197**, 132.
15. A.H. Weerkamp, N. Klijn, R. Neeter and G. Smit, *Neth. Milk Dairy J.*, 1996, **50**, in press.

THE CO-OXIDATION OF CAROTENOIDS BY LIPOXYGENASE IN TOMATOES

C.L. Allen and J.W. Gramshaw

Procter Department of Food Science, University of Leeds, Leeds, LS2 9JT, U.K.

1 INTRODUCTION

Lipoxygenase (EC 1.13.11.12, LOX) catalyses the oxygenation of polyunsaturated fatty acids containing a (Z),(Z)-1,4-pentadiene system to conjugated hydroperoxy acids (HPO), which are further metabolized by hydroperoxide lyase (HPO lyase). HPO lyase catalyses the cleavage of HPO to aldehydes, such as 3(Z)-nonenal and hexanal from linoleic acid HPO and 3(Z),6(Z)-nonadienal and 3(Z)-hexenal from linolenic acid HPO. Tomato fruit possess a lyase activity that cleaves only the 13-HPO and thus produces only C_6-aldehydes,[1] which contribute to the characteristic aroma.

Many volatiles present in the tomato fruit[2] are thought to be formed as secondary products. Thus, the LOX reaction involving the oxidation of fatty acids produces a range of free radicals, such as fatty acid radicals,[3] alkyl radicals[4] and peroxyl radicals,[4,5] as well as highly active species, such as superoxide[5] and singlet oxygen.[6] These reactive species readily attack unsaturated compounds, e.g., carotenoids, to produce volatiles that play a key role in the aroma profile of the fruit.[7] Non-volatile degradation products of β-carotene formed during enzymic co-oxidation with lipoxygenase and linoleic acid have also been identified.[8]

Our objective was to isolate, purify and partially characterize LOX from tomato fruit and use the purified LOX to catalyse the oxidation of linoleic acid in the presence of β-carotene. Volatiles produced were determined by gas chromatography and mass spectroscopy. Rates of carotene bleaching in the presence of antioxidants were determined by spectrometry, using two isoenzymes of LOX; L-1 from soyabean and L-2 from tomato. Hydroperoxides produced by the isoenzymes were analysed using HPLC.

2 EXPERIMENTAL SECTION

2.1 Lipoxygenase

2.1.1 Extraction. Ripe tomato fruits were washed, cooled and sliced, the seeds removed, and the flesh homogenized in Tris-HCl buffer (pH 8.0) containing glycerol at 4 °C. The homogenate was centrifuged and calcium chloride added to the supernatant to precipitate pectin. After 2 hours, the suspension was centrifuged and the supernatant filtered and stored (−20 °C) until purified.

2.1.2 Purification. The depectinized solution was treated with ammonium sulfate (35%–60%), the precipitate dissolved in Tris-HCl (pH 8.0) containing glycerol, applied to a G-100 Sephadex column and eluted with the same buffer. Active fractions were pooled, placed onto a DEAE–cellulose column in the same buffer, and eluted using a linear gradient (0 M to 0.5 M NaCl) in the buffer. Active fractions were collected and applied to a Phenyl Sepharose CL-4B column and eluted with Tris-HCl buffer (pH 8.0) containing glycerol. Active fractions were collected and applied to a hydroxyapatite chromatography column. After washing with Tris-HCl buffer (pH 8.0) containing glycerol, a linear gradient (10 mM to 0.4 M phosphate, pH 8.0) was applied to the column. Active fractions eluted from the last column were collected and concentrated.

2.1.3 Enzyme assay. The enzymic activity was determined by measuring the increase in absorbance at 234 nm due to hydroperoxide production. Phosphate buffer (pH 6.4) was used for the isoenzyme isolated from tomatoes (L-2) and borate buffer (pH 9.0) was used for soyabean isoenzyme (L-1). Linoleic acid solution (1.5 mM) was prepared by adding linoleic acid (4.7 μl) and ethanol (50 μl) to the appropriate buffer (total volume 10 ml).

Each assay used oxygen-saturated buffer (2 ml), linoleic acid (1 ml) and enzyme preparation (5–200 μl) in a 1 cm cell. The reaction was started by addition of enzyme and the absorbance was followed using a Pye-Unicam SP8-200 recording spectrophotometer, fitted with a temperature controlled cell (25 °C).

2.1.4 Characterisation. The protein concentrations of the LOX preparations were measured, using a Bio-Rad protein assay kit and BSA as standard.

The apparent molecular mass was determined by gel permeation chromatography on a Sephadex G-100 column eluted with Tris-HCl (pH 8.0), containing glycerol, and calibrated using alcohol dehydrogenase, BSA, carbonic anhydrase and cytochrome c.

The purity of the enzyme extracts was determined by analytical isoelectric focusing (IEF), using a polyacrylamide gel (110 × 240 × 0.5 mm) and carrier ampholytes in the pH range 3.5–10. Proteins were stained with Coomassie Brilliant Blue and presence of LOX in the purified sample was determined with an *o*-dianisidine stain.[9] The isoelectric point (pI) of LOX was estimated by (a) measuring the pH gradient across the gel and (b) using standard pI markers. The pI value was also determined by preparative IEF in a liquid medium.

2.2 Analysis of Hydroperoxides

Phosphate buffer (pH 6.4) was used for the isoenzyme isolated from tomatoes (L-2) and borate buffer (pH 9.0) for that from soyabean (L-1). A solution of linoleic acid (10 mM) was prepared by adding linoleic acid (77 μl) and Tween-20 (77 μl) to the appropriate buffer (total volume 25 ml).

The depectinized enzyme solution (Section 2.1.1) (4 ml) was incubated with linoleic acid solution (16 ml) for 15 min (25 °C). Hydroperoxides were extracted with diethyl ether (2 × 25 ml), dried with anhydrous sodium sulphate, the diethyl ether was evaporated, and then products were dissolved in hexane and analysed by HPLC, using a Zorbax SIL column (250 × 4.6 mm) eluted (0.7 ml min⁻¹) with hexane/isopropanol/acetic acid (98:2:0.05 v/v/v); the separation was followed by monitoring absorbance at 234 nm.

2.3 Co-oxidation of Carotenoids

2.3.1 Bleaching. A stock solution of β-carotene (Sigma) was prepared by dissolving the carotenoid (1 mg) in chloroform (1 ml) containing Tween 80 (36 μl). A phosphate or borate buffer was used as appropriate (Section 2.1.3). Each buffer was used under (a) anaerobic (degassed and N₂ bubbled through), (b) aerobic (O₂ bubbled through) or (c) semi-aerobic (partially degassed) conditions.

Chloroform was evaporated from β-carotene stock solution (0.1 ml) and the residue suspended in appropriate buffer (17.5 ml) containing EDTA. Linoleic acid (2.5 ml, Section 2.2) was added and the solution was stored in the dark and on ice until use. Each bleaching assay contained linoleate/β-carotene (2 ml) and LOX (0.1 ml). The absorbance was followed at 460 nm over 10 min (25 °C).

Caffeic acid, chlorogenic acid, (±)-α-tocopheryl acetate, propyl gallate, butylated hydroxyquinone (BHQ), butylated hydroxyanisole (BHA) and ascorbic acid were tested as antioxidants. Each antioxidant was used at a concentration of 1×10^{-5} M in the final solution. Ascorbic acid was dissolved in distilled water, whilst the others were dissolved in ethanol. This experiment was repeated using lycopene.

2.3.2 Formation of volatiles. Chloroform was evaporated from β-carotene stock solution (0.1 ml, Section 2.3.1) and the residue suspended in borate buffer (19 ml, pH 9.0), containing EDTA and linoleic acid (1 ml, section 2.2). L-1 from soyabean (75.73 μmol min^{-1} mg^{-1}) was added and the reaction mixture allowed to react for 10 min (25 °C). The headspace was continuously swept with air onto a Tenax trap (80 × 4 mm) containing Tenax TA (250 mg) until the colour had disappeared. The headspace from ten replicates was collected on the same trap.

The trap was eluted with distilled diethyl ether (4 ml) and the extract evaporated to *ca.* 200 μl using N$_2$. The extract was analysed by GLC and by coupled GLC–MS (Kratos MS80 RFA mass spectrometer), using fused-silica open tubular columns, BP1 (25m × 0.33 mm; 0.5 μm df) and BP20 (50 m × 0.33 mm; 0.5 μm df), and helium at 1.2 m s^{-1}. The temperature programme was 50 °C for 5 min rising by 5 °C min^{-1} to 280 °C. EI mass spectra were generated using an energy of 70 eV at a source temperature of 150 °C and collected over the range 600–20 a.m.u.

Table 1   *Purification of Lipoxygenase from Tomatoes*

Sample	Volume (ml)	Protein (mg/ml)	Total activity (μmol/min)	Specific activity (μmol min^{-1} mg^{-1})	Yield (%)	Purification
Crude Ext.	250	1.10	31.62	0.12	100	—
Depect. Ext.	167	0.71	15.32	0.13	48.5	1.1-fold
35-60% A.S.	10	5.20	13.05	0.25	41.3	2.2-fold
Gel Filtration	60	0.36	10.62	0.49	33.6	4.3-fold
Anion Exc.	40	0.107	8.17	1.91	26.0	16.6-fold
HIC	40	0.025	6.70	6.70	21.2	58.3-fold
HC	32	0.0082	3.25	12.40	10.3	107.8-fold

Ext. = extract; Depect. = depectinated; A.S. = ammonium sulfate saturation; Exc. = exchange; HIC = hydrophobic interaction chromatography; HC = hydroxyapatite chromatography.

3 RESULTS AND DISCUSSION

3.1 Lipoxygenase

3.1.1 Purification. Typical values obtained during purification of tomato lipoxygenase are given in Table 1. Lipoxygenase was purified 107.8-fold, but the loss in total activity during purification was 89.7%. Buffers used during the extraction and purification of lipoxygenase were maintained at pH 8.0 throughout since this minimised loss of activity. Glycerol was added where possible to buffers used in the purification since this had a stabilising effect on the enzyme.

The removal of pectins from the crude extract by the addition of $CaCl_2$ removed over 50% of the protein present in the initial sample. However, the total activity of the enzyme was also almost halved so that little purification was achieved. Recent reports[10,11,12] have shown the existence of a membrane-associated lipoxygenase in tomato fruit, similar to soluble L-2 found in tomato, and with a similar optimum pH and molecular mass, but with a different pI value. Difficulty can arise in a crude extract for the determination of LOX activity as there are many oxidising enzymes present which can interfere with specific enzyme assays. It appears that during the removal of pectins by $CaCl_2$, the membrane-associated lipoxygenase was also removed.

The 35%–60% saturation cut selected matched that of other workers[13] and produced a 2.2-fold purification. Loss of the enzyme through the four chromatographic procedures was related to the length of time the enzyme remained on the column and was largest using the hydroxyapatite column. This column showed a low flow rate and the enzyme remained on the column much longer than with the other columns; in addition, glycerol could not be used to stabilise the enzyme here.

One major protein band was seen upon staining with Coomassie Brilliant Blue on an analytical IEF gel. This was confirmed as LOX by use of *o*-dianisidine. The purification technique was repeated several times and the appearance of a minor band was seen on one occasion only. This could be due to a contaminating enzyme but, most probably, since the sample was two days old, was due to denatured enzyme.

3.1.2 Characterisation. The molecular mass, estimated by gel permeation chromatography of L-2 from tomatoes, was 94,500 Da. This figure is in agreement with those previously reported.[10,12] The pI of the protein, determined by analytical isoelectric focusing, was found to be 5.1 and that determined by preparative IEF was 5.06 (published value[11] 5.1). The pH optimum was found to be between 6.3 and 6.35; however, the activity was found to be fairly constant over the pH range 6.25–6.4 (published value[13] 6.3).

3.1.3 Ratio of Hydroperoxides. Lipoxygenase from tomatoes (L-2) produced a ratio of 9- to 13-HPO of 24:1, whereas soyabean (L-1) showed a value of 9:91 (value given for L-2 in tomato fruit is 24:1[14] and for soyabean L-1[15], 5:95).

3.2 Co-oxidation Reactions of Carotenoids

3.2.1 Bleaching. Optimum bleaching conditions for β-carotene were found to differ for each isoenzyme. Tomato lipoxygenase (L-2) showed the highest bleaching activity under aerobic conditions, whilst L-1 from soyabean reacted fastest under semi-aerobic conditions.

Antioxidants led to the inhibition of bleaching and three groups of antioxidants were recognised (Figure 1). Compounds in Group 1, butylated hydroxyquinone (BHQ) and butylated hydroxyanisole (BHA), gave complete inhibition of the bleaching reaction. Inhibition of 50%–70% occurred with Group 2 compounds – caffeic acid, chlorogenic acid, propyl gallate and (±)-α-tocopheryl acetate. Ascorbic acid (Group 3) showed no inhibition. The two isoenzymes of lipoxygenase gave similar results with both β-carotene and lycopene (a typical result is shown in Figure 1).

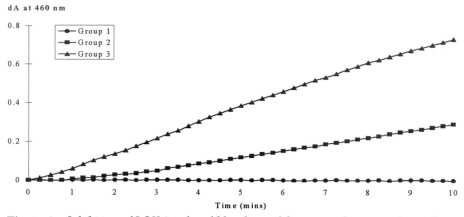

Figure 1   *Inhibition of LOX-2 induced bleaching of β-carotene by antioxidants. Group 1 = BHA, BHQ; Group 2 = caffeic acid, chlorogenic acid, propyl gallate, α-tocopheryl acetate; Group 3 = control, ascorbic acid*

The antioxidants, BHQ and BHA, are renowned for their high antioxidant effect and it is not unusual for complete inhibition of bleaching to take place in their presence. Ascorbic acid was the only hydrophilic antioxidant used and lack of activity may be due to lack of solubility in the bleaching medium. Since lipoxygenase acts at a micellular surface in model systems, ascorbyl palmitate would probably have been a better choice.

3.2.2 *Volatiles*. Table 2 shows the degradation products of β-carotene, identified using gas chromatography–mass spectrometry and retention indices. The three compounds were found in the headspace above the reaction mixture of β-carotene, linoleic acid and LOX (L-1 from soyabean). These preliminary experiments were chosen as L-1 is commercially available and β-carotene is easily accessible and, in our hands, more stable than lycopene.

Table 2   *Major Co-oxidation Products of β-Carotene*

Compound	Molecular Mass	Structure	Identity
β-Ionone	192	$C_{13}H_{20}O$	a, b
β-Cyclocitral	152	$C_{10}H_{16}O$	a, b
5,6-Epoxy-β-ionone	208	$C_{13}H_{20}O_2$	a

a   MS;
b   Retention Index of known standard on 2 columns of different polarity.

Figure 2 shows the structure of β-carotene in relation to the three volatiles produced. Co-oxidation of squalene and lycopene using LOX-1 from soyabean and LOX-2 from tomatoes is presently being studied.

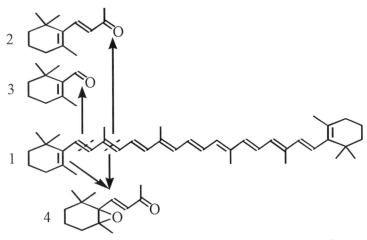

Figure 2   *Oxidative fission of β-carotene to give volatile products. 1 = β-carotene; 2 = β-ionone; 3 = β-cyclocitral; 4 = 5,6-epoxy-β-ionone*

ACKNOWLEDGEMENTS

The authors would like to thank Mr. I.D. Boyes for assistance with GC–MS analysis and the B.B.S.R.C. and Unilever p.l.c. for generous financial support.

REFERENCES

1. T. Galliard and J.A. Matthew, *Phytochem.*, 1977, **16**, 339.
2. R.G. Buttery and L.C. Ling, A.C.S. Symposium Series, 1993, **525**, 23.
3. J.J.M.C. Groot, G.J. Garssen, J.F.G. Vliegenthart and J. Boldingh, *Biochim. Biophys. Acta*, 1973, **326**, 279.
4. M.J. Nelson, R.A. Cowling and S.P. Seitz, *Biochem.*, 1994, **33**, 4966.
5. W. Chamulitrat, M.F. Hughes, T.E. Eling and R.P. Mason, *Arch. Biochem. Biophys.*, 1991, **290**, 153.
6. J.R. Kanofsky and B. Axelrod, *J. Biol. Chem.*, 1986, **261**, 1099.
7. M.A. Stevens, *J. Am. Soc. Hort. Sci.*, 1970, **95**, 461.
8. C. Zinsou, *Physiol. Veg.*, 1971, **9**, 149.
9. M.O. Funk, M.A. Whitney, E.C. Hausknecht and E.M. O'Brien, *Anal. Biochem.*, 1985, **146**, 246.
10. C.G. Bowsher, B.J.M. Ferrie, S. Ghosh, J. Todd, J.E. Thompson and S.J. Rothstein, *Plant Physiol.*, 1992, **100**, 1802.
11. M.J. Droillard, M.A. Rouetmayer, J.M. Bureau and C. Lauriere, *Plant Physiol.*, 1993, **103**, 1211.
12. J.F. Todd, G. Paliyath and J.E. Thompson, *Plant Physiol.*, 1990, **94**, 1225.
13. J.L. Bonnet and J. Crouzet, *J. Food Sci.*, 1977, **42**, 625.
14. J.A. Matthew, *Lipids*, 1977, **12**, 324.
15. B. Axelrod, T.M. Cheesebrough and S. Laakso, 'Methods of Enzymology' (Vol. 71), Academic Press, New York, 1981, 441.

NEW SULFUR-BEARING COMPOUNDS IN BUCHU LEAF OIL

Gerhard E. Krammer, Hans-Jürgen Bertram,[*] Jürgen Brüning, Matthias Güntert,[*] Stefan Lambrecht, Horst Sommer, Peter Werkhoff and Johannes Kaulen

Haarmann & Reimer GmbH, Corporate Research, D-37603 Holzminden, Germany

[*]Haarmann & Reimer GmbH, Flavor Division, D-37603 Holzminden, Germany

1 INTRODUCTION

Buchu leaf oil is obtained by standard steam-distillation from the leaves and stems of different *Agathosma* species belonging to the family of *Rutaceae*. The most important species *Agathosma betulina* (Berg.) Bartl. *et* H.L. Wendl. (syn. *Barosma betulina*) is a shrub endemic to the Western Cape mountains of South Africa. The essential oil is commercially obtained from wild and cultivated plants of *A. crenulata* (L.) Hook and of *A. betulina*. Recent studies[1] showed that the yield from *A. betulina* plant material is significantly higher during the hot dry summer period (2%) than in winter (1%). Previous studies on the chemistry of buchu leaf oil reported clear differences between the essential oil of the different species.[2–4] Among major components (+)-limonene and (–)-pulegone, the presence of (+)-menthone (**1**) and (–)-isomenthone (**2**) has been described[2,5] (Figure 1). The configurations of these compounds in other essential oils such as peppermint oil, in contrast, are always those of (–)-menthone and (+)-isomenthone.[6] These facts and the occurrence of the buchu camphors, diosphenol and ψ-diosphenol, stress the interest of the flavour chemistry of buchu leaf oil.

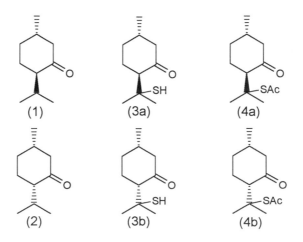

Figure 1 *Important* p-*menthan-3-one derivatives in buchu leaf oil*

Because of its typical 'cassis' flavour note, buchu leaf oil is frequently used as a valuable ingredient in flavours and fragrances. This powerful and characteristic odour impression is attributed to *p*-menthan-8-thiol-3-one (**3**) and its thiol acetate (**4**).[2,7,8] Recent studies reported the synthesis and the analysis of the optically pure stereoisomers of both compounds.[9–12] Varying diastereomeric ratios of *trans*-(1*S*, 4*S*) and *cis*-(1*S*, 4*R*) thiols (**3a**, **3b**) in authentic oils have been reported.[13,14] The corresponding acetates *trans*-(1*S*, 4*S*) and *cis*-(1*S*, 4*R*) (**4a**, **4b**) have also been found to belong to the small group of (1*S*)-configured monoterpenes (Figure 1).

These findings indicate that there are differences in the biosynthesis of monoterpenes in buchu compared to other plants. Several *S*-prenyl thioesters, *e.g.*, *S*-(3-methylbut-2-enyl) ethanethioate, *S*-(3-methylbut-2-enyl) 2-methylpropanethioate and *S*-(3-methylbut-2-enyl) 3-methylbutanthioate, have been found in related species (*A. apiculata* G.F.W. Mey, *A. clavisepala* R.A. Dyer, *A. puberula* Fourcade, A. *rosmarinifolia* (Bartl.) Williams).[15] From the perspective of biosynthesis, this is an interesting finding, since the biosynthesis of terpenes involves isoprene building blocks such as isopentenyl pyrophosphate.

In order to learn more about the sulfur chemistry in buchu leaf oil, the aim of the work reported here was the isolation and structure elucidation of sulfur-bearing compounds present only in trace levels. Parallel to the analytical work, we tried to correlate the sensory characteristics of new compounds to the general sensory impression of buchu leaf oil.

2 EXPERIMENTAL SECTION

Commercial buchu leaf oil of *Agathosma betulina* from South Africa was used. Distillation was performed in two steps. For preconcentration, a 80 cm packed column was used. Final fractionation was achieved on a 50 cm slit-tube distillation column (Fischer, Germany). Instrumentation (capillary gas chromatography, spectroscopy) as well as analytical and preparative conditions have been described in previous publications.[16,17] For GC–FTIR analyses, a Bio-Rad Digilab FTS-45A spectrometer, connected to the Bio-Rad Tracer (Bio-Rad, Krefeld, Germany) equipped with a liquid nitrogen cooled narrow-band MCT detector and coupled to an HP 5890 series II gas chromatograph (Hewlett-Packard, Waldbronn, Germany), was applied. The samples were separated on a J&W DB-1 column (30 m × 0.25 mm; 0.25 μm film thickness) with helium as carrier gas (split injection mode). Deposition tip and transfer line were held above 200 °C. Absorbance spectra were recorded from 4000 to 700 cm^{-1} at a spectral resolution of 1 cm^{-1}. For chiral separations, two fused-silica columns (25 m × 0.25 mm internal diameter, film thickness 0.25 μm) from MEGA capillary columns laboratory (Legnano, Italy) were used. Column 1 was coated with a solution of 30% diacetyl-tert.-butylsilyl-γ-cyclodextrin and 70% OV-1701. Column 2 was coated with a solution of 30% diethyl-tert.-butylsilyl β-cyclodextrin and 70% PS086.

2.1 Analysis of Trace Compounds

For the study of sulfur-containing compounds at trace levels, 85 g of preconcentrated buchu leaf oil were separated into ten fractions by distillation on a slit tube column. After further separation of two selected fractions on silica gel (pentane/diethyl ether), all final fractions were checked by capillary gas chromatography (GC) with FID and sulfur-specific detection (FPD). Subsequently the different fractions were analysed by capillary gas chromatography–mass spectrometry (GC–MS). Specific sulfur-bearing unknowns were enriched by preparative multi-dimensional gas chromatography (MDGC). For further structure elucidation, complementary analyses using GC–MS and capillary gas chromatography–Fourier transform infrared spectroscopy (GC–FTIR) as well as different NMR pulse experiments, such as ^{1}H, ^{13}C, COSY and HMQC (heteronuclear multiple quantum coherence), were applied.

3 RESULTS AND DISCUSSION

The present study revealed a variety of new sulfur-containing terpenoids, which can be divided into different groups.

3.1 Sulfur-bearing *p*-Menthane Derivatives

A well known representative of this group is menthothiophene (**5**), which is described here for the first time as a constituent of buchu leaf oil (Figure 2). The sweet, malty, cocoa-like odour impressions of (**5**) differ significantly from the other three compounds 3,8-epidithio-3-*p*-menthene (**6**) and 3,8-methylenedithia-3-*p*-menthene (**7**) and 3,8-epidithio-*p*-cymene (**8**). Most likely, (**6**) and (**7**) are derived from the same progenitor with *p*-menthene structure and two free thiol groups in the 3- and 8- positions. If so, the epidithio derivative (**6**) is a direct product of oxidation, whereas (**7**) represents a dithioacetal formed with formaldehyde. In order to confirm the structure of (**8**), the compound was synthesized from pulegone using a modification of the method of Bertaina *et al.*[18] There is some evidence that molecules like (**6**) and (**8**) are responsible for the unpleasant rubbery note, which is inherent in some commercial qualities of oil. In this study, we identified the derivatives (**5–8**) as new constituents in buchu leaf oil.

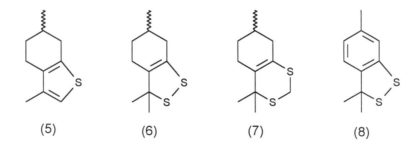

(5) (6) (7) (8)

Figure 2   *Sulfur containing* p-*menthane derivatives in buchu leaf oil*

3.2 *p*-Menthan-2-one Derivatives

8-Epidithio-3-*p*-menthene (**6**) on the one hand and different carvone derivatives on the other were the basis for the identification of a series of new flavour compounds with *p*-menthan-2-one skeleton (Figure 3). In comparison with the structure of (**6**), 3,8-epidithio-*p*-menthan-2-one, (**9**), lacks the ring double bond, but it has a carbonyl group at position 2. These changes result in completely different sensory characteristics which are described in Table 1 as pepper and woody. Similar sensory impressions are observed from 7-acetylthio-*p*-menthan-2-one (**10**). The position of sulfur in this molecule was confirmed by COSY and HMQC NMR pulse experiments. Bearing in mind the above mentioned sulfur chemistry, the presence of 2-oxomenthothiophene (**11**) is not really surprising.

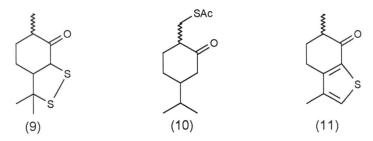

Figure 3 p-*Menthan-2-one derivatives in buchu leaf oil*

3.3 Diosphenol Derivatives

This group is represented by formal derivatives of diosphenol (**12**) and ψ-diosphenol (**13**) (Figure 4). The enantiomeric composition of diosphenol in buchu leaf oil was determined recently by Werkhoff *et al.*,[19] revealing an enantiomeric excess of 43.4% of the laevorotatory form. The odour descriptions for the optically pure stereoisomers were summarized as woody, camphor-like, menthofuran, minty, isoeugenol, clove, buchu, menthone/pulegone, dusty, jasmon-like for (–)-diosphenol. The enantiomer is not as strong. It is characterised as woody, camphor-like, dusty, medicinal, isoeugenol, *cis*-jasmon-like, terpene-like.

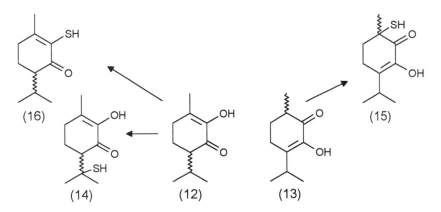

Figure 4 *Sulfur-bearing diosphenol derivatives in buchu leaf oil*

The S-containing diosphenol derivative, 8-mercaptodiosphenol (**14**), is possibly derived from diosphenolen, which is a minor constituent of buchu leaf oil. In Table 1 the sensory characteristics of (**14**) together with 1-mercapto-ψ-diosphenol (**15**) and thiodiosphenol (**16**) are outlined. In comparison to diosphenol, the sensory characteristics of (**14**) and (**15**) show the strong influence of the thiol group.

On the basis of MS data, Kaiser and co-workers proposed (**15**) to be a mercapto derivative of ψ-diosphenol.[2] We have been able to determine its complete structure using an HMQC NMR pulse experiment. This heteronuclear [1]H–[13]C experiment identifies bonded [1]H–[13]C pairs with excellent sensitivity. So it was possible to assign the SH-function to position 1. Compound (14) had been tentatively identified as 8-mercaptodiosphenol.[2] From our own investigations we have been able to verify the proposed structure.

The occurrence of thiodiosphenol (16) in buchu leaf oil is a new finding. The mass spectral data of (16) are shown in Table 1. In addition, the structure of (16) was checked by synthesis starting from diosphenol with Lawesson's reagent as an S-donor.[20] Subsequent chirospecific analysis of (16) was performed by GC, using diacetyl-tert.-butylsilyl-γ-cyclodextrin in OV-1701 (column 1) as a chiral stationary phase. The stereodifferentiation of the isolated compound from buchu leaf oil indicated a racemic mixture, which is probably due to isomerisation under acidic conditions. The sensory evaluation of (16) revealed interesting odour impressions (Table 1) similar to menthofuran together with a sulfur note.

3.4 *p*-Menthan-3-one Derivatives

Our investigations led us to a group of four diastereomers with a relative molecular mass of 186. After GC–MS analysis, the two major isomers (17a) and (17b) were isolated by means of preparative MDGC. Subsequently, the characterization of the isolated compounds was extended by GC–FTIR, as well as ^{1}H- and ^{13}C-NMR spectroscopy. On the basis of NMR data, the relative configuration was assigned. Additionally, both isomers were synthesized starting from (–)-menthone and (+)-isomenthone, as well as from (+)-menthone and (–)-isomenthone isolated from buchu leaf oil. After bromination, the α-haloketone was treated with thioacetic acid, yielding the corresponding thioacetate, which was hydrolysed using a standard technique.[21,22]

Finally, stereodifferentiation of (17a, 17b) and all synthesized isomers was achieved on a diacetyl-tert.-butylsilyl-γ-cyclodextrin column (column 1) and a diethyl-tert.-butylsilyl-β-cyclodextrin column (column 2). These results clearly showed the presence of the 1*S*, 2*S*, 4*R* isomer (17a) with an enantiomeric purity of 88% and of the 1*S*, 2*R*, 4*S*-diastereomer (17b) with a purity of 84% (Figure 5). Since we observed four almost identical mass spectra in buchu leaf oil, one has to conclude that the other two isomers have to be described as 2-mercapto-*p*-menthan-3-one, 1*S**, 2*R**, 4*R** and 2-mercapto-*p*-menthan-3-one, 1*S**, 2*S**, 4*S**.

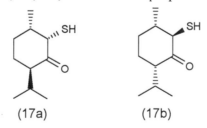

(17a)   (17b)

Figure 5   p-*Menthan-3-one derivatives*

3.5 Miscellaneous Sulfur Compounds

Besides the above mentioned new sulfur-bearing monoterpenoids, we also identified mercaptodihydromethyleugenol (18), a product from the phenylpropanoid pathway. Methyl eugenol, a well known constituent of the oil, is a possible building block for the formation of (18). The racemate, however, was synthesized from methylisoeugenol by addition of thioacetic acid. After hydrolysis of the *S*-acetate, the sensory properties of the free thiol were described as meaty and terpene-like (Table 1).

Table 1 *Sensory Descriptions for New Sulfur-bearing Compounds in Buchu Leaf Oil obtained by Gas Chromatography–Olfactometry (GC–O) and MS Data*

No.	Compound name	Sensory description	RI (DB-1)	MS-data (m/z, %)
6	3,8-epidithio-3-*p*-menthene	sulfur-like, off-flavour, licorice, musty, rubber, oily	1475	185 (100), **200** (19), 186 (13), 143 (11), 187 (10), 41 (7), 39 (6), 77 (5), 111 (5), 59 (5)
7	3,8-methylenedithia-3-*p*-menthene	weak, woody, herbal note	1635	199 (100), 167 (78), **214** (67), 111 (32), 41 (27), 153 (18), 168 (19), 77 (15)
8	3,8-epidithio-*p*-cymene	reminiscent of sulfur, rubber	1515	181 (100), **196** (36), 148 (21), 147 (13), 182 (13), 163 (11), 183 (10), 197 (5), 115 (5), 97 (4)
9	3,8-epidithio-2-*p*-menthanone	sulfur-like, herbal note, pepper, grapefruit, terpeny, woody	1658	152 (100), **216** (89), 123 (77), 41 (63), 110 (62), 95 (51), 81 (49), 39 (42), 109 (38), 67 (37)
10	7-acetylthio-2-*p*-menthanone	caryophyllene oxide, pepper, herbal note, nootkatone-note	1726	153 (100), 43 (72), 143 (27), 41 (21), 69 (21), 55 (19), **228** (17), 81 (16), 109 (15), 154 (11)
11	2-oxo-menthothiophene	gasoline, herbal note, tarry, menthofuran-note	1555	138 (100), **180** (66), 110 (49), 165 (22), 151 (20), 139 (10), 181 **(8)**, 70 (8), 66 (8), 39 (8)
14	8-mercapto-diosphenol	oil, tarry, tropical	1486	125 (100), 124 (79), 41 (56), 83 (38), 69 (37), 75 (36), 126 (32), 107 (31), 82 (28), **200** (27)
15	1-mercapto-ψ-diosphenol	grapefruit, sweet	1408	124 (100), 125 (73), 166 (72), 167 (48), 123 (46), 43 (44), 41 (41), 69 (28), 55 (24), 107 (19), **200** (4)
16	thiodiosphenol	menthofuran-like, sulfur note, chemical note	1428	**184** (100), 142 (89), 141 (74), 86 (34), 41 (33), 169 (29), 108 (28), 85 (25), 79 (20), 127 (18)
17a	2-mercapto-3-*p*-menthanone, 1*S*, 2*S*, 4*R*-	buchu, off-flavour, tropical, cedar wood, herbal note, passion fruit, fruity	1352	69 (100), 41 (92), 100 (74), 55 (67), 74 (56), 27 (47), 115 (45), 43 (44), **186** (42), 81 (38), 158 (18)
17b	2-mercapto-3-*p*-menthanone, 1*S*, 2*R*, 4*S*-	woody, terpeny, balsamic	1345	41 (100), 69 (98), 100 (77), 55 (72), 74 (61), 115 (50), 81 (40), 27 (40), 43 (39), 102 (38), **186** (37), 158 (23)
18	mercaptodihydromethyl-eugenol	fatty, terpeny, metallic, sulfur-like, meaty	1620	151 (100), **212** (28), 152 (13), 107 (5), 91 (4), 213 (4), 77 (3), 108 (3), 106 (3), 179 (3)
19	8-(3-oxo-2-*p*-menthanylthio)-3-*p*-menthanone	grapefruit, nootkatone, dusty, fermentation note, woody, vetiver-like	2261, 2324	153 (100), 186 (91), 69 (66), 109 (44), 41 (39), 81 (31), 55 (28), 152 (26), 43 (20), 154 (18), **338** (12)

Furthermore, two major peaks in a group of isomers were observed with a relatively high retention index in chromatograms. GC–MS analysis revealed a molecular ion at m/z 338. On the basis of GC–FTIR measurements, as well as COSY and HMQC NMR experiments, we assigned the structure (**19**). Formation is probably due to the condensation of 8-mercapto-*p*-menthan-3-one with 2-mercapto-*p*-menthan-3-one (**17**). Finally, we found for the first time in buchu leaf oil mintsulfide (**20**), an interesting sulfur-containing sesquiterpene, which has previously been found in several essential oils, *e.g.*, peppermint[23] (Figure 6).

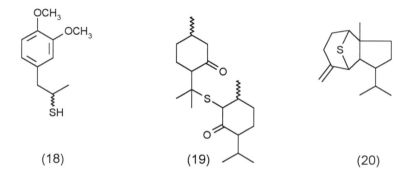

     (18)   (19)   (20)

Figure 6   *Miscellaneous sulfur containing compounds in buchu leaf oil*

4 CONCLUSION

The complex flavour of buchu leaf oil is mainly attributed to its unusual sulfur chemistry. In this study, we isolated and identified thirteen new flavour compounds, eleven of which are described for the first time. In addition, the structures of two diosphenol derivatives (**14, 15**), tentatively described in literature, were characterised.

The stereodifferentiation of thiodiosphenol revealed racemisation after storage under acidic conditions. The chirality evaluation of the two diastereomeric 2-mercapto-*p*-menthan-3-ones showed the presence of the 1*S*, 2*S*, 4*R* configured isomer (**17a**) with an enantiomeric purity of 88% and of the 1*S*, 2*R*, 4*S*-diastereomer (**17b**) with a purity of 84%.

On the one hand, the occurrence of sulfur-bearing diosphenol derivatives (**14, 15, 16**) and of the diastereomeric 2-mercapto-*p*-menthane-3-ones (**17a, 17b**) indicates a biosynthetic origin. On the other, the formation of some dithio-*p*-menthane derivatives, (**6, 7, 8**) and other compounds (**19**) is attributable to thermal treatment during oil production.

The sensory properties of each compound were evaluated and correlated with the complex flavour impression of the oil.

ACKNOWLEDGEMENT

We would like to thank the management of Haarmann & Reimer for permission to publish this paper and the staff of the H & R research department for their valuable and skilful work. We are indebted to Dr. H. Surburg for helpful discussions. Furthermore, we are grateful to Mr. R. Hall for the distillation of buchu leaf oil and to Mr. G. Kindel for sensory evaluations.

REFERENCES

1. N.F. Collins, Poster Presentation P-34, 26th International Symposium on Essential Oils, Hamburg, Germany, 1995.
2. R. Kaiser, D. Lamparsky and P. Schudel, *J. Agric. Food Chem.*, 1975, **23** (5), 943.
3. M.A. Posthumus, T.A. van Beek, N.F. Collins and E.H. Graven, *J. Essent. Oil Res.*, 1996, **8**, 223.
4. N.F. Collins, E.H. Graven, T.A. van Beek and G.P. Lelyveld, *J. Essent. Oil Res.*, 1996, **8**, 229.
5. E. Klein and W. Rojahn, *Dragoco Rep.*, 1967, **9**, 183.
6. P. Werkhoff, S. Brennecke, W. Bretschneider, M. Güntert, R. Hopp and H. Surburg, *Z. Lebensm. Unters. Forsch.*, 1993, **196**, 307.
7. D. Lamparsky and P. Schudel, *Tetrahedron Lett.*, 1971, **36**, 3323.
8. E. Sundt, B. Willhalm, R. Chappaz and G. Ohloff, *Hel. Chim. Acta*, 1971, **54**, 1801.
9. T. Köpke and A. Mosandl, *Z. Lebensm. Unters. Forsch.*, 1992, **194**, 372.
10. T. Köpke, H.-G. Schmarr and A. Mosandl, *Flav. Frag. J.*, 1992, **7**, 205.
11. P. Werkhoff, S. Brennecke, W. Bretschneider and K. Schreiber, *H&R CONTACT*, 1995, **63**, 5.
12. R. Fellous, G. George, L. Lizzani, M. Loiseau, Ch. Schippa and S. Rochard, *Rivista Italiana Eppos*, 1993, Numero Speciale Febbraio, 66.
13. K. MacNamara, P. Brunerie, S. Keck and A. Hoffmann, in 'Food Science and Human Nutrition', ed. G. Charalambous, Elsevier, Amsterdam, 1992, p. 351.
14. T. Köpke, A. Dietrich and A. Mosandl, *Phytochemical Analysis*, 1994, **5**, 61.
15. W.E. Campbell and B.K. Williamson, *Flav. Frag. J.*, 1991, **6**, 113.
16. M. Güntert, J. Brüning, R. Emberger, M. Köpsel, W. Kuhn, T. Tielmann and P. Werkhoff, *J. Agric. Food Chem.*, 1990, **38**, 2027.
17. M. Güntert, J. Brüning, R. Emberger, R. Hopp, M. Köpsel, H. Surburg and P. Werkhoff, in 'Flavor Precursors – Thermal and Enzymatic Conversions', eds. R. Teranishi, G.R. Takeoka and M. Güntert, A.C.S. Symposium Series 490, American Chemical Society, Washington, D.C., 1992, p. 140.
18. C. Bertaina, F. Cozzolino, G.G. Fellous and E. Rouvier, *Parfums, Cosmétiques, Arômes*, 1986, **71**, 69.
19. P. Werkhoff, S. Brennecke, W. Bretschneider and K. Schreiber, *H&R CONTACT*, 1995, **65**, 5.
20. M.P. Cava and M.I. Levinson, *Tetrahedron*, 1985, **41**, 5061.
21. P. Dubs and R. Stüss, *Synthesis*, 1979, 696.
22. J.H. Chapman and L.N. Owen, *J. Chem. Soc.*, 1950, 579.
23. K. Takahashi, S. Muraki and T. Yoshida, *Agric. Biol. Chem.*, 1981, **45**, 129.

COMPOSITION OF VOLATILE CONSTITUENTS IN TARRAGON (*ARTEMISIA DRACUNCULUS* L.) AT DIFFERENT VEGETATIVE PERIODS

R. Venskutonis[*], J.W. Gramshaw[†], A. Dapkevicius[*] and M. Baranauskienė[§]

[*]Department of Food Technology, Kaunas University of Technology, Radvilenu pl. 19, Kaunas, LT-3028, Lithuania

[†]Procter Department of Food Science, University of Leeds, Leeds, LS2 9JT, U.K.

[§]Lithuanian Institute of Horticulture, Babtai, Lithuania

1 INTRODUCTION

Two broad divisions of tarragon are recognised: French (*A. dracunculus*) and Russian (*A. dracunculoides*). The main commercial sources of tarragon are southern France and the United States. It is also successfully grown in Lithuania, although industrial cultivation is not widely organised.

The essential oil of tarragon is widely used in food flavouring formulations. However, the components of tarragon essential oil have been studied much less intensively than those of many other aromatic plants.[1] Thieme and Tam identified α- and β-pinenes, camphene, limonene, *cis-* and *trans-* ocimenes, methyl chavicol, *p*-methoxy cinnamic aldehyde, δ-4-carene, β-phellandrene and linalool in tarragon.[2] Later, Zarghami and Russell complemented this list with *cis-* and *trans-* alloocimenes, butyric acid, geraniol, 1,2-dimethoxy-4-allyl benzene and eugenol.[3] Frattini *et al.* additionally identified menthone, *o*-methylisoeugenol, elemicin, *p*-methoxybenzaldehyde, *p*-methoxycinnamic alcohol, isobornyl acetate and two γ-lactones in the essential oil of Piedmontese tarragon.[4] Vostrowsky *et al.* reported, in addition, α-thujene, sabinene, myrcene, α- and γ-terpinenes *p*-cymene, sabinene hydrate, terpinolene, terpinen-4-ol, α-terpineol, farnesene and bornyl, citronellyl, geranyl and cinnamyl acetates.[5]

Data published on tarragon flavour composition reveal an extreme variability in its quality. The content of methyl chavicol, which imparts to French tarragon its preferred anise-like aroma, can vary in a large range depending on the plant chemotype, harvesting time and even day length.[6] For instance, in the total volatile content of seed-grown tarragon, it constituted 0.16%, in Russian 1.29%, in French 52.83% and in commercial oil 82.29%.[7] Investigations on the biotechnological production of flavour compounds by callus and suspension cultures of tarragon have been carried out recently.[8,9]

The seasonal dependence of the content of essential oil in tarragon was determined by Vostrowsky *et al.*[5] The maximum peaks in essential oil content were observed in tarragon herb harvested in the middle of August and later in the beginning of September. Considerable changes in the concentration of some individual compounds were also found: the early harvested herb (June 22) possessed high elemicin (27.7%) and low methyl chavicol (0.6%) concentration, whereas at the delayed harvest time, sabinene (38.81%), methyl eugenol (28.87%) and methyl chavicol (17.26%) dominated.

In view of the prized status of tarragon oil and the variability in its composition, which may stem from the taxonomic peculiarities and/or long-term instability, it was of interest to investigate the composition of flavour of some *A. dracunculus* cultivars cultivated in Lithuania.

2 MATERIALS AND METHODS

Stock Plants. Three types of tarragon plants cultivated in the experimental garden of the Lithuanian Institute of Horticulture were investigated:
1. The plants cultivated from the seeds obtained from Denmark in 1991, and referred to as 'Danish' tarragon (DT); these plants were propagated by sowing in the beginning of April.
2. The plants cultivated from the seedlings obtained from Latvia in 1987, conditionally named 'Siberian' tarragon (ST); these plants were propagated vegetatively.
3. The plants cultivated from the seedlings obtained from Latvia in 1987, conditionally named 'Georgian' tarragon (GT); these plants were also propagated vegetatively.

It should be noted that 'French'-type tarragon also was introduced into the Lithuanian Institute of Horticulture from the seedlings obtained from Latvia in 1987. However, the plants were destroyed by the frost during the first winter.

Tarragon herb was harvested in the 1992 season at three periods: June 22 (beginning of flower formation); July 8 (advanced flower formation); July 27 (intensive flowering). Also, it should be mentioned that climatic conditions in 1992 were very specific because of the severe drought. Therefore, the flowering period for the most aromatic plants started earlier than usual.

Freshly harvested herb was sorted and dried naturally at ambient temperature, protected against direct sunlight. The dried herb was packaged into paper bags and stored at room temperature before analysis.

Isolation of Volatile Compounds. Essential oil was hydrodistilled from 150 g of dried herb during 2.5–3 hours by the standard AOAC method using a Clevenger apparatus and stored in the refrigerator. Simultaneous distillation-extraction (SDE) was carried out in a Likens–Nickerson apparatus using a mixture of n-pentane and diethyl ether (1:1) as the extraction solvent. Both solvents were freshly distilled before use, 25 cm^3 of the solvent were used in each extraction. One gram of the shredded dried plant material was diluted with 500 cm^3 of purified water and distilled for 1.5 h. Two cm^3 of the internal standard (IS) solution (0.005% 4-methyl-pentan-1-ol in diethyl ether) were added at the beginning of the SDE procedure. The extracts were dried over 2 g of anhydrous sodium sulphate, filtered and concentrated to 200 mg under the nitrogen stream.

GC and GC-MS Conditions. Samples of essential oil diluted in diethyl ether (25 mm^3 in 1000 mm^3) and concentrated extract samples (0.5 mm^3) were chromatographed under the following conditions:
1. Fisons/Carlo Erba 8261 instrument with HP 3395 integrator and FID; fused silica BP-1 column (25 m; internal diameter: 0.32 mm; film thickness: 0.5 mm); helium as a carrier gas at a mean flow rate of 1.6 cm^3 min^{-1}; detector temperature: 300 °C; oven temperature programme: 50 °C for 10 min, increasing by 3 °C min^{-1} to 250 °C; split injection: 1:15.
2. Carlo Erba 4200 series instrument with FID; fused silica BP-20 column (25 m; internal diameter: 0.32 mm; film thickness: 0.25 mm); helium as a carrier gas at a mean flow rate of 2 cm^3 min^{-1}; detector temperature: 190 °C; oven temperature programme: 50 °C for 5 min, increasing by 2 °C min^{-1} to 190 °C; split injection: 1:20.

Table 1   *Constituents Identified in Tarragon (A. dracunculus) and Not Previously Reported* [1–12,14]

Compound	RI on BP-1	Georgian id	Georgian %	Danish id	Danish %	Siberian id	Siberian %
(E)-Hex-2-enal	822	a,b,c	0.05	a,b,c	0.06	a,b,c	0.03
(Z)-Hex-3-en-1-yl acetate	987			a,b,c	0.02		
Δ-3-Carene	1000			a,b	0.01	a,c	0.02
(4-Methyl-3-pentenyl)-furan	1063			a,c	0.02		
(E)-*p*-Menth-2-en-1-ol	1103			a,b,c	0.11	a,c	0.05
p-Menthadiene-7-ol[a]	1112			a,c	0.03	a,c	0.08
(Z)-*p*-Menth-2-en-1-ol	1118			a,b,c	0.07	a,c	0.04
Alloocimene	1118			a,b	0.02		
p-Menthadiene-7-ol[a]	1122			a,c	0.07	a,c	0.08
Citronellal	1131			a,c	0.04		
Menthol	1157			a,b	0.02	a,b,c	0.02
Pulegone	1174			a,b,c	0.03		
cis-Carveol	1207	a,b,c	0.04			a,b	0.04
Neral	1212			a,b,c			
Geranial	1235	a,b,c	0.01	a,b,c	0.03	a,c	0.03
Carvacrol	1279					a,c	0.04
Methyl geranate	1301			a,c	0.40		
Neryl acetate	1341			a,b,c	0.05		
α-Cubebene	1347			a,b,c	0.01	a,b,c	0.03
β-Caryophyllene	1414	a,b,c	0.11	a,b,c	0.07	a,b,c	0.04
α-Humulene	1447			a,b,c	0.04	a,c	0.04
β-Ionone	1462			a,b,c	0.02		
Germacrene D	1473			a,b,c	0.28	a,b,c	0.08
Germacrene B	1488					a,c	0.08
γ-Elemene	1489			a,b,c	0.09		
β-Bisabolene	1501			a,b,c	0.06		
(E)-Nerolidol	1547			a,b,c	0.36	a,b,c	0.08
Germacrene-D-4-ol[a]	1563	a,b		a,b,c		a,b	
Caryophyllene oxide	1568	a,b,c	0.24	a,b,c	0.14	a,b,c	0.20
T-Cadinol	1624			a,c	0.16	a,c	0.11
α-Cadinol	1640			a,c	0.22	a,c	0.02
Farnesol[b]	1652	a,c	0.04	a,b	0.04	a,c	0.11
α-Bisabolol	1667			a,b,c	0.03		

a   Match with RI on BP-1;
b   Match with RI on BP-20;
c   Match with MS of the reference compound;
[a]   For concentration see Table 3;
[b]   Unknown isomer.

GC-MS was performed on a coupled Carlo Erba 4200 series GC with a Kratos MS 80 RFA mass spectrometer under the following conditions: fused silica open tubular columns (BP-1, 30 m, internal diameter: 0.5 mm; film thickness: 0.5 mm; and BP-20, 50 m, internal diameter: 0.5 mm; film thickness: 0.5 mm); helium as a carrier gas at a mean flow rate of 3 cm^3 min^{-1}; oven temperature programme: 50 °C for 5 min, increasing by 5 °C min^{-1} to 180 °C (BP-20) or 240 °C (BP-1); on-column injection. Mass spectra were obtained by electron ionization (voltage: 70 eV; beam current: 100 mA; source temperature: 150 °C) and chemical ionisation with isobutane.

Identification. The identification was based on matching of the retention indexes (RI) and comparison of the obtained mass spectra with those of authentic compounds and reported in the literature. The amount of the individual constituents was expressed as a peak area percent and in arbitrary units (a.u.), calculated on the basis of the peak area ratio with IS. The absolute concentration, in mg kg^{-1} of dried herb, was determined for certain compounds in the samples isolated by SDE.

3 RESULTS AND DISCUSSION

A total of 77 compounds were identified positively or tentatively in *A. dracunculus*: 32 in 'Georgian', 57 in 'Danish' and 56 in the 'Siberian' type. Among these, 33 compounds were not found to have previously been reported in the literature[1-12,14] (Table 1).

3.1 Composition of Essential Oils

The differences in the total essential oil content, in most cases, were not very high for tarragon cultivars (Table 2). The exception was 'Georgian' tarragon, which yielded 4.27% of oil during the period of flower formation (July 8). In all cases, the largest yield of oil was obtained at this time. During further growth, the content of oil decreased considerably in GT and ST (by 44% and 39%) respectively.

The percentage content of the major constituents in the essential oils is presented in Table 2. The composition of GT was typical of the French variety (methyl chavicol is the dominant constituent – approximately 80%). ST cultivar was rich in elemicin and sabinene, and seems to be a typical Russian tarragon. The herb which originated from Denmark (DT) possessed a different, and the most complex, composition: methyl eugenol, sabinene, isoelemicin and elemicin were the most abundant constituents. Phenyl propene compounds constitute a considerable part of the essential oil for all *A. dracunculus* cultivars analysed.

Very slight differences were found in the percentage composition of essential oil of GT harvested on June 22, July 8 and July 27. The main compound, methyl chavicol, constituted 78.7%, 79.4%, and 81.2%, respectively; the concentration of β-ocimenes and limonene in the essential oils was also similar at all harvesting periods.

The fluctuations in the composition of distilled oils from DT and ST harvested at different periods were considerable. The concentration of some quantitatively important aliphatic mono-terpenes (sabinene, ocimenes and terpinolene) in ST was the highest in the herb collected on July 8, whereas the content of oxygenated terpenes (terpinen-4-ol) and phenyl propenes (methyl eugenol and elemicin) at the same time was significantly lower than in the herb harvested on June 22 and July 27. Similar changes in the concentration of elemicin were observed in DT essential oil. However, a different trend was seen in the amounts of sabinene and methyl eugenol when DT was compared with ST. The concentration of sabinene gradually increased from June 22 until July 27, that of methyl eugenol significantly decreased during the same period, from 24.4% to 8.9%. The percentage content of another important DT constituent, isoelemicin, was lower on June 22. The changes in the content of β-ocimenes in DT were similar to those in ST.

Table 2 *Essential Oil Composition (%) of Tarragon at Different Vegetative Periods: a June 22; b July 8; c July 27 (AOAC Method)*

Compound	GTa	GTb	GTc	DTa	DTb	DTc	STa	STb	STc
Sabinene	0.73	0.86	0.71	29.96	38.90	44.57	17.29	48.58	29.96
Myrcene	0.53	0.18	0.25	0.65	1.59	1.62	0.85	1.92	0.95
Limonene	2.02	1.73	1.60	0.36	0.46	0.28	1.12	1.57	1.02
(Z)-β-Ocimene	5.56	5.53	4.98	1.46	5.75	3.58	0.98	1.86	1.13
(E)-β-Ocimene	5.32	6.21	5.74	2.44	5.42	3.05	7.45	10.45	3.95
Terpinolene	—	—	—	0.35	0.27	0.43	5.18	7.33	2.46
Terpinen-4-ol	0.23	—	—	1.80	1.96	1.76	2.65	0.89	2.05
Methyl chavicol	78.7	79.4	81.2	0.36	0.26	0.27	0.63	0.75	1.54
Geranyl acetate	—	—	—	0.69	0.66	1.03	0.85	1.03	2.31
Methyl eugenol	0.49	0.44	0.27	24.4	18.0	8.92	3.66	1.84	3.24
Germacrene D	—	—	—	0.72	1.05	0.95	0.04	0.23	0.20
Elemicin	0.43	0.45	0.28	5.29	1.23	4.85	54.9	16.1	40.8
Methyl isoeugenol	0.18	0.13	0.08	1.25	1.04	1.06	0.51	0.90	0.39
Isoelemicin	0.14	0.12	0.17	11.3	13.7	18.7	0.31	0.11	0.33
Total essential oil, ml per 100 g	2.97	4.27	2.37	2.19	2.71	2.54	2.25	2.60	1.59

3.2 Composition of Volatile Compounds Isolated by the SDE Method

The compositions of (main compounds only) SDE samples (June 22) is presented in Table 3. In the SDE preparation of DT, the percentage concentration of isoelemicin increased from 8.2% to 16.2% in the period between June 22 and July 8. The percentage contents of the other quantitatively important compounds, sabinene (24.7% and 21.7%), methyl eugenol (35.9% and 35.6%) and elemicin (9.1% and 9.5%), were almost equal for the herbs harvested at both periods. However, the absolute concentration of those constituents increased significantly between June 22 and July 8. Some changes in the concentration of minor compounds were also detected, *e.g.* the content of methyl chavicol decreased from 1.1% to 0.5%.

The differences in the composition between essential oils and SDE samples were quite considerable, especially for DT and ST. Sabinene was the prevailing component in the essential oils from these herbs, whereas in SDE extracts its content was much lower. On the contrary, the concentration of phenyl propene compounds (methyl eugenol in DT and elemicin in ST) was considerably higher in SDE samples than in the essential oils. In the case of GT, considerable differences were found only in the concentration of β-ocimenes which were found to be several times higher in the essential oil. Two factors could be the likely reasons for these differences. One could be the very procedure of the isolation of the flavour compounds. Some amount of more polar phenyl propene compounds could be lost with water during hydrodistillation, whereas non-polar terpene hydrocarbons could be separated with smaller losses. The second could be the changes in volatiles during the storage of the herbs: essential oils were hydrodistilled approximately two months before SDE procedures were carried out. Huopalahti *et al.* applied solvent extraction, hydrodistillation, CO_2 extraction and head space technique in the analysis of dill compounds and each method gave a different composition for the volatiles of this herb.[13] In another study, the differences determined between solvent extract and essential oil of tarragon were also significant.[14]

Table 3 *Major Volatile Compounds in Tarragon (*Artemisia dracunculus*) Isolated by SDE*

Compound	GT (June 22)		DT (June 22)		ST (June 22)	
	$mg\ kg^{-1}$	%	$mg\ kg^{-1}$	%	$mg\ kg^{-1}$	%
Sabinene*	27.7	0.45	2170.1	24.75	1325.2	14.28
Limonene	87.3	2.40	24.6	0.36	42.1	0.56
(Z)-β-Ocimene	40.7	0.70	40.2	0.36	2.1	0.02
(E)-β-Ocimene	27.4	0.47	41.3	0.37	5.2	0.04
Terpinolene	5.6	0.14	26.0	0.35	20.4	0.24
Terpinen-4-ol	11.7	0.23	177.6	1.80	617.4	5.41
Methyl chavicol	4036.0	82.06	111.7	1.08	69.0	0.63
Geranyl acetate	—	—	102.1	0.69	150.5	0.85
Methyl eugenol	113.5	1.24	4978.9	39.35	575.4	3.39
Methyl isoeugenol*	9.8	0.18	152.2	1.87	5.7	0.06
Germacrene D-4-ol*	123.7	1.52	127.8	1.41	356.5	3.29
Elemicin*	—	—	844.1	10.37	6281.6	57.21
Isoelemicin*	—	—	753.0	8.15	34.2	0.31

* Response factor in relation to the IS was not determined, but taken from literature.[1]

REFERENCES

1. 'The Complete Database of Essential Oils', BACIS, The Netherlands, 1991–1995.
2. H. Thieme, and N.T. Tam, *Pharmazie*, 1972, **27**, 255.
3. N.S. Zarghami and G.F. Russell, *Chem. Mikrobiol. Technol. Lebensm.*, 1973, **2**, 184.
4. C. Frattini, F. Belliardo, C. Reyneri and C. Bicchi, *Riv. Ital. Essenze Profumi Piante Off.*, 1978, **60**, 286.
5. O. Vostrowsky, K. Michaelis, H. Ihm, R. Zintl and K. Knobloch, *Z. Lebensm. Unters. Forsch.*, 1981, **173**, 365.
6. K. Suchorska, B. Jedraszko and I. Olszewska-Kaczynska, *Annals of Warsaw Agricultural University, Horticulture*, 1992, No. **16**, 79.
7. C.M. Cotton, L.V. Evans and J.W. Gramshaw, *J. Experim. Botany*, 1991, **42**, 365.
8. C.M. Cotton, J.W. Gramshaw and L.V. Evans, *J. Experim. Botany*, 1991, **42**, 377.
9. J.W. Gramshaw, C.M Cotton and L.V. Evans, in 'Bioflavour 87', ed. P. Schreier, Walter de Gruyter Co., Berlin, 1988, p. 341.
10. B.M. Lawrence and R.J. Reynolds, *Perf. & Flav.*, 1990, **15**, 75.
11. B.M. Lawrence and R.J. Reynolds, *Perf. & Flav.*, 1995, **20**, 38.
12. 'Volatile Components in Food – Qualitative and Quantitative Data', Vol. 2. eds. H. Maarse and C.A. Visscher, TNO-CIVO, Zeist, Netherlands, 1989, p. 915.
13. R. Huopalahti, R. Lahtinen, R. Hiltunen, and I. Laakso, *Flav. Fragr. J.*, 1988, **3**, 121.
14. E. Werher, E. Putievsky, U. Ravid, N. Dudai and I. Katzir, *J. Herbs, Spices, Med. Plants*, 1994, **2**, 19.

BIOCHEMICAL PATHWAYS FOR THE FORMATION OF ESTERS IN RIPENING FRUIT

S. Grant Wyllie, D.N. Leach, H.N. Nonhebel and I. Lusunzi

Centre for Biostructural and Biomolecular Research, University of Western Sydney, Hawkesbury, Richmond, New South Wales, Australia, 2753

1 INTRODUCTION

The aroma profiles of many ripe fruits are notable for the presence of a range of esters whose qualitative and quantitative composition makes a significant contribution to the characteristic odour and quality perception of the fruit. The production of these esters appears to be under genetic control[1] which results in marked differences in aroma profile even between cultivars of the same species. For example, we have examined the aroma profiles of a considerable number of the cultivars of the *Cucumis melo* reticulatis group. These profiles are dominated by the esters with only small amounts of other functional groups being represented.[2,3] It was found that subtle changes in the aroma of individual cultivars are reflected in small but important changes in the composition of the esters. It has been shown that most of these esters are biosynthesised from either lipid or amino acid precursors[4,5] by a series of enzyme-mediated steps. It can be postulated that the qualitative and quantitative composition of the esters finally produced will depend on the properties of these enzymes and/or the supply of the substrates required by them. The transformations of the branched chain amino acids offer a suitable starting point to investigate these questions since their characteristic carbon skeletons can be identified and quantified in the resulting esters with comparative ease. For example, examination of the ester profiles of bananas shows that most of the branched chain derived esters originate from leucine and valine.[5] On the other hand, those from melons[3,6] are derived from isoleucine and valine whilst those from strawberries come from leucine, isoleucine and valine.[7] In the case of banana, this predominance could be explained by increased substrate availability since it has been shown[5,8] that the concentration of leucine rises markedly during ripening. However, in melons, both leucine and isoleucine are present in comparable concentrations[2] (at least at a gross cell level) and it would appear that the almost exclusive use of isoleucine must be a result of the selectivity properties of the biosynthesis pathway. An overall scheme depicting the steps in this pathway is shown in Scheme 1.

The enzymes of importance in this pathway are (a) α-aminotransferase, (b) α-ketoacid decarboxylase, (e) α-ketoacid dehydrogenase, (c) alcohol dehydrogenase and (d) alcohol acyltransferase. This last enzyme (AAT) appears to be the only one of these enzymes to have been investigated in any detail in ripe fruit. It has been isolated and partially purified from a small number of fruit.[9-13] In the most recent and extensive of these studies,[13] AAT from strawberries was prepared in a high state of purity and the selectivity determined for a range of alcohols and acyl coenzyme A derivatives. The authors concluded that the composition of the aroma profile for the strawberries investigated did indeed reflect the selectivity of the AAT isolated. However, none of these investigations has demonstrated that the AAT enzyme is capable of discriminating between 2- and 3-methylbutanol and/or 2- and 3-methylbutyryl CoA, the products of leucine and isoleucine bioconversion, during ester formation. This is a key factor in determining whether the selectivity referred to above for bananas and melons is expressed at this point of the biosynthetic pathway. The work reported here describes the properties of this and other enzymes of the pathway with regard to their ability to participate in the transformation of selected amino acids and hence control the ester profile generated in the ripening fruit.

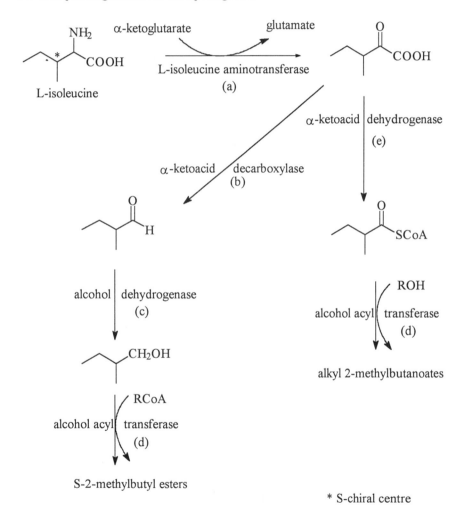

Scheme 1 *Biosynthetic pathways for esters from branched chain amino acids*

2 RESULTS AND DISCUSSION

2.1 Properties of Some Fruit Alcohol Acyl Transferases

Results obtained by measuring the esters formed by the addition of exogenous substrates to fruit tissue slices or homogenised tissues are complicated by the almost universal presence of extracellular esterase enzymes in these preparations which prevents an accurate assessment of the quantities of ester formed and hence enzyme activity. Previous workers[13] have overcome this difficulty by isolating the protoplasts before extraction of the enzymes. This procedure was adopted for most of the work reported here. Alcohol acyl transferase enzyme from bananas (*Musa sapientum* L.), strawberries (*Fragaria ananassa*) and melons (*Cucumis melo*) was isolated and partially purified, essentially according to the procedure of Perez *et al.*[13] The activity of the resulting enzymes was determined using a similar headspace method to that described by these workers. The relative activities of AAT from these three fruit tissues towards a range of alcohol substrates are shown in Table 1.

Table 1   *Relative Activity of AAT from Fruit Protoplasts towards a range of Alcohols (Acyl donor Acetyl CoA)*

Alcohol substrate	Strawberry tissue	Strawberry enzyme	Banana tissue	Melon tissue
Methanol	0	nd	0	0
Ethanol	0	100	0	0
Propanol	8	5	0	7
Butanol	54	40	47	57
2-Methylpropanol	24	7	65	50
2-Methylbutanol*	100	84	93	88
3-Methylbutanol	95	75	100	100

nd not determined;
*   S enantiomer.

These values are generally in accord with those found by others, although the variability of the results appears to indicate that factors, such as cultivar and physiological status of the fruit and the preparation and analytical method used, may exert a strong influence on the outcomes. Two points are of particular interest. First, the tissue samples show no apparent activity towards either methanol or ethanol whereas the enzyme preparations from all three fruit exhibit considerable activity towards these substrates (data not shown). This supports previous suggestions of the presence of a very active and specific extracellular esterase in the tissue which is eliminated during enzyme extraction. Secondly, these data demonstrate that fruit tissue and partially purified AAT enzyme extracts were able to convert both (S-)-2- and 3-methyl butanol to their corresponding acetates with almost equal facility. Fruit protoplasts also showed this ability. Therefore it may be concluded that the predominance of 3-methylbutyl acetate found in bananas and of 2-methylbutyl acetate in melons does not arise from the selectivity characteristics of the AAT enzyme in these fruit. However these results support the previous observations that there are considerable differences in the activity of AATs towards different alcohols and these properties could, in part, influence the composition of the esters produced.

Although there is a considerable body of evidence that the amino acids are transformed into esters in fruits,[2] chiral analysis of the volatile products provides further support for the integrity of these transformations. The 2-methylbutyl moiety is particularly useful for this analysis as it should exhibit the S configuration if it is derived stereospecifically from isoleucine. It has been shown that the ethyl 2-methylbutanoate from pineapple is exclusively in the S(+) configuration and that the 2-methylbutanoic acid from strawberries has an S:R ratio of 91.3: 8.7.[17] Similarly the 2-methylbutyl esters from Granny Smith apples are of the S configuration and of high optical purity.[18] In this investigation, chiral analyses of 2-methylbutyl acetate and ethyl 2-methylbutanoate obtained from melons showed that they were also essentially enantiomerically pure and of the S configuration. These data therefore provide further confirmation of the validity of the postulated biogenetic pathway.

2.2 Alcohol Dehydrogenase Activity

The reduction of the intermediate aldehydes to their corresponding alcohols represents another enzyme-mediated step at which substrate selectivity could be expressed. The ability of some fruits[10] to convert aldehydes to alcohols has been demonstrated and the responsible enzyme, alcohol dehydrogenase, measured in ripening fruit.[14,15] The alcohol dehydrogenase from strawberry fruit achenes and receptacles has been examined for selectivity[14] and shown to exhibit quite marked selectivity towards some alcohols although the pair of alcohols central to this study, *i.e.*, 2- and 3-methylbutanol, were not investigated.

In our hands, both crude enzyme extracts and tissue slices from melons, strawberries and bananas showed that the alcohol dehydrogenase from all of these fruit, under the conditions used, reduced 3-methylbutanal more extensively than 2-methylbutanal by a factor of two or three (data not shown). This is not consistent with, for example, the predominance of the 2-methylbutyl skeleton in the aroma profile of melons but could be explained if the biogenetic step where the 2- and 3-methylbutyl skeletons are differentiated precedes the reduction step.

2.3 Properties of the Aminotransferases and Ketoacid Decarboxylases

Since neither the AAT nor the ADH enzymes demonstrated selectivity properties that suggested that they determined the final production of 2- or 3- methylbutyl esters, it is likely that any control exerted lies with preceding enzymes in the pathway (see Scheme 1) *i.e.* α-amino transferase and α-ketoacid decarboxylase. The amino transferases from fruit have not been extensively investigated. Yu and Spencer[16] have demonstrated the conversion of leucine to 3-methylbutanal by enzyme preparations from tomatoes. However, in our study, the conversion of added valine, leucine and isoleucine to esters by either fruit slices, fruit protoplasts or enzyme extracts could not be demonstrated. While the appropriate aldehydes were found in the headspace of some preparations, it has not been possible to determine unequivocally whether these arise from enzymatic or non-enzymatic reactions. Similar results were obtained when α-ketoacids were used as substrates and the known instability of these compounds in biological media adds a further element of difficulty.

3 CONCLUSIONS

Based on the experimental evidence discussed above it appears that the selectivity properties of the enzymes controlling the reduction and ester formation steps (see steps c and d, Scheme 1) of ester biogenesis do not provide the specificity required to explain the composition of the esters derived from the branched chain amino acids observed in many fruits. Therefore, if the transformation of branched chain amino acids is controlled by the enzymes depicted in Scheme 1, then either or both of the aminotransferase or α-ketoacid decarboxylase enzymes must be implicated.

4 EXPERIMENTAL

4.1 Preparation of Enzyme Extract

Fruit tissue (200 g) was added to a solution (pH 5.6, 400 ml) containing glycerol (0.7 M), pectinase (0.7% w/v) mercaptoethanol (50 mM), potassium dextran sulfate (5%), and 2-(N-morpholino)ethane-sulfonic acid (MES) (100 mM, 20 ml). The tissue was vacuum infiltrated for 10 min then incubated at 27 °C for one hour. Protoplasts were collected by filtration through cheesecloth followed by centrifugation (1000g for 10 min).

Protoplasts were ruptured by sonication (3 min) in a solution of Tris–HCl buffer (50 mM, pH 8.0) containing mercaptoethanol (50 mM) and Triton X-100 (0.1%). After centrifugation (1200g, 10 min), the supernatant was treated with ammonium sulfate until its concentration reached 80%. The resulting pellet was dissolved in buffer (pH 8.0) and the solution concentrated using a Centriplus concentrator 30 (Amicon Inc., Beverly, MA, U.S.A.) to a volume of about 20 ml to give the crude enzyme extract.

4.2 Analysis of Alcohol Acyl Transferase (AAT) Activity

A mixture of Tris–HCl buffer (200 µl, pH 8.5), acetyl CoA (200 µl, 0.2 mg ml^{-1}), dithiothreitol (DTT) (100 µl, 0.2 mg ml^{-1}), bovine serum albumin (BSA) (5 mg), the alcohol (100 µl) and the enzyme solution (1.0 ml) was incubated at 35 °C for 30 min in a sealed headspace vial (10 ml). The headspace composition was then analysed according to the procedure described in Section 4.5.

4.3 Incubation of Amino Acids

A mixture of α-ketoglutarate (0.4 ml, 6.6 mM), pyridoxal-5-phosphate (0.2 ml, 200 mM), DTT (0.1 ml, 0.05 mM) and the amino acid (0.5 ml, 100 mM) was added to the enzyme extract (1.0 ml), the solution was sealed in a headspace vial (20 ml) and incubated with gentle agitation for 4 and 24 hours before headspace analysis.

4.4 Incubation of α-Ketoacids

This followed essentially the same protocol as above but with the amino acids replaced by the α-ketoacids.

4.5 HS–GC Analysis of Volatile Compounds

The HS–GC system consisted of a Hewlett-Packard 5890 gas chromatograph, equipped with a Hewlett-Packard 19395A headspace sampler. Samples were separated on a fused silica BP1 capillary column (Scientific Glass Engineering, Australia, 25 m × 0.22 mm internal diameter 1.0 µm film thickness). The chromatographic conditions were: FID temperature, 240 °C; injector temperature, 230 °C; carrier gas (nitrogen) pressure, 10.0 psi. The oven temperature was held at 50 °C for 5 min after injection, then increased to 60 °C at 2 °C min^{-1}, held for 2 min and then increased to 120 °C at 10 °C min^{-1} and then held for a further 7 min. Data were acquired with HP 3365 series II ChemStation software (Hewlett Packard). The components were identified by comparison of retention times with those of authentic standards and confirmed by GC–MS.

The following conditions were used for the headspace sampler: carrier gas pressure, 0.9 bar; auxiliary pressure, 1 bar; servo air, 3.3 bar; equilibration time 25 min.; bath temperature, 80 °C; valve/loop temperature, 90 °C; pressurise time, 10 s; vent/fill loop time, 20 s; inject time, 100 s.

4.6 GC–MS Chiral Analysis

Extracts obtained from melons by simultaneous distillation extraction[6] were analysed using a Hewlett-Packard 5890 Series II gas chromatograph coupled to a Hewlett-Packard 5971A Mass Selective Detector operating in the EI mode at 70 eV. The gas chromatograph was fitted with a fused silica capillary column (J&W Cyclodex-β, 30 m, 0.25 mm internal diameter, 0.25 μm film thickness). The carrier gas was helium set at a pressure of 4 psi. The oven temperature programme was: initial temperature 50 °C for 5 min, 2 °C min^{-1} to 200 °C then held for 10 min. The injector was maintained at 220 °C and the split ratio was 100:1.

REFERENCES

1. S.G. Wyllie and D.N. Leach, *J. Agric. Food Chem.*, 1992, **40**, 253.
2. S.G. Wyllie, D.N. Leach, Y Wang and R.L. Shewfelt, in 'Fruit Flavors', eds. R.L. Rouseff and M.M. Leahy, A.C.S. Symposium Series 596, American Chemical Society, Washington, DC, 1995, p. 248.
3. V. Homatidou, S. Karvouni and V. Dourtoglou, *J. Agric. Food Chem.*, 1992, **40**, 1385.
4. P. Schreier, 'Chromatographic Studies of Biogenesis of Plant Volatiles', Dr. Alfred Huthig Verlag, Heidelberg, 1984, p. 53.
5. R. Tressl and F. Drawert, *J. Agric. Food Chem.*, 1973, **21**, 560.
6. S.G. Wyllie, D.N. Leach, Y Wang and R.L. Shewfelt, in 'Sulfur Volatiles in Foods', eds. C.J. Musssian and M.E. Keelan, A.C.S. Symposium Series 564, American Chemical Society, Washington, DC, 1994, p. 36.
7. A.G. Perez, J.J. Rios, C. Sanz and J.M. Olias, *J. Agric. Food Chem.*, 1992, **40**, 2232.
8. I. Yamashita, Y. Nemoto and S. Yoshikawa, *Agr. Biol. Chem.*, 1975, **39**, 2303.
9. H. Yoshioka, Y. Ueda and T. Iwata, *Nippon Shokuhin Kogyo Gakkaishi*, 1982, **29**, 333.
10. I. Yamashita, K. Iino, Y. Nemoto and S. Yoshikawa, *J. Agric. Food Chem.*, 1977, **25**, 1165.
11. Y. Ueda, A. Tsuda, J. Bai, N. Fujishita and K. Chachin, *Nippon Shokuhin Kogyo Gakkaishi*, 1992, **39**, 183.
12. M. Harada, Y. Ueda and T. Iwata, *Plant Cell Physiol.*, 1985, **26**, 1067.
13. A.G. Perez, C. Sanz, J.M. Olias, *J. Agric. Food Chem.*, 1993, **41**, 1462.
14. W.C. Mitchell and G. Jelenkovic, *J. Amer. Soc. Hort. Sci.*, 1995, **120**, 798.
15. D. Ke, E. Yahia, M. Mateos and A.A. Kader, *J. Amer. Soc. Hort. Sci.*, 1994, **119**, 976.
16. M.H. Yu and M. Spencer, *Phytochemistry*, 1969, **8**, 1173.
17. G. Takeoka, R.A. Flath, T.R. Mon, R.G. Buttery, R. Teranishi, M. Guntert, R. Lautamo and J. Szejtli, *J. High Res. Chromatography*, 1990, **13**, 202.
18. A. Mosandl, K. Fischer, U. Hener, P. Kreis, K. Rettinger, V. Schubert and H.-G. Schmarr, *J. Agric. Food Chem.*, 1991, **39**, 1131.

VOLATILE COMPONENTS OF SPOILED DRY-CURED HAMS

M. Acilu, J. Font, A. Garmendia, I. Susaeta and A Azpiroz

GAIKER, Technological Center, Parque Tecnológico, Edificio 202, 48170-Zamudio, Bizkaia, Spain

1 INTRODUCTION

Spanish dry-cured ham is a food product with an estimated market of more than 27 million hams in 1994. Of the total production volume, 1%–2% results in spoiled hams (normally referred to as 'cala' hams) due to some putrefaction processes that take place during the ripening process. This problem generates great economic loss and decreases the overall quality of the product.

In recent years, papers dealing with flavours of different kind of hams have been published,[1-5] but few, if any, dealing with the volatiles from spoiled hams. The aim of this paper is the study of volatile compounds from spoiled hams and the selection of an adsorbent material capable of concentrating the target volatile compounds.

2 MATERIALS AND METHODS

2.1 Materials

Hams were supplied by a Spanish producer who was responsible for the identification of spoiled and good hams. The samples were taken from different parts of the whole piece at the same stage of maturity (five months) and were sent vacuum-packed, and then stored at −20°C until analysis.

The hams used as control were from the same ham piece as the bad ones, parts in which 'cala' defect was not detected. In order to avoid the possibility that volatiles diffuse through the piece of ham from the bad part to the good one, some parts from the whole good hams were sent by the producer and analysed. No differences between these two kinds of hams were found, so the work was carried out with the good parts of the spoiled hams, called 'control', and the bad parts called 'cala'. The samples were cut and ground in the frozen state using liquid nitrogen. The equipment used was a cryogenic grinder (Retsch) with a sieve of 2 mm. The assays were carried out using 50 g of ground hams.

2.2 Study of the Concentration Material

Due to the variability of different batches and hams, an homogeneous sample – enough for carrying out all the assays – was obtained by mixing spoiled hams of the same batch. The control sample was prepared in the same way.

The adsorbent materials used for the assays were Tenax-TA, Chromosorb 106, Activated Carbon and GraphTrap. All the assays were carried out with cartridges filled with 90 mg of the material to be tested.

2.3 Isolation and Analysis of Volatiles

Samples of ground ham (50 g) were held at 35 °C in a flask and flushed with nitrogen for three and a half hours at a flow rate of 100 ml min^{-1}. The outlet of the flask was connected to a cartridge filled with 90 mg of activated Tenax-TA. The cartridge was thermally desorbed onto a CDS-Peak Master and directly injected into the gas chromatograph coupled with a mass selective detector. The mass spectra were obtained by electron impact at 70 eV and the gas chromatograms were recorded by monitoring the total ion current in the 30–200 a.m.u. range. Chromatographic separations were performed with an HP-5 (5% phenylmethyl silicone) column (60 m × 0.32 mm internal diameter × 1.0 μm film thickness). The oven temperature was held at 50 °C for 5 min and then programmed at a rate of 3 °C min^{-1} to 170 °C where it was held for 2 min. The mass spectral library used for the identification of the compounds was the Wiley library.

3 RESULTS AND DISCUSSION

3.1 Study of Volatiles

Before the identification of the volatile profile, a simple and non-destructive method for the concentration of volatiles was developed. This simple methodology (described in Section 2) allows us to concentrate them in sufficient quantity to get a significant ion chromatogram in full-scan mode.

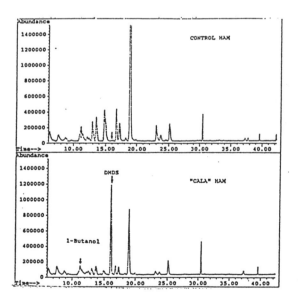

Figure 1 *Characteristic chromatograms of control and 'cala' hams showing compounds contributing to the difference*

Three different batches, from a total amount of 64 hams, were studied in order to find if there was any difference in the volatile profile between 'cala' and control hams.

Figure 1 shows the characteristic chromatograms of 'cala' and control hams. All components appeared in similar proportions in both chromatograms, except dimethyl disulfide (DMDS) and 1-butanol, which clearly appeared in higher concentration in the 'cala' hams. The relative intensity of the DMDS was always higher than that of the 1-butanol in 'cala' hams.

In the three batches studied, DMDS was not present in any of the control hams, but in the 'cala' hams, DMDS was clearly detected. In batch 1, it was found in 33% of 'cala' hams, in batch 2 in all them, and in batch 3 in 71% of them.[6]

It has been shown[7] that, in spoiled beef, the concentration of DMDS increases continuously during the storage and spoilage process and it has been suggested that it might be useful in serving as a spoilage index.

Sulfur compounds may derive from the degradation of S-amino acids and, in particular, DMDS may be obtained from methionine via Strecker degradation.[5] However, in spoiled samples it is probable that this compound arises from microbial activity.[7]

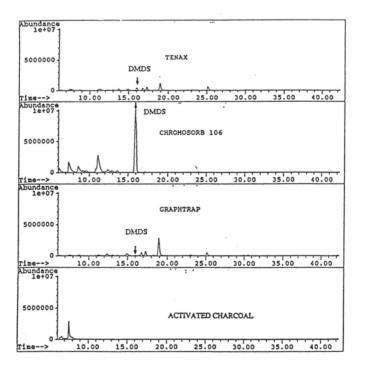

Figure 2   *Adsorbed volatiles of 'cala' hams onto the four different adsorbents showing the relative abundance of the DMDS*

3.2 Study of Adsorbents

After identification of the differences in volatile compounds between 'cala' and control ham, and their relative importance, a study of the best adsorbent material for this compound was carried out. When control samples were analysed, no DMDS was found in any of the concentrating materials, as was expected. In Figure 2 the adsorbed volatiles of 'cala' hams onto different matrices are presented. DMDS was detected when adsorbed onto Tenax-TA, Chromosorb 106 and GraphTrap. Activated charcoal was not able to adsorb DMDS in the studied samples.

Comparing the adsorption capacity of different adsorbents for DMDS and using Tenax-TA as reference adsorbent, and based on total ion chromatogram areas, the adsorption capacity was ten times higher for Chromosorb than for Tenax-TA. On the other hand GraphTrap has 75 times less adsorption capacity for the DMDS than Tenax-TA.[6]

In summary, the results of this work indicate that the presence of DMDS in 'cala' hams can be used as an indicator of spoilage and Chromosorb 106 as the best adsorbent material for concentrating this sulfur compound.

ACKNOWLEDGEMENTS

The work reported here was carried out in the course of the HAMS project. This project is partially funded by the ESPRIT Programme of the Commission of the European Community as project number 8095. The partners in the project are: GAMESA (Spain), Conservera Campofrio S.A. (Spain), De Montfort University (U.K.), GAIKER Centro Tecnologico (Spain), Stazione Esperimentale Per l'Industriadelle Conserve Allimentari (Italy), EFFE SPA (Italy) and the University of the West of England (U.K.). This paper represents the author's point of view and does not necessarily reflect that of the HAMS Consortium.

REFERENCES

1. J.L. Berdagué, C. Denoyer, J.L. Le Quére and E. Semon, *J. Agric. Food Chem.*, 1991, **39**, 1257.
2. J.L. Berdagué, N. Bonnaud, S. Rousset and C. Touraille, 'Proceedings of the 37th International Congress on Meat Science and Technology', Bundesanstal für Fleischforschung, Kulmbach, 1991, Vol. 3, p. 1135.
3. C. García, J.L. Berdagué, T. Antequera, C. López-Bote, J.J. Córdoba and J. Ventanas, *Food Chem.*, 1991, **41**, 23.
4. M.O. López, L. de la Hoz, M.I. Cambero, E. Gallardo, G. Reglero and J.A. Ordoñez, *Meat Science*, 1992, **31**, 267.
5. G. Barbieri, L. Bolzoni, G. Parolari, R. Virgili, R. Buttini, M. Careri and A. Mangia, *J. Agric. Food Chem.*, 1992, **40**, 2389.
6. M. Acilu, J. Font, A. Garmendia, I. Susaeta and A. Azpiroz, *J. Agric. Food Chem.*, 1996, in press.
7. H.K. Stutz, G.J. Silverman, P. Angelini and R.E. Levin, *J. Food Sci.*, 1991, **56**, 1147.

ANALYSIS OF VOLATILES OF *KECAP MANIS* (A TYPICAL INDONESIAN SOY SAUCE)

A. Apriyantono, E. Wiratma, H. Husain, Nurhayati, L. Lie, M. Judoamidjojo, N.L. Puspitasari-Nienaber, S. Budiyanto and H. Sumaryanto

Department of Food Technology and Human Nutrition, Bogor Agricultural University, Kampus IPB Darmaga, P.O. Box 220, Bogor 16002, Indonesia

1 INTRODUCTION

In general, soy sauce is classified into two types, Japanese and Chinese type. The main difference between the sauces is their principal raw materials. Japanese type soy sauces usually use a mixture of yellow soybean and wheat (1:1) as the main raw materials, whereas the Chinese type soy sauces use soybean and either no wheat at all or only very little.[1] However, the preparation of both types of sauce is similar. The preparation of soy sauce involves a fermentation process, *i.e.*, a mould and brine fermentation stage, with *moromi* as the final fermentation product.

Traditional *kecap manis* is different from these two types of soy sauce, and is made from whole black soybean as the principal raw material but the main difference is that coconut sugar and spices are added in the final stage of *kecap manis* preparation, *i.e.* cooking all raw materials (*moromi*, coconut sugar and spices) for 1–2 hours. The amount of coconut sugar used is about 45%–50% of the total raw materials. Unlike Japanese or Chinese soy sauces, the flavour of *kecap manis* is sweet, salty, spicy and umami, whereas that of Japanese or Chinese soy sauces is mainly salty and umami. Similar types of soy sauce to *kecap manis* are also produced in Thailand (*see-iew*), Malaysia (*kicap*) and Singapore, but caramel is used instead of coconut sugar and spices are not added. In addition, their preparation does not involve cooking for a long time.[1] Therefore, *kecap manis* is considered as a typical Indonesian soy sauce.

This is the first in-depth study on the flavour of *kecap manis*, focusing on the composition of its volatiles and the volatiles of its raw materials, as well as on the effect of brine fermentation time on the formation of volatiles.

2 EXPERIMENTAL

Seven samples of *kecap manis* were obtained from local markets in Bogor, Indonesia. Other raw materials used for *kecap manis* preparation, *i.e.*, coconut sugar, chinese star anise, fennel and *moromi*, were obtained from local markets or traditional *kecap manis* industries located in Bogor and Karawang. *Moromi* was also prepared in the laboratory using *Aspergillus sojae* as the starter for the mould fermentation stage, and brine fermentation was done spontaneously. The *moromi* obtained from brine fermentation times of 0, 1, 2, 3 and 4 months were filtered, pressed, and used further. To prepare *kecap manis*, coconut sugar (1 kg) was dissolved in water (300 ml) and boiled for 10 min. The solution was mixed with *moromi* filtrate (750 ml), boiled for 50 min, and then mixed with 4 ml spice extract (prepared by heating until boiling and filtering a mixture of 25 g fennel and 6 g chinese star anise in 100 ml water). Finally, the mixture was boiled for another 10 min, filtered, cooled and bottled.

Volatiles of *kecap manis* and its raw materials were extracted using Likens-Nickerson apparatus with diethyl ether as the extraction solvent. The extracts were then dried with anhydrous sodium sulfate, concentrated using a rotary evaporator followed by flushing using nitrogen until the volume was about 0.5 ml. The extracts were analysed using GC–MS and identification was done by matching the mass spectra obtained with those present in the NIST Library or published literature. Identification was confirmed by matching their LRI values with those reported in the literature. Quantification of volatiles was carried out using GC–MS with 1,4-dichlorobenzene as the internal standard.

3 RESULTS AND DISCUSSION

3.1 Volatiles Composition of *kecap manis* and its Raw Materials

More than 70 volatiles were identified in seven samples of commercial *kecap manis*. They included esters, aldehydes, alcohols, ketones, carboxylic acids, furans, pyrazines, pyrroles, cyclopentenones and benzene derivatives. The major volatiles identified were 2- and 3-methylbutanal, ethanol, 2-methyl-3(2*H*)-furanone, 2,6-dimethylpyrazine, acetic acid, 2-furfural, 2-acetylfuran, 5-methyl-2-furfural, butanoic acid, phenylacetaldehyde, 2-furanmethanol, 2- and 3-methylbutanoic acid, anethole and dodecanoic acid (see Table 1 for volatiles composition of a representative sample of commercial *kecap manis*).

Among the major volatiles identified, those likely to be derived from the Maillard reaction, such as furans, pyrazines, pyrroles, cyclopentenones and acetic acid, dominated the volatile composition. The volatiles of the Maillard reaction products apparently come from coconut sugar, one of the raw materials of *kecap manis* (Table 1), since many volatiles typical of the Maillard reaction products present in coconut sugar were also present in *kecap manis*. They were also likely to be formed during the cooking step as indicated by the increase of the proportion of these volatiles present in *kecap manis* compared to those present in coconut sugar (Table 1).

Most carboxylic acids, esters, aldehydes, alcohols and benzene derivatives are apparently derived from *moromi*, since they were also identified in *moromi*; not many were identified in other raw materials (Table 1). Major volatiles of the two-month-old *moromi* prepared in the laboratory were 2 and 3-methylbutanal, ethanol, phenylacetaldehyde, 2-furanmethanol, 3-methyl-1-butanol and 2-furfural (Table 1). The volatiles present in *moromi* are likely either to have been derived from soybean or were formed during the fermentation.

The major volatiles of fennel and chinese star anise (the spices mostly used in the preparation of *kecap manis*) were similar, *i.e.* anethole at an amount of 69.5% and 59.2%, respectively (data is not shown in Table 1). Anethole was identified either in samples of commercial *kecap manis* or *kecap manis* prepared in the laboratory (Table 1).

The major volatiles of coconut sugar were dodecanoic acid, acetic acid, 2-undecanone, decanoic acid, 2-nonanone, and 2-furfural. In addition, qualitatively many volatiles of coconut sugar were detected in *kecap manis* as has been mentioned above. The volatiles of coconut sugar were most probably derived from its raw material, *i.e.* coconut sap, and many of which are formed during its preparation , *i.e.*, cooking the sap coconut for about four hours.

Table 1   *Selected Volatiles Identified in* kecap manis *and its Raw Materials*

Component	A (%)ᵃ	B (%)ᵃ	C (%)ᵃ	D (ppm)
Acetic acid	21.2	14.9	nd	32.1
2-Methylpropanoic acid	nd	< 0.5	nd	1.6
Butanoic acid	<0.5	< 0.5	nd	1.0
2 and 3-Methylbutanoic acid	4.6	0.3	nd	3.5
Decanoic acid	< 0.5	5.5	< 0.5	0.9
Undecanoic acid	nd	nd	0.8	nd
Dodecanoic acid	1.3	42.5	nd	3.8
Ethanol	1.8	2.9	9.0	< 0.5
3-Methylbutanol	0.9	< 0.5	5.5	< 0.5
2-Octanol	nd	0.6	nd	nd
1-Octen-3-ol	nd	nd	0.5	nd
Phenylethanol	0.5	nd	1.2	< 0.5
Propylene glycol	nd	nd	2.0	nd
2- and 3-Methylbutanal	5.2	0.5	36.2	17.3
Hexanal	nd	nd	0.6	nd
Ethyl acetate	1.5	0.9	nd	< 0.5
Ethyl hexanoate	0.5	< 0.5	nd	< 0.5
Ethyl linoleate	0.5	nd	nd	< 0.5
trans-β-Terpinyl butanoate	nd	nd	1.0	nd
2,3-Butanedione	1.2	< 0.5	1.0	< 0.5
3-Hydroxy-2-butanone	< 0.5	< 0.5	1.4	< 0.5
2-Nonanone	nd	3.3	nd	nd
2-Undecanone	< 0.5	7.4	nd	< 0.5
4-Cyclopenten-1,3-dione	0.9	< 0.5	nd	< 0.5
Benzaldehyde	< 0.5	< 0.5	1.0	< 0.5
Phenylacetaldehyde	3.9	0.5	6.3	3.0
Anethole	3.9	nd	nd	< 0.5
Naphthalene	nd	nd	1.0	nd
2-Furanmethanol	3.8	1.4	6.3	2.8
2-Furfural	7.3	3.1	2.2	9.1
5-Methyl-2-furfural	1.9	< 0.5	0.8	1.3
2-Acetylfuran	2.0	< 0.5	nd	2.3
2-Methyl-3(2*H*)-furanone	2.6	nd	0.5	4.7
3-(2-Furanyl)-3-penten-2-one	nd	nd	0.7	nd
5-(2-Octenyl)-2(3*H*)-furanone	< 0.5	nd	1.6	< 0.5
Furan MW 124	< 0.5	nd	nd	9.1
2-Methylpyrazine	< 0.5	2.2	nd	< 0.5
2,5-Dimethylpyrazine	0.8	0.9	nd	< 0.5
2,6-Dimethylpyrazine	2.7	2.1	nd	< 0.5
2-Acetylpyrrole	0.9	< 0.5	nd	0.7
A sulfur compound	nd	nd	3.0	nd

[a]   Relative percentage area of total ion chromatogram;
      A: *kecap manis* prepared from the 2-month-old *moromi* and the coconut sugar; B:
      Coconut sugar; C: 2-month-old moromi; D: A commercial sample of *kecap* manis;
nd Not detected.

Compared to the volatiles of Japanese soy sauce, which are mainly derived from soybean and formed during fermentation, such as alcohols, acids, esters, aldehydes and ketones,[2,3] the composition of volatiles of *kecap manis* is quite different. The main difference is the proportion of the volatiles of the Maillard reaction products present in *kecap manis* is much higher than those present in Japanese soy sauce. This is due to the different methods of preparation. For preparing *kecap manis*, 45%–50% of coconut sugar is added to the *moromi* and cooked for about 1–2 hours, whereas for preparing Japanese soy sauce either no sugar or very little at all is added to *moromi* and pasteurisation, not boiling for a long time, is done in the final stage of preparation.

3.2 Effect of Brine Fermentation Time

It was surprising to observe that volatiles composition of *kecap manis* prepared from *moromi* with different brine fermentation times, *i.e.*, 0, 1, 2, 3 and 4 months, was similar (data is not shown, but the volatiles composition is similar to those of *kecap manis* prepared in the laboratory, see Table 1). It is therefore likely that many volatiles and precursors of most volatiles of *kecap manis* (*e.g.*, amino acids and peptides of *moromi*) formed during brine fermentation do not play a major role in the volatiles of *kecap manis*. In Japanese soy sauce, brine fermentation time plays a major role, *i.e.* the longer the time of fermentation the better the flavour of the soy sauce. However, the flavour of *kecap manis* is apparently more dependent on its main raw material (coconut sugar) which gives the sweet aroma and taste. Since the taste of *kecap manis* is not only sweet, but also umami, which is not the main taste of coconut sugar, brine fermentation is likely to play an important role in forming the non-volatile components responsible for the umami taste. Therefore, attempts will be made to investigate the taste components formed during brine fermentation stage of *kecap manis* preparation.

ACKNOWLEDGEMENT

The authors wish to thank the Directorate General of Higher Education, Ministry of Education and Culture of Indonesia for the financial support for this project via University Research Graduate Education (URGE) project. We thank the Department of Food Science and Technology, University of Reading (U.K.) for providing GC–MS for analysis of some samples.

REFERENCES

1. W. Röling, 'Traditional Indonesian Soy Sauce (Kecap): Microbiology of the Brine Fermentation', Ph.D. thesis, Free University of Amsterdam, 1995.
2. N. Nunomura and M. Sasaki, in 'Legume-Based Fermented Foods', eds. N.R. Reddy, M.D. Pierson and D. K. Salunkhe, CRC Press, Florida, 1986.
3. N. Nunomura and M. Sasaki, in 'Off-flavors in Foods and Beverages', ed. G. Charalambous, Elsevier, Amsterdam, 1992.

DYNAMIC HEADSPACE GAS CHROMATOGRAPHY OF DIFFERENT BOTANICAL PARTS OF LOVAGE (*LEVISTICUM OFFICINALE* KOCH.)

E.Bylaitë,* A.Legger,† J.P.Roozen† and P.R.Venskutonis*

*Department of Food Technology, Kaunas University of Technology, Radvilёnø pl. 19, Kaunas 3028, Lithuania

†Department of Food Science, Wageningen Agricultural University, P.O. Box 8129, 6700 EV Wageningen, The Netherlands

1 INTRODUCTION

Lovage (*Levisticum officinale* Koch.) is a widely used aromatic plant characterised as a medium aromatic material[1] with a pronounced seasoning-like flavour.[2] Extracts of lovage roots, seeds and leaves have been used in the food, beverage, perfumery and tobacco industries for a long time.[3] Volatile compounds in lovage have been studied previously and more than 190 compounds have been reported in its roots, seeds and leaf essential oil, as well as in root and leaf solvent extracts.[2-5] In addition, the dynamic head space method has been applied to the analysis of volatiles in lovage root[3] and fresh leaves[4] and, as a result, 20 main compounds characterised in the former and 9 in the latter. The importance of these compounds to the aroma has not been investigated greatly because the evaluation of the odour has only recently started, *e.g.* sotolon was found to be the key aroma compound of the acidic fraction of lovage extract.[2]

The present study deals with volatile compounds of different botanical parts of lovage and their contribution to the release of flavour. Volatile compounds were analysed by dynamic headspace gas chromatography (DH–GC), identified by combined gas chromatography–mass spectrometry (GC–MS) and their odour activity was evaluated by a sniffing panel (GC–SP).

2 MATERIALS AND METHODS

Plant material was collected in the experimental garden of the Lithuanian Institute of Horticulture during the blooming period (blossoms) and after formation of seeds (leaves, stems and seeds). It was air-dried at 30 °C, packed in glass containers and stored at room temperature in the absence of light until sampled.

Prior to analysis, 10 ml of 100 °C distilled water were added to 40 mg of homogenised sample and cooled down to room temperature for 5 minutes. The sample flask (40 ml) was transferred into a water bath at 37 °C in which purified nitrogen gas (40 ml min⁻¹) was flushed for 5 min in order to trap volatile compounds in 0.1 g Tenax, TA 35/60 mesh.

Combined GC–SP analysis was performed with a Carlo Erba MEGA 5300 gas chromatograph equipped with a flame ionisation detector at 275 °C. The volatiles were desorbed by a thermal desorption (210 °C, 5 min) / cold trap (−120 °C/240 °C) device (Carlo Erba TDAS 5000) and injected into a Supelcowax 10 capillary column (60 m × 0.25 mm internal diameter, df 0.25 μm). The oven temperature was held at 40 °C for 4 min and then programmed to 92 °C at a rate of 2 °C min^{-1} and then to 272 °C at 6 °C min^{-1} with a final hold of 5 min. The GC effluents were split in the ratio of 1:2:2 for the FID and the two sniffing ports respectively, and assessed by a sniffing panel[6] consisting of ten assessors who had been selected and trained. After three preliminary sessions, a list of 17 descriptors was prepared for describing odour components isolated from lovage. Besides 'Other', these descriptors were used for each component detected by the ten assessors at the sniffing port. The background noise of the sniffing port panel was determined by analysing clean Tenax tubes.

Table 1   *Volatile Compounds of Different Parts of Lovage Isolated by DH–GC Analysis, their Average Peak Areas and Coefficients of Variance*

No.	Compound	KI	GC peak area[a]			
			Blossoms	Seeds	Leaves	Stems
1	[Unknown]	763	tr	1.40	nd	nd
2	[Unknown]	827	0.63	tr	0.40	0.20
3	Butanal	878	0.45	0.35	0.08	0.29
4	2-Methyl-2-propenal	880	0.22	nd	nd	tr
5	2-Methylbutanal	921	tr	nd	tr	tr
6	3-Methylbutanal	923	tr	nd	tr	tr
7	Pentanal	985	tr	nd	tr	tr
8	α-Pinene	1022	1.20	1.90	0.70	2.00
9	α-Thujene	1028	0.08	0.24	0.20	0.14
10	Camphene	1063	0.22	0.75	0.27	0.53
11	β-Pinene	1102	0.26	0.90	0.37	0.45
12	Sabinene	1117	0.73	1.10	1.70	1.90
13	α-Phellandrene	1166	3.14	11.91	2.37	0.92
14	Myrcene	1172	4.30	4.20	3.30	2.00
15	α-Terpinene	1180	0.70	4.80	0.80	0.29
16	Limonene	1198	5.60	2.60	3.00	3.60
17	β-Phellandrene	1228	36.30	209.90	38.3	18.20
18	(Z)-β-ocimene	1247	0.60	1.70	1.20	0.93
19	γ-Terpinene	1255	1.40	2.11	2.60	0.28
20	(E)-β-ocimene	1263	0.55	0.98	tr	tr
21	p-Cymene	1282	2.33	3.50	7.11	0.90
22	Terpinolene	1289	1.50	0.86	tr	0.45
23	Pentylcyclohexadiene	1334	0.56	1.66	tr	0.07
24	Alloocimene	1383	0.40	tr	0.315	0.17
25	Pentylbenzene	1428	0.48	0.26	0.30	0.64
26	Bornyl acetate	1562	0.42	0.70	0.38	0.48
27	α-Terpinyl acetate	1690	4.70	0.80	2.60	6.16
	Total peak area		70.97	266.10	68.20	44.01
	CV[b] [%]		26	38	28	30

[a]   Average peak areas of six replicates (Vs);
[b]   CV, average coefficient of variance of individual compounds.

The volatiles trapped on Tenax were identified by Dr. M.A. Posthumus (Department of Organic Chemistry, Wageningen Agricultural University) using combined GC–MS (Varian 3400, Finnigan MAT 95) equipped with a thermal desorption/cold trap device (TCT injector 16200). The capillary column and oven temperature programme were the same as those used in the GC–SP analyses. Mass spectra were obtained with 70 eV electron impact ionisation and scanned from $m/z = 300$ to 24 at a scan speed of 0.7 s dec^{-1}.

Odour-active compounds were identified by comparison of Kovats Indices (KI), obtained from GC–SP responses, with KIs of compounds identified on GC–MS or compounds described in the literature.

3 RESULTS AND DISCUSSION

Volatile compounds isolated from four different parts of the lovage plant (leaves, stems, blossoms and seeds) were identified by GC–MS and characterised by their GC peak areas (Table 1). All of them have been reported in previous studies[3–5] as the constituents of lovage.

In Table 1, the major compounds (on a quantitative basis) were found to be monoterpene hydrocarbons (β-phellandrene, limonene, myrcene) and α-terpinyl acetate. β-Phellandrene appeared to be the most abundant compound in all parts of lovage ranging from 41.3% in stems to 78.6% in seeds. However, phthalides were almost absent in our DH–GC analysis, although these compounds have been reported to play a major role in the celery-like odour note of lovage. This confirms the results of De Pooter et al.[5] who found the headspace of fresh lovage to be composed almost exclusively of mono- and sesquiterpenes and free of phthalides. But it is in contrast with the results of Jian-Qin Cu et al.[3] who reported ligustilide as one of the major components in DHS of lovage root. It is understandable that the trace of phthalides (high KI) was not detected in our DH–GC. Our sampling was performed at 37 °C for a short time (5 min) which explains why volatile compounds with a higher KI, such as phthalides, might not be released.

Odour descriptors were provided by the assessors at the sniffing port (Table 2). Sniffing port analysis of blank samples showed that detection of an odour by fewer than two out of the ten assessors could be considered as background 'noise'. GC–SP revealed that 11, 7, 7 and 8 volatile compounds isolated from lovage blossoms, seeds, leaves and stems possessed detectable odours, respectively. Two odour-active components (α-pinene, α-phellandrene/myrcene) were detected by GC–SP in all the samples. Both of these have a 'green' note (pine, grassy, floral–chemical) and were detected by the greatest number of assessors. Blossoms and seeds possessed more volatile compounds at smaller KIs than leaves and stems. One of the most important lovage compounds, α-terpinyl acetate, was detected by GC–SP only in blossoms and stems .The concentrations of other components most probably did not exceed their threshold values even though their DH–GC responses were high. Some of the odour-active components (KI = 700, 763, 827) could not be identified by GC–MS nor detected by DH–GC due to extremely low concentrations; however, their contribution to the lovage flavour is obvious.

Table 2 *Odour Active Compounds (GC–SP) of Different Botanical Parts of Lovage: Blossoms (BL), Seeds (SE), Leaves (LE) and Stems (ST); Max snf – Number of Assessors Detecting the Smell Simultaneously*

No. [a]	KI	Component	Odour descriptor	Max snf			
				BL	SE	LE	ST
1a	700	[Unknown]	Grassy, spicy	3	3	—	—
1	763	[Unknown]	Fruity, sweet, butter	3	4	—	—
2	827	[Unknown]	Chocolate, grassy, chemical	6	—	4	3
3	878	Butanal	Chocolate, chemical	6	5	—	3
4	880	2-Methyl-2-propenal	Grassy, spicy	6	—	—	—
5/6	923	2/3-Methylbutanal	Chocolate, spicy, chemical	5	—	3	5
7	985	Pentanal	Caramel, butter, sour	6	—	3	5
8	1022	α-Pinene	Pine, grassy, floral	5	5	4	5
13/14	1166	α-Phellandrene/	Pine, grassy, chemical	7	6	6	6
	1172	myrcene					
17	1228	β-Phellandrene	Grassy, chemical	4	5	4	—
18	1247	(Z)-β-Ocimene	Mushrooms, musty, chemical	—	4	3	3
27	1690	α-Terpinyl acetate	Floral, sweet	3	—	—	3

[a] Numbers refer to Table 1.

It can be concluded from the results shown in Tables 1 and 2 that several volatile compounds isolated from different botanical parts of lovage contribute to its flavour. However, none of the volatile compounds identified in this study possesses a particular 'lovage' character. Therefore, we expect that some of the as yet unidentified odour-active compounds might play an important role in the aroma complex of lovage. Further research is required to elucidate these compounds.

REFERENCES

1. H.B. Heath and G. Reineccius, 'Flavor Chemistry and Technology', Macmillan Publishers, U.S.A., 1986, p. 176.
2. I. Blank and P. Schieberle, *Flavor Fragr. J.*, 1993, **8**, 191.
3. Jian-Qin Cu, Fan Pu, Yan Shi, F. Perineu, M. Delmas and A. Gaset, *J. Ess. Oil Res*, 1993, **2**, 53.
4. B. Toulemonde and I. Noleau, in 'Flavors and Fragrances: a World Perspective', eds. B.M. Lawrence, B.D. Mookherjee and B.J. Willis, Elsevier Science Publishers B.V., Amsterdam, 1988, p. 641.
5. H.L. De Pooter, B.A. Coolsaet, P.J. Dirinck and N. Schamp, in 'Essential Oils and Aromatic Plants', eds. A. Baerheim Svendsen and J.J.C. Scheffer, Martinus Nijhoff / Dr. W. Junk Publ., Dordrecht, The Netherlands, 1985, p. 67.
6. J.P.H. Linssen, J.L.G.M. Janssens, J.P. Roozen and M.A. Posthumus, *Food Chem.*, 1993, **46**, 367.

NEW FLAVOUR COMPOUNDS OF *CUCUMIS MELO* L.

Cornelius Nussbaumer and Bernhard Hostettler

Givaudan-Roure Research Ltd., CH-8600 Duebendorf, Switzerland

1 INTRODUCTION

Much work has been done on the identification of the volatile flavour compounds of sweet melon (*Cucumis melo* L.)[1] which is one of the largest fruit crops cultivated anywhere in the world. To date, approximately two hundred components have been found in the many different varieties and are compiled in the TNO list.[2]

More than twenty years ago, Kemp[3] showed that certain C_9-aldehydes and alcohols, *e.g.*, (*Z*)-6-nonenal and (*Z,Z'*)-3,6-nonadienol, which are compounds with low odour thresholds, are important and characteristic flavour ingredients of both musk and watermelons. The fruity odour of sweet melons is mainly due to esters, which are qualitatively and quantitatively dominant.[2] Among these, ethyl isobutyrate, ethyl butyrate and methyl 2-methylbutyrate possess the highest odour values.[4] Recent work by Homatidou[5] and Wyllie[6] revealed the importance of several methylthio esters for the aroma of certain melon varieties. Highest aroma values were found for ethyl (methylthio)acetate, 2-(methylthio)ethyl acetate and 3-(methylthio)propyl acetate.

Regarding the creation of new flavours for melons of the 'Cavaillon' type (*Cucumis melo* L. var. Cantaloupe cv. Charentais), which is very popular in Switzerland, we undertook a detailed analysis of that aroma. To the best of our knowledge, no previous determination of an accurate flavour profile of 'Cavaillon' melons has been carried out.

2 EXPERIMENTAL

2.1 Melons

Cucumis melo var. Cantaloupe cv. Charentais originated from Cavaillon (France) and were purchased from local supermarkets (July 1992).

2.2 Isolation of Headspace Volatiles

A ripe, intact Cavaillon melon (*ca.* 900 g) and also the orange-fleshed fruit pulp (200 g) were placed separately in conical flasks and the headspace trapped on Porapak Q® (50–80 mesh) microfilters under quasi-static conditions by sucking air with the aid of a personal air sampler (SKC Inc., Eighty Four, PA 15330 U.S.A., Model 222-4) for six hours as described by Neuner-Jehle and Etzweiler.[7] Adsorbed volatiles were released with 50 µl of hexane/acetone 4:1 and then analysed by GC and combined GC-MS.

2.3 Isolation of Volatiles by Solvent Extraction

The fruit pulp (100g, freed from seeds) was homogenized in a Waring blender, 2-hexanol as an internal standard was added, followed by extraction with 200 ml of pentane/dichloromethane 2:1 for six hours in a Kutscher-Steudel perforator. The extract was concentrated to a volume of *ca.* 10 ml by careful evaporation of the solvent at ordinary pressure passing a 10 cm Vigreux column.

2.4 Identification of Compounds by Combined GC-MS

Gas Chromatography. Carlo Erba HRGC 4160 with fused silica capillary columns (DB-Wax and DB-1701 J&W, 30 m × 0.32 mm); on column injection; He as gas carrier (70 kPa); FID detection; temperature programme: 50 °C for 3 min, then 4 °C min^{-1} to 230 °C.

Mass Spectroscopy. Varian MAT, models CH-5 and 212 (70 eV), connected to a Finnigan INCOS data system.

3 RESULTS AND DISCUSSION

About one hundred components were identified in the headspace of our melons. The most abundant compounds were 2-methylbutyl acetate and hexyl acetate (40%–50% combined). The spectrum of esters comprised the whole series of ethyl, propyl, butyl, isobutyl and 2-methylbutyl esters of acetic, propionic, butyric, isobutyric and 2-methylbutyric acid.

In gradually diluting the headspace concentrate followed by GC sniffing, four esters were found to have the highest odour values: ethyl isobutyrate, methyl 2-methylbutyrate, ethyl butyrate and ethyl 2-methylbutyrate.

Besides these volatile esters, a trace component was encountered in the very early part of the gas chromatogram exhibiting an ethereal, fruity and diffusive odour at the sniffing device. Combined GC-MS revealed a molecular weight of 115 for this unknown substance. Based on the mass fragmentation pattern, the structure 2-methylbutanal oxime O-methyl ether (1), was tentatively assigned (Figure 1).

(1)   (2)

Figure 1 *New compounds detected in the headspace of Cavaillon melons*

The (*E*)-isomer of a synthetic sample of (1), prepared according to Karabatsos and Hsi[8] showed the same retention time, MS and odour as the compound found in melon. Furthermore, (*Z*)-(1) and *E/Z*-(2) (Figure 1) were identified in the headspace concentrate. The (*E/Z*) ratio of (1) isolated from melon is *ca.* 15:1, while synthetic (1) exhibits a 73:27 ratio, which corresponds to the equilibrium value given by Karabatsos.[8] The (*E/Z*) ratio of (2) in the headspace extract could not be determined due to the overlapping signal of the (*Z*)-isomer with other peaks.

(*E/Z*)-Isomers of (1) and (2) showed comparable odour qualities and strengths as judged from GC sniffing. The threshold value of (*E/Z*)-(1) was *ca.* 1 ng l^{-1} air, whereas (*E/Z*)-(2) was perceived at somewhat higher concentrations. Thus, (1) with its special diffusive odour note may contribute to the overall olfactory impression of melons.

Table 1   *Approximate Concentrations (± 20%) of Heavier Volatiles in* Cucumis melo *L.*
var. Cantaloupe cv. Charentais determined by Solvent Extraction and GC

Compound	Concentration (mg kg^{-1})
1-Butanol	1
2-Methylbutanol	1
Hexyl acetate	1
Acetoin	80
1-Hexanol	1
Acetoin acetate	5
Ethyl (methylmercapto)acetate	1
2,3-diacetoxy butane (2 isomers)	5
3-acetoxy-2-butanol (2 isomers)	20
Benzyl acetate	1
Benzyl alcohol	2
Phenylethanol	0.5
Phenylpropanol	1
2-Ethyl-4-hydroxy-5-methyl-3(2*H*)-furanone	10
5-Ethyl-4-hydroxy-2-methyl-3(2*H*)-furanone	
4-Hydroxy-5-methyl-3(2*H*)-furanone	4
Cinnamic alcohol	0.2
Dihydroactinidiolide	0.3

Components are arranged in order of increasing retention time on a DB-wax column.

Oxime ethers have been recently found in several flower scents by Kaiser[9] and, without specifying (*E/Z*)-isomers, in headspace volatiles of spider mite infested young and old cucumber leaves.[10]

As (1) and (2) are most likely biosynthesized from isoleucine and leucine, respectively, they might be present in other fruits as well.

Among the sulfur compounds, S-methylthio 2-methylbutyrate dominated in the headspace of intact melons (*ca.* 2%). Its odour was perceived as sweaty and musty at the sniffing port. It has previously been found in hop and beer,[2] but not yet in fruits. Interestingly, several S-methylthio alkanoates were recently described by Wyllie *et al.*[11] to occur in *Cucumis melo* L. cv. Makdimon and to contribute to the musky aroma, but S-methylthio 2-methylbutyrate was not identified.

The solvent extract, which contained the more polar and less volatile components was dominated by acetoin and 3-acetoxy-2-butanol. Acetoin, found 1957 in high concentrations by Serini[12], was recently mentioned as a flavour constituent of *Cucumis melo* L. by Homatidou *et al.*[5]

Other major constituents of the solvent extract were 2-ethyl-4-hydroxy-5-methyl-3(2*H*)-furanone (3a), 5-ethyl-4-hydroxy-2-methyl-3(2*H*)-furanone (3b) (homofuraneol) and 4-hydroxy-2-methyl-3(2*H*)-furanone (4) (Figure 2). Homofuraneol (3) has previously been found in several food systems[13], but not previously in fruits. Due to its strong sweet, caramel-like, fruity and bread-like flavour and high concentration (*ca.* 10 mg kg^{-1}) it may be an important contributor to the complete flavour profile of Cavaillon melons. The high concentration was confirmed by direct HPLC analysis of the fruit juice,[14] where a (**3a**):(**3b**) ratio of 3:1 was determined. The well-known 2,5-dimethyl-4-hydroxy-3(2*H*)-furanone (furaneol) was only detected in sub ppm levels by GC and HPLC methods (*ca.* 0.1 ppm).

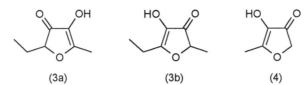

| (3a) | (3b) | (4) |

Figure 2   *Major hydroxyfuranones found in Cavaillon melons*

REFERENCES

1. K.-H. Engel, D. Heidlas and R. Tressel, in 'Food Flavours, Part C: The Flavour of Fruits', eds. I.D. Morton and A.J. MacLeod, Elsevier, 1990, p. 195.
2. H. Maarse and C.A. Visscher, 'Volatile Compounds in Food', Supplement 3, TNO-CIVO Food Analysis Institute, Zeist, The Netherlands, 6th edition, 1989.
3. T.R. Kemp, *Phytochemistry*, 1975, **14**, 2637.
4. P. Schieberle, S. Ofner and W. Grosch, *J. Food Science*, 1990, **55**, 193.
5. V.I. Homatidou, S.S. Karvouni, V.G. Dourtoglou and C.N. Poulos, *J. Agric. Food Chem.*, 1992, **40**, 1385.
6. S.G. Wyllie and D.N. Leach, *J. Agric. Food Chem.*, 1992, **40**, 253.
7. N. Neuner-Jehle and F. Etzweiler, in 'Perfumes – Art, Science and Technology', eds. P.M. Müller and D. Lamparsky, Elsevier, London, 1991, p. 153.
8. G.J. Karabatsos and N. Hsi, *Tetrahedron*, 1967, **23**, 1079.
9. R. Kaiser, in 'Flavours, Fragrances and Essential Oils', Proceedings of the 13th International Congress of Flavours, Fragrances and Essential Oils, 1995, ed. K.H.C. Baser, AREP Publ. 1995, Vol. 2, p.135.
10. J. Takabayashi, M. Dicke, S. Takahashi, M.A. Posthumus and T.A. Van Beek, *J. Chem. Ecol.*, 1994, **20** (2), 373.
11. S.G. Wyllie, D.N. Leach, Y. Wang and R.L. Shewfelt in 'Sulfur Compounds in Foods', eds. C.J. Mussinan and M.E. Keelan, A.C.S. Symposium Series 564, Washington, D.C., 1994, p. 36.
12. G. Serini, *Ann. Sper. Agr.*, 1957, **11**, 583.
13. U. Huber, *Perfumer & Flavourist*, 1992, **7/8**, 15.
14. B.I. Magyar and L.Tollsten, Givaudan-Roure Research Ltd., 1994, unpublished results.

THE VOLATILE CONSTITUENTS OF THE FLOWERS OF THE 'SILK TREE', *ALBIZZIA JULIBRISSIN*

Hazel R. Mottram[*] and Ivon A. Flament

Firmenich SA, Corporate Research Division, P.O. Box 239, CH-1211 Geneva 8, Switzerland

1 INTRODUCTION

The 'silk tree', *Albizzia julibrissin*, is a small deciduous tree, originating from Asia and tropical Africa. It is named after Filippo degli Albizzi, the Florentine nobleman who first cultivated the plant in the mid 18th century.[1] The word *julibrissin* is derived from the Persian *gul-i-abrischim*, meaning a woven silky fabric[2] and refers to the delicate appearance of the pink and white flowers, which are seen from July to September. The leaves are bipinnate with 5–15 pairs of pinnae, each consisting of between 18–40 pairs of leaflets. Although *Albizzia julibrissin* is popular as an ornamental tree, especially in Japan (Nemo-No-Ki, sleeping tree), it is not used commercially to any great extent.

Very little work has been performed on the perfume of the flower. Li *et al.*[3] analysed the headspace volatiles of the flower and identified, or tentatively identified, twenty five components. Two other references to the fragrance of the flower are as patents, one concerned with the use of silk flower extract as an ingredient in fragranced inks,[4] the other with its use in sustained release perfumes.[5]

2 EXPERIMENTAL

2.1 Sample Preparation

Collection of the headspace volatiles of the living flowers was achieved by enclosing several flowers in a plastic beaker fitted with a Teflon sleeve. Over a period of five to six hours, carbon-filtered air was drawn over the flowers and through a cartridge filled with Tenax (100 mg), by means of a Supelco SP-13 pump. The cartridges were dried with nitrogen before analysis of the adsorbed volatiles by GC–MS. The headspace of picked flowers was sampled in the laboratory using similar apparatus.

A solvent extract of picked flowers (*ca.* 100 g) was prepared by allowing the flowers to stand in redistilled dichloromethane (500 ml) for 24 hours. The extract was dried over MgSO₄ before concentration by distillation to *ca.* 0.5 ml. Absolute ethanol (10 ml) was added and the extract heated under Argon (1 hour, 50 °C). After storage at −20 °C for twelve hours, the extract was filtered and reduced to a final volume of 0.5 ml by distillation under vacuum (100 mbar, 30 °C).

---

[*]Present address: School of Chemistry, Cantock's Close, University of Bristol, Bristol, BS8 1TS, U.K.

2.2 GC–MS Analysis

Analyses were performed on a Carlo-Erba HRGC 5300 gas chromatograph coupled to a Finnigan ITD-800 mass spectrometer. The headspace volatiles were introduced to the GC column by thermal desorption, using a FLACHE5000 desorber (Firmenich Model, manufactured by Brechbühler SA, CH-1226, Plan-les-Ouates, Geneva). The volatile components were desorbed by slow heating of the cartridge (2 min at 220 °C) and transferred into a loop cooled with liquid nitrogen. The loop was then closed, progressively heated (8 min up to 220 °C), re-opened and purged in a helium stream, so that the vapours were instantaneously and completely injected onto the GC column. Analysis of the solvent extract was achieved by injection of the extract (1 μl) through the FLACHE5000 unit.

A fused silica capillary column coated with apolar SPB-1 stationary phase (60 m; 0.25 mm internal diameter; 0.1 μm film thickness) was used. The GC temperature program consisted of 5 min at 60 °C followed by an increase in temperature from 60 to 120 °C at 3 °C min^{-1} then from 120 to 280 °C at 5 °C min^{-1}. The temperature was then held at 280 °C for 30 min. Helium was used as a carrier gas at a linear velocity of 20 cm sec^{-1}. The column effluent was split equally between the mass spectrometer, a flame ionisation detector, a nitrogen phosphorus detector, a flame photometric detector and an odour port.

A second analysis was performed using a fused silica capillary column coated with a polar CPWax52CB stationary phase (25 m, 0.53 mm internal diameter, 2.0 μm film thickness, Chrompak Ltd.) and a temperature programme consisting of an increase in temperature from 60 to 250 °C at 5 °C min^{-1} followed by 25 min at 250 °C. Identification of compounds was achieved by comparison of mass spectra and linear retention indices with those of authentic standards.

2.3 GC-FTIR Analysis

Analysis of the solvent extract by GC–FTIR was achieved using a Hewlett Packard 5890 gas chromatograph coupled to a HP 5965B infra-red detector. A fused silica capillary column with a Supelcowax stationary phase (25 m, 0.53 mm internal diameter, 2.0 μm film thickness) was used with a temperature program consisting of 7 min at 50 °C followed by an increase in temperature from 50 to 230 °C at 5 °C min^{-1}, before a second isothermal at 230 °C for 30 minutes.

2.4 Oxime Synthesis

The 2- and 3-methylbutanal oximes were synthesised from their related aldehydes following the procedure described in Fieser and Fieser.[6] An aqueous solution of sodium carbonate (5 M, 25 ml) was added, with stirring, to a mixture of the aldehyde (0.2 moles) and an aqueous solution of hydroxylamine hydrochloride (8.3 M, 30 ml). After stirring overnight, the oily upper layer was separated, washed twice with water then purified by distillation under reduced pressure (*ca.* 25 mbar).

3 RESULTS

The headspace volatiles of *Albizzia julibrissin* were collected as described above, using 22.2 litres of air. Analysis of the volatiles by GC–MS provided identification, or tentative identification, of over 75 compounds, 25 of which had been previously reported by Li *et al.*[3] The relative percentages of the major constituents are given in Table 1. Ethanol was present at 26.1% in the headspace sample, but this was excluded from the calculations. The majority of compounds identified were oxygen-containing. Nine nitrogen-containing compounds

Table 1   *Major Compounds Identified in the Headspace Sample and Solvent Extract of A.* julibrissin *(expressed as a percentage of the total GC peak area)*

Compound	Headspace % Area	Solvent Extract % Area	Linear Retention Index [a]
2-Butanone	5.9	—	569
2-Methyl-3-buten-2-ol	—	0.73	618
3-Methylbutanal	1.1	0.25	629
3-Methyl-3-buten-2-one	—	0.69	646
3-Methyl-1-butanol	0.8	1.9	696
3-Methylbutanenitrile			697
+ 2-methylbutanenitrile	22.5	0.18	698
2-Methyl-1-butanol	0.42	—	699
(Z)-3-Hexenal + hexanal	0.04	—	774
Butyl acetate	0.4	—	795
(E)-2-Hexenal	0.5	—	826
(Z)-3-Hexen-1-ol	0.02	—	831
(E)-3-Hexen-1-ol	0.7	—	840[b]
(E)-2-Methylbutanal oxime			840
+(E)-3-Methylbutanal oxime	2.8	3	843
1-Hexanol	1.3	—	845
(Z)-2-Methylbutanal oxime	2.1	1.8	851/859
+(Z)-3-Methylbutanal oxime	—	—	
3-Methylbutyl acetate	2.6	—	857
2-Methylbutylacetate	0.3	—	860
Myrcene	0.5	—	986
(Z)-3-Hexenyl acetate	1.4	—	987
Hexyl acetate	1.3	—	994
Decane	0.4	—	1000
cis-Ocimene	0.3	—	1033
trans-Ocimene	14.5	0.28	1041
Linalool oxide (furanoid)	13.7	35.7	1064/1077
Linalool	14.5	4.6	1083
Benzonitrile	0.3	—	1096
(E)-4,8-Dimethyl-1,3,7-nonatriene	1.1	—	1107
allo-Ocimene	0.4	—	1121
Linalool oxide (pyranoid)	2.3	30.7	1151/1156
Phenyl acetaldehyde oxime	—	0.12	1227
α-Gurjunene	0.6	—	1415[b]
(Z)-7-Decen-5-olide (jasmin lactone)	—	2.0	1454
α-Farnesene	0.6	—	1505
Farnesol	—	2.6	1580[b]

[a]   SPB-1 stationary phase;

[b]   Retention Index not confirmed by direct injection of the pure compound.

were noted, including nitriles and oximes, as well as various terpenes and both aliphatic and aromatic hydrocarbons.

Of the compounds identified in the headspace volatiles, the oximes were of particular interest. The headspace of *A. julibrissin* was found to contain 2- and 3-methylbutanal oxime, both of which were present as the (*E*) and (*Z*) isomer. The identification of these compounds was confirmed by synthesis from the related aldehydes then characterisation by ^{1}H-NMR, ^{13}C-NMR and mass spectrometry. These compounds were first reported in flower headspace by Kaiser and Lamparsky,[7] who identified 2- and 3- methylbutanal oxime in the headspace of *Lonicera* (honeysuckle). These have since been found in the headspace of several flowers including *Angraecum, Citrus, Coffea* and *Hedychium*.[8] Phenylacetaldehyde oxime[8,9] and 2-methylpropanal oxime[7,8] have also been identified in certain flowers, although no oximes have previously been reported in *A. julibrissin*.

A solvent extract of the flowers of *Albizzia julibrissin* was prepared as described above and analysed by GC–MS. The major components identified are shown in Table 1 and were similar to those seen in the headspace. The four isomers of methylbutanaloxime were present, as well as phenylacetaldehyde oxime. Analysis of the solvent extract by GC–FTIR allowed IR spectra to be obtained for the peaks suspected to be oximes. Absorption bands for oximes are expected at 3600–3130 cm^{-1} (v O–H, s), 1685–1610 cm^{-1} (v C=N, s), 1475–1315 cm^{-1} (δ O–H, m) and 1070–800 cm^{-1} (v N–O).[10] These are all observed in the spectra, except for the C=N stretch. However this band is also absent in the library spectrum of 4-methyl-2-pentanone oxime.

REFERENCES

1. P. Lanzara and M. Pizzetti, 'Simon and Schuster's Guide to Trees', Simon and Schuster, New York, 1977.
2. E. Walker, 'Flora of Okinawa and the Southern Ryukyu Islands', Smithsonian Institute Press, Washington, 1976.
3. C. Li, Y. Zheng, Y. Sun, Z. Wu. and M. Liu, *Fenxi Huaxue*, 1988, **16**, 585, Chem. Abs. 110:92059s.
4. J. Zhao, Chinese patent 1069042 (01.08.92), Chem. Abs. 120:P109665t.
5. Z. Yang, Chinese patent 1034754 (03.02.88), Chem. Abs. 113: P103240j.
6. L. Fieser and M. Fieser, 'Reagents for Organic Synthesis', Wiley, New York, 1967.
7. R. Kaiser and D. Lamparsky, in 'Proceedings of the 8th International Congress of Essential Oils, Cannes', 1980, ed. FEDAROM, Grasse, 1982, p. 287.
8. R. Kaiser, in 'Perfumes – Art, Science and Technology', eds. P.M. Müller and D. Lamparsky, Elsevier Applied Science, London, 1991, p. 213.
9. D. Joulain, in 'Progress in Essential Oil Research', ed. E.-J. Brunke, 1986, p. 57.
10. R. Miller and H. Willis, 'Infra Red Structural Correlation Tables and Datacards (IRSCOT)', Heydon and Son Ltd., New York, 1969.

COMPARISON OF FLAVOUR VOLATILES IN A SOMATIC HYBRID FROM WEST INDIAN LIME (*CITRUS AURANTIFOLIA*) AND VALENCIA ORANGE (*C. SINENSIS*) WITH ITS PARENTS

Russell L. Rouseff, Jude W. Grosser, Harold E. Nordby and David H. Powell

University of Florida, IFAS, Citrus Research and Education Center, 700 Experiment Station Road, Lake Alfred, FL 33850, U.S.A

1 INTRODUCTION

The West Indian or Key lime (*Citrus aurantifolia*, Swing.) is a tropical small fruited acidic lime. The trees are medium in size and very sensitive to cold. Fruit are relatively small (*ca.* 3–5 cm diameter), moderately seedy with a thin peel and high juice content. The peel oil of this cultivar is highly prized for its distinctive flavour and aroma characteristics.

Valencia sweet orange (*Citrus sinensis* (L.) Osbeck) is moderately cold hardy and is extensively cultivated because it is widely considered to be the orange cultivar with the best flavour and aroma characteristics. Fruit are medium in size (10–15 cm), and contain few seeds.

In normal sexual hybridization half of the genetic material comes from each parent in the form of haploid (.5×) pollen or ovules producing progeny with the normal (1×) complement of genes. In somatic hybrids the entire genetic code of each parent is combined to produce a progeny (tetraploid) that contains twice (2×) the normal complement of genes. Autotetraploids are observed on rare occasions. Barrett[1] reported finding a Key lime autotetraploid which exhibited true Key lime flavour and aroma although was horticulturally less vigorous. Biotechnological processes have been recently developed to produce a tetraploid hybrid between Valencia orange and Key lime.[2] This unique hybrid could not be developed using conventional breeding techniques. Most citrus cultivars go through an extended juvenility period (5–8 years) before producing fruit. The juvenility period for the above mentioned hybrid is now over and the fruit can be evaluated. Whereas most citrus tetraploids are low in vigour and horticulturally inferior, this hybrid is unusual in its vigour and production of fruit. Since both parents are significant sources of natural flavours, and since it is not possible to predict dominant or recessive flavour characteristics in tetraploid hybrids, the purpose of this study was to compare the volatile peel oil components of the hybrid to that of its parents. A related goal was to determine overall flavour potential for the new hybrid.

2. MATERIALS AND METHODS

2.1 Chromatography

Samples were analysed on a Hewlett-Packard 5890 high resolution capillary gas chromatograph with a 0.32 mm internal diameter × 30 m DB-5 column (Restek, Bellefonte, PA). The initial oven temperature was 35 °C, held for 5 min, then ramped at 6 °C min^{-1} until reaching 220 °C and held there for 5 min. The injector temperature was 250 °C, the detector 320 °C and the helium carrier gas flow was 2.3 ml min^{-1}. Kovat's retention index values were established using a C_{10}–C_{25} alkane standard. Peaks were identified using both retention index and mass spectral information obtained from standards and the literature. Literature retention index values were from Adams[3].

2.2 Sample Preparation

The volatile components in the intact oil glands were extricated with a syringe immediately prior to GC analysis. Each injection consisted of the contents of 6–10 oil glands taken from the equatorial region of the fruit. To minimize peel oil compositional changes due to maturity, all samples were taken from fruit that were fully mature but not overmature.

3 RESULTS AND DISCUSSION

The somatic hybrid fruit were intermediate in size between the larger orange and smaller lime parents. The peel colour resembled a lime at early maturity but a lemon in late maturity. The overall physical appearance favoured the lime parent. The overall sensory impact of the freshly disrupted oil glands was almost indistinguishable from the lime parent. Essentially no orange aroma character was noted.

3.1 Terpenes

As shown in Table 1, terpene composition (area %) was lowest in the Key lime and highest in Valencia orange. These results are similar to those reported in other citrus oil compositional studies for lime [4,5] and orange.[6,7] The component in greatest concentration in both parents and hybrid was (+)-limonene.

Table 1 *Terpene Composition*

Retention Index	Lit. Value	Compound	Key Lime (area %)	Hybrid (area %)	Valencia orange (area %)
927	931	Thujene	0.55	0.77	0.01
933	939	α-pinene	2.68	3.89	1.28
944	953	Camphene	0.12	0.03	0.01
975	980	β-pinene	20.93	2.79	0.66
978	976	Sabinene	0.00	0.00	0.01
995	991	Myrcene	1.22	8.07	4.19
1006	1005	α-phellandrene	0.03	0.87	0.34
1016	1018	α-terpinene	0.32	0.47	0.00
1030	1031	(+)-limonene	39.89	58.12	89.34
1052	1050	β-ocimene	0.45	0.25	0.36
1064	1062	γ-terpinene	12.39	10.62	0.35
1086	1088	Terpinolene	0.70	1.11	0.01
Totals			79.29	86.99	96.68

The hybrid contained intermediate levels of camphene, β-pinene and γ-terpinene. It should be noted that several terpenes were present in relatively higher percentages in the hybrid than either parent. These include terpinolene and α-pinene but especially myrcene, a key terpene aroma impact compound. Myrcene has an aqueous odour threshold of 0.042 µg ml^{-1}.[8] Even though the hybrid favoured the lime parent in sensory characteristics, it did not have the exceptionally large proportion of the bi-cyclic terpene, β- pinene, that usually characterizes limes.

3.2 Sesquiterpenes

Shown in Figure 1 is the sesquiterpene region of the chromatograms for the hybrid and both parents. Many sesquiterpenes were common to all three cultivars. The major interest for this study was to identify those compounds found only in one parent and the hybrid and not the other parent. Valencia orange was characterized by valencene, which could not be identified in the lime. West Indian lime was characterized by germacrene B, one of the putative key flavour impact compounds of lime.[9] Even though the hybrid's physical and overall sensory qualities resembled the West Indian lime, it lacked germacrene B. The hybrid also lacked valencene.

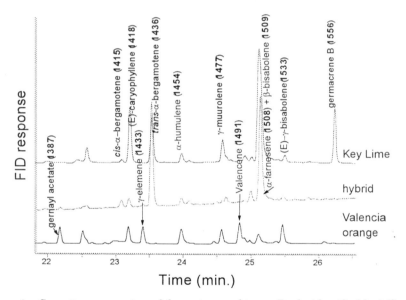

Figure 1 *Sesquiterpene region of three citrus cultivars. Peaks identified by MS and includes literature retention index values*

3.3 Alcohols, Aldehydes, Esters and Ketones

Shown in Table 2 are some of the oxygenated terpenes and esters. It is well known that this fraction contains most of the key aroma impact compounds in citrus fruits. However, it can be seen that there are not appreciable differences in the three cultivars for these compounds. Elevated citral (neral + geranial) has traditionally been associated with lemon and lime flavours. Both the lime parent and hybrid have elevated levels of these compounds compared to the orange. It should be noted that the total peak area represents both identified and unidentified peaks. (Unidentified peaks have been omitted from the table). The hybrid contained a unusually large proportion of oxygenated compounds that would explain the strong pleasant lime-like aroma. The hybrid was also characterized by having an unusually large proportion of α-terpineol, which is usually described as being floral.

4 CONCLUSIONS

Whereas the hybrid contains similarities from both parents, the physical and chemical properties of the somatic hybrid appears to favour the Key lime parent to a greater degree than the Valencia orange. Even though larger in size, with less smooth peel, the hybrid appears in many respects to be a large Key lime. Its peel oil contains elevated levels of some highly desirable aroma chemicals and would be an excellent candidate for commercial exploitation. Finally it should be noted that although the hybrid's aroma was very similar to the Key lime, it did not contain germacrene B, the reported key aroma impact compound of Key lime.

Table 2 *Oxygen Containing Compounds*

Retention Index	Lit. Value	Compound	Key Lime (area %)	Hybrid (area %)	Valencia Orange (area %)
1002	1001	Octanal	0.10		0.14
1072	1070	Octanol	0.59	0.43	0.46
1100	1098	Linalool	0.40	0.18	0.42
1103	1102	Nonanal	0.08	0.05	0.04
1157	1153	Citronellal		0.21	
1182	1177	Terpin-4-ol	0.26	0.17	0.12
1200	1189	α-terpineol	0.14	0.60	0.05
1235	1240	Neral	0.10	0.16	0.05
1245	1242	Carvone	1.07	0.84	0.07
1275	1270	Geranial	1.76	0.78	0.22
1369	1365	Neryl Acetate	0.52	0.03	0.08
1388	1383	Geranyl Acetate	0.32	0.03	0.02
Totals			7.44	8.81	2.73

REFERENCES

1. H.C. Barrett, *Fruit Varieties Journal*, 1992, **46**, 166.
2. J. Grosser, G. Moore and F.G. Gmitter, *Scientia Horticulturae*, 1989, **39**, 23.
3. R.P. Adams, 'Identification of Essential Oil Components by Gas Chromatography–Mass Spectrometry', Allured Publishing Corporation, Carol Stream, IL, 1995.
4. B.C. Clark and T.S. Chamblee, in 'Off-flavors in Foods and Beverages', ed. G. Charalambous, Elsevier Publishers, New York, 1992, p. 229.
5. P.E. Shaw, in 'Volatile Compounds in Foods and Beverages', ed. H. Maarse, Marcel Dekker, New York, 1991, p. 764.
6. P.E. Shaw and R.L. Coleman, *J. Agric. Food Chem.*, 1974, **22**, 785.
7. P.E. Shaw and R.L. Coleman, *International Flavours and Food Additives*, 1975, **6**.
8. E.M. Ahmed, R.A. Dennison, R.H. Daugherty and P.E. Shaw, *J. Agric. Food Chem.*, 1987, **26**, 187.
9. B.C. Clark, T.S. Chamblee and G.A. Iacobucci, *J. Agric. Food Chem.*, 1987, **35**, 514.

CHANGES IN LEMON FLAVOURING COMPONENTS DURING UV-IRRADIATION

Hideki Tateba, Yuko Iwanami, Nobuko Kodama and Katsumi Kishino

Ogawa & Co. Ltd., 6-32-9 Akabanenishi, Kita-ku, Tokyo 115, Japan

1 INTRODUCTION

Lemon oil dissolved in aqueous solutions (lemon flavouring) is one of the most useful flavouring materials for foods, such as soft drinks, candies and cakes. However, it is well known that the flavouring often deteriorates with age because of various factors. Among these, two major ones are heat and light. The effects of heat on the deterioration of lemon flavouring has been the subject of much research.[1] However, few reports of the effect of light on lemon flavouring are to be found except for photo-oxidation.[2] Many soft drinks, packaged in clear bottles under reduced oxygen pressure, are manufactured commercially. This paper reports how lemon flavouring deteriorates on UV-irradiation under strictly controlled conditions and discusses how the photo-deterioration can be suppressed.

2 EXPERIMENTAL

Lemon flavourings (0.1 g), made from either Californian lemon oil (FCC III grade)[3] or citral (**1** and **2**) (5 mg), were each dissolved in a 0.05 M phosphate buffer (pH 6)–ethanol (35/65) (1 l) by stirring vigorously, and the upper oil layer was removed. Each residual solution was purged with nitrogen gas, and was then sealed into clear glass bottles.

These bottles were exposed to UV-light ($\leq$ 400 nm; Toshiba FL20BLBx2 + FL20SEx2) from a 50 cm height for four days at 30 °C. Control samples were kept in the dark. The content of unreacted citral was determined by HPLC using an HP1090M instrument equipped with a reverse phase column (Capcell Pak C-18 SG120, 4.6 mm internal diameter × 250 mm, Shiseido). The volatile components were recovered from the lemon flavouring by column chromatography using Porapak Q as an absorbent and methylene chloride as the eluent.[4] Each component was identified by GC–MS by comparing its Kovat's Index and mass spectrum with those of standard compounds using an HP5971A mass spectrometer interfaced to an HP5980 series II gas chromatograph equipped with a fused silica capillary column (SupercoWaxTM10, 60m × 0.25 mm internal diameter, 0.25 μm film thickness, Supelco Inc.).

The amount of each component was calculated by computing the GC peak area against an internal standard (2-octanol). Photoreaction products of citral were isolated from a concentrate of the reaction mixture by column chromatography (n-hexane–ethyl acetate) and preparative GC on a fused silica capillary column (DB-Wax, 30 m × 0.53 mm internal diameter, 1 μm film thickness, J & W Scientific).

3 RESULTS AND DISCUSSION

3.1 UV-Irradiation of Lemon Flavouring

Lemon oil is mainly composed of terpene hydrocarbons, such as pinenes, limonene and terpinenes, and oxygen-containing compounds, such as nonanal, citronellal, linalool and citral. Many of these are major contributors to the odour of lemon flavouring as determined by AEDA (aroma extract dilution analysis),[5] and the odour quality of lemon flavouring can be evaluated by quantitative analysis of these aroma components. Because most of these components in lemon flavouring are easily degraded by heat in the presence of acid, the heat effects were controlled in investigating the UV effects on flavouring deterioration. Even though lemon flavourings are often used in soft drinks together with lemon or other fruit juices, which contain natural or added acids, the examination of UV-irradiation effects on lemon flavouring was carried out in pH 6 phosphate buffer–ethanol solution, to prevent the acid-catalysed reactions.

Figure 1 shows the quantitative and sensory results from UV-irradiation of the lemon flavouring. The analyses were repeated five times and the quantitative data were expressed as mean values. The coefficient of variation (CV) was no more than 23% for any data. Citral, which provides the typical lemon-like odour, decreased rapidly and then new peaks, such as **I**, **II** and **III**, appeared. Monoterpene hydrocarbons, such as limonene and γ-terpinene, and nonanal decreased, while *p*-cymene increased.

Other components, such as sesquiterpene hydrocarbons, esters, alcohols, changed only slightly. The fresh, sweet, lemon-like odour mainly characterised by citral disappeared, while a dusty and herbaceous odour increased. Peaks **IV**, **V** and **VI**, shown in Figure 1, were observed in trace quantities and have not yet been identified. However, **IV** was confirmed to come from citral by UV-irradiation. Peak **V**, found to contribute to the off-odour, was obtained indirectly by photoreaction of citral. Therefore, photoreaction of citral was suggested to be a very important factor in the photo-deterioration of lemon flavouring.

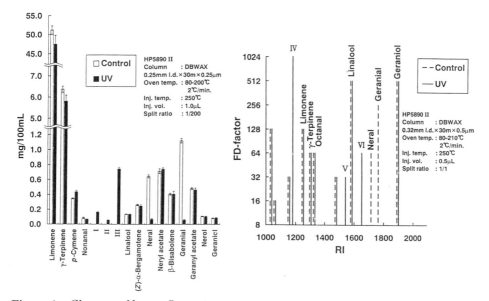

Figure 1 *Changes of lemon flavouring components*

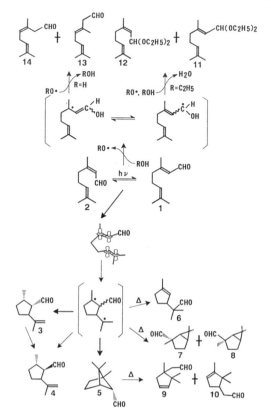

Figure 2   *Photoreduction of citral in protic solvents*

3.2 Photoreaction of Citral in Aqueous Solutions

It has been reported that citral is converted to photocitral A (**3** in Figure 2), epiphotocitral A (**4**) and photocitral B (**5**) at room temperature[6] and to the products (**7** and **8**) from 1,2-migration of the formyl group at higher temperature ($\geq 80\ °C$) in an aprotic solvent under nitrogen.[7] No other compounds from the photoreaction of citral in a protic solvent have been reported.[8] In this present study, **6–14** were produced from the photolysis of citral, in addition to **3–5**, (Figure 2). α-Campholenealdehyde (**10**) and compounds **11–14** were identified by comparing the GC–MS data with that of known compounds. Photocitrals (**3–5**) and other products(**7** and **8**) were identified by comparing the spectral data with that reported in the literature.[7,9] Compound **6** was confirmed by comparing the spectral data of an authentic sample synthesised according to a previously reported procedure.[10] In the photoreaction of citral, **6** was a new product whose formation requires the 1,3-migration of the formyl group. Small amounts of products, **7** and **8**, whose formation required the 1,2-migration of the formyl group, were obtained. The yields of **7** and **8** increased at 80 °C as in an aprotic solvent.[6] Diethyl acetals (**11** and **12**) and aldehydes (**13** and **14**) might be obtained from the hydrogen-abstraction reaction of excited citral because they were not produced in the dark. The structure of **9**, which is a new compound, was deduced from various spectroscopic data. Aldehydes **9** and **10** might be derived from **5**. The peaks **I**, **II** and **III**, shown in Figure 1, were confirmed to be **5**, **6** and **3**, respectively.

3.3 Prevention of Photo-deterioration of Lemon Flavouring

The odour deterioration of lemon flavouring under UV-light was found to be caused mainly by photolysis of citral. It was reconfirmed that citral was mainly converted to intramolecular cycloaddition products, such as photocitrals. The reactivity of citral in UV-light was investigated under various conditions of solvents, pH and acid sodium salts, appropriate to the composition of commercial soft drinks. Figure 3 shows a solvent effect, and suggested that the photoreaction could be promoted by including more water, such as soft drinks. The reactivity increased at lower pH, but salts hardly influenced the reactivity.

As the reactions had been postulated to proceed via radical mechanisms,[7] we tried to control the reaction using radical scavengers, such as antioxidants. Ascorbic acid showed a clear preventive effect on photo-reaction of citral in lemon flavouring which increased with concentration (Figure 4). Therefore, photo-deterioration of lemon flavouring was suggested to be stabilised not only by excluding UV-light, but also by the addition of a radical scavenger.

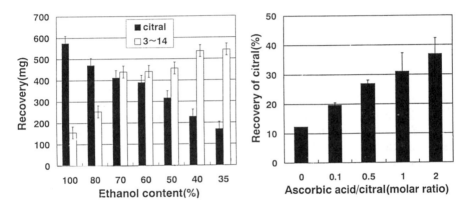

Figures 3 and 4 *Effect of ethanol and ascorbic acid on the photoreduction of citral*

REFERENCES

1. B.C. Clark, Jr. and T.S. Chamblee, in 'Off-flavours in foods and beverages', ed. G. Charalambous, Elsevier, Amsterdam, 1992, p. 229.
2. P. Schieberle and W. Grosch, *Z. Lebensm. Unters. Forsch*, 1989, **189**, 26.
3. Committee on Codex Specifications, 'Food Chemicals Codex', National Academy Press, Washington, D.C., 3rd edition, 1981, p. 168.
4. M. Shimoda, K. Hirano and Y. Osajima, *Bunseki Kagaku*, 1987, **36**, 792.
5. P. Schieberle and W. Grosch, *J. Agric. Food Chem.*, 1988, **36**, 797.
6. R.C. Cookson, J. Hudec, S.A. Knight and B.R.D. Whitear, *Tetrahedron*, 1963, **19**, 1995.
7. S. Wolff, F. Barany and W. C. Agosta, *J. Am. Chem. Soc.*, 1980, **102**, 2378.
8. G. Büchi and H. Wüest, *J. Am. Chem. Soc.*, 1965, **87**, 1589.
9. R. Kaiser and D. Lamparsky, *Helv. Chim. Acta*, 1976, **59**, 1797.
10. M.B. Erman, O.O. Volkova, G.V. Cherkaev, I.M. Pribytkova, M. Yu. Antipin, Yu.T. Struchkov and V.B. Mochalin, *Zh. Org. Khim.*, 1986, **22**, 2508.

EXPLORATION OF AN EXOTIC TROPICAL FRUIT FLAVOUR – OPTIMISATION OF *CEMPEDAK* (*ARTOCARPUS INTEGER* (THUNB.) MERR.) FLAVOUR EXTRACTION

C.H. Wijaya, T.A. Ngakan, I. Utama, E. Suryani and A. Apriyantono

Department of Food Technology and Human Nutrition, Bogor Agricultural University, Kampus IPB Darmaga, P.O. Box 220, Bogor 16002, Indonesia

1 INTRODUCTION

Cempedak (*Artocarpus integer* (Thunb.) Merr.), one of Indonesia's exotic tropical fruits, has a strong characteristic aroma, reminiscent of durian and jackfruit.[1,2] Due to its unique aroma, there is a high demand for finding a suitable method to obtain an aroma extract of the fruit for both direct applications and analytical purposes. The studies on the flavour components of *cempedak*, however, are very limited,[2] although some studies have been done on jackfruit,[2,3] which is a fruit similar to *cempedak*.

The aim of this research was to find which part of the fruit possesses the highest intensity of the fruit aroma, the form of material most suitable for extraction, and the most suitable extraction method. The most suitable method has been further optimised by selecting the appropriate solvents and their appropriate ratio with respect to the amount of material to be extracted, as well as the optimum conditions for each extraction step. In addition, the composition of the volatiles of the extract was also analysed.

2 EXPERIMENTAL

Cempedak fruits were obtained from a local market in Bogor, West Java-Indonesia. The fruits were separated into three parts: rind, flesh and fibrous layer. The intensity of the aroma of each part was then compared. The most intense part was then treated by either slicing it 3 mm thick or pulping it using a blender. Four methods commonly used in flavour extraction[2-4] were tried in order to determine the most suitable. The method that gave the extract with the aroma which most resembled the aroma of the fresh fruit was further optimised by varying the condition of each extraction step. The possibility of prolonging the shelf life of the flesh before extraction was also studied by keeping this material at 15 and 25 °C for 2, 4 and 6 days.

The analyses done included sensory evaluations (ranking test, multiple comparison test, aroma description and hedonic test)[5,6] and GC and GC–MS analysis.[2] Sensory evaluations were performed by trained assessors, except for the hedonic test. Determination of the aroma threshold value of the best extract and its application in a soft drink model were also carried out.

3 RESULTS AND DISCUSSION

3.1 Material Preparation

3.1.1 Selection of Fruit Part with the Highest Aroma Intensity. The multiple comparison test for the aroma showed that the flesh of *cempedak* has a significantly higher aroma intensity than the rind or the fibrous layer.

3.1.2 Storage Effect of Raw Flesh before Extraction. The result of this study showed that it is possible to prolong the shelf life of *cempedak* flesh for 2–4 days at 15 °C without losing aroma intensity. If the temperature were increased and the period of storage extended, however, the intensity of the aroma decreased and the formation of off-flavour increased. The activity of the enzymes present in the plant tissue, such as lypoxygenase, apparently plays a role in promoting the oxidation reaction that produces the unpleasant odour.[7,8]

3.2 Selection of Extraction Method

Based on hedonic and multiple comparison tests (Figure 1), and supported by the aroma description of each extract, the extract obtained by maceration using dichloromethane as the solvent possessed the aroma most resembling fresh *cempedak* in comparison with those obtained by vacuum or steam distillation or simultaneous distillation and extraction methods.

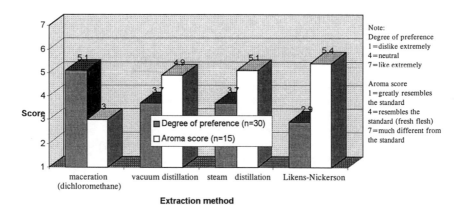

Figure 1 *Quality of flavour extract of* cempedak *extracted by various methods*

Several other solvents with various polarities had been tried for the maceration, *viz* hexane, pentane, diethyl ether, ethanol and a mixture of diethyl ether-pentane (2:1, v/v). Of the solvents used, dichloromethane, the mixture of diethyl ether-pentane, and ethanol yielded extracts with good *cempedak* aroma quality. The ethanol extract, however, was difficult to separate from the pulp because of its water miscibility. It also gave a cloudy and rather viscous extract. Diethyl ether apparently extracted more acidic aroma compounds, since the extract had a more acidic note. Hexane and pentane gave extracts with an unpleasant aroma (resembling deteriorated coconut milk). Based on the quality of the extract obtained, its stability during storage and its ease of evaporation, dichloromethane was chosen as the extraction solvent.

The results indicated that volatile components responsible for typical *cempedak* aroma are susceptible to heat. Extracts obtained by the distillation methods were described as having boiled *cempedak* aroma or overripe aroma with burnt characteristics. Special attention should be paid to the extract obtained by steam distillation, since it had also an aroma typical of *pandan* (a shrub common in South East Asia). This aroma in *pandan* is believed to be due to 2-acetyl-1-pyrroline.[2] In processed foods, 2-acetyl-1-pyrroline is usually produced from the reaction of proline and carbonyl compounds at elevated temperatures.

3.3 Optimisation of Maceration Condition

3.3.1 Selection of the Form of the Material. The aroma of the extract prepared from the flesh pulp of *cempedak* possessed a higher intensity compared to that prepared from the sliced flesh. This may be due to the smaller particle size, giving a greater surface area in contact with the solvent[9] and, accordingly, a higher number of components extracted.

3.3.2 Ratio of Solvent to Material. No significant differences in either aroma quality or intensity among the extracts (after the volume of the extract had been adjusted to the same volume for all the extracts) was found for ratios of 1:1, 1:2, 1:3 and 1:4.

3.3.3 Influence of Stirring and Freezing Times. In order to facilitate the separation of the solvent extract from the residue, the extract was frozen at –20 °C.[4] The extracts prepared with stirring times of 5, 30 and 60 min and freezing times of 12, 24 and 48 hours gave no significant differences in aroma intensity. A stirring time of 5 min and a freezing time of 12 hours were considered optimal.

3.4 Aroma Threshold Value and Application of the Extract in a Soft Drink Model

The aroma threshold value of the evaporated dichloromethane extract prepared under optimum conditions was 163 ppm. This suggests that the extract could be used as a flavouring. The sensory evaluation of the soft drink model (a mixture of 528.6 ml sucrose syrup (67° Brix), 2 g sodium benzoate, 8 g citric acid, 2 ml tartrazine, and 0.5 ml evaporated dichloromethane extract, diluted to 1000 ml with water) showed that the aroma was slightly weaker than the standard (fresh flesh of *cempedak*), whereas the average assessor score in the hedonic test was 5.5 which lies between 'Like slightly' and 'Like'.

Table 1　　*Positively Identified Volatiles of* cempedak *Fruit Extract*

Volatile compound	ppm	Volatile compound	ppm
Methyl 3-methylbutanoate	0.1	Butyl pentanoate	0.1
Toluene	0.1	1-Hexanol	0.2
Propyl 3-methylbutanoate	0.1	Butyl hexanoate	0.3
3-Hexanol	0.5	Acetic acid	0.2
3-Methyl-1-butanol	1.1	2,5-Dimethyl-4-methoxy-3(2*H*)-furanone	0.1
2-Hexanol	0.8	Isoamyl octanoate	0.2
Butyl 3-methylbutanoate	0.5	3-Methylbutanoic acid	0.9
3-Methylbutyl 2-methylpropanoate	0.3	Phenylethanol	0.1
Octanal	0.1	Octanoic acid	0.7
3-Methylbutyl 3-methylbutanoate	3.0	Ethyl hexadecanoate	0.2

3.5 Composition of Volatile Compounds

About 64 volatile compounds of the *cempedak* extract were detected by GC–MS, 20 of which were positively identified (Table 1). Except for 2-acetyl-1-pyrroline and 2,5-dimethyl-4-hydroxy-3(2*H*)-furanone, the major volatile compounds reported in the previous report for a different variety of *cempedak*,[2] such as 3-methylbutanoic acid and 3-methylbutanol, were also identified in the extract. GC–Odour analysis indicated 3-methylbutyl 3-methylbutanoate and butyl 3-methylbutanoate to be the character-impact compounds of *cempedak*.

ACKNOWLEDGEMENT

This experiment was financially supported by the Indonesian National Research Council via the Integrated Outstanding Research (RUT) project. The authors are indebted to Dr. Slamet Budiyanto as the team leader of this project and to Mr. Han Han Hariyanto for preparing the figures.

REFERENCES

1. Widyastuti, Nangka dan Cempedak: Ragam Jenis dan Pembudayaannya, Penebar Swadaya, Jakarta, 1993.
2. K.C. Wong, C.L. Lim and L.L. Wong, *Flavour and Fragrance J.*, 1992, 7, 307.
3. G. Swords, P.A. Bobbio and G.L.K. Hunter, *J. Food Sci.*, 1978, **43**, 639.
4. M. Larsen and L. Poll, in 'Flavour Science and Technology', eds. Y. Bessiere and A. F. Thomas, John Wiley and Sons, Chichester, 1990, p. 209.
5. E. Larmond, 'Methods for Sensory Evaluation of Foods', Department of Agriculture, Ottawa, 1975.
6. W. Grosch, P. Schieberle and S. Ofner, *J. Food Sci.*, 1990, **55**, 193.
7. B. Nijssen, in 'Volatile Compounds in Foods and Beverages', ed. H. Maarse, Marcel Dekker, New York, 1991, p. 689.
8. R.G. Buttery, in 'Flavor Science: Sensible Principles and Techniques', eds. T.E. Acree and R. Teranishi, A.C.S. Professional Reference Book, Washington, D.C., 1993.
9. G.A. Reineccius and S. Anandaraman, in 'Food Constituents and Food Residues: Their Chromatographic Determination', ed. J.F. Lawrence, Marcel Dekker, New York, 1984.

THE BIOSYNTHESIS OF 2,5-DIMETHYL-4-HYDROXY-2H-FURAN-3-ONE AND ITS DERIVATIVES IN STRAWBERRY

I. Zabetakis, J.W. Gramshaw and D.S. Robinson

Procter Department of Food Science, University of Leeds, Leeds, LS2 9JT, U.K.

1 INTRODUCTION

Current research focuses on the production of natural flavour compounds. 2,5-Dimethyl-4-hydroxy-2H-furan-3-one (furaneol) is one of the most important components of strawberry flavour,[1,2] but it is also present in other fruits such as pineapples[3] and mango.[4] In strawberries, two derivatives of furaneol are also present: 2,5-dimethyl-4-hydroxy-2H-furan-3-one-glucoside (furaneol-glucoside) and 2,5-dimethyl-4-methoxy-2H-furan-3-one (mesifuran).[5] Lactaldehyde (2-hydroxypropanal) and dihydroxyacetonephosphate (DHAP) have been proposed as precursors of furaneol-glucoside.[6] It is suggested that the aldol condensation of these two compounds yields 6-deoxy-D-fructose and this deoxysugar in turn may be the main precursor of furaneol-glucoside.[7] The natural occurrence of isomers of 6-deoxy-D-fructose in strawberries[8] (i.e., rhamnose and fucose) prompted us to investigate whether feeding of strawberry cells with rhamnose (6-deoxymannose) or fucose (6-deoxygalactose) enhances the formation of furaneol and its derivatives.

2 MATERIALS AND METHODS

Strawberry callus cultures have been established and cultured as described elsewhere.[7] These calluses have been cultured in control medium and 6-deoxysugar-fed medium for four weeks. The control medium consisted of Murashige and Skoog[9] (MS) basal salt mixture, supplemented with agar (1% w/v), sucrose (2% w/v), benzylaminopurine (2.22 μM) and 2,4-dichlorophenoxyacetic acid (2.26 μM). The pH of the medium was adjusted to 5.7 and the medium was sterilised.[7] The precursor-fed culture medium contained either the 6-deoxysugar (6-deoxy-D-galactose or 6-deoxy-L-galactose or 6-deoxy-L-mannose) (0.5% w/v) or furaneol (0.5% w/v), as well as sucrose (1.5% w/v) and all the other chemicals at the above levels and was filter-sterilised. All chemicals were purchased from Sigma, U.K. The purity of all 6-deoxysugars, purchased from Sigma, was claimed to be at least 99%. Their purity was tested by thin layer chromatography (TLC): a unique spot was obtained for each sugar. At the end of the culture period, the calluses were homogenised and analysed for furaneol and its derivatives as described elsewhere.[7]

3 RESULTS AND DISCUSSION

3.1 Precursor Feeding

The aim of these studies was to determine the biosynthetic pathway for furaneol. The quantitative effect of each of four different 6-deoxysugars on the formation of furaneol-glucoside is shown in Table 1. The highest amount of furaneol-glucoside was formed with 6-deoxy-D-fructose, whereas with 6-deoxy-D-galactose approximately only half the amount of furaneol-glucoside was observed. For control experiments, where the supplemented sugar was only sucrose, furaneol-glucoside was not detected. Consequently, it is suggested that it is the aldose or ketose nature of the sugar, and not the configuration (D- or L-), which mainly governs product formation. The levels of furaneol-glucoside obtained with 6-deoxy-L-galactose and 6-deoxy-L-mannose, as carbon sources, were similar. Likewise, these results indicate that the configuration at C2 and C4 of the deoxyhexose has no effect on the formation of furaneol-glucoside.

Table 1 *Furaneol and its Derivatives (mg per g of fresh weight of tissue) in Control and Precursor Fed Cultures*

	Furaneol	*Furaneol-glucoside*	*Mesifuran*
Control	n.d.[a]	n.d.[a]	n.d.[a]
6-deoxy-D-galactose	n.d.[a]	0.51 ± 0.02[b]	n.d.[a]
6-deoxy-L-galactose	n.d.[a]	0.89 ± 0.16[b]	n.d.[a]
6-deoxy-D-fructose[c]	n.d.[a]	1.03 ± 0.22[b]	n.d.[a]
6-deoxy-L-mannose	n.d.[a]	0.85 ± 0.06[b]	n.d.[a]

[a] n.d.: not detected;
[b] Each figure (95% confidence limits) represents the mean of three analyses;
[c] The 6-deoxy-D-fructose results are reproduced from a previous paper[7] in order to render easier the evaluation of the effect of the different deoxysugars on the formation of furaneol-glucoside.

Interestingly, only furaneol-glucoside, and not free furaneol, was detected in the deoxysugar-fed cultures, indicating that the glucoside may be the main metabolite formed from the deoxysugars. This proposal is also supported by the fact that, when furaneol alone was fed to the callus, furaneol-glucoside was not detected in the callus cultures after a culture period of four weeks. Since no furaneol-glucoside was formed, assuming that the substrate was assimilated, it is suggested that furaneol-glucoside is the main product of the biosynthetic pathway and the precursor of free furaneol (Figure 1). However, the fate of the added furaneol is not known, since this also was not detected after the four-week culture period.

3.2 The Biosynthetic Pathway of Furaneol

It seems that the 6-deoxyhexoses with an aldo group (6-deoxy-D-galactose, 6-deoxy-L-galactose and 6-deoxy-L-mannose) have a smaller effect on the amount of furaneol-glucoside formed than the 6-deoxyhexose with a keto group (6-deoxy-D-fructose). This may be because 6-deoxyketoses predominantly exist in solution as a five-membered ring which is similar to furaneol, whereas 6-deoxyaldoses predominantly exist as a six-membered ring.

During fruit ripening, the cell wall polysacharides are hydrolysed, monosacharides (*e.g.*, rhamnose, fucrose) are liberated and may provide a source of furaneol precursors.[8,10]

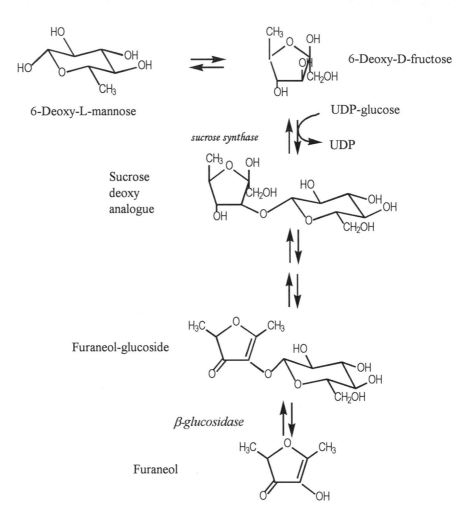

Figure 1   *Proposed outline for the biosynthesis of furaneol*

ACKNOWLEDGEMENT

The authors are grateful to the Hellenic State Scholarships Foundation for a Research Studentship to one of us (I.Z.).

REFERENCES

1. E. Honkanen and T. Hirvi, in 'Food Flavours, Part C: The Flavour of Fruits', eds. I.D. Morton and A.J. MacLeod, Elsevier, Amsterdam, 1990, p. 125.
2. M. Larsen, L. Poll and C.E. Olsen, *Z. Lebensm. Unters. Forsch.*, 1992, **195**, 536.
3. J.O. Rodin, C.M. Himel, R.M. Silverstein, R.W. Leeper and W.A. Gortner, *J. Food Sci.*, 1965, **30**, 280.
4. C.W. Wilson, P.E. Shaw and R.J. Knight, *Dev. Food Sci.*, 1988, **18**, 283.
5. A. Latrasse, in 'Volatile Compounds in Foods and Beverages', ed. H. Maarse, Marcel Dekker, New York, 1991.
6. I. Zabetakis and M.A. Holden, in 'Bioflavour 95: Analysis – Precursor Studies – Biotechnology', ed. P. Eteivant, INRA Editions, 1995, p. 211.
7. I. Zabetakis and M.A. Holden, *Plant Cell Tiss. Org. Cult.*, 1996, **45**, 25.
8. A.F. Pisarnitskii, A.G. Demechenko, I.A. Egorov and R.K. Gvelesiani, *Appl. Biochem. Microbiol.*, 1992, **28**, 97.
9. T. Murashige and F. Skoog, *Physiol. Plant.*, 1962, **15**, 473.
10. P. Muda, G.B. Seymour, N. Errington and G.A. Tucker, *Carbohydr. Polym.*, 1995, **26**, 255.

BIOTECHNOLOGICAL PRODUCTION OF FLAVOUR

GENESIS OF AROMA COMPOUNDS IN PHOTOTROPHIC CELL CULTURE OF GRAPEFRUIT, *CITRUS PARADISI* CV. *WHITE MARSH*

G. Reil* and R.G. Berger

*Technische Universität München, Institut für Lebensmitteltechnologie und Analytische Chemie, D-85350 Freising Weihenstephan, Germany

Universität Hannover, Institut für Lebensmittelchemie, D-30453 Hannover, Germany

1 INTRODUCTION

In vitro plant cells have lost their previous position in the intact plant and are forced by growth regulators, incorporated into the nutrient medium, to divide continuously. A pronounced primary metabolism ensures survival in the synthetic environment; very low levels of expression of secondary pathways are usually observed as a result. As early as the occasion of the fifth Weurman Symposium, however,[1] a regreened callus culture of *Ruta graveolens* was shown to contain significant amounts of volatile flavour compounds. Evidence for light induced generation of volatiles is now accumulating.[2]

2 RESULTS

2.1 Cell Cultures

Photomixotrophic calli of *C. paradisi* on different media formed soft, dark green, aggregates (< 1 cm), and contained 7 to 9 wall associated chloroplasts, irrespective of the actual chlorophyll content. Thorough light microscopic investigation gave no hint of the presence of any secretory structures, but showed a heterogeneous cellular morphology. Organogenesis was not observed.

2.2 Chemical Analysis of *C. paradisi*

Chlorophyll and chlorophyll a/b ratios reached stable values after five subcultivations of *C. paradisi* under 3000 lux. A combination of a modified Murashige and Skoog (MSL) medium,[3] light tubes of the emission colour 12, a 24 hour photoperiod and two-fold phytohormone concentrations, as compared with the original medium, yielded maximum amounts. The calli grown on MSL medium exhibited a typical *Citrus* aroma, and these cells were examined for their volatile composition.

2.3 Volatiles in Fruit and Cultured Cells of *C. paradisi*

Citrus essential oils are complex mixtures, mainly composed of oligoisoprenoids and fatty acid degradation products. The sensory impression is determined by (+)-*4R*-limonene, and quantity and ratio of side and trace components impart species character. Composition and accumulation of volatile constituents in *C. paradisi* cell cultures growing under different environmental conditions were compared with the essential oil of the mature fruit (Table 1; Figures 1 and 2). GC-MS examination of the essential oil of *C. paradisi* reconfirmed the presence of forty-nine constituents, mainly monoterpenes, in the flavedo–albedo layer (exocarp) and fleshy (mesocarp–endocarp) proportion of the fruit.[4] Limonene, the progenitor of the oxygenated p-menthanes, accounted for 88% of the volatile fraction in the peel. Samples of similar size were taken from the juice vesicles. Comparatively few volatiles were found, but they included 2(*E*)-hexenal and nootkatone.

Formation of volatiles was not observed in any of the heterotrophically initiated cells nor in phototrophic cells that contained less than 140 mg chlorophyll per kg. When callus was grown under intense light and phytoeffector conditions as indicated in Figure 2A, the identities of the volatiles identified in the solvent extract matched almost completely those in the cold pressed grapefruit peel oil; moreover, the quantitative distribution was very similar. All components extracted from the cell culture were also found in the peel oil, apart from n-tri- and n-tetradecanal. Some minor constituents of the peel oil, such as acetic acid esters, citronellol, and the carvones, were not found in cell cultures, probably as a result of a lack of analytical sensitivity under the conditions used. Experiment **B** (Table 1) used elevated concentrations of phytoeffectors which resulted in maximum formation of limonene and total volatiles at the expense of diversity of the volatiles. This accumulation must represent true *de novo* synthesis since carry-over from the primary callus was negligible because composition of volatiles reproducibly depended on the subcultivation conditions and because suspension cultures that did not accumulate any volatile constituents recovered the full synthetic potential upon re-transfer to solidified medium.

Distinct quantitative differences existed between accumulating fruit tissues and the photomixotrophic callus culture: The highest yielding cell culture (culture **B** in Table 1) contained about 5% (186 mg kg^{-1} wet weight) of the volatiles found in peel tissue (exo/mesocarp section), and about twenty times that found in the fleshy endocarp. A direct comparison of the aroma producing potentials of both differentiated plant parts and cell culture is complicated by the heterogeneous morphology and the obvious concentration gradients in the fruit. If the peel proportion of the whole fruit is estimated as about 5% of the total weight, the wet-weight based figures of whole fruit tissue and cell culture are not too different (0.029% *vs.* 0.019% w/w); on a time base, the productivity of the cultured cells is superior (13 months from anthesis to mature fruit *vs.* 4 to 5 weeks of subcultivation). Small amounts (< 10%) of volatiles leaked into the surrounding agar zones, but were not included in these calculations. A systematic approach to improve terpene yields was not within the scope of this work. Production of volatile aroma compounds has now been observed during more than two years of continued sub-culturing.

The variation of physical and chemical parameters of cultivation modulated the pattern and yield of volatiles. Permanent illumination instead of a long-day photoperiod was superior for the formation of volatiles. A mercury high-pressure lamp, frequently used for the cultivation of plant cell cultures, was inferior to a neon tube type that emitted a continuous, uniform, spectrum. An increased concentration of the gelling agent (9 g l^{-1} gellan gum) induced an increase in the concentrations of the saturated aldehydes from six to fourteen carbons, while limonene remained the major compound. Chlorophyll concentration and accumulation of volatiles were correlated. In the range from 150 to 300 mg chlorophyll kg^{-1}, the data fitted a linear regression with a coefficient of $R^2 = 0.912$. The same linearity of chlorophyll and terpene accumulation was observed for callus of *C. limon* from 15 to 320 mg chlorophyll kg^{-1} ($R^2 = 0.9849$), whereas data obtained from *C. aurantifolia* were logarithmically correlated (25 to 390 mg chlorophyll kg^{-1}, $R^2 = 0.9775$).

Table 1 *Volatiles (mg kg⁻¹ wet wt) in Photomixotrophic callus of* Citrus paradisi *(conditions see Figure 2)*

No.	Compound	Flavedo/ Albedo	Juice Vesicles	Tissue Culture A	Tissue Culture B	Identn.[a]
1	α-Pinene	19	0.03	0.09	0.58	MS;RT
2	β-Pinene	trace	n.d.	trace	trace	MS;RT
3	Sabinene	26	n.d.	0.22	0.67	MS;RT
5	Myrcene	0.42	n.d.	0.74	3.75	MS;RT
6	(+)-Limonene	3814	0.12	48.35	176.73	MS;RT
9	(E)-β-Ocimene	6.75	n.d.	0.06	0.28	MS;RT
11	n-Octanal	83	n.d.	1.17	0.41	MS;RT
13	3(Z)-Hexenal	3.58	0.15	0.11	n.d.	MS
14	n-Nonanal	10	n.d.	0.41	n.d	MS;RT
15	Limonen-1,2-epoxide	3.42	n.d.	0.02	n.d.	MS
16	(Z)-Linalooloxide	1.92	n.d.	0.02	n.d.	MS
17	α-Pineneoxide	4.05	n.d.	0.05	n.d.	MS
18	(E)-Sabinenehydrate	20.75		0.20	0.32	MS;RT
19	Citronellal	43.2	n.d.	0.95	0.59	MS;RT
21	α-Copaene	2.66	n.d.	0.02	0.16	MS
22	n-Decanal	33.9	n.d.	0.35	0.23	MS;RT
25	Linalool	6.94	n.d.	0.07	0.16	MS;RT
26	Octan-1-ol	6.86	0.21	0.08	0.65	MS;RT
27	β-Caryophyllene	2.93	n.d.	0.19	n.d.	MS;RT
28	2-Undecanone	Internal	Standard			
29	Terpinen-4-ol	3.75	0.13	0.24	n.d.	MS;RT
31	(E)-2,8-p-Menthadien-1-ol	2.95	n.d.	0.03	n.d.	MS;RT
32	α-Humulene	1.04	n.d.	n.d.	0.08	MS;RT
33	Neral	23.45	0.03	0.24	0.15	MS;RT
34	(E)-Piperitol	3.55	n.d.	0.04	0.17	MS
35	α-Terpineol	21.5	n.d.	0.22	0.27	MS;RT
36	n-Dodecanal	1.72	0.76	0.76	0.17	MS;RT
38	Geranial	27.57	n.d.	0.29	0.17	MS;RT
39	δ-Cadinene	3.03	n.d.	n.d.	0.19	MS;RT
40	Geranylacetate	3.02	n.d.	0.03	0.16	MS;RT
41	Perillaldehyde	4.42	n.d.	0.05	n.d.	MS
44	Nerol	3.02	0.12	0.03	n.d.	MS;RT
45	n-Tridecanal	n.d.	n.d.	0.20	n.d.	MS;RT
46	(E)-Carveol	1.78	n.d.	0.03	n.d.	MS;RT
47	Geraniol	3.01	n.d.	0.04	n.d.	MS;RT
48	n-Tetradecanal	n.d.	n.d.	0.20	n.d.	MS;RT
51	Nerolidol	2.23	n.d.	0.06	n.d.	MS;RT
53	Nootkatone	26.25	3.71	0.30	0.64	MS;RT

[a] Identification based on mass spectral (MS) and gas chromatographic (RT) data of authentic reference compounds on CW 20 M.

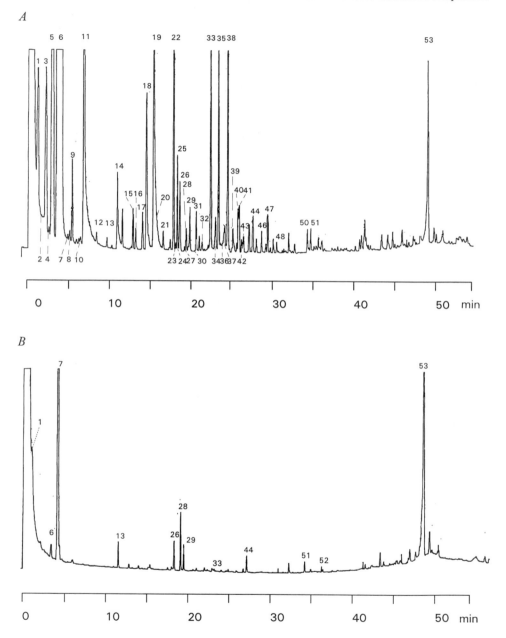

Figure 1 *Gas chromatograms of the essential oil of berries of* C. paradisi: *A: Flavedo plus Albedo tissue; B: Endocarp with mesocarp proportion (juice vesicles); CW 20 M fused silica 25 m × 0.32 mm, 0.5 μm film, 60 °C isothermal 5 min, to 200 °C at a rate of 2 °C min⁻¹, injector and detector temperature 225 °C*

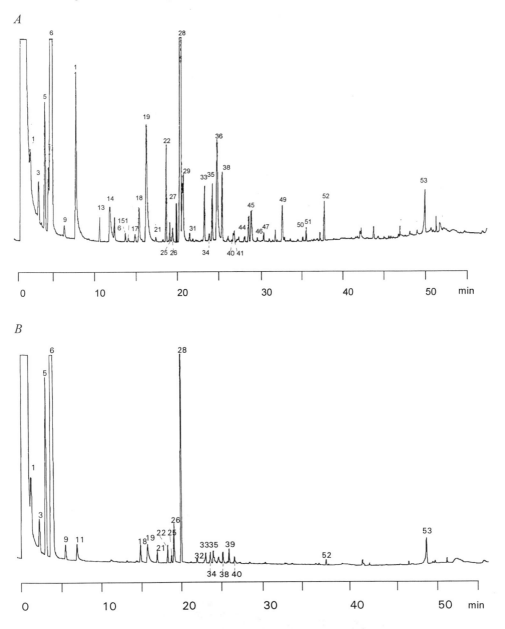

Figure 2 *The essential oil of photomixotrophic callus cultures of* C. paradisi: *MSL-medium, 3 g l^{-1} gellan gum, 3000 lux, 24 hour photoperiod, light colour 12:*
A: 2 mg of indoleacetic acid, 1 mg of indolebutanoic acid, 0.5 mg of naphthylacetic acid, and 1 mg l^{-1} of kinetin;
B: halved auxin concentrations, and 0.25 mg l^{-1} of kinetin.
Chromatographic conditions as in Figure 1.

2.4 Bioconversion of Exogenous Valencene

In all callus cultures of *C. paradisi* the synthesis of nootkatone, one of the character impact components of grapefruit flavour, was linked with chlorophyll and, thus, with limonene formation (Figure 3). Photomixotrophic suspension cells, however, did not accumulate any volatiles. A subsequent transfer of the suspended microcalli to plate cultivation resulted in a full recovery of terpene formation upon six subcultivations.

Incubation of suspended cells with exogenous valencene led to the intermediate formation of the 2-hydroxy-derivative, followed by conversion to the 2-oxo-compound, nootkatone (68% in 24 hours; Figure 4). Once reached, the concentration (*ca.* 0.7 mg l^{-1}) was maintained for another 48 hours, after which the experiment was terminated. Chemical oxidation was negligible under these conditions. In their heterotrophic counterparts, the conversion was much faster and reached a maximum accumulation after 6 hours (66% yield), but the nootkatone formed was rapidly and completely degraded within 24 hours.[5]

3 DISCUSSION

Plant cells are distinguished by their unique ability to synthesize complex compounds using light and carbon dioxide. Although this is the basis of existence of all higher life forms, commercial applications of cell systems to generate phytochemicals in a sterile and confined environment remain rare. Photoheterotrophic cells at high densities, the standard *in vitro* approach, experience an environment resembling those of root cells. Aroma compounds and oligoprenoids in particular, however, are usually formed in green parts of the plant.

Accordingly, many observations with non-illuminated cultures of essential oil plants indicated that the cells produced only very low amounts of terpenoids or completely fail to accumulate, whereas light may induce biochemical conditions that are more suitable for the generation of oligoisoprenoids.[2]

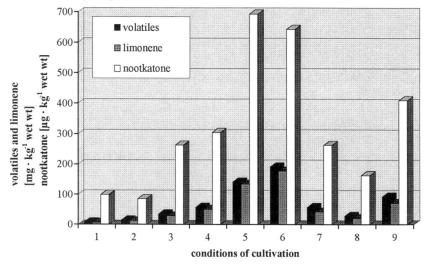

Figure 3   *Accumulation of limonene, nootkatone and total volatiles by* in vitro *cells of C. paradisi; 6: conditions as in Figure 2A, 4: as in Figure 2B, 9: as in case 6 with 9 g l^{-1} gellan gum*

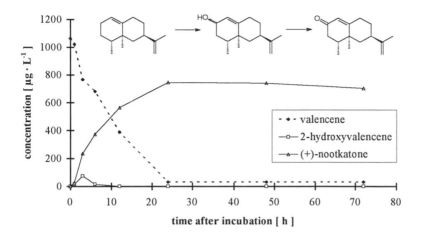

Figure 4 *Bioconversion of exogenous valencene by suspended cells of* C. paradisi *(MSL-medium, HQL at 3000 lux, 1μM 150 ml⁻¹ of precursor, end of log phase cells)*

3.1 The Effect of Light

Quality and quantity of light affect the synthesis of terpenes in intact plants by regulating the key enzyme, 3-hydroxy-3-methylglutaryl-CoA reductase.[6] Analogous reports on cell cultures refer to the formation of leaf-like terpenoids in photomixotrophic callus of *Coleonema album,*[7] and to cultures of *Pelargonium fragrans*, in which a 200-fold stimulation of the accumulation of essential oil was induced by a transfer into a long-light photoperiod.[8] Callus of cotton lavender contained himachalene-type sesquiterpenes when grown under continuous illumination of 600 lux.[9] The concentration of a precursor of the norisoprenoid, β-damascenone, in *Concord* grape callus was raised by light.[10] Electron microscopy of dark grown, valencene – and nootkatone – producing grapefruit cells revealed that plastids were present, possibly from the light grown preculture.[11]

The data obtained with callus cells of *Citrus paradisi* show that the phototrophic state is indispensable for the generation of monoterpenes. Secondary parameters affect composition and yield. A (-)-4S-limonene synthase was recently purified from spearmint, and the enzyme expressed in *E. coli*.[12] This recombinant limonene cyclase was catalytically active, producing about the same spectrum and proportions of monoterpene hydrocarbons as the *C. paradisi* callus enzyme. The cytoplasmic sesquiterpene pathway appears to compete poorly for the available isopentenyl diphosphate relative to the plastidic monoterpene pathway, as this class of oligoisoprenoids was, as in peel oil, not well represented in callus extracts.

3.2 Role of Phytoeffectors

Phytoeffectors may in some way affect or interact with the processes by which light controls gene expression. Their application to *in vitro* cells has led to differentiation, organogenesis, and, ultimately, to regenerated plantlets that usually produce the same essential oil as the parent plant.[13] It was concluded that some degree of specialization must be present in order for both synthesis and accumulation of oligoprenoids to be observed. In the present study, the number of chloroplasts per cell remained constant under different light and phytoeffector regimes, while concentrations of chlorophyll and terpenes changed. It is supposed that plastid differentiation, *i.e.* formation and orientation of enzymes and possibly transporters, but not cytodifferentiation or organogenesis, is crucial.

3.3 Toxicity of Oligoisoprenoids

Higher plants excrete mono- and sesquiterpenes into specialized storage sites. In tissue cultures that do not contain an appropriate metabolic sink, essential oil components may severely disturb respiration, photosynthesis, and membrane functions. It was estimated that suspended cells could withstand < 100 mg l^{-1} of certain monoterpenoids of medium before toxicity would occur. Valencene concentrations > 1 mg l^{-1} resulted in decreased conversion capacity of photoheterotrophic *C. paradisi*.[5]

The cells of *C. paradisi* accumulated significant amounts of intracellular oligoprenoids. While growth rates and oligoprenoid formation were inversely related in all experiments, the physiological state of photomixotrophic cells may contribute to an improved tolerance to oligoprenoids. This view is supported by the lower tolerance towards exogenous valencene of heterotrophic suspension cultures of *Citrus*.[5] A second aspect may be a reduced access to, or activity of, endogenous degrading enzymes. A remarkably high tolerance towards the monoterpenoid precursor Δ^2-carene and stable accumulation of transformation products has also been reported for illuminated callus cells of *Myrtillocactus geometrizans*.[14]

REFERENCES

1. R.G. Berger and F. Drawert, in 'Flavour Science and Technology', eds. M. Martens, G.A. Dalen and H. Russwurm Jr., Wiley, Chichester, 1988, p. 199.
2. Y.P.S. Bajaj (ed.), 'Biotechnology in Agriculture and Forestry' Series, 'Medicinal and Aromatic Plants' I to VII (1988 to 1995), Springer, Berlin.
3. R. Agrawal, M.V. Patwardhan and K.N. Gurudutt, *Biotechnol. Appl. Biochem.*, 1991, **14**, 265.
4. C.W. Wilson and P.E. Shaw, *J. Agric. Food Chem.*, 1987, **26**, 1432.
5. F. Drawert, R.G. Berger and R. Godelmann, *Plant Cell Rep.*, 1984, **3**, 37.
6. B.A. Stermer, G.M. Bianchini and K.L. Korth, *J. Lipid Res.*, 1994, **35**, 1133.
7. R.G. Berger, Z. Akkan and F. Drawert, *Z. Naturforsch.*, 1990, **45c**, 187.
8. B.V. Charlwood, C. Moustou, J.T. Brown, P.K. Hegarty and K.A. Charlwood, in 'Prim Sec Met Plant Cell Cult', ed. W.G.W. Kurz, Springer, Berlin, 1989, p. 73.
9. D.V. Banthorpe, in 'Biotechnology in Agriculture and Forestry' 28, ed. Y.P.S. Bajaj, Springer, Berlin, 1994, p. 412.
10. K.B. Shure and T.E. Acree, *Plant Cell Rep.*, 1994, **13**, 477.
11. J.A. Del Rio and A. Ortuño in 'Biotechnology in Agriculture and Forestry' 28, ed. Y.P.S. Bajaj, Springer, Berlin, 1994, p. 123.
12. S.M. Colby, W.R. Alonso, E.J. Katahira, D.J. McGarvey and R. Croteau, *J. Biol. Chem.*, 1993, **268**, 23016.
13. A.A. Abou-Mandou, T. van den Berg and F.-C. Czygan, *Angew. Bot.*, 1994, **86**, 163.
14. G. Gil, P. Ferreira dos Santos and C. Bullard, *Phytochemistry*, 1994, **38**, 629.

MICROBIAL PRODUCTION OF BIOFLAVOURS BY FUNGAL SPORES

Jan C.R. Demyttenaere, Ilse E.I. Koninckx and Arvid Meersman

Department of Organic Chemistry, Faculty of Agricultural and Applied Biological Sciences, Coupure Links 653, B-9000 Gent, Belgium

Dedicated to Prof. Dr. Herman L. De Pooter (*obiit* 23 December 1995)

1 INTRODUCTION

Because of the public's fear of the words 'chemical' and 'synthetic', there is a trend to produce more natural flavours, the so-called 'bioflavours'. These can be obtained from plant sources, but also by biotechnological processing.[1]

Methyl ketones, for example, constitute a very important group of flavour compounds. They are responsible for the characteristic flavour of Blue-type cheeses.[2,3] The most important of them are 2-pentanone, 2-heptanone, 2-nonanone and 2-undecanone, and their corresponding alcohols. They are derived by the partial oxidation of fatty acids, which are released by the lipase-catalysed hydrolysis of triacyl glycerides.

Methyl ketones have a fruity to spicy aroma and have a low odour threshold value. For 2-pentanone and 2-heptanone this is 0.5 ppm and 0.7 ppm, respectively.[4] Hence, their concentrations in Blue-type cheese are very low: 1.52, 3.48, 3.31 and 0.85 mg per 100 g cheese for 2-pentanone, 2-heptanone, 2-nonanone and 2-undecanone, respectively.[2]

A more specific example is 6-methyl-5-hepten-2-one (MHO), a naturally occurring ketone, originally identified in lemongrass and found in the essential oils of palmarosa, lemon, citronella, *etc.* It is used in non-alcoholic beverages, ice cream, candy, *etc.*, at about 1 ppm. It has a strong, fatty, green, citrus-like odour.[5]

It was found that spores of *Penicillium digitatum* (and not *P. italicum* as was believed earlier) carry out the biotransformation of both geraniol and nerol to MHO.[6,7]

This article deals with the microbial production of bioflavours by fungal spores. The degradation of triglycerides to methyl ketones by spores of *P. roqueforti* and *A. niger* is discussed in the first part. The second part describes a comparative study of the biotransformation of the terpenoid alcohols, geraniol, nerol, their mixture, citrol, and the mixture of the aldehydes, citral, to MHO by spores of *P. digitatum*.

2 EXPERIMENTAL

2.1 Micro-Organisms

For the degradation of triglycerides, three cultures of *Aspergillus niger* and one of *Penicillium roqueforti* were used: one culture, *A. niger* DSM 821 (designated 'ANA'), was bought from DSM (Deutsche Sammlung von Mikroorganismen). The other two *A. niger* strains were isolated from plant material ('ANG' and 'COR'). The culture of *Penicillium roqueforti* was isolated from commercially available Roquefort cheese('ROQ'). For the biotransformation of terpenes, a culture of *Penicillium digitatum* was used, which had been isolated from a spoiled tangerine. It was identified by the MUCL (Mycothèque de l'Université Catholique de Louvain, Laboratoire de Mycologie Systématique et Appliquée) as *P. digitatum* (Persoon: Fries) Saccardo.

2.2 Growth Medium

For the isolation, growth and conservation of the fungi in petri dishes, one and the same solid medium was used: Malt Extract Agar (MEA) (OXOID®) – malt extract (2%), bacteriological peptone (0.1%), glucose (2%) and agar (2%). For cultivation in liquid media, broth was used (YMPG) – yeast extract (0.5%), malt extract (1%), bacteriological peptone (0.5%), glucose (1%).

2.3 Cultivation

For the production of methyl ketones by shaken liquid cultures, the fungi were cultivated in 500 ml conical flasks, filled with 100 ml of YMPG medium. Before sterilization 100 µl Miglyol 812 was added. The flasks were shaken on a rotary shaker at 120 rpm. Three days after inoculation, full mycelial growth had taken place and the cultures were sampled by dynamic headspace. At different intervals, further additions of Miglyol were made. For the production of methyl ketones by spores, spore suspensions were made from three two-week old cultures on petri dishes, completely covered with spores. To each petri dish, 10–15 ml of a 0.1% Tween 80 solution were added and the spores brought into suspension. The suspensions were adjusted to 100 ml and brought into 500 ml conical flasks and shaken at 120 rpm. To these suspensions, 1 ml Miglyol was added. After three days, the suspensions were sampled by dynamic headspace.

For the biotransformation of terpenoids by fungal spores, six batches of sporulated surface cultures of *Penicillium digitatum* were compared. Two batches were treated with pure terpenoid alcohols: one (Flask 1) with geraniol and the other (Flask 2) with nerol; two batches (Flask 3 and Flask 4) were treated with a mixture of the alcohols, *i.e.*, citrol; and two batches (Flask 5 and Flask 6) were treated with the mixture of the aldehydes, geranial and neral, *i.e.*, citral. The biotransformation capacity of the six cultures was followed over prolonged periods (three weeks) with dynamic headspace techniques as described earlier[7]. During this period, three substrate additions were made (100 µl per week), each time followed by four or five headspace collections of 24 hours or longer.

2.4 Chemical Compounds

For the degradation of triglycerides to methyl ketones, Miglyol 812 (Federa) was used as substrate. It has to be pointed out that Miglyol is a semi-synthetic triglyceride and is not natural, and cannot be used as a precursor for natural methyl ketones. In this study, it is only used as a test substrate, because it is commercially available at a low price and it has a well defined composition: 56% octanoic acid, 42% decanoic acid and <2% dodecanoic acid. The control compounds were 2-heptanone (Fluka), 2-heptanol, 2-nonanone and 2-nonanol (Janssen Chimica).

For the biotransformation of terpenoids, the substrates used were nerol (Aldrich), geraniol (Fluka) and citral (Janssen Chimica). The control compound used was 6-methyl-5-hepten-2-one (Janssen Chimica).

2.5 Recovery and Analysis of the Flavours

For the degradation of triglycerides by liquid cultures, at the end of the experiments, liquid samples were taken: 5 ml culture medium. They were extracted with Et_2O. For the production of flavours by surface cultures, the flasks were sampled by dynamic headspace with Tenax. All analyses were carried out with GC and GC–MS, as described earlier.[7]

3 RESULTS AND DISCUSSION

3.1 Degradation of Triglycerides to Methyl Ketones

In the first experiment, the degradation of Miglyol by five different fungal cultures was examined. Five 500 ml conical flasks were filled with 100 ml of liquid medium (YMPG). To each culture medium were added 100 µl of Miglyol and the culture flasks sterilized. After cooling of the media, four flasks were inoculated with 1 ml of a different spore suspension of the following fungal strains: *A. niger* DSM 821, two *A. niger* strains isolated in the laboratory, marked here as 'ANG' and 'COR', and *P. roqueforti* isolated from a piece of Roquefort cheese. The fifth flask was not inoculated, but kept in sterile conditions and used as a control – it was treated in exactly the same conditions as the culture flasks. The fungi were grown for three days as submersed mycelial cultures and then headspace samples were taken during 24 hours. After the first headspace sample, 100 µl sterile Miglyol were added to the flasks. During an additional period of two weeks, five more additions of 100 µl Miglyol were made and seven more headspace samples of 24 hours each were taken. Only to the culture with *A. niger* DSM 821, were two more additions of 100 µl Miglyol each made and four more headspace samples taken during an additional period of one week.

Results for the four cultures after treatment with 700 µl Miglyol are shown in Table 1.

Table 1 *Production of Methyl Ketones by Fungi from 700 µl Miglyol*

Strain	2-Heptanone (µl)	2-Nonanone (µl)
A. niger DSM 821	99.98	36.56
A. niger 'ANG'	15.10	0.13
A. niger 'COR'	37.84	3.70
P. roqueforti	14.56	3.73

The production of flavours by the spore suspension of *A. niger* DSM 821 after treatment with 900 µl Miglyol was 156.15 µl 2-heptanone, 3.03 µl 2-heptanol, 49.95 µl 2-nonanone and 1.75 µl 2-nonanol. No methyl ketones were found in the control flask.

From these data, it is clear that *A. niger* DSM 821 produces more methyl ketones than *P. roqueforti* and the other *A. niger* strains. These data are comparable with literature data. Chalier and Crouzet (1993)[8] obtained 15.1 mg l^{-1} 2-heptanone, 1.3 mg l^{-1} 2-heptanol, 11.7 mg l^{-1} 2-nonanone and 0.6 mg l^{-1} 2-nonanol with *P. roqueforti* in a modified Czapek medium containing sucrose and urea and Miglyol 812 (1 g l^{-1}). Van der Schaft *et al.* (1992)[9] used the strain *Fusarium poae* and they obtained 91 mg l^{-1} 2-heptanone, 36 mg l^{-1} 2-heptanol, 61 mg l^{-1} 2-nonanone and 6 mg l^{-1} 2-nonanol after five days with 1% Miglyol.

In the next experiment, two cultures of *A. niger* DSM 821 and two cultures of *P. roqueforti* were inoculated in conical flasks containing 100 ml liquid medium at 0.1 % (v/v) Miglyol in the same way as done in the previous experiment. After three days, headspace analyses were started. During a period of three weeks, seven more headspace samples of 24 hours each were taken and three more additions of 100 μl Miglyol were made. After the last headspace sample, liquid samples were taken and extracted with ether.

The results of the headspace samples for the four cultures are shown in Table 2.

Table 2   *Production of Blue-Cheese Flavours by Fungi from 400 μl Miglyol (in μl)*

Strain	2-Heptanone	2-Heptanol	2-Nonanone	2-Nonanol
A. niger DSM 821: 1	73.93	1.21	31.61	0.73
A. niger DSM 821: 2	54.16	2.84	15.96	0
P. roqueforti: 1	8.26	1.45	2.36	1.69
P. roqueforti: 2	3.99	2.42	0.70	0.76

In the liquid extracts, only small amounts of 2-heptanone were found (7.5 μl for culture 1, 4.9 μl for culture 2), whereas the amounts of octanoic and decanoic acid were still high (13.0 μl, resp., 5.2 μl for culture 1 and 27.0 μl, resp., 13.3 μl for culture 2).

From these data, it can be concluded that most of the volatile Blue-cheese flavours (methyl ketones and secondary alcohols) are found in the headspace samples and that high amounts of non-converted fatty acids still remain in the liquid phase. Again it is clear that *A. niger* converts Miglyol better than *P. roqueforti*.

In the next experiment, the degradation of Miglyol was carried out with spore suspensions. Therefore the spores of three petri dishes of ANA were recovered with Tween 80 solution and diluted to obtain a spore suspension of 100 ml. The same procedure was followed with ROQ. No nutrients were added to the spore suspensions. To each suspension, 1 ml sterile Miglyol was added, the flasks were closed with a cotton plug and shaken on a rotary shaker (120 rpm). After three days, the flasks were equipped with a headspace aeration glass piece and the first headspace was taken for 24 hours. During an additional period of two weeks, three more 24-hour headspace samples were taken. The agitation was then stopped and the cultures stored as standing cultures for 12 weeks. After this period, four more headspace samples of 48 hours each were taken in a total period of two weeks. The results of the headspace samples for both cultures are shown in Table 3.

Table 3   *Production of Blue-Cheese Flavours by Spores of ANA and ROQ (in μl)*

Headspace	2-Heptanone		2-Heptanol		2-Nonanone		2-Nonanol	
	ANA	ROQ	ANA	ROQ	ANA	ROQ	ANA	ROQ
1–4	3.91	2.46	0.50	0.44	0.25	0.60	0.03	0.11
5	15.07	11.53	1.21	2.03	21.49	6.49	0.70	3.01
6	6.83	6.86	1.89	1.54	9.58	4.51	0.61	2.85
7	6.01	8.97	1.18	0.94	7.81	5.02	0.53	2.62
8	6.33	5.00	1.23	0.36	6.93	3.43	0.46	1.24
Total	38.15	34.82	6.01	5.31	46.06	20.05	2.33	9.83

It can be concluded that spores are indeed able to catalyse the degradation of fatty acids to methyl ketones, but that this reaction only takes place after very long periods. It is assumed that germination of the spores occurs and that the spores only reach their maximum activity while germinating.

In the final experiment, the influence of certain nutrients on spore activity was tested. It was believed that addition of glucose and L-leucine to the spore suspensions would promote the germination of the spores and stimulate the methyl ketone formation.[10,11] A spore suspension was made using three two-week old petri dishes of ANA completely covered with spores. The suspension was diluted to obtain a total volume of 100 ml. To this suspension was added 1 ml Miglyol. After five days, a headspace sample of 24 hours was taken. After three days, 360 mg glucose and 131 mg L-leucine were added to obtain a concentration of 20 mM and 10 mM, respectively. After shaking the suspension for 4 hours, the second headspace was taken during 48 hours, immediately followed by a similar third headspace. Three days later, two more headspaces of 24 hours each were taken with a time interval of 24 hours. The results of the headspace analyses of this experiment are shown in Table 4.

Table 4 *Production of Methyl Ketones by a Spore Suspension of ANA (in µl)*

Headspace no.	1	2	3	4	5	Total
2-Heptanone	8.17	21.68	25.56	11.17	10.99	77.57
2-Nonanone	3.42	3.67	8.43	3.07	5.44	24.03

From these data, it can be concluded that, during the early stages of cultivation, glucose and L-leucine indeed enhance the degradation of fatty acids by fungal spores: the production of methyl ketones increases after addition of the nutrients. More ketones are formed in a shorter period.

3.2 Bioconversion of Terpenoids by Fungal Spores

A comparative study was undertaken of the biotransformation of geraniol, nerol, citrol, and citral to 6-methyl-5-hepten-2-one (MHO) by fungal spores.

The first addition of 100 µl substrate was carried out on a Friday evening. After the addition, the flasks were closed during the week-end and the first 24-hour headspace was started on the next Monday morning. Three more headspace samples were taken of 24, 52 and 65 hours, respectively, the last headspace being taken from the following Friday evening until the next Monday morning.

The second addition of 100 µl substrate took place on Monday morning. The flasks were closed for 8 hours. The first headspace sample was taken from Monday evening until Tuesday morning during 16 hours. The next two headspace samples were taken during 24 hours, the fourth during 30 hours and the fifth and last headspace from Friday evening until Monday morning (66 hours).

The third addition of 100 µl substrate was carried out on Monday morning. The first headspace was taken immediately from Monday morning until Tuesday morning during 24 hours. All the next headspace samples were taken as in the previous week. The results for each headspace sample of the third week are shown in Table 5. The final results and yields for the three weeks are shown in Table 6.

It was found that citral was converted faster than the other substrates and gave an overall yield of 65%, whereas the individual alcohols and the mixture of alcohols, citrol, were converted more slowly, but gave yields of 70%.

From the headspace analyses, it was found that the substrate losses were minimal. Only in the third week were small amounts of terpenoids found in the first headspace sample.

Flavour Science: Recent Developments

Therefore, the best method seems to be the one applied during the second week: after addition of the substrates, the flasks are closed for 8 hours, after which headspace samples are taken.

Table 5 *Bioconversion of Terpenoids to Methylheptenone by spores of* P. digitatum *(in μl): Week 3*

Headspace	Geraniol	Nerol	Citrol	Citrol	Citral	Citral
1: 24 h	28.1	29.8	40.1	38.7	48.3	46.3
2: 24 h	26.3	20.7	17.8	19.1	11.6	12.4
3: 24 h	10.6	8.8	8.1	8.4	5.8	6.1
4: 30 h	2.7	2.6	2.4	2.3	1.8	1.9
5: 66 h	7.2	7.1	6.0	6.8	4.2	4.2

Table 6 *Production of Methylheptenone by Six Batches of Fungal Spores over Three Weeks*

Yield	Geraniol	Nerol	Citrol	Citrol	Citral	Citral
Week 1 (μl)	59.4	54.4	58.4	59.2	54.2	53.0
Week 2 (μl)	75.4	93.1	77.1	76.2	72.0	71.2
Week 3 (μl)	74.9	69.0	74.4	75.3	71.7	70.9
Total (μl)	209.7	216.5	209.9	210.7	197.9	195.1
Yield (%)	70.0%	72.2%	70.0%	70.3%	66.0%	65.2%

In a further step, the same cultures were tested for their continuous production of MHO during an additional period of five weeks. During the first three weeks, the cultures were continuously treated with substrate (20 μl each day during five days to give 100 μl per week) and headspaces of 24 hours were taken each day. During the last two weeks, three more headspaces were taken of 48 hours each to strip all the MHO produced. It was found that the overall yield was 80%–90%, depending on the substrate used. Again citral was converted faster than the alcohols, but gave a lower yield.

REFERENCES

1. P. Schreier, *Food Rev. Int.*, 1989, **5**, 289.
2. B.K. Dwivedi and J.E. Kinsella, *J. Food Sci.*, 1974, **39**, 620.
3. P.V. Hatton and J.L. Kinderlerer, *J. Appl. Bacteriol.*, 1991, **70**, 401.
4. M. Groux and M. Moinas, *Le Lait*, 1974, **531**, 44.
5. R.L. Hall and B.L. Oser, *J. Food Tech.*, 1965, **19**, 253.
6. J.C.R. Demyttenaere and H.L. De Pooter, *Med. Fac. Landbouww. Rijksuniv. Gent*, 1995, **60**, 1961.
7. J.C.R. Demyttenaere and H.L. De Pooter, *Phytochemistry*, 1996, **41**, 1079.
8. P. Chalier and J. Crouzet, *Flavour and Fragrance J.*, 1993, **8**, 43.
9. P.H. van der Schaft, N. ter Burg, S. van den Bosch and A.M. Cohen, *Appl. Microbiol. Biotechnol.*, 1992, **36**, 709.
10. R.C. Lawrence, *J. Gen. Microbiol.*, 1976, **46**, 65.
11. J.E. Kinsella and D. Hwang, *Biotechnol. Bioeng.*, 1976, **18**, 927.

BIOTECHNOLOGICAL PRODUCTION OF VANILLIN

Ivan Benz and Andreas Muheim

Givaudan-Roure Research Ltd., 8600 Dubendorf, Switzerland

1 INTRODUCTION

Natural flavours represent a growing demand within the food industry. In order to expand the range of natural flavours, biotechnological methods are used for the production of natural aroma chemicals. Vanillin is one of the most commonly consumed flavour chemicals (5,550 t/a worldwide).[1] At present, 97% of the world vanilla flavour market is synthetic vanillin, the remaining 3% (weight basis), consisting of vanilla extracts. The limited supply and high price for this botanical extract led to the necessity of using large amounts of synthetic material, not only in vanilla, but also in fruit, chocolate, caramel and many other flavours.

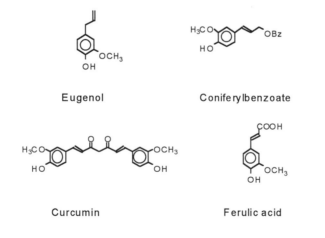

Eugenol   Coniferylbenzoate

Curcumin   Ferulic acid

Figure 1 *Precursors for the production of natural vanillin*

Various patents recently filed for the production of natural vanillin indicate the great interest in this compound and the efforts undertaken for its biotechnological production.[2-4] To produce natural vanillin, a natural precursor is needed. Several raw materials extracted from plant sources are suited to be converted to vanillin. These include lignin, eugenol, ferulic acid, curcumin and benzoe siam resin (coniferyl benzoate) (Figure 1). In general, microbial turnover rates of 30% and production levels below 1 g l^{-1} were reported when using these precursors. The toxicity of both the precursor and the vanillin formed represents a major obstacle. An elegant method of avoiding this problem is to use enzymes, such as lipoxygenases, which oxidatively cleave double bonds conjugated with the aromatic ring as a side reaction.[5] Ferulic acid is the least toxic of the chemicals suggested above, which makes it suitable for microbial conversion. This acid can be extracted from several waste materials such as molasses or husks of different corn brans.[6] The latter were reported to contain up to 6% (w/w) of ferulic acid. The acid is covalently linked to the xylan residue present in all corns and is easily liberated by the action of specific ferulic acid esterases. Several commercial enzyme preparations were shown to contain such esterase activity.[7]

We have developed an industrial process for the production of vanillin *ex* ferulic acid. In order to authenticate natural flavour chemicals, modern isotopic methods, such as natural-abundance ^{2}H-NMR spectroscopy, are employed.[8] These techniques allow the determination of the origin (synthetic, natural) as well as the determination of the starting materials. This, however, requires an appropriate reference material of which the starting material and generation pathway are known. In this report we discuss several microbial routes to vanillin starting from ferulic acid and also describe the novel vanillin reference sample that has been produced by fermentation of ferulic acid on an industrial scale.

2 EXPERIMENTAL

2.1 Materials

Different samples of vanillin were used for the isotopic measurements. Vanillin *ex* eugenol and *ex* ferulic acid were prepared by Givaudan-Roure Ltd. Vanillin *ex* guaiacol was purchased from Rhône-Poulenc, Geneva, Switzerland. Vanillin *ex* vanilla pods was extracted from authentic vanilla beans by supercritical CO_2 extraction, followed by preparative HPLC. Ferulic acid was purchased from Fluka for analytical purposes or extracted from rice husks for the synthesis of natural vanillin.

2.2 Analysis of the Metabolites

Compounds from the fermentation were separated by reverse phase HPLC using a 4.6 × 250 mm C$_8$ column (Brownlee column, Applied Biosystems) with a flow rate of 1 ml min^{-1} and a linear gradient from 0%–100% in 15 min using aqueous 0.1% TFA as solution A and aqueous 80% acetonitrile supplemented with 0.1% TFA as solution B. The absorbance was read at 254 nm. Commercial vanillin (Fluka) and ferulic acid were used as standards for quantification.

2.3 Isotope Ratio Mass Spectrometry (IRMS)

Vanillin samples were submitted to Krueger Food Laboratories Inc. (Cambridge, MA, U.S.A.), where IRMS measurements were performed by GC–combustion–MS.

2.4 Natural Abundance ^{2}H-NMR Spectroscopy

Natural abundance ^{2}H-NMR spectra of vanillin were obtained at 308 K on a Bruker AVANCE DPX-400 MHz NMR spectrometer.

3 RESULTS AND DISCUSSION

3.1 Vanillin *ex* Lignin and Eugenol

Every year, several tons of synthetic vanillin are produced by alkaline oxidation of lignin, which accumulates as a waste material in the paper and pulp industry.[9] In an analogous manner, vanillin can be obtained by forced air oxidation starting from, *e.g.*, eugenol or curcumin. Eugenol is available in large quantities from clove oil, whereas roots of the turmeric plant contain high amounts of curcumin. Production by air oxidation results in a vanillin that is considered natural according to the U.S.A.'s FDA legislation, whilst according to the Council directive of the European Community this vanillin must be considered as nature-identical. Production of a vanillin that meets the E.C. guidelines for naturalness can be achieved either by microbial or enzymic transformation of a precursor or by microbial *de novo* synthesis.

3.2 *De novo* Synthesis of Vanillin

Submerged growing cells of *Vanilla planifolia* were shown to produce vanillin. An improved system consisting of hairy root cells of the vanilla plant has been described.[10] Upon feeding these cells with ferulic acid, higher amounts of vanillin were determined than without feeding. These systems, however, are handicapped by the fact that culturing plant cells is rather expensive and time-consuming.

As an alternative, *de novo* synthesis by white-rot fungi has been suggested. During secondary metabolism, these fungi produce a variety of aromatic compounds derived via the shikimate pathway.[11] 3,4-Dimethoxybenzyl alcohol and the corresponding aldehyde were found as major metabolites in the culture supernatant.[12] Vanillin was detected in trace amounts, suggesting the need for major improvements before contemplating future industrial application.

3.3 Ferulic Acid as a Precursor for Natural Vanillin

Ferulic acid is abundantly available from waste materials and is an excellent precursor for the production of vanillin as there is no need for a substitutional change in the aromatic ring. Different microbial pathways have been reported for the conversion of ferulic acid: first, the reduction to the corresponding alcohol; secondly, the demethylation to caffeic acid; thirdly, the metabolism via hydration of the acid; and, fourthly, the decarboxylation to vinylguaiacol.

3.3.1 Reduction of Ferulic Acid. The first route was reported to occur within wood-rotting fungi.[13] These fungi reduce ferulic acid predominantly to dihydroferulic acid, coniferylaldehyde and dihydroconiferyl alcohol. Further cleavage of the C_α–C_β bond was observed as vanillic acid was formed. Due to the reductive potential of these fungi, vanillic acid is reduced to vanillin and vanillyl alcohol. The formation of vanillin from ferulic acid was observed in the case of the white-rot fungus *Pycnoporus cinnabarinus* I-937.[14] The low yields of vanillin formed are possibly due to the rather inefficient cleavage of the side chain, and, additionally, phenoloxidases may convert vanillic acid to methoxyhydroquinone sidestepping the formation of vanillin.

3.3.2. Demethylation of Ferulic Acid. A rather uninteresting route for the production of vanillin is the demethylation of ferulic acid to caffeic acid.[15] This acid can either be ring-opened by dioxygenases or further degraded to dihydroxybenzoic acid and then ring-opened.

3.3.3. Hydration of Ferulic Acid. We have isolated and studied several soil bacteria which grew on ferulic acid as the sole carbon and energy source and also degraded the acid efficiently. Degradation of ferulic acid **1** to vanillin **3** appeared to be the rate-limiting reaction as vanillin **3** was very rapidly further oxidized to vanillic acid **4** (Figure 2). In line with this observation, cell extracts were shown to contain an extremely powerful vanillin oxidoreductase. This NAD^+-dependent enzyme was responsible for the rapid oxidation of vanillin **3**. Cells incubated with ferulic acid **1** produced vanillic acid **4** in yields up to 70%, but no vanillin as an intermediate was detected. The vanillic acid **4** accumulated was further degraded to 3,4-dihydroxybenzoic acid after ferulic acid **1** had been depleted. The micro-organisms were identified as *Pseudomonas* spp. From the literature, it is known that *Pseudomonas acidovorans* hydrates the C_α–C_β double bond of ferulic acid **1**, leading to a rather unstable intermediate, β-hydroxy-α-hydroferulic acid **2**.[16] Lyase cleavage of this compound results in the direct formation of vanillin **3** and acetic acid. A vanillin oxidoreductase-negative *Pseudomonas* strain obtained by random mutagenesis was recently patented.[17] As vanillin oxidation was blocked, vanillin **3** accumulated in the medium.

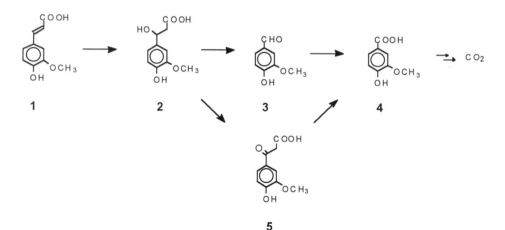

Figure 2   *Hypothetical degradation pathway of ferulic acid 1 via hydration of the acid. Depending on the further degradation route vanillin 3 may be formed. The upper pathway via vanillin 3 has been observed within* Pseudomonas *spp.*

Interestingly, some plants, *e.g.*, *Lithospermum erythrorhizon,* were reported to contain an additional pathway.[18] The intermediate hydrated ferulic acid **2** was not cleaved, but further oxidized to the keto compound **5**. Cleavage of this compound resulted in the formation of vanillic acid **4** and acetic acid circumventing the formation of vanillin **3**. This pathway resembles the oxidative degradation of fatty acids. Obviously such a metabolic route cannot be used for the production of vanillin **3** and is not likely to occur within the vanilla pod.

3.3.4. Decarboxylation of Ferulic Acid. The fourth degradation route of ferulic acid involves vinylguaiacol **6** as an intermediate (Figure 3). The oxidative decarboxylation of ferulic acid **1** was reported in various organisms, such as *Pseudomonas* spp. and various yeast strains.[19] The latter are responsible for the production of vinylguaiacol **6** as an off-note in orange juice and beer. Further degradation of vinylguaiacol **6** can either proceed via the corresponding phenethyl alcohol or by cleavage of the double bond, leading to the formation of vanillin **3** and formaldehyde. The two enzymes involved in the vanillin-forming pathway, ferulic acid decarboxylase and the dioxygenase, have been cloned.[20,21] Heterologous co-expression of these two enzymes should ultimately result in the formation of vanillin upon feeding the host with ferulic acid. Liberation of formaldehyde might cause cell metabolism to cease before high amounts of vanillin accumulate, due to its toxicity. This problem, however, could be circumvented by additional co-expression of formaldehyde-converting enzymes, such as formaldehyde dehydrogenase, producing the less toxic and further degradable formate.

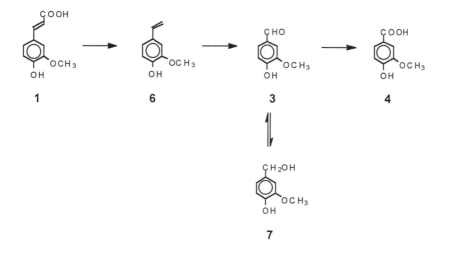

Figure 3 *Hypothetical degradation route of ferulic acid 1, as reported to occur in micromycetes.[22] The two enzymes responsible for the decarboxylation of ferulic acid 1 and the subsequent oxidative cleavage of the vinyl double bond in compound 6 have been cloned from independent hosts. Heterologous co-expression might lead to an efficient vanillin 3 producing micro-organism.*

3.4 Microbial Production of Vanillin and its Isotopic Characterisation

We have established an industrial production route to natural vanillin. A carefully selected micro-organism was found to accumulate vanillin upon feeding with ferulic acid, allowing the efficient bioconversion and production of vanillin. The vanillin produced was purified by counter-current extraction, followed by crystallization, and the pure vanillin was then used as a reference sample for authenticity experiments.

IRMS measurements allow easy distinction of the way vanillin has been formed. Vanillin from vanilla pods exhibits a very characteristic PDB value in the range of −19‰ to −20‰ (Table 1).[23] This value results from the unique metabolic pathway of vanillin formation within the orchid. Vanillin from ferulic acid showed a different value of −35.5 ‰.

Table 1   *Mean Values of IRMS Data from Various Vanillin Samples*

Vanillin Source	$\delta^{13}C$ [‰PDB]
Madagascar vanilla beans	−20.2
Reunion vanilla beans	−19.4
Comores vanilla beans	−18.9
Natural vanillin *ex* ferulic acid	−35.5
Synthetic vanillin *ex* lignin	−27.0
Synthetic vanillin *ex* guaiacol	−31.0

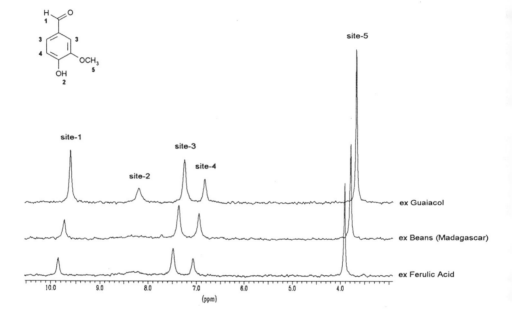

Figure 4   *Natural abundance ^{2}H-NMR spectra of three different vanillin samples (dissolved in acetone) were recorded on a Bruker AVANCE DPX-400 MHz NMR spectrometer operating at 61.4 MHz with ^{19}F-lock, ^{1}H-decoupled, using a 10 mm BBO probe at a temperature of 308 K, and TMU as internal standard*

As vanillin samples can be adulterated in order to meet required PDB values, natural abundance ^{2}H-NMR spectroscopy (SNIF-NMR®) has been successfully used for authenticity control. In this method, in contrast to IRMS measurements, not only is the relative amount of carbon-13 of the entire molecule measured, but also the relative amount of deuterium at each individual position.[24] A characteristic ^{2}H-NMR spectrum was obtained for vanillin from ferulic acid (Figure 4). Comparison of this spectrum with others, *e.g.*, those obtained for synthetic vanillin from guaiacol and natural vanillin from Madagascar beans, underlined the novel production route to natural vanillin from fermentation. For example, the relative deuterium concentration of the aldehyde group was much higher in the synthetic vanillin than in the biotechnological one. A possible explanation for this phenomenon is the way the aldehyde proton is incorporated, as enzymic and chemical reactions have different isotopic preferences and kinetics. The small differences between the individual spectra can be better illustrated by using statistical operations, such as principal component analysis. The different ^{2}H-NMR spectra, however, demonstrate the powerful application of the SNIF-NMR$^®$ technique to the determination of the origin of aroma chemicals.

ACKNOWLEDGEMENTS

We are grateful to Dr. C. Selden and Mr. J. Märki for the NMR spectra. We also thank Mrs. N. Silke and Dr. A. Häusler for helpful advice in the preparation of this manuscript.

REFERENCES

1. L.P. Somogyi, *Chem. Ind. L*, 1996, **5**, 170
2. I. Labuda *et al.*, US patent 5'279'950, Jan. 18, 1994
3. B. Gross *et al.*, EP 0 453 368 A1, Oct. 23, 1991
4. J. Rabenhorst and R. Hopp, EP 0 405 197 A1, Jan. 2, 1991
5. P.H. Markus *et al.*, EP 0 542 348 A2, May 19, 1993
6. L.P. Christov and B.A. Prior, *Enzyme Microb. Techn.*, 1993, **15**, 460
7. V. Micard *et al.*, *Lebensm.-Wiss. u. -Technol.*, 1994, **27**, 59
8. G. Martin *et al.*, *Flav. Fragr. J.*, 1993, **8**, 97
9. A.L. Mathias *et al.*, *J. Chem. Tech. Biotechnol.*, 1995, **64**, 225
10. R.J. Westcott *et al.*, *Phytochemistry*, 1994, **35**, 135
11. R.G. Berger, *Flav. Fragr. J.*, 1986, **1**, 181
12. K.A. Jensen *et al.*, *Appl. Environm. Microbiol.*, 1994, **60**, 709
13. J.K. Gupta *et al.*, *Arch. Microbiol.*, 1981, **128**, 349
14. B. Falconnier *et al.*, *J. Biotechn.*, 1994, **37**, 123
15. R. Tillet and J.R.L. Walker, *Appl. Microbiol.*, 1990, **154**, 206
16. A. Toms and J.M. Wood, *Biochemistry*, 1970, **2**, 337
17. Takasago Co Ltd., Japanese patent application 35338, Sept. 7, 1993
18. K. Yazaki *et al.*, *Phytochemistry*, 1991, **30**, 2233
19. Z. Huang *et al.*, *Appl. Environm. Microbiol.*, 1993, **59**, 2244
20. A. Zago *et al.*, *Appl. Environm. Microbiol.*, 1995, **61**, 4484
21. S. Kamoda and Y. Saburi, *Biosci. Biotech. Biochem.*, 1993, **57**, 926
22. M. Rahouti *et al.*, *Appl. Environm. Microbiol.*, 1989, **55**, 2391
23. G. Lamprecht *et al.*, *J. Agric. Food Chem.*, 1994, **42**, 1722
24. G. Martin *et al.*, *Ital. J. Food. Sci.*, 1993, **3**, 191

THE DETERMINATION OF DIACETYL IN CULTURED BUTTERMILK IN THE PRESENCE OF THE UNSTABLE PRECURSOR α-ACETOLACTIC ACID

D.A. Cronin and E. Rispin

Department of Food Science, University College Dublin, Belfield, Dublin 4, Ireland

1 INTRODUCTION

The mechanisms responsible for the production from citrate of diacetyl, the key aroma compound in mesophilic cultured milks, have long been a matter of controversy. An enzymic route involving the condensation of acetyl-coenzyme A and an 'active acetaldehyde' moiety was proposed in 1968,[1] while others[2] claimed that the compound was produced chemically by the oxidative decarboxylation of α-acetolactic acid (ALA), an unstable intermediate of citrate metabolism. While more recent research is tending to highlight the importance of the latter route, there are still conflicting reports on the ability of different cultures to produce ALA.[3,4,5,6] The lack of reliable methodology for the measurement of ALA in cultures and, in particular, for the determination of diacetyl in its presence has proved a major impediment in much of the work published in this area.

Diacetyl, normally present in cultures at only a few mg kg^{-1}, is usually determined either by steam-distillation–colorimetry or by headspace gas chromatographic methods. During analysis, breakdown of the labile ALA to diacetyl can occur and, although decomposition yields of the latter as low as 2% have been reported,[7,8] there is frequently doubt as to whether the diacetyl measured in a fresh culture is actually preformed or 'true diacetyl', or an artefact of the method of analysis, especially if the culture produces significant amounts of ALA. One of the more satisfactory approaches to this problem was that of Veringa et al.[9] who measured yields of diacetyl produced by distillation at a range of pH values of synthetic ALA added to skim-milk–buttermilk matrices, and derived simple equations to calculate both the 'true diacetyl' and ALA contents of the samples from the data at pH values of 4.6 and 9.0. The present study examined the decomposition of synthetic ALA to diacetyl under more strongly acidic conditions than previously used, and describes how the resulting methodology could be used to follow the course of diacetyl and ALA development during the preparation and storage of a buttermilk culture.

2 EXPERIMENTAL

2.1 Effect of pH on Decomposition of ALA to Diacetyl

Oxidative decarboxylation of ALA was examined in the following matrices: skim milk, heat treated (90 °C, 10 min) skim milk artificially coagulated to pH 4.6 by the addition of δ-gluconolactone (1.25% w/v), and one-day-old buttermilk cultures. ALA stock solutions (1.05 g l^{-1}) were freshly prepared by hydrolysing α-methyl-α-acetoxy-ethylacetate (Oxford

Organic Chemicals Ltd., Brackley, Northamptonshire, U.K.) with stoichiometric amounts of 0.1 M NaOH for 30 min under nitrogen.[7] Previously determined quantities of 1 M and 0.1 M H_2SO_4 and 0.2 M NaOH required to give final pH values in the range 1.0 to 9.0 were transferred to 150 ml distillation flasks followed by 10 ml aliquots of cooled (0 °C) skim-milk–buttermilk, using a modified pipette widened at the outlet to facilitate rapid delivery of the sample. For coagulated products, the exact quantity of material delivered was predetermined by weighing. The ALA stock (1 ml) was then quickly added, the samples thoroughly mixed and immediately steam distilled. Diacetyl was determined colorimetrically in the first 10 ml of distillate by the method of Walsh and Cogan[9] and acetoin in the second 10 ml by the Westerfeld method.[10] Unsupplemented buttermilks were analysed under the same conditions as the spiked samples and the differences between the diacetyl levels at the respective pH values were used to determine the percentage decomposition of ALA to diacetyl during the analysis.

2.2 Diacetyl Development in Experimental Buttermilk

Separate samples (250 ml) of heat treated skim milk (90 °C, 10 min) were cultured at 22 °C with a 1% inoculum of a commercial DL culture in screw-cap glass bottles. After approximately 14 hours and at 2-hour intervals thereafter, the bottles were tightly capped to prevent ingress of air and cooled to 0 °C in an ice bath for an hour. The curd was then broken by shaking the bottles and aliquots were quickly removed for analysis of diacetyl after acidification with 1 M and 0.1 M H_2SO_4 to pH values of 1.0 and 2.0, respectively. Residual citrate was also determined in each sample.[9] After cooling and uncapping the bottles, culture samples in separate bottles were thoroughly agitated to induce aeration. The outlets were covered with air-permeable polythene film and the cultures placed in storage at 5 °C.

Table 1 *Diacetyl Production from Distillation of Samples spiked with ALA (70 mg l^{-1})*

pH	Diacetyl yield(%)		
	Skim milk	*Artificially coagulated skim milk*	*Buttermilk*
1.0	2.5	0.4	0.2
2.0	23.2	15.9	14.0
3.5	48.3	45.2	35.5
4.6	12.4	11.9	10.2
9.0	1.9	1.9	1..8

2.1 Results and Discussion

ALA is an unstable compound which decarboxylates non-oxidatively to acetoin and oxidatively to diacetyl when heated in aqueous solution. The data in Table 1 show that during distillation of ALA solutions the amount of diacetyl produced was profoundly influenced by the pH of the medium and, to some extent also, by the nature of the matrix. Maximum production of diacetyl occurred at around pH 3.5, decreasing on either side of this to low levels at both high and low pH values. Acetoin production at 50%–70% (data not shown) was much less pH dependent, but theoretical recoveries (diacetyl + acetoin > 98%) were only achieved at pH 3.5, suggesting the possible production of additional undetected decomposition products at the other pH values. Diacetyl yields were generally somewhat higher in fresh unheated skim milk than in the heated artificially coagulated milk or 'simulated buttermilk'. However, under acidic conditions, diacetyl yields from decomposition of ALA were consistently lower in the fresh buttermilks than in the other matrices, possibly due to the cultures having lower dissolved oxygen levels and/or redox potentials.

The most significant feature of this experiment was the very low production of diacetyl from ALA when the media were adjusted to pH 1.0 by the addition of 1M H_2SO_4. A diacetyl yield of only 0.2% in buttermilks is approximately an order of magnitude lower than has been reported in previous studies.[7,8] Thus, it offers the possibility of using a single distillation analysis to measure the preformed or 'true diacetyl' content of a culture even in the presence of significant amounts of ALA. The concentration of the latter can be calculated from the steam distillation data at pH 1.0 and 2.0 using the following equation, and assuming a 14% decomposition of ALA to diacetyl at the latter pH:

$$ALA = \frac{M_{ALA}}{M_D} \times \frac{D_{2.0} - D_{1.0}}{0.14}$$

where M_{ALA} and M_D are the molecular weights of ALA and diacetyl, respectively, and $D_{2.0}$ and $D_{1.0}$ are the diacetyl concentrations at pH 2.0 and 1.0.

When this equation was used to calculate the ALA content of various buttermilks supplemented with known amounts of the compound and steam distilled at pH 1.0 and 2.0, the values obtained were usually within ±5% of the actual levels.

Table 2   *The Diacetyl and ALA Status of a Buttermilk Culture During Fermentation and Storage at 5 °C for Six Days*

Fermentation time (hours)	pH	Diacetyl (mg/l)[a]		ALA (mg/l)[a]		Citric Acid (mg per 100 ml)
14	4.80	0.20	(3.7)	64.5	(6.2)	62.5
16	4.70	0.45	(3.2)	73.3	(3.4)	16.5
18	4.65	0.42	(3.5)	54.3	(3.9)	0
20	4.58	0.45	(1.8)	26.5	(8.2)	0
22	4.55	0.30	(2.0)	6.05	(8.6)	0

[a] Figures in brackets represent aerated samples stored for six days.

It is generally accepted that a good 'buttery' aroma in buttermilks requires diacetyl levels of around 1 mg l^{-1} or higher. From a survey of the diacetyl status of a range of commercial buttermilks available on the Irish market, a particular (DL) culture which had a strong aroma and consistently high diacetyl levels in the range 2–4 mg l^{-1} was selected for a study of the aroma of the product. Experimental buttermilks prepared with this culture were monitored for diacetyl, ALA and citrate at the later stage of the fermentation when the pH had fallen below 5.0. Results for a typical culture are presented in Table 2, together with the diacetyl and ALA levels in the cultures after aeration and storage at 5 °C for six days. The distillation data at pH 1.0 showed clearly that diacetyl levels only slightly above the limit of detection of the colorimetric method used (0.2 mg l^{-1}) were present throughout the duration of anaerobic fermentation up to the disappearance of citrate. ALA concentration increased to a maximum at 16 hours when most of the citrate had been consumed and thereafter decreased quite rapidly. The crucial role of oxidative decarboxylation of ALA as the principal mechanism of aroma production in this culture was demonstrated by a rapid increase in diacetyl in the aerated stored samples over a 2-day period to levels which remained relatively unchanged after a further four days in storage; the latter values are shown in Table 2. Beneficial effects on diacetyl production resulting from post-culture oxygenation has also been shown for other cultures,[12] where the effect was assumed to have resulted from an enhancement of the oxidative decarboxylation of ALA. The data in Table 2 show that low levels of ALA were still present in the culture after six days and, in the case of the 22 hour sample, had actually increased somewhat over the initial value. This rather odd phenomenon of a decrease in ALA to very low or even zero concentrations, followed by a small subsequent increase during storage, has been observed in several buttermilks prepared with this culture. It may be indicative of the presence of some membrane-bound or immobilised form of ALA within the bacterial cell which is released as the latter gradually undergoes lysis with time.

The fact that the primary mechanism of aroma production in this culture appears to be via the oxidative decarboxylation of ALA does not preclude the possibility that synthesis of diacetyl by the enzymic route of Speckman and Collins[1] is also taking place during the active phase of the anaerobic fermentation. However, the low levels observed under these conditions may be due to a high activity of the enzyme diacetyl reductase which converts diacetyl to acetoin and which requires NADH as the electron donor. The environment of cultures towards the end of fermentation has been shown to be strongly reducing with a redox potential of the order of –340 mV,[13] conditions which would favour the reduction of diacetyl, as well as inhibiting the oxidative decarboxylation of ALA.

In conclusion, this study has shown that a simple protocol involving a steam distillation carried out at pH 1.0 could be used to follow the course of diacetyl production during the fermentation and storage of a DL buttermilk culture. Anomalous results arising from decomposition of ALA during analysis were avoided, even though the culture contained significant amounts of this compound, which was the main precursor of the diacetyl present in the stored product.

REFERENCES

1. R.A. Speckman and E.B. Collins, *Can. J. Bacteriol.*, 1968, **26**, 744.
2. J. de Man, *Rec. Trav. Chim.*, 1959, **78**, 480.
3. K.N. Jordan and T.M. Cogan. *J. Dairy Res.*, 1988, **55**, 227.
4. K.N. Jordan and T.M. Cogan, *Irish J. Agric. Food Res.*, 1995, **34**, 39.
5. W.M. Verhue and F.S.B. Tjan, *Appl. Envir. Microbiol.*, 1991, **57**, 3371.
6. N. Bassit, C.Y. Boquie, D. Picque and G. Corrieu, *Appl. Envir. Microbiol.*, 1993, **59**, 1893.
7. H.A. Veringa, E.H. Verburg and J. Stadhouders, *Netherlands Milk & Dairy J.*, 1984, **38**, 251.
8. C. Monnet, P. Schmitt and C. Divies, *J. Dairy Sci.*, 1994, **77**, 809.
9. B. Walsh and T.M. Cogan, *J. Dairy Res.*, 1974, **41**, 31.
10. W.W. Westerfeld, *J. Biol. Chem.*, 1945, **161**, 494.
11. J.C.D. White, *J. Dairy Sci.*, 1963, **30**, 171.
12. F.M. Driessen and Z . Puhan, *IDF Bulletin*, 1988, **227**, 75.
13. H. Jönsson and H.E. Pettersson, *Milchwissenschaft*, 1977, **32**, 587.

THE PRODUCTION OF NATURAL FLAVOURS BY FERMENTATION

H. Stam, A.L.G.M. Boog and M. Hoogland

Quest International, Food Science and Technology Centre, P.O. Box 2, 1400 CA Bussum, The Netherlands

1 INTRODUCTION

The use of biotransformations to produce flavour topnotes and flavour building blocks is increasing in order to meet consumers demand for natural products. Natural ingredients with improved functionalities, produced in an environmentally friendly way, can be manufactured using biocatalysis. By means of fermentation specific compounds (topnotes) or unique complex building blocks can be produced.

Research on fermentative production of natural flavours is focused on three areas: flavour topnotes, flavour building blocks and yeast extracts.

2 FLAVOUR TOPNOTES

In this class of compounds, a precursor is often converted by micro-organisms to the desired molecule with a known molecular structure. In cases where precursor and/or product are hydrophobic, a two (oil–water) phase fermentation is frequently the preferred production method. After fermentation, cells are removed using centrifugation or microfiltration and the flavour compounds are purified from the remaining broth. A high degree of purity is often requested in order to avoid the influence of so called off-flavours. The purified compounds are often mixed with other natural flavour ingredients.

2.1 Production of δ-Decalactone

A range of natural flavour compounds can be derived using the fatty acid metabolism from yeast or filamentous fungi. Well known in the literature is the production of γ-decalactone starting from the precursor ricinoleic acid, abundantly present in commercially available castor oil.[1,2,3]

We have studied the production of the peach-like flavour, δ-decalactone (DDL), by *Saccharomyces cerevisiae* with 11-hydroxy-hexadecanoic acid as precursor.[4,5] 11-Hydroxy-hexadecanoic acid was isolated from the roots of a Mexican plant (Mexican jalap, *Ipomea orizabensis*). An example of a typical fermentation is given in Figure 1. In a medium containing peptone, yeast extract and the precursor, three different phases could be observed: a first phase in which sugars present in the yeast extract are converted into ethanol, a second phase where the ethanol is consumed completely and a third phase in which the precursor is oxidised to 5-hydroxydecanoic acid. During this conversion, two molecules of 11-hydroxy-hexadecanoic acid yielded one molecule of 5-hydroxydecanoic acid. The various phases are characterised by different RQ (rCO_2/rO_2) values. The sequence of metabolic events suggest a strong catabolic control by the yeast. An RQ of about 0.7 during the stage that the precursor is converted to 5-hydroxydecanoic acid is in accordance with the expected RQ value during metabolism of long chain (hydroxy) fatty acids by means of beta oxidation as can be seen in the following formulae:

1. Production of 5-hydroxydecanoic acid:
$$C_{16}H_{32}O_3 + 9\,O_2 \longrightarrow C_{10}H_{18}O_2 + 6\,CO_2 + 7\,H_2O \quad (RQ=0.67)$$

2. Complete beta oxidation:
$$C_{16}H_{32}O_3 + 22.5\,O_2 \longrightarrow 16\,CO_2 + 16\,H_2O \quad (RQ=0.711)$$

At the end of the fermentation the broth is acidified and heated to 100 °C for 10 min to perform the lactonisation of 5-hydroxydecanoic acid to DDL. After extraction with butylacetate and distillation, pure δ-decalactone was obtained in a yield of 90%.

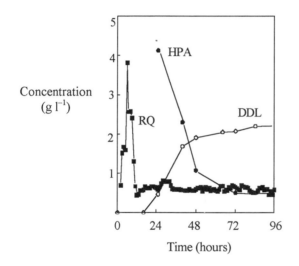

Figure 1 *Profile of hydroxyhexadecanoic acid (HPA), DDL and RQ value during a fermentation with* Saccharomyces cerevisiae

2.2 Production of Sulfur Containing Topnotes

An interesting new area is the production of sulfur-containing flavour topnotes starting with dimethylsulfoniumpropionate (DMSP) as precursor. Natural available sulfur precursors for fermentations have been limited to cysteine, methionine and thiamine so far. Research has indicated that 3-dimethylsulfonium propionate (DMSP) from algae can also be used to generate flavour molecules such as dimethylsulfide (DMS), methylmercaptan, 3-S-methyl-3-mercaptopropionic acid (MTPA), 3-mercaptopropionic acid (MPA) and esters thereof.

The alga *Ulva*, a common and widespread species, contains considerable quantities of DMSP (up to 2% by dry weight). In some waters, *Ulva* can become a nuisance and will give a stink during massive starvation at the end of summer. DMSP in the algae is thought to play a role in osmoregulation. DMSP can be isolated by extraction using alcohols. In isolated form it falls easily apart into DMS and acrylic acid.

DMSP was isolated to serve as a raw material for sulfurous products made by micro-organisms. Some anaerobic bacteria (*Desulfobacterium* species) were identified, which were able to demethylate DMSP into MTPA. The advantage of these biotransformations is a conversion yield of 100% combined with economically acceptable levels of MTPA. The disadvantage is the slow generation time of the bacteria used combined with a low final biomass concentration, resulting in long fermentation and production times.

Some aerobic strains were recognised as producers of MTPA, starting from DMSP. One main disadvantage of these strains is the possibility of degrading MTPA further, giving a highest recorded yield of 50%. Other bacterial strains were identified, able to convert MTPA into MPA. The identified methanogenic bacteria produced acceptable levels of MPA (> 0.5%) in about 14 days.

Further research is being directed to an acceleration of growth rate and an increased biomass production of the identified bacteria in order to reduce the present long fermentation times to more acceptable production times.

3 FLAVOUR BUILDING BLOCKS

This is an area of research in which food fermentation plays an important role. Raw materials which are used in well known processes such as baking, brewing, cheese making *etc.* are transformed in a fermentor using micro-organisms and/or enzymes which normally play a role in these processes. In fact, it is an acceleration of the processes and a concentration of the products which are formed during the baking, brewing *etc.*

Although the principle is simple, the final product is often a mixture of ingredients, whose quality is strongly dependent on the choice of raw materials, enzymes and micro-organisms, the fermentation type (submerged, slurry or solid state), the conditions and downstream processing.

3.1 Complex Bakery Ingredient

Cereals as raw materials provide most of the necessary precursor molecules for full (bio)generation of flavour in baked goods when fermented with specific micro-organisms (*e.g.*, yeast) either with or without enzymes (*e.g.*, amyloglucosidase).

Enzymes facilitate the (partial) breakdown of starchy and proteinaceous material into mono-, di-, tri- and oligosaccharides, amino acids and peptides. These molecules may act as flavour precursor as such (*e.g.*, in crust formation) or after conversion by microbial metabolic processes (*e.g.*, crumb flavours).

In order to generate a complex bread flavour a high protein-containing wheat fraction is fermented in a slurry-type of fermentation with specific enzymes, yeasts and lactic acid bacteria. After fermentation, the matrix is subjected to conventional downstream processing such as spray drying. The resulting multifunctional preparation (HYBAKE) can be used in bakery goods (*e.g.*, Premium Crackers, French Baguette) to improve flavour and texture (resembling more traditionally baked goods) and to replace laborious sponge handling in S&D processes (see Figure 2).

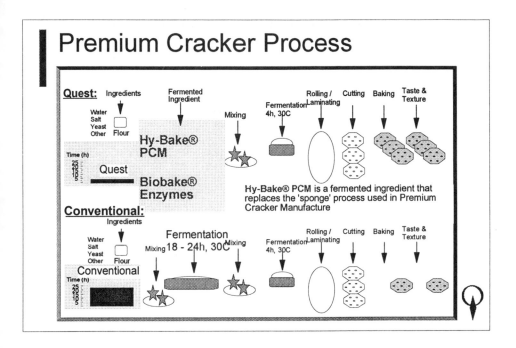

Figure 2 *Effect of Bread improver (HyBake PCM) on cracker manufacturing*

4 YEAST EXTRACT

Yeast extracts can be very powerful savoury ingredients. In yeast extract manufacturing, large scale yeast propagation is followed by yeast autolysis. In this process the yeast cells are subjected to enzymic breakdown by yeast enzymes and/or added enzymes. The specific cell composition of the yeast cell determines the quality of the final product.

In general, a high protein content, resulting in high amounts of free amino acids after autolysis, is desirable. Therefore, often proteases and peptidases are added during the autolysis procedure. High RNA levels can result in increased levels of 5-GMP or 5-IMP. The tripeptide glutathione, an important compound in maintaining the right redox balance in the yeast cell, also contributes to the savoury character of some yeast extracts. Research is focusing on influencing the specific cell composition of yeast by fermentation strategy or genetic engineering (metabolic reprogramming).

REFERENCES

1. M. Farwood and B.J. Willis, 1983, US Patent 4560656.
2. A.L.G.M. Boog *et al.*, 1989, EP 89203041.2.
3. K.A. Maume *et al.*, *Biocatalysis*, 1991, **5**, 79.
4. A.L.G.M. Boog *et al.*, submitted for publication.
5. A.L.G.M. Boog *et al.*, 1990, EP 90201882.9.

VOLATILES PRODUCED BY *STAPHYLOCOCCUS XYLOSUS* GROWING IN MEAT/FAT SYSTEMS SIMULATING FERMENTED SAUSAGE MINCES WITH DIFFERENT INGREDIENT LEVELS

Louise H. Stahnke

Department of Biotechnology, Building 221, Technical University of Denmark, DK-2800 Lyngby, Denmark

1 INTRODUCTION

Several studies have shown that the profile of volatiles of fermented sausage is very complex, including more than a hundred different compounds from many classes of components.[1,2] Microbial growth in the sausage mince together with the activity of enzymes from the meat and fat are responsible for many of these components. Autoxidative reactions are also of great importance.[4]

It is not known which processes play the major part in the development of the characteristic, cured flavour and how this flavour is related to microbial growth, ingredient levels and other production parameters. In one study, correlation of sensory data and the profile of volatiles and content of the starter culture *Staphylococcus xylosus* indicated that salami aroma was related to the content of ethyl esters, certain short-chain, branched aldehydes, methyl ketones and *Staphylococcus xylosus*. However, it was not possible to establish whether this bacterium produced the compounds or whether other micro-organisms or mechanisms were involved.[2,3]

The purpose of the present study was to investigate the production of major volatile compounds by *Staphylococcus xylosus*. In order to avoid interference from the background flora, the bacterium was grown in aseptic model sausage minces. The experimental design was set up as a fractional factorial design, examining the influence of temperature, pH, oxygen access, and the concentrations of salt, nitrite/nitrate, glucose and ascorbate. The factor levels were chosen to cover a wide spectrum of production parameters of different types of fermented sausages.

2 MATERIALS AND METHODS

The experimental design was set up as a seven factor fractional design at two levels with resolution IV (2^{7-3} structure). Four centre points were included giving a total of 20 sausage minces. Main effects were confounded with three-factor interactions, two-factor interactions were confounded with each other.

Model minces were made of 45% lean pork, 15% pork back fat and 40% w/w water including ingredients according to the experimental design. Low and high levels were: pH (5 and 6.5), temperature (17 and 33 °C), salt (2.4 and 9.0% w/w in water), nitrite:nitrate (150 ppm:0 and 0:0.2 %w/w), glucose (0 and 0.5 %w/w), ascorbate (0 and 500 ppm), oxygen (growth flask closed and open). The minces were produced aseptically by immersing pork back in boiling water for 2 min, removing the cooked surface and cutting, mincing the meat and fat by sterile instrumentation on a sterile bench. Ingredients were dissolved in water and sterilised by filtration. At this point, the viable, aerobic and anaerobic count was less than 10 g^{-1} of mince. 5×10^7 cfu g^{-1} *Staphylococcus xylosus* (Chr. Hansen's A/S, Denmark) was added to each mince.

Volatiles from the minces were collected over a five-day period by diffusive sampling in the following manner: The minces were placed in Erlenmeyer flasks fitted with two short, horizontal glass tubes closed with Swagelok fittings. First, the minces were incubated for 4 days at the conditions specified in the experimental design. Secondly, all flasks were closed and minces equilibrated at 17 °C for 18 hours. Finally, Tenax TA tubes (200 mg, 60/80 mesh, Buchem, Holland) were fixed onto the glass tubes and headspace volatiles from the minces collected by passive diffusion for 5 days at 17 °C. This procedure simulated the fermentation and drying periods of a fermented sausage. Duplicate tubes were made for each mince. The volatiles on the Tenax TA tubes were thermally desorbed (ATD50, Perkin-Elmer) and analysed by GC–FID and GC–MS on Hewlett-Packard instrumentation. The separation was performed on a DB-1701 capillary column (J&W Sci., U.S.A.). The desorption procedure was calibrated by desorbing five calibration tubes containing 5 µl of a 0.01% octane in methanol solution, prepared according to Perkin-Elmer.[5] Headspace volatiles were semi-quantified by dividing peak areas with the averaged area of the octane peaks from the five calibration tubes. Identification was based on Kovat's retention indices of authentic compounds and of MS spectra compared to the NIST database. The results were analysed using multiple linear regression and analysis of variance (MODDE version 3.0, UMETRI AB, Umeå, Sweden). The peak areas were transformed into the logarithmic$_{10}$ scale prior to analysis.

3 RESULTS AND DISCUSSION

Table 1 lists all the identified compounds except alkanes, alkenes, benzene, methylbenzenes and furanes. Many branched alkanes appeared to be contaminants and it is very likely that none of those compounds have a significant impact on the aroma. The listed compounds have all been detected in earlier studies of fermented sausages.[1,2] Some of them are most certainly produced by *Staphylococcus xylosus*, *e.g.* diacetyl and ethanol, while others are more likely to arise from lipid autoxidation, *e.g.* hexanal.

The volatiles were collected by diffusive sampling over a five-day period. This method has been used earlier for collecting secondary, volatile metabolites from fungi growing in Petri dishes.[8] The major advantage of diffusive sampling is the ease of collecting volatiles from a large number of samples simultaneously without the tedious and slow job of analysing the samples one by one. In experiments investigating the effect of many factors a large number of samples is usually required. In addition, losses of highly volatile compounds due to breakthrough of adsorbent traps or due to reconcentration steps are avoided. The major disadvantage is the low sensitivity, though sampling over several days compensates for that.

Table 2 shows the results from the regression analysis on a few of the major headspace volatiles. The analysis indicates that the level of acetonitrile was increased by increasing the amounts of nitrate. The same relationship was shown between heptanonitrile and nitrate in fermented sausage in an earlier study.[2] The occurrence of nitriles in cured meat has also been reported by Mottram *et al.*[6] It is not known what the influence of nitriles on aroma is and whether the nitriles are formed by *Staphylococcus xylosus* or by a chemical reaction. Mottram *et al.*[6] proposed that nitriles are formed during reactions involving nitrite and lipid oxidation products at the expense of the corresponding aldehydes. This may be possible. It seems more likely though, that nitriles are formed during reactions directly between aldehydes and nitrite (or decomposition products of nitrite), not involving lipid oxidation since, for example, 2-methylpropanal and benzaldehyde (corresponding to 2-methylpropanonitrile and benzonitrile) are not products of lipid oxidation (*cf* Table 1).

Table 1   *Components identified in headspace above minces*

Component	Reliability of identification	Component	Reliability of identification
Aldehydes		*Ketones*	
Acetaldehyde	rt/ms	2-Propanone = acetone	rt/ms
2-Methylpropanal	rt/ms	2-Butanone	rt/ms
Butanal	rt/ms	2,3-Butadione = diacetyl	rt/ms
2-Methylbutanal	rt/ms	2-Pentanone	rt/ms
3-Methylbutanal	rt/ms	1-Penten-3-one	ms
Pentanal	rt/ms	4-Methyl-2-pentanone	rt/ms
Hexanal	rt/ms	Cyclohexanone	rt/ms
Heptanal	rt/ms	2-Heptanone	rt/ms
Nonanal	rt/ms	2-Undecanone	rt/ms
Decanal	rt/ms		
Benzaldehyde	rt/ms	*Acids*	
		Acetic acid	rt/ms
Alcohols		Propanoic acid	rt/ms
Ethanol	rt/ms	2-Methylpropanoic acid	rt/ms
2-Propanol	rt/ms	Butanoic acid	rt/ms
2-Methyl-1-propanol	rt/ms	3-Methylbutanoic acid	rt/ms
2-Butanol	rt/ms	2-Methylbutanoic acid	ms
3-Methyl-1-butanol	rt/ms	Hexanoic acid	rt/ms
3-Hydroxy-2-butanone	ms	Heptanoic acid	rt/ms
1-Pentanol	rt/ms	Octanoic acid	rt/ms
1-Penten-3-ol	rt/ms		
		Nitriles	
Esters, Sulfides		Acetonitrile	ms
Ethylacetate	rt/ms	2-Methylpropanonitrile	ms
3-Methylbutylacetate	rt/ms	Hexanonitrile	ms
Dimethyldisulfide	rt/ms	Heptanonitrile	ms
Dimethyltrisulfide	ms	Benzonitrile	ms

The amount of 2- and 3-methylbutanal was increased by high temperature, low amount of ascorbate and, most strongly, by low oxygen level during growth. These findings indicate that the aldehydes are produced by fermentative oxidation from the amino acids isoleucine and leucine, probably by oxidative deamination to the corresponding α-keto acid followed by decarboxylation. It is often suggested that 2- and 3-methylbutanal in dried, fermented meat products originate from non-enzymatic Strecker degradation.[1] This is not likely in this study since the water content of the minces was quite high (75%–80%). The content of 2-

and 3-methylbutanoic acid was increased by increasing the amount of nitrate. This indicates that the acids were produced from 2- and 3-methylbutanal by chemical oxidation. 3-Methyl-1-butanol probably arises from microbial reduction of 3-methylbutanal. The pH-effect reconfirms that *Staphylococcus xylosus* grows better at high pH (*cf* Table 2).

Table 2 *Significant Main Effects[a]*

Compound	pH	Temp	Salt	Nitrite: nitrate	Glucose	Ascorbate	Oxygen access
Acetonitrile				+			
2- and 3-Methylbutanal	+					–	–
2- and 3-Methylbutanoic acid				+			
Diacetyl	+			+	+	–	
2-Pentanone	+	+			–		+
3-Methyl-1-butanol	+						

[a] Significance level $p \leq 0.05$; + (–) : concentration of compound increases (decreases) by increasing the factor level from low to high level. Factor levels are shown under Materials and Methods.

The level of diacetyl was increased by an increasing amount of glucose, indicating that *Staphylococcus xylosus* produces diacetyl from glucose. Nitrate and ascorbate increased and decreased, respectively, the amount of diacetyl. This shows that high redox potential raises the content of diacetyl. Otherwise, most of the diacetyl would probably convert into acetoin by acetoin dehydrogenase as in the case of lactic acid bacteria fermentation of diacetyl.[7]

The amount of 2-pentanone was increased by access to oxygen, increasing pH and salt content and by decreasing glucose content. 2-Alkanones may be formed by β-oxidation of fatty acids or by decarboxylation of β -keto acids. The glucose effect implies that *Staphyloccus xylosus* only forms 2-pentanone when no easily degradable carbon and energy source is present (in this case glucose). The effect of salt is puzzling.

REFERENCES

1. J.L. Berdagué, P. Monteil, M. Montel and R. Talon, *Meat Sci.*, 1993, **35**, 275.
2. L.H. Stahnke, *Meat Sci.*, 1995, **41**, 193.
3. L.H. Stahnke, *Meat Sci.*, 1995, **41**, 211.
4. K. Incze, *Fleischwirtsch.*, 1992, **72**, 58.
5. Perkin-Elmer, Thermal Desorption Data Sheet No. 9, Perkin-Elmer Ltd., Beaconsfield, Buckinghamshire, U.K., 1991.
6. D.S. Mottram, S.E. Croft and R.L.S. Patterson, *J. Sci. Food Agric.*, 1984, **35**, 233.
7. G. Gottschalk, 'Bacterial Metabolism', Springer-Verlag, New York, 1986, 2nd ed., p. 208.
8. T.O. Larsen and J.C. Frisvad, *J. Microb. Methods*, 1994, **19**, 297.

ENZYMES INVOLVED IN THE METABOLIC PATHWAY LEADING TO 3-METHYLBUTANAL IN TOMATO FRUIT

H.T.W.M. van der Hijden and I.J. Bom

Unilever Research Laboratory, P.O. Box 114, 3130 AC Vlaardingen, The Netherlands

1 INTRODUCTION

Perception of flavour is the result of a complex interplay of many volatile and non-volatile compounds. Some 400 volatiles[1] along with such compounds as sugars, amino acids and free acids (mainly malic and citric acid) are responsible for the typical tomato flavour. The characteristic tomato flavour is mainly a result of *in vivo* and *in vitro* enzymatic reactions. We have studied one of the metabolic pathways in tomato in which leucine is converted into the very characteristic tomato volatiles 3-methylbutanal and 3-methylbutanol. In a patent application[2] Unilever has described the use of antibody fragments against off-flavour components.

In the regular catabolic pathway of amino acids, leucine is converted to α-ketoisocaproic acid (4-methyl-2-oxopentanoic acid) by deamination. α-Ketoisocaproic acid is finally converted to acetoacetate and acetyl-CoA. The latter compound can enter the TCA cycle. As an alternative to the major catabolic route, α-ketoisocaproic acid can be used as the starting molecule for the generation of the branched aldehyde, 3-methylbutanal. α-Ketoisocaproic can be converted into the aldehyde via at least two biochemical pathways. One possible route, which is catalysed by a branched-chain 2 oxo-acid dehydrogenase[3] complex, results in 3-methylbutyryl-CoA ester which is converted into 3-methylbutanal by a coenzyme A-acylating aldehyde dehydrogenase[4]. In an alternative pathway, an α-keto acid decarboxylase[5] competes with the branched-chain 2 oxo-acid dehydrogenase complex for 2-ketoisocaproic acid. The latter route leads to 3-methylbutanal in one single enzymatic step. The aldehyde in its turn can be converted into the corresponding alcohol by alcohol dehydrogenase.

In order to elucidate the pathway of the formation of 3-methylbutanal and manipulate it by, for example, a transgenic approach, we have identified enzymes from tomato which are involved in this pathway. An α-keto acid decarboxylase capable of converting α-ketoisocaproic acid to 3-methylbutanal has been isolated from tomato and characterized.

2 ENZYMES INVOLVED IN THE GENERATION OF 3-METHYLBUTANAL

There are several routes by which volatile flavour compounds are formed. Some volatiles are formed by the endogenous metabolism of the fruit and are present in the intact tomatoes. Alternatively, other volatile compounds are formed enzymatically or chemically after disruption of the fruit. Volatility measurements have indicated that 3-methylbutanal is present in intact tomatoes[6]. On the other hand, it was qualitatively demonstrated long ago that crude extracts from tomato are capable of converting L-leucine into 3-methylbutanal and 3-methylbutanol *in vitro*[7]. In order to determine the relative contribution of the *in vivo* versus the *in vitro* production of 3-methylbutanal, we quantified the amount of this compound in the intact fruit and at several time intervals after disruption. It appeared that there was no significant increase of the leucine-derived volatiles, 3-methylbutanal and 3-methyl-1-butanol, after disruption of the tomato fruit. From these measurements it can be concluded that 3-methylbutanal in the tomato and in the tomato homogenate originate completely from endogenous metabolism in the intact fruit.

In order to identify enzymes which are involved in the *in vivo* generation of leucine-derived aldehydes and alcohols, we have measured enzyme activities in several developmental stages of the tomato fruit, in order to correlate them with the levels of 3-methylbutanal. The maximal 3-methylbutanal concentration in the pink tomato corresponded with a maximum in an α-keto acid decarboxylase enzyme activity. The latter activity, tested with pyruvate as the substrate, was approximately 1.7 times higher in the pink tomato than in the green stage.

Table 1   *Levels of 3-Methylbutanal and 3-Methyl-1-Butanol in Normal and Anaerobically Stressed Pink Tomatoes*

Compound	Intact (non-stress) tomatoes ng g^{-1}	Intact (anaerobically stressed) tomatoes ng g^{-1}
3-Methylbutanal	78	32
3-Methyl-1-butanol	326	1046
Total	404	1078

In order to test the correlation between the α-keto acid decarboxylase activity and the leucine-derived volatiles further, we looked for a means of manipulating the decarboxylase activity in a non-invasive way. We found that incubation of the pink tomato fruit under anaerobic conditions increased the α-keto acid decarboxylase activity. After one hour of anaerobic incubation, the enzyme activity was twice as high as that measured in the untreated pink tomato fruit. Measurements of volatiles showed that, in intact tomatoes, anaerobic stress leads to an increase of leucine-derived volatiles (Table 1). It should be noted that the total of 3-methylbutanal and 3-methylbutanol has to be taken to monitor the conversion of leucine into volatile compounds since it is known that ADH is also induced by anaerobiosis[8]. The above findings led to the hypothesis that an α-keto acid decarboxylase is involved in the generation of 3-methylbutanal. For this reason we decided to purify and characterise the α-keto acid decarboxylase of tomato fruit.

3 ENZYME PURIFICATION

The results of the purification of α-keto acid decarboxylase activity from tomato pericarp of *Lycopersicon esculentum* cv. Ferrari are shown in Table 2. The enzyme was purified approximately 900-fold over the crude extract and was obtained in 20% overall yield. The α-keto acid decarboxylase is a labile enzyme. Extraction of the tomato pericarp in the presence of 10% glycerol, 5 mM phenyl methylmethanesulphonyl fluoride, 1 μM pepstatin, 1 μM leucopeptin, 0.1 mM EDTA, 0.1 mM EGTA, 1 mM MgCl₂ and 1 mM DTT was necessary to retain enzyme activity as much as possible, especially after ammonium sulphate precipitation. To achieve maximum stability of the enzyme preparation, 5% glycerol, 1 mM DTT and 1 mM $MgCl_2$ were included in all chromatographic buffers. In spite of these additions, the stability of the decarboxylase was considerably less after the gel filtration step. Only hydroxyapatite was useful to concentrate the diluted enzyme solution for further purification.

Table 2   *Purification of α-Keto acid Decarboxylase Obtained from Tomato Pericarp from* L. esculentum *var.* Ferrari

Step	Volume (ml)	Protein (mg)	Activity U (total)	Spec. Activity U/mg	Yield %	Step purification
Crude extract	470	1800.0	43.0	0.023	100.0	1.0
Am. sulphate prec. (30–55%)	52	550.0	37.0	0.070	86.0	3.0
Gel filtration (Sepharose CL-6B)	400	12.5	32.0	2.560	74.4	36.6
Hydroxyapatite	380	2.1	23.0	10.950	53.5	4.3
Hydrophobic interaction chromatography	18	1.2	14.0	11.700	32.6	1.0
Hydroxyapatite	18	0.4	8.3	20.800	20.8	1.8

4 CHARACTERISATION OF THE TOMATO α-KETO ACID DECARBOXYLASE

The α-keto acid decarboxylase purified in this manner shows two bands of slightly different molecular weight in SDS–PAGE electrophoresis. The protein bands have sizes of 60 and 62 kD approximately. The double bands corroborate earlier results using a tomato α-keto acid decarboxylase[9]. The native molecular weight was estimated by gel filtration on a Superose 12 column with standard proteins of known molecular weight. A native molecular weight of 240 kD was found. From this, together with its chromatographic behaviour on the Sepharose CL-6B column, it may be concluded that the native α-keto carboxylase consisted of four subunits.

The kinetics of pyruvate α-keto-carboxylase were measured on several α-keto acids, *viz* pyruvate, α-ketovaleric acid, α-ketoisovaleric acid and α-ketoisocaproic acid. Among these α-keto acids, pyruvate showed the highest rate of conversion. The substrate concentrations at which the enzyme shows half maximal activity, the K_m values, for these compounds are presented in Table 3.

Table 3   *Kinetic Parameters of Tomato α-Keto acid Decarboxylase*

	Pyruvate	α-Ketovaleric acid	α-Ketoisovaleric acid	α-Ketoisocaproic acid
K_m (mM)	1.06	2.33	2.92	3.41
V_{max} (U/mg protein)	23.5	1.45	0.43	0.23

The pH optimum of the pyruvate α-keto carboxylase was found to be 6.2. The product derived from the incubation of α-ketoisocaproic acid and purified α-keto acid decarboxylase were analysed on the GC. The chromatographic analysis showed only one volatile which was identified as 3-methyl butanal. From the product specificity and kinetic data on the α-keto acids, it appears that the decarboxylase shows the highest catalytic power (K_{cat}/K_m where K_{cat} represents the turnover number (s^{-1}) of the enzyme) on pyruvate and hence can be defined as tomato *pyruvate decarboxylase* which is able to convert α-ketoisocaproic acid into 3-methylbutanal.

According to the above data it is very likely that pyruvate decarboxylase is involved in the *in vivo* generation of 3-methylbutanal and 3-methylbutanol in tomato. During the purification of pyruvate decarboxylase on the Sepharose CL-6B column, however, another α-keto acid decarboxylase activity was determined. The latter enzyme had an optimum pH at 7.5 and showed a preference for branched chain keto acids. Unfortunately, this enzyme activity was at least a factor of ten less than that measured for PDC. The possibility that this enzyme activity is also involved in the generation of 3-methylbutanal can not be excluded. The origin and biological function of this enzyme in tomato with a preference for branched chain keto acids in tomato is unknown.

REFERENCES

1. M. Petro-Turza, *Food Rev. Int.*, 1987, **2**, 309.
2. Patent application WO-A-94/14934 (published 7 July 1994).
3. J.R. Dickinson and I.W. Dawes, *J. Gen. Microbiol.*, 1992, **138**, 2029.
4. R.-T. Yan and J.-S. Chen, *Appl. & Environ. Microbiology*, 1990, **56**, 2591.
5. S. Derrick and P.J. Large, *J. Gen. Microbiol.*, 1993, **139**, 2783.
6. R.G Buttery and L. Ling, in 'Progress in Flavour Studies – Proceedings of the International Conference, Würzburg, Germany', eds. P. Schreier and P. Winterhalter, Allured Publishing Corporation, U.S.A., 1992
7. M.H. Yu, D.K. Salunkhe and L.E. Olson, *Plant & Cell Physiol.*, 1968, **9**, 633.
8. A.-R.S. Chen and T. Chase, Jr., *Plant Physiol. Biochem.*, 1993, **31**, 875.
9. A. Foster and T. Chase, Jr., *The FASEB Journal*, 1995, **9**, A1 292.

BAKER'S YEAST REDUCTION OF 2,3-PENTANEDIONE TO NATURAL 3-HYDROXY-2-PENTANONE

P.H. van der Schaft, H. de Goede and N. ter Burg

Tastemaker B.V., P.O. Box 414, 3770 AK Barneveld, The Netherlands

1 INTRODUCTION

An increasing consumption of processed foods is accompanied by a growing demand for flavours. Due to consumer preferences, it is important that these flavours can be designated as natural. This implies that both the starting materials *and* the processing are natural; physical and enzymatic/microbiological methods are regarded as such. 2,3-Pentanedione, 2-hydroxy-3-pentanone, 3-hydroxy-2-pentanone and 2,3-pentanediol are all nature-identical flavour ingredients according to European legislation and used in aromas by the flavour industry. In the U.S.A., 2,3-pentanedione and 2-hydroxy-3-pentanone are considered as GRAS and used in commercial flavours. These flavour ingredients can, for instance, be applied in dairy flavours. The only commercially available natural ingredient of this group of ingredients is 2,3-pentanedione, which can be recovered from the pyroligneous juices of wood or from fermentation broths. It was shown earlier,[1] that wine yeast is able to form dicarbonyl compounds, including 2,3-pentanedione and converts these compounds into their reduction products which are present in wine below their odour threshold value. The reduction of 2,3-pentanedione leads to unique products, but they have a much lower flavour intensity than the diketone. Microbial reduction of this 2,3-diketone may result in 2-hydroxy-3-pentanedione, 3-hydroxy-2-pentanedione and 2,3-pentanediol as shown by others.[2] This paper presents a study on the development of a laboratory scale process for the production of 3-hydroxy-2-pentanone.

2 MATERIALS AND METHODS

2.1 Micro-organism Screening

Twenty-one micro-organisms were obtained from microbial culture collections; *Saccharomyces cerevisiae* is commercially available in bulk quantities ('Koningsgist') from Gist-brocades, Delft, Holland (Table 1). During the screening experiment, all microbes were cultured on appropriate media in the presence of 4 g l^{-1} 2,3-pentanedione under conditions prescribed by the supplier.

Table 1 *Screening of micro-organisms for the reduction of 2,3-pentanedione to 3-hydroxy-2-pentanone (3-hyp-2), 2-hydroxy-3-pentanone (2-hyp-3) and 2,3-pentanediol (2,3-diol)*

Micro-organism	Code number	3-hyp-2	2-hyp-3	2,3-diol
Aspergillus oryzae	ATCC 14895	++	+++	0
Bacillus licheniformis	DSM 13	++++	+++	0
Bacillus polymyxa	DSM 36	+++	++	+
Bacillus subtilis	ATCC 8188	+++	++	0
Bacillus subtilis	ATCC 15245	++++	+++	0
Candida guillermondii	CBS 5660	+	+	+++
Candida maltosa	ATCC 20184	+	+	++++
Dipodascopsis uninucleata	CBS 190.37	++++	++++	+
Enterobacter cloacae	NCIMB 10101	++	++++	+
Enterobacter aerogenus	DSM 30053	++	++++	++
Fusarium poae	CBS 317.73	+++	+++	+++
Geotrichum candidum	CBS 615.84	++	++++	+++
Klebsiella oxytoca	DSM 4798	+	++++	++
Lactobacillus acidophilus	ATCC 4356	+	++	-
Lactobacillus lactis	DSM 20661	+	-	-
Leuconostoc mesenteroides	ATCC 8293	0	++	-
Listeria innocua	ATCC 33090	+++	++++	-
Penicillium roqueforti	CBS 221.30	+++	++++	++
Saccharomyces cerevisiae		+++	++++	++
Serratia marcescens	DSM 30121	++	++++	++
Sporobolomyces odorus	CBS 6781	++	++++	++
Yarrowia lipolytica	CBS 6124	+++	+++	+

No production (–);
1–10 mg l^{-1} product (0);
10–50 mg l^{-1} product (+);
50–250 mg l^{-1} product (++);
250–750 mg l^{-1} product (+++);
750 mg l^{-1} and more product (++++).

2.2 Conversion Studies using Baker's Yeast

Bioconversion studies were carried out using 28 g l^{-1} baker's yeast (dry weight) in 250 mM phosphate buffer (pH 7) at 28 °C and a substrate concentration of 4 g l^{-1} 2,3-pentanedione in shaking flasks or in an aerated fermenter.

2.3 Analysis

Analysis was performed using GC or HPLC. In both cases cells were first removed by centrifugation. For GC analysis, the supernatant was extracted three times with an equal volume diethylether. The extract was dried over Na_2SO_4. Heptanol-1 was added as an internal standard and the extract was concentrated by distilling off most of the solvent. The concentrated extract was analysed by GC. HPLC analysis of the supernatant was done on a Hypercarb column using methanol/water (15:85) as eluent. External standards were used for the quantitation of all reaction products.

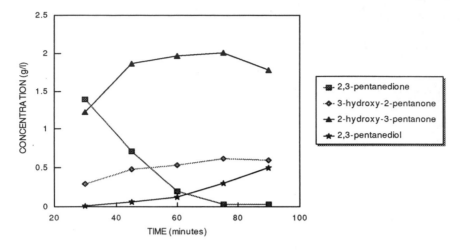

Figure 1　*Baker's yeast reduction of 2,3-pentanedione without addition of a co-substrate*

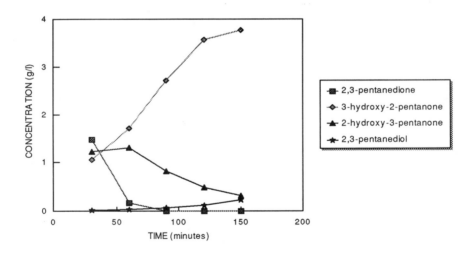

Figure 2　*Baker's yeast reduction of 2,3-pentanedione in the presence of 0.5% glucose as a co-substrate*

3 RESULTS AND DISCUSSION

3.1 Screening

The screening of twenty-two micro-organisms from very different origin and classification, including bacteria, yeasts and fungi showed that reduction of 2,3-pentanedione to its hydroxyketone and diol derivatives is a general feature in the microbial world (Table 1). Baker's yeast also showed a good performance with regard to this reduction and this food-grade catalyst was chosen for the further development of this bioconversion.

3.2 Role of Temperature and pH

The optimal temperature and pH for the bioconversion were 30 °C and 6.5, respectively. Good conversions were obtained in the temperature range from 21 to 37 °C and in the pH range from 6 to 7.5.

3.3 Substrate and Product Inhibition

According to the literature, micro-organisms reduce dicarbonyl compounds to diols because the reduced compounds are less toxic to micro-organisms.[3] The optimal substrate concentration for bioconversion is 5 g l^{-1} according to literature.[4] In our studies 4 g l^{-1} substrate was found to cause no inhibition; at 5 g l^{-1} some inhibition was observed.

3.4 Effect of Co-substrate Addition

The conversion of 2,3-pentanedione by baker's yeast results in a mixture of 2,3-pentanediol, 3-hydroxy-2-pentanone and 2-hydroxy-3-pentanone in a ratio of 1:1:3 (Figure 1). Addition of 0.5% glucose to the bioconversion medium resulted in a quite different profile of reduced products (Figure 2). 2,3-Pentanediol was only formed in small amounts. After a reaction time of 40 minutes, about equal amounts of the two hydroxyketones were formed. During the next 110 minutes, 3-hydroxy-2-pentanone was formed at the expense of 2-hydroxy-3-pentanone. Comparable results are obtained when 0.5% ethanol or 0.5% fructose are used as a co-substrate.

No studies were undertaken to explain this phenomenon. Several individual or combined issues may play a role.[4,5] The presence of a co-substrate may facilitate co-factor regeneration which may result in another product pattern; also product inhibition may play a role. The addition of glucose may also induce other dehydrogenases that can catalyze the reduction resulting in another product pattern.

REFERENCES

1. G. de Revel and A. Bertrand, in 'Trends in Flavour Research', eds. H. Maarse and D.G. van der Heij, Elsevier, Amsterdam, 1994, p. 353.
2. R. Bel-Rhlid, A. Fauve, M.F. Renard and H. Veschambre, *Biocatalysis*, 1992, **6**, 319.
3. J. Heidlas and R. Tressl, *Eur. J. Biochem.*, 1990, **188**, 165.
4. P. Besse, J. Bolte, A. Fauve and H. Veschambre, *Bioorganic Chemistry*, 1993, **21**, 342.
5. A. Fauve and H. Veschambre, *Tetrahedron Letters*, 1987, **28**, 503.

CHIRALITY AND FLAVOUR

FORMATION OF γ- AND δ-LACTONES BY DIFFERENT BIOCHEMICAL PATHWAYS

R. Tressl, T. Haffner, H. Lange and A. Nordsieck

Institut für Biotechnologie / Chemisch-Technische Analyse, Technische Universität Berlin, Seestraße 13, D-13353 Berlin, Germany

1 INTRODUCTION

Chiral γ- and δ-lactones are known as important flavour compounds and are widely distributed in nature. The chirospecific analysis of these lactones (by enantioseparation on chiral gas-chromatographic capillary columns) is an effective tool for proving the authenticity of natural flavourings in food. Therefore, knowledge of the biosynthetic pathways to lactone enantiomers is of great importance. The lactone-producing yeast, *Sporobolomyces odorus*, is an appropriate model system for biosynthetic studies of fatty acid-derived γ- and δ-lactones. *Sporobolomyces* red yeasts possess alternative cyanide and antimycin A-insensitive respiratory systems and regenerate coenzymes by enzymically catalysed hydroxylations, desaturations, epoxidations, peroxidations and Baeyer–Villiger-type oxidations.[1]

In cultures of *S. odorus*, the dominant pathway to (*R*)-γ-decalactone was characterized as an (*R*)-12-hydroxylation of oleic acid, followed by β-oxidation. In addition, oleic acid was transformed into linoleic acid, which was subsequently degraded to (*Z*)-6-γ-dodecenolactone, possessing an enantiomeric ratio of 92:8 (*R*):(*S*).[2] The biosynthesis of (*R*)-δ-decalactone ((*R*) > 99% ee) in *S. odorus* is initiated by a lipoxygenation at C-13 of linoleic acid to (*S*)-13-hydroxy-(*Z,E*)-9,11-octadecadienoic acid, which is subsequently converted to (*R*)-δ-decalactone.[3] The stereochemistry of the peroxysomal β-oxidation system in yeasts possesses a strict D-stereospecificity of the involved D-3-hydroxyacyl-CoA dehydrogenase and 2-enoyl-CoA hydratase, as demonstrated by Filippula and co-workers.[4] The β-oxidation of unsaturated fatty acids with odd- and even-numbered double bonds was recently elucidated. Odd-numbered double bonds are isomerized and reductively removed by NADH-dependent 2,4-dienoyl-CoA reductases.[5] These enzyme systems are involved in the biosynthesis of γ- and δ-lactones.

The biochemical pathways to lactone enantiomers were investigated using synthesized, labelled precursors, supposed to act as intermediates in the existing pathways. The metabolites and lactones, which were formed from labelled precursors in cultures of *S. odorus* were analysed by established gas-chromatographic mass-spectrometric methods. Microsomal pathways leading to lactones are very sensitive to high substrate concentrations and free fatty acids, which are normally not released to the culture medium. Spinnler *et al.*[6] investigated the microsomal β-oxidation in *S. odorus* in organic solvents and traced intermediates derived from the metabolic degradation of ricinoleic acid. This method uses the same advantages as enzymic catalysis of hydrophobic substrates in organic solvents.[7]

These methods may be used for the elucidation of biosynthetic pathways after administration of appropriately labelled substrates, followed by regio- and stereochemically specific analysis of the course of the reaction by achiral and chiral GC–MS.

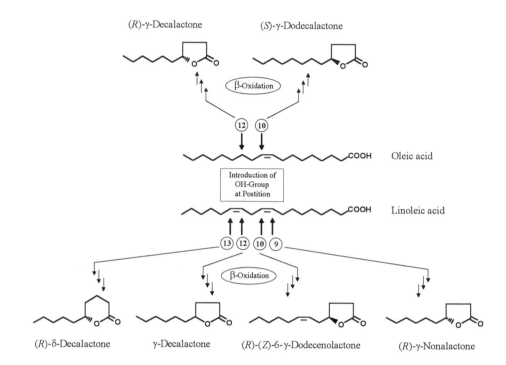

Scheme 1 *Biosynthesis of γ- and δ-lactone by initial hydroxylation of oleic and linoleic acid in the yeast, S. odorus*

However, the initial step in the lactone biosynthesis is always the regio- and enantioselective hydroxylation of an unsaturated fatty acid (Scheme 1). In cultures of *S. odorus*, unsaturated fatty acids are synthesized and oxidatively degraded in a cyclic manner. Therefore, the elucidation of biochemical pathways and of the enzymes involved can only be achieved by the administration of labelled precursors.

2. BIOSYNTHESIS OF δ-LACTONES IN *SPOROBOLOMYCES ODORUS*

2.1 Biosynthesis of (*R*)-δ-Decalactone in *S. odorus*

Biotransformations of [9,10,12,13-²H₄]-linoleic acid and its derivatives [9,10,12,13-²H₄]-(*S*)-13-hydr(per)oxy-(*Z*,*E*)-9,11-octadecadienoic acid ((*S*)-13-H(P)OD) and [13-²H₁]-(*R/S*)-13-hydroxy-(*Z*,*E*)-9,11-octadecadienoic acid into optically pure (*R*)-δ-decanolactone, catalysed by *S. odorus*, demonstrated an inversion of the initial configuration of the secondary hydroxy group during the β-oxidative degradation to the lactone.[3,8] Additionally, [13-¹⁸O]-labelled (*S*)-13-HOD was administered to cell suspensions of the yeast and transformed into the δ-lactone. These experiments confirmed a 13-lipoxygenase–peroxidase–β-oxidation pathway for (*R*)-δ-decalactone in this micro-organism. The inversion step of this degradation was further investigated by addition of [9,10-^{2}H$_2$]-13-oxo-(Z)-9-octadecenoic acid (13-KOE) to growing cultures of the yeast.

Table 1 *Incubation Experiments for the Elucidation of the Biosynthesis of δ-Decalactone in the Yeast,* Sporobolomyces odorus

Precursor	Product	Enantiomeric Purity [% ee]	Conversion [%]
[^{2}H$_4$]-(S)-13-HOD	[^{2}H$_2$]-(R)-δ-Decalactone	(R) > 98	15.0
[^{2}H]-(R,S)-13-HOD	(R)-δ-Decalactone	(R) > 98	14.0
13-KOD	(R)-δ-Decalactone	(R) > 98	10.8
[^{2}H$_2$]-13-KOE	[^{2}H]-(R)-δ-Decalactone	(R) > 98	12.0

As summarized in Table 1, [9,10-^{2}H$_2$]-13-KOE is efficiently transformed into (R)-δ-decanolactone. The obligatory reduction of 13-oxo-9,11-octadecadienoic acid to 13-KOE in the biochemical pathway is catalysed by enonereductases, which were recently characterized in *Saccharomyces cerevisiae*.[9] During β-oxidation of 13-KOE, we observed no direct reduction of the oxo-compound to 13-hydroxy-(Z)-9-octadecenoic acid (13-HOE). Nevertheless, we identified di-deuterated 9-oxo-(Z)-5-tetradecenoic acid and monodeuterated 5-oxodecanoic acid and 5-hydroxydecanoic acid as intermediates. Some of the results are summarized in Scheme 2.

These labelling experiments demonstrate an effective β-oxidation pathway for (S)-13-HOD in the yeast via the oxidation to 13-KOD and subsequent reduction of the α,β-double bond by enonereductases to 13-KOE. The 13-KOE undergoes four cycles of β-oxidation to the 5-oxodecanoic acid, which is reduced stereospecifically by a 5-oxodecanoyl-CoA reductase and transformed finally into the δ-decalactone.[10] The further degradation of this lactone goes via (R)-3-hydroxyoctanoic acid. Therefore, the inversion of the stereocentre of the (S)-configured 13-HOD is a prerequisite for the successful β-oxidation of this compound. The peroxysomal β-oxidation system of yeast possesses D-3-hydroxyacyl-CoA dehydrogenase and 2-enoyl-CoA hydratase 2 and no epimerase system for the inversion of the configuration like protcaryotes, plant and animal cells do.[4,11] The biosynthesis of (R)-δ-decanolactone (Scheme 2) is in agreement with the stereochemical course found for the β-oxidation of oleic acid in various organisms.[5]

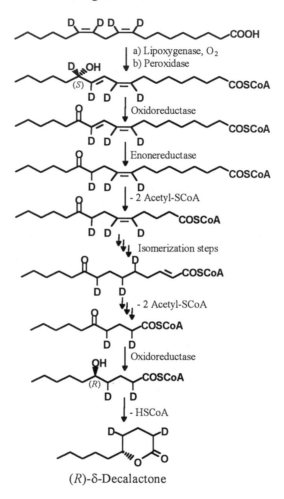

(*R*)-δ-Decalactone

Scheme 2 *Pathway to δ-decalactone from linoleic acid*

2.2 Biosynthesis of (*R*)-δ-Jasmin Lactone in *S. odorus*

δ-Jasmin lactone ((*Z*)-7-δ-decenolactone) possesses a fruity, sweet floral aroma and occurs in a variety of different foods, for example, cheddar cheese, tea, peaches and mangoes, as well as in flower extracts from *Jasminum*, *Gardenia*, *Tuberosa* and *Mimosa* species. Extracts of *Jasminum* flowers contain the lactone in the pure (*R*)-configuration, whereas in *Tuberose* species only the (*S*)-enantiomer is biosynthesized. In most plants and micro-organisms, this lactone is produced in distinct enantiomeric ratios. In several strains of *S. odorus*, δ-jasmin lactone was identified as a mixture of both antipodes ((*R*):(*S*) 73:27). The influence of the ω-3-double bond in linolenic acid-derived lactones was investigated by a series of labelling experiments.[12] The stereocentre of the resulting labelled δ-jasmin lactone corresponds to the β-oxidation of (*S*)-13-hydroxy-(*Z*,*E*,*Z*)-9,11,15-octadecatrienoic acid ((*S*)-13-HOT). The identification of a 13-lipoxygenase–peroxidase pathway in the yeast by administration of [9,10,12,13, 15,16-^{2}H$_6$]-linolenic acid and [9,10,12,13, 15,16-^{2}H$_6$]-(*S*)-13-HOT is summarized in Table 2.

Table 2 *Relative distribution of the isotopomers of δ-jasmin lactone in cultures of S. odorus formed 96 hours after addition of 53 mg [^{2}H$_6$]-linolenic acid ([^{2}H$_6$]-LA) or formed 102 hours after addition of 80 mg [^{2}H$_6$]-(S)-13-hydroxyoctadecatrienoic acid ([^{2}H$_6$]-HOT (capillary column: DB wax)*

Isotopomer	Represented by Mass Fragment [*m/z*]	Substrate	
		[^{2}H$_6$]-LA [%]	[^{2}H$_6$]-HOT [%]
Unlabelled δ-jasmin lactone	99	28.1	32.8
[^{2}H$_3$]-δ-Jasmin lactone	100	9.9	8.1
[^{2}H$_4$]-δ-Jasmin lactone	101	25.7	23.8
[^{2}H$_5$]-δ-Jasmin lactone	102	36.3	35.3

The quantitative mass-spectrometric investigations of the ring-labelled δ-jasmin lactone isotopomers showed similar deuterium distributions coming from [^{2}H$_6$]-linolenic acid and [^{2}H$_6$]-(*S*)-13-HOT, respectively, indicating a common pathway. The enantioseparation of the deuterated lactones indicated a pure [^{2}H$_5$]-(*R*)-enantiomer and a ratio of 89.1:10.9 (*R*):(*S*) for the [^{2}H$_4$]-labelled jasmin lactones. These results support a pathway from (*S*)-13-HOT to (*R*)-δ-jasmin lactone without inversion of the stereocentre (Scheme 3). The loss of one deuterium atom during the β-oxidative degradation of the hydroxy fatty acid cannot be explained by an oxidation–reduction step at the stereocentre at C-5 of the lactone. The D–H-exchange may result by a 1,5-H–D-shift during $\Delta^{3,5}$,$\Delta^{2,4}$-dienoyl-CoA isomerizations, leading to a D-loss at C-2 of the lactone.[5] Only 12% of the total amount of the deuterated δ-jasmin lactones possessed the inverted (*S*)-configuration. Thus this inversion constitutes only a minor pathway during the formation of δ-jasmin lactone. The chiral discrimination of the (*R*)- and (*S*)-enantiomer occurs during the β-oxidation of the (*S*)-configured 13-HOT.

3 BIOSYNTHESIS OF γ-LACTONES IN *SPOROBOLOMYCES ODORUS*

3.1 Epoxide Pathway to γ-Lactones in *S. odorus*

The successful degradation of (*E*)-3,4-epoxydecanoic acid by *S. odorus* to γ-decalactone demonstrated an epoxide pathway to γ-lactones in yeasts. On the contrary, the isomeric (*Z*)-3,4-epoxydecanoic acid was not a precursor for *S. odorus*.[13] This pathway is operative in the formation of γ-decalactone and γ-dodecalactone in fruits such as strawberries, peaches and apricots. Recently, Schöttler and Boland[14] confirmed this pathway by administration of deuterated (*Z*)- and (*E*)-configured 3,4-epoxyundecanoic acids to fruit tissues and identified labelled (*R*)-γ-undecanolactones as main products. In *S. odorus*, we showed this pathway by incubation experiments with [9,10,12,13-^{2}H$_4$]-linoleic acid to be a minor one. Less than 2% of γ-decalactone was found to be labelled, whereas more than 20%

of (*Z*)-6-γ-dodecenolactone was detected as labelled, coming from linoleic acid. Incubation experiments with [9,10,12,13-^{2}H$_4$]-(*S*)-9-hydroxy-(*E*,*Z*)-10,12-octadecadienoic acid ((*S*)-9-HOD) and cultures of the yeast resulted in labelled γ-decalactone isotopomers and unlabelled γ-dodecenolactone. The enantiospecific investigations of the [^{2}H$_3$]- and [^{2}H$_2$]-γ-decalactones showed, that this epoxide pathway is not enantioselective in *S. odorus*. Part of [9,10,12,13-^{2}H$_4$]-(S)-9-HOD is not degraded via β-oxidation by the yeast. After addition of this compound to cultures of *S. odorus*, tri-deuterated (*E*,*E*)-2,4-nonadien-1-ol, (*E*)-4-nonen-1-ol, (*Z*)-3-nonen-1-ol, the corresponding fatty acids and unlabelled azelaic acid were identified. Therefore, for the first time we have evidence for a Baeyer–Villiger-type reaction in a yeast. The β-oxidation and the Baeyer–Villiger reaction of 9-functionalized fatty acids were further investigated by synthesis and administration of a [9,10-^{2}H$_2$]-9-oxo-(*Z*)-12-octadecenoic acid ([^{2}H$_2$]-9-KOE). Analogously to the β-oxidation of (*S*)-13-HOD, the (*S*)-9-HOD isomer is transformed into 9-KOE during its metabolism. This compound is subsequently degraded by two complete cycles of β-oxidation to 5-oxo-(*Z*)-8-tetradecenoic acid. After 48 hours of fermentation, the corresponding aldehyde was identified as a labelled intermediate and after 72 hours doubly labelled γ-nonalactone accumulated in the culture broth.

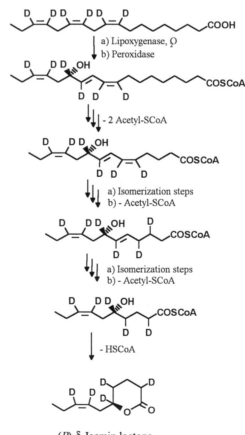

(*R*)-δ-Jasmin lactone

Scheme 3 *Pathway to jasmin lactone from linoleic acid*

The administration of $[12,13-^2H_2]$-9-oxo-(Z)-12-octadecenoic acid to cultures of *S. odorus* resulted in the corresponding di-deuterated intermediates and lactones and confirmed a combined pathway for the degradation of 9-HOD by (I) β-oxidation and (II) Baeyer–Villiger oxidation in the yeast. The results of these labelling experiments will be published in detail.[15]

3.2 Biosynthesis of (R)-γ-Dodecenolactone in *S. odorus*

Labelling experiments with $[^2H_4]$-linoleic acid demonstrated the incorporation of four deuterium atoms into (R)-γ-dodecenolactone. For the introduction of a hydroxyl group at C-10 of the fatty acid, there are three plausible microbial pathways: (I) 10-lipoxygenase–peroxidase; (II) 9,10-epoxygenase–epoxide hydrolyse; and (III) 10-hydratase. The enantiomeric composition of this γ-lactone ((R):(S) = 92:8) indicates different pathways for each enantiomer. The formation of the minor (S)-configured enantiomer was proved by fermentation of the yeast in 10% $H_2^{18}O$. The analysis of the $[4-^{18}O]$-labelled lactone by chiral capillary GC–MS revealed only labelling of (S)-γ-dodecenolactone and (R)-γ-dodecalactone, respectively. The (R)-γ-dodecenolactone was unlabelled and synthesized by the introduction of an (S)-configured hydroxyl group as the initial step. Therefore (S)-10-lipoxygenase–peroxidase or 9,10-epoxygenase–epoxide hydrolase systems can be expected. The β-oxidation of $[9,10,12,13-^2H_4]$-(S)-10-hydr(per)oxy-(E,Z)-8,12-octadecadienoic acid ((S)-10-H(P)OD) and of $[9,10,12,13-^2H_4]$-(9,10)-dihydroxy-(Z)-12-octadecenoic acid result in the same (R)-configured $[^2H_4]$-(Z)-6-γ-dodecenolactone. Discrimination between the two pathways by such labelling experiments is therefore not possible.

Addition of $[9,10-^2H_2]$-oleic acid to growing cultures of *S. odorus* resulted in $[3,4-^2H_2]$-(S)-γ-dodecalactone and $[3,4,6,7-^2H_4]$-(R)-γ-dodecenolactone, respectively (Table 3).

These investigations indicate similar pathways for both γ-C_{12}-lactones. Therefore, a lipoxygenation of the linoleic acid to the unsaturated lactone can be excluded as the initial step.

Table 3 *Yield, Configuration and Labelling Pattern of Two Lactones from Labelled Oleic and Linoleic Acid*

Precursor				Precursor			
$[9,10-^2H_2]$-Oleic Acid				*$[9,10,12,13-^2H_4]$-Linoleic Acid*			
Labelling pattern	*Relative amount (%)*	*Configuration*		*Labelling pattern*	*Relative amount (%)*	*Configuration*	
		(R) (%)	(S) (%)			(R) (%)	(S) (%)
$[^2H_2]$-Labelled	33.4	7.5	92.5	$[^2H_4]$-Labelled	24.7	89.4	10.6
$[^2H_1]$-Labelled	15.8	79.2	20.8	$[^2H_3]$-Labelled	1.8	66.4	33.6
Unlabelled	50.5	32.4	67.6	Unlabelled	73.5	86.2	13.8

4 OLEIC ACID AS PRECURSOR OF γ- AND δ-LACTONES IN *SPOROBOLOMYCES ODORUS*

In a series of labelling experiments, we identified an (R)-hydroxylase–Δ^{12}-desaturase system for oleic acid in *S. odorus*. Both enzyme activities are cytochrome b_5-dependent and hydroxylate and desaturate oleic acid in parallel to (R)-12-hydroxy-(Z)-9-octadecenoic acid and linoleic acid in a correlated reaction. The hydroxylation product, (R)-ricinoleic acid, is subsequently β-oxidized to (R)-γ-decalactone.[2] A proposed biochemical pathway to this compound is presented in Scheme 4.

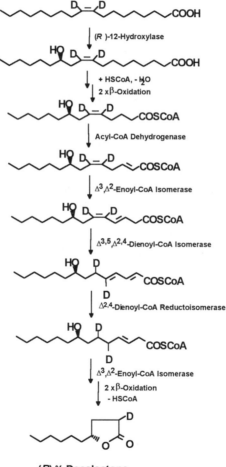

(R)-γ-Decalactone

Scheme 4 *Conversion of oleic acid to (R)-γ-decalactone*

Oleic acid, as well as linoleic acid, can also be epoxidized and degraded to (S)-γ-dodecalactone and (R)-(Z)-6-γ-dodecenolactone, respectively. The three lactones are synthesized in distinct isotopomeric and enantiomeric compositions. The biosynthesis of γ-decalactone was further investigated using appropriately labelled C_{14}- and C_{16}-monoenoic acids. These experiments demonstrate strict substrate-, regio- and enantioselectivities of the key enzymes (R)-12-hydroxylase, Δ^{12}-desaturase, 9,10-epoxygenase, and 13-lipoxygenase), which are involved in the biosynthesis of (R)-γ-decanolactone, (R)-(Z)-6-γ-dodecenolactone, and (R)-δ-decanolactone in *S. odorus* (Scheme 5).

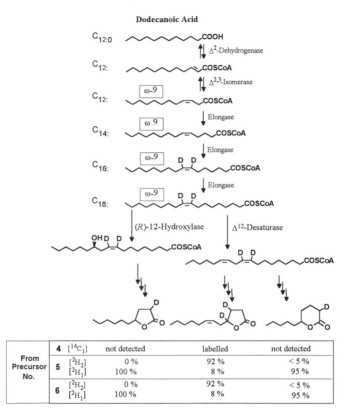

Dodecanoic Acid

4	$[^{14}C_1]$	not detected		labelled	not detected
From Precursor No.	**5**	$[^2H_2]$	0 %	92 %	< 5 %
		$[^2H_1]$	100 %	8 %	95 %
	6	$[^2H_2]$	0 %	92 %	< 5 %
		$[^2H_1]$	100 %	8 %	95 %

Scheme 5 *Production of lactones from dodecanoic acid*

[9,10-^{2}H$_2$]-Oleic acid and [7,8-^{2}H$_2$]-(Z)-7-hexadecenoic acid are comparatively transformed into ring-labelled γ- and δ-lactones. These results are evidence for an elongation system for the C_{16}- to C_{18}-monoenoic acids. No indications were found for a hydroxylation of the (ω-9)-hexadecenoic acid. In addition, the (ω-7)-C_{14} and (ω-)7-C_{16} monoenoic acids, which are known as intermediates of the anaerobic fatty acid biosynthesis in microorganisms and plants, act as lactone precursors.[16,17] The labelling patterns of the ring moieties of the γ- and δ-lactones derived from the (ω-7)-monoenoic acids are different from those obtained from the (ω-9)-acid pathway. An explanation for these unexpected results may be the action of a (Z)-11-octadecenoic acid (vaccenic acid)–oleic acid isomerase, a system identified by Shibahara *et al.* in mango and kaki pulp.[18,19] By the assumption of the existence of a vaccenic acid–oleic acid isomerase system in *S. odorus*, the transformation of double-bond labelled vaccenic acid into [11,12-^{2}H$_2$]-oleic acid as labelled lactone precursor is possible. After addition of these (ω-7)-precursors to growing cells of the yeast, γ-decanolactone and δ-decanolactone were found to be ring-labelled.

REFERENCES

1. A. Shiraishi and H. Fuiji, *Agric. Biol. Chem.*, 1986, **50**, 447.
2. T. Haffner and R. Tressl, *J. Agric. Food Chem.*, 1996, **44**, 1218.
3. W. Albrecht, M. Schwarz, J. Heidlas and R. Tressl, *J. Org. Chem.*, 1992, **57**, 1954.
4. S. Fillippula, R.T. Sormunen, A. Hertig, W.-H. Kunau and K. Hiltunen, *J. Biol. Chem.*, 1995, **270**, 27453.
5. T.E. Smeland, M. Nada, D. Cuebas and H. Schulz, *Proc. Natl. Acad. Sci. U.S.A.*, 1992, **89**, 6673.
6. H.E. Spinnler, C. Gines, J.A. Khan, E.N. Vulfson, *Proc. Natl. Acad. Sci. U.S.A.*, 1996, **39**, 3373.
7. A. Zaks and A.M. Klibanov, *Science*, 1984, **224**, 1249.
8. W. Albrecht, J. Heidlas, M. Schwarz and R. Tressl, 'Flavor Precursors', eds. R. Teranishi, G.R. Takeoka and M. Günthert, American Chemical Society, Washington, D.C., 1992, A.C.S. Symposium Series 490, p. 46.
9. P. Wanner and R. Tressl, in preparation.
10. A. Francke, *Biochem. J.*, 1965, **95**, 633.
11. J.K. Hiltunen, B. Wenzel, A. Beyer, R. Erdmann, A. Fossa and W.-H. Kunau, *J. Biol. Chem.*, 1992, **267**, 6646.
12. T. Haffner, A. Nordsieck and R. Tressl, *Helv. Chim. Acta*, submitted.
13. W. Albrecht and R. Tressl, *Z. Naturforsch.*, 1990, **45c**, 207.
14. M. Schöttler and W. Boland, *Helv. Chim. Acta*, 1995, **78**, 847.
15. T. Haffner, H. Lange and R. Tressl, in preparation.
16. H. Hollberg and S.J. Wakil, *J. Biol. Chem.*, 1964, **230**, 2489.
17. J.M. Schwab, A. Habib and J.B. Klassen, *J. Am. Chem. Soc.*, 1986, **108**, 5304.
18. A. Shibahara, K. Yamamoto, M. Takeoka, A. Kinoshita, G. Kajimoto, T. Nakayama and M. Noda, *Lipids*, 1989, **24**, 488.
19. A. Shibahara and K. Yamamoto, *FEBS Lett.*, 1990, **264**, 228.

THE EFFECT OF PROCESSING ON CHIRAL AROMA COMPOUNDS IN FRUITS AND ESSENTIAL OILS

B.D. Baigrie,* M.G. Chisholm† and D.S. Mottram§

*Reading Scientific Services Limited, Lord Zuckerman Research Centre, The University of Reading, Whiteknights, Reading, RG6 6LA, U.K.

†The Pennsylvania State University, Division of Science, Behrend College, Erie, PA 16563, U.S.A.

§The University of Reading, Department of Food Science, and Technology, Whiteknights, Reading, RG6 6AP, U.K.

1 INTRODUCTION

The determination of the enantiomeric distribution of chiral compounds found in fruits and essential oils has become increasingly important in determining the origin of natural and processed foodstuffs. The method depends in part on a knowledge of the enantiomeric distributions found in nature (and their consistency) and also the changes that may occur with the extraction and processing methods used.

The enantiomeric distribution of linalool in nature demonstrates the variation which may be found for chiral compounds. In bergamot oil[1] and basil herb,[2] linalool occurs as the almost pure R-(–) isomer; in lavender flowers it occurs as 95%–98% of the R-(–) isomer[2,3] and in bitter orange oil it occurs as 79%–87% of the R-(–) isomer.[4] However in sweet orange oil the S-(+) isomer predominates, occurring as 94%–96% of the total,[5,6] and in coriander oil 86%–87% is found as the S-(+) isomer.[2,5] In passion fruit and apricot, linalool occurs as the racemate.[5]

Another group of compounds that may occur naturally in the racemic form are the γ- and δ-lactones. Their enantiomeric distribution in many fruits has been extensively studied[7–9] and summarized by Casabianca et al.[10] The enantiomeric distributions of γ- and δ-lactones vary among different fruits and with the origin for the same fruit. It has been noted that despite its relatively high abundance in many fruits, γ-hexalactone is useless for assessing the origin of flavours because of the wide range of enantiomeric distributions found.[11]

Racemates also occur naturally when fermentation takes place in the production of the foodstuff. Racemic linalool and α-terpineol are both found in bergamot tea,[12] and in wine, α-terpineol has been found in the racemic form.[13] This is the result of fermentation which occurs in the production of both tea and wine.

The R-(+) form of α-terpineol occurs frequently in essential oils,[14,15] but it rarely occurs in the enantiomerically pure form. Naturally occurring racemates have been reported in geranium oil[14,16] and yellow passion fruit.[6] Since α-terpineol lies further along the biosynthetic pathway than linalool, and it is also an artefact whose abundance depends upon the extraction and processing methods used, a wide range of enantiomeric distributions would be expected.

Racemic compounds may occur in essential oils and foods as a result of the extraction and processing methods used in their isolation or production. Partial racemisation may also occur depending on the procedures used for chiral analysis. It is well known that linalool and α-terpineol are very labile in acidic media,[17,18] so it is important to know the pH at which the oil was isolated, before making an assessment of its origin. Varying tendencies towards the racemisation of limonene, linalool and α-terpineol have been reported for lavender oils, depending upon the method of extraction, and the pH of steam distillation.[15] In contrast, model studies on linalool under simultaneous distillation/extraction conditions at pH 7 showed no racemisation.[5] Kreis *et al.* found up to 8% racemisation of linalool isolated from *Flores Lavandulae* by hydrodistillation at pH 5 over differing time periods.[19] Solvent extraction produced negligible racemisation. These results, together with the variation in enantiomeric distribution of key monoterpenes in different fruits, which could be used to determine the origin of essential oils and foods, provide a complex picture when using chirospecific analysis as a tool to determine authenticity.

A preliminary investigation of changes in the enantiomeric distribution of limonene, linalool and α-terpineol in cherries caused by canning, showed significant shifts towards racemisation for all three compounds.[20] To investigate the effect of processing conditions on the racemisation of chiral aroma compounds further, a simpler investigation was undertaken, which examined the effect of temperature and pH used in the extraction of some selected citrus oils from different regions. Pinene (α- and β-), limonene, linalool and α-terpineol were the chiral aroma compounds selected for analysis.

2 MATERIALS AND METHODS

2.1 Samples

Bergamot (*Citrus aurantium* (L.) Bergamia), lemon (*Citrus limon*, (L.) Burman) and Persian lime (*Citrus latiofolia*, Tanaka) were obtained as tree fruit from several regions of the world. Commercial oils from the same fruits were supplied by two flavour houses in the U.K. either in the expressed or in the distilled forms.

2.2 Sample Preparation

2.2.1 *Hydrodistillation.* Samples of the expressed commercial oils were hydrodistilled for an hour at pH 2 and 6. The pH of 2 was achieved by carrying out the distillation in a 5% solution of citric acid.

2.2.2 *Extraction of Fruit.* The skin of each fruit was scraped with a zesting tool to remove a thin layer of the oil-rich material. Care was taken to keep the skin separated from the pulp and juice by ensuring that the fruit was not broken. The zest from an individual fruit was stirred with 30 ml of a 1:1 mixture of pentane–ether for 30 min. The solvent was filtered from the peel residue and diluted by a factor of 10 with pentane for chiral analysis.

2.2.3 *Simulation of Extraction in Contact with Juice.* Authentic bergamot oil (1 ml) was stirred with a 5% citric acid solution for 72 hours at room temperature. Samples were removed at regular intervals and prepared for chiral analysis.

2.3 Chiral Analysis using Multidimensional Gas Chromatography (MDGC)

MDGC was carried out on a dual oven linked system comprised of a Carlo Erba 5160 Mega Series gas chromatograph (GC 1) containing the pre-column and a Carlo Erba Fractovap 4200 series GC (2) containing the analytical column. Heart cutting was achieved

Table 1 The Effect of Distillation on the Enantiomeric Distribution of Chiral Aroma Compounds. All Samples are Commercial. The Values are the Average of at least Two Runs

	α-Pinene			β-Pinene			Limonene			Linalool			α-Terpineol		
	S(−)	R(+)	wt %[a]	R(+)	S(−)	wt %	S(−)	R(+)	wt %	R(−)	S(+)	wt %	S(−)	R(+)	wt %
Lemon															
Distilled	77.3	22.7		2.9	97.1		7.5	92.5		47.9	52.2		57.7	42.3	
Cold expressed	66.3	33.7	3.4	7.6	92.4	11.6	1.3	98.7	66.3	32.8	67.2	0.22	60.7	39.3	0.33
Hydrodist. pH 6	66.2	33.9	3.8	7.4	92.6	16.7	1.3	98.7	50.1	34.3	65.7	0.44	60.5	39.5	0.69
Hydrodist. pH 2	66.0	34.0	5.2	7.6	92.4	18.9	1.4	98.6	51.0	40.0	60.0	0.43	45.1	54.9	0.70
Lime															
Cold expressed	69.1	30.9	3.7	8.8	91.2	15.8	7.9	92.1	38.5	63.0	37.0	0.31	74.6	25.4	0.56
Hydrodist. pH 6	68.8	31.2	3.3	8.8	91.2	15.1	8.0	92.0	38.1	62.8	37.0	0.35	70.5	29.5	0.63
Hydrodist. pH 2	65.6	34.4	4.6	9.4	90.6	14.8	8.8	91.2	38.7	54.1	45.9	0.90	62.0	38.0	1.19
Bergamot															
Cold pressed	70.4	29.6	1.5	8.5	91.5	6.8	1.9	98.1	31.4	77.7	23.3	16.7	38.0	62.0	0.31
Hydrodist. pH 6	69.0	31.1	1.0	8.8	91.2	5.7	1.8	98.2	25.2	73.3	26.7	22.5	35.1	64.9	1.54
Hydrodist. pH 2	69.1	30.9	1.3	9.4	90.6	6.9	1.8	98.2	27.2	63.7	36.3	16.0	49.6	50.4	6.74

[a] Quantity of compound (both enantiomers) expressed as % total as measured by GC.

by using a **MU**ltiple **S**witching **I**ntelligent **C**ontroller (MUSIC) system (Chrompack U.K. Ltd.) which is based on the Deans pressure switching concept.[21] The heart cut sample was trapped and cooled by liquid nitrogen. Injector temperature: 200 °C; detector temperature: 250 °C (both); injection mode: on-column; sample size 2 µl; solution concentration: 0.2% in pentane.

2.3.1 Precolumn. Stabilwax-DA chemically bonded 30 m × 0.53 mm internal diameter, 1.0 µm film thickness (Restek Corp). Carrier gas: He 10 ml min^{-1} flow rate; temperature programme: 30 °C for 30 s, 30 °C to 60 °C at 40 °C min^{-1}, 60 °C to 200 °C at 3 °C min^{-1}.

2.3.2 Analytical Column. Chirasil Dex CB chemically bonded (heptakis-(2,3,6-tri-O-methyl)-β-cyclodextrin in 10% OV 1701) 25 m × 0.25 mm, 0.25 µm film thickness (Chrompack, U.K. Ltd.). Carrier gas: 0.88 bar He; temperature programme: α- and β-pinene: 40 °C for 15 min then 4 °C min^{-1}; limonene and linalool: 70 °C for 15 min then 4 °C min^{-1}; α-terpineol: 120 °C for 15 min then 4 °C min^{-1}.

2.3.3 Identification of Enantiomers. The order of elution for each compound was assigned by comparison to standards of known optical purity. The order of elution was found to be: α-pinene: *S*-(–), *R*-(+); β-pinene: *R*-(+), *S*-(–); limonene: *S*-(–), *R*-(+); linalool: *R*-(–), *S*-(+); α-terpineol: *S*-(–), *R*-(+).

3 RESULTS AND DISCUSSION

The monoterpene hydrocarbons underwent little change in their enantiomer distribution upon hydrodistillation in neutral or acid solution, as shown in Table 1, in agreement with previous work.[5,15] Linalool showed only a slight change in neutral solution, but significant racemisation occurred at pH 2. Kreis *et al.* found a 5% change for lavender oil under similar conditions[19] and Weinreich reported a 2%–15% change.[15] α-Terpineol is the least stable of these monoterpenes in acid solution,[22] and underwent the most racemisation upon hydrodistillation at pH 2. Since it is formed as an artefact in citrus oils upon heating at pH 2, then the higher level of racemisation is not unexpected.

The racemisation of linalool which occurred while bergamot oil was in contact with cold 5% citric acid, shown in Table 2, suggests that if good manufacturing procedures permit the extraction of citrus oils from the whole fruit, then in the case of bergamot oil, linalool may not be present in the pure *R*-(–) form. Under hydrodistillation conditions, most of the oil is removed from the vicinity of the acid in the first few minutes, so prolonged distillation under these conditions will not increase the amount of racemisation found in Table 1.

The variation in enantiomeric distribution that occurs with fruits from different regions are shown in Table 3. These results point to some of the difficulties in using known values for determining the origin and processing methods used in the production of citrus oils. The range of values for linalool in lemon oil, where the enantiomeric excess ranges from 0% to

Table 2 *The Effect of Acid Contact at pH 2 at 25 °C on the Enantiomeric Distribution of Linalool in Bergamot Oil*

Enantiomer	Cold pressed	2 hours	4 hours	24 hours	72 hours
R (–)	100.0	92.2	93.0	79.9	53.0
S (+)	0.0	7.8	7.0	20.1	47.0

Table 3 *Enantiomeric Distribution of Chiral Aroma Compounds for Authentic Citrus Oils*

	α-Pinene		β-Pinene		Limonene		Linalool		α-Terpineol	
	S(−)	R(+)	R(+)	S(−)	S(−)	R(+)	R(−)	S(+)	S(−)	R(+)
Lemon										
Cyprus	60.1	39.9	7.7	92.3	1.2	98.9	39.7	60.3	71.2	28.8
Israel	72.2	27.9	6.2	93.8	1.4	98.6	45.0	55.0	80.0	20.0
U.S.A. 1	66.4	33.6	6.7	93.3	1.3	98.7	50.8	49.2	78.8	21.2
U.S.A. 2	73.3	26.7	4.7	95.3	1.6	98.4	72.0	28.0	88.8	11.2
Lime										
Cuba	70.1	29.9	10.0	90.0	2.7	97.3	62.1	37.7	79.8	20.2
Mexico	68.6	31.4	8.9	91.1	2.2	97.8	64.6	35.4	77.2	22.8
Bergamot										
U.K. unripe	71.2	28.8	8.2	91.8	2.3	97.7	98.3	1.7	61.3	38.7
U.K. ripe	68.2	31.7	9.4	90.6	1.7	98.3	98.7	1.3	37.4	62.6

Table 4 *Enantiomeric Distributions of Chiral Compounds in Commercial Citrus Oils. All oils were cold pressed except those marked*[*]

	α-Pinene		β-Pinene		Limonene		Linalool		α-Terpineol	
	S(−)	R(+)	R(+)	S(−)	S(−)	R(+)	R(−)	S(+)	S(−)	R(+)
Lemon										
Sicily	69.2	30.8	6.1	93.4	1.5	98.5	61.0	39.0	80.4	19.6
California	70.2	29.8	5.1	94.9	1.4	98.6	58.9	41.1	77.6	22.4
Spain	69.8	30.2	6.1	93.9	1.6	98.4	48.3	51.7	78.5	21.5
Argentina	69.4	30.6	5.1	94.9	1.3	98.7	46.8	53.2	76.8	23.2
Unknown 1	66.3	33.7	7.5	92.5	1.3	98.7	51.8	48.2	77.2	22.8
Unknown 2	79.7	20.3	4.9	95.1	1.2	98.8	32.8	67.2	60.7	39.3
Distilled[*]	77.3	22.7	2.9	97.1	7.8	92.2	47.9	52.1	57.7	42.3
Lime										
Unknown	69.1	30.9	8.8	91.2	7.9	92.1	63.0	37.0	74.6	25.4
Distilled[*]	70.9	29.1	3.5	96.5	7.4	92.6	47.7	52.3	51.5	48.5
Bergamot										
Italy 1	68.0	32.0	8.6	91.4	1.6	98.4	70.7	29.3	23.9	76.1
Italy 2	69.2	30.8	8.5	91.5	4.9	95.1	100	0.0	51.4	48.6
Italy 3	69.8	30.2	8.5	91.5	1.9	98.1	77.7	23.3	38.0	62.0
Sicily	68.8	31.2	8.3	91.7	1.8	98.2	68.6	31.4	23.3	76.7
Argentina	63.0	37.0	8.7	91.3	1.3	98.7	100	0.0	64.0	36.0
'Compound'[*]	70.7	29.3	5.2	94.8	1.3	98.7	64.2	35.8	36.5	63.5

45%, makes linalool a poor choice as an indicator compound for lemon oil production. The range of values found for α-terpineol show a narrower range for lemon and lime oils than those of linalool, but the values for bergamot oil show no pattern with both the R-(+) and the S-(–) isomer being present in excess, which is further supported by the data for the commercial oils shown in Table 4. α-Terpineol is clearly not a good indicator compound to determine the origin of an oil, particularly for bergamot oil. However, linalool and α-terpineol are minor components in lemon and lime oils, although they are major odourants.

The enantiomeric distribution of limonene shows good consistency for all the authentic citrus oils examined, and agrees with reported values, and lies within the range shown by a wide range of citrus oils.[23,24] The commercial samples shown in Table 4, which contain more than 3% of S-(–) limonene (both lime, the distilled lemon and the Italian No. 2 bergamot samples), clearly lie outside the expected range for citrus oils and suggest an anomaly in their production.

The one good indicator compound that emerges from these data is linalool in bergamot oil. If the oil is cold pressed, and has been isolated without juice contact, then close to 100% of the R-(–) isomer should be present.[1] However, the commercial bergamot oil samples in Table 4 show an additional dilemma. The Italian No. 2 sample has acceptable values for linalool, but those of both limonene and α-terpineol lie outside the expected ranges for a cold pressed citrus oil isolated without juice contact. The samples from Sicily and Italy No. 1 show a high percentage of S-(–)-linalool for a cold pressed oil, but if good manufacturing processes permit juice contact, then this value may be acceptable. The sample from Argentina appears to be an authentic cold pressed bergamot oil, on the basis of the enantiomeric distribution data.

4 CONCLUSION

The data in Tables 1–4 show that it would be almost impossible to use chiral analysis alone, particularly on a single compound, to specify the origin of citrus oils. The range of values found for authentic citrus oil samples together with the loose definition of 'good manufacturing procedures' make the task of establishing acceptable values for enantiomeric distributions very difficult. The best way to assess the origin of a natural product would seem to be to adopt a more comprehensive approach and use chiral analysis in conjunction with quantitative analysis of chiral indicator compounds and other instrumental methods such as isotope ratio mass spectrometry or site-specific natural isotope fractionation nuclear magnetic resonance spectroscopy (SNIF–NMR). In all cases a comprehensive data base of acceptable values is needed to effectively establish the origin of a natural product.

REFERENCES

1. A. Crotoneo, I. Stagno d'Alcontres and A. Trozzi, *Flavour Fragr. J.*, 1992, **7**, 15.
2. V. Schubert and A. Mosandl, *Phytochem. Anal.*, 1991, **2**, 171.
3. P. Kreis and A. Mosandl, *Flavour Fragr. J.*, 1992, **7**, 187.
4. P. Dugo, L. Mondello, E. Cogliandro, A. Verzera and G. Dugo, *J. Agric. Food Chem.*, 1996, **44**, 544.
5. A. Bernreuther and P. Schreier, *Phytochem. Anal.*, 1991, **2**, 167.
6. P. Werkoff, S. Brennecke, W. Bretschneider, M. Güntert, R. Hopp and H. Surburg, *Z. Lebensm. Unters. Forsch.*, 1993, **196**, 307.
7. A. Bernreuther, N. Christoph and P. Schreier, *J. Chromatogr.*, 1989, **481**, 363.

8. S. Nitz, H. Kollmansberger and F. Drawert, *Chem. Mikrobiol. Technol. Lebensm.*, 1989, **12**, 75.
9. E. Guichard, A. Kustermann and A. Mosandl, *J. Chromatogr.*, 1990, **498**, 396.
10. H. Casabianca, J.-B. Graff, P. Jame and C. Perrucchietti, *J. High Resol. Chromatogr.*, 1995, **18**, 279.
11. A. Mosandl, *J. Chromatogr.*, 1992, **624**, 267.
12. H. Casabianca and J.-B. Graff, *HRC. J. High Resolut. Chromatogr*, 1994, **17**, 184.
13. C. Askari, U. Hener, H. Schmarr, A. Rapp and A. Mosandl, *Fresenius J. Anal. Chem.*, 1991, **340**, 768.
14. U. Ravid, E. Putievsky and I. Katzir, *Flavour Fragr. J.*, 1995, **10**, 281.
15. B. Weinreich and S. Nitz, *Chem. Mikrobiol. Technol. Lebensm.*, 1992, **14**, 117.
16. P. Kreis and A. Mosandl, *Flavour Fragr. J.*, 1993, **8**, 161.
17. G. Schmaus and K.H. Kubeczka, in 'Essential Oils and Aromatic Plants', eds. A. Baerheim-Svendsen and J.J. Scheffer, Nijhoff/Junk Publ., Dordrecht, Netherlands, 1985, p. 127.
18. T.S. Chamblee, B.C. Clarke, Jr., G.B. Brewster, T. Radford and G.A. Iacobucci, *J. Agric. Food Chem.*, 1991, **39**, 162.
19. P. Kreis, R. Braundorf, A. Dietrich, U. Hener, B. Maas and A. Mosandl, in 'Progress in Flavour Precursor Studies', eds. P. Schreier and P. Winterhalter, Allured Publishing Co., Carol Stream, IL, 1993, p. 77.
20. B.D. Baigrie, D.S. Mottram and K.J. Pierce, in 'Chemical Markers for Processed and Stored Foods', eds. T.C. Lee and H.J. Kim, American Chemical Society, Washington, D.C., 1996, A.C.S. Symposium Series 631, in press.
21. D.R. Deans, *Chromatographia*, 1968, **5–6**, 187.
22. B.C. Clark, Jr. and T.S. Chamblee, in 'Off-flavors in Foods and Beverages' ed. G. Charalambous, Elsevier Science Publishers, Amsterdam, 1992, p. 229.
23. A. Mosandl, U. Hener, P. Kreis and H.-G. Schmarr, *Flavour Fragr. J.*, 1990, **5**, 193.
24. U. Hener, A. Hollnagel, P. Kreis, B. Maas, H.-G. Schmarr, V. Schubert, K. Rettinger, B. Weber and A. Mosandl, in 'Flavour Science and Technology', eds. Y. Bessière and A.F. Thomas, John Wiley and Sons, Chichester, 1990, p. 25.

THE USE OF CHIRAL ANALYSIS FOR THE CHARACTERIZATION OF CERTAIN AROMATIC AND MEDICINAL PLANTS

F. Tateo,* L.F. Di Cesare,† M. Bononi,* S. Cunial* and F. Trambaiolo*

*Dipartimento di Fisiologia delle Piante Coltivate e Chimica Agraria, Università di Milano, Via Celoria 2, 20133, Milano, Italy

†Istituto per la Valorizzazione dei Prodotti di Trasformazione Agricola, Via Venezian 26, 20133, Milano, Italy

1 INTRODUCTION

Teucrium chamaedrys (L.) is a Labiatae found all over Italy and in other Mediterranean countries. This plant is used in the preparation of hydroalcoholic extracts of aromatic herbs employed in the production of many alcoholic and non-alcoholic beverages in everyday use. It is a medicinal plant chosen for its slightly bitter flavour, though the aromatic note of its more volatile fraction is also significant.

Analytical characterization of *Teucrium chamaedrys* based on modern techniques involving GC equipment is limited to the contributions of Chialva.[1] The bibliography does not, however, contain any information on the enantiomeric distribution of the chiral compounds contained in the plant in question. Further analytical research in GC-MS and chiral GC has now become necessary so as to have more complete analytical data available for identification of the use of *Teucrium chamaedrys* in fragrancy additives and in drinks available on the market. At the same time, in a series of studies of which this is the first, it is proposed to investigate the influence of a number of extraction and concentration techniques on the quantitative analysis of the volatile fraction of various aromatic and medicinal plants.

2 MATERIAL AND METHODS

2.1 Extraction of Volatile Compounds

A Likens-Nickerson extractor-concentrator was used to extract about 30 g of dried herb with 30 ml of hexane as the extractor solvent. The extract obtained was analysed as indicated in section 2.3, both immediately on production and after two months. In one case, these extracts were used to produce a concentrate by low-temperature evaporation of the solvent, equal to 15 ml in terms of quantity. In a second case, a concentrate was produced by employing the method already used for other applications.[2] Here, the hexane extract produced with the Likens-Nickerson equipment, was percolated on a Sep-Pak Silica (Millipore) microcolumn, and eluted with 3×1 ml fractions of 96% ethanol, only the second fraction being taken for analysis. The two concentrates were analysed in accordance with the procedures described in sections 2.2 and 2.3. For the evaporated sample, the analyses were carried out both immediately and two months after concentration.

2.2 HRGC–MS Analysis

The analyses were performed on an SPB-5 silica capillary column (30 m × 0.32 mm internal diameter, 0.25 μm film). The oven temperature was programmed as follows: 80 °C for 10 min, then 1°C min^{-1} to 120 °C, 120°C for 10 min, then 3 °C min^{-1} to 230°C. The GC-MS analyses were carried out on an HP 5971 A MSD, equipped with an HP 5890 Series II GC unit, under conditions analogous to those given for the HRGC analyses.

2.3 Enantioselective GC Analysis

HRGC chiral analyses were performed on a silica capillary column β-Dex™ (30 m × 0.25 mm internal diameter, 0.25 μm film). The oven temperature was programmed as follows: 60 °C for 10 min, then 1.5 °C min^{-1} to 120 °C, followed by 3 °C min^{-1} to 220 °C.

3 RESULTS

Table 1 contains the composition data for the evaporated extracts and shows the differences in composition attributable to two different concentration methods. Table 2 contains the enantiomeric distribution values.

Table 1 *Essential Oil Composition of* Teucrium chamaedrys *(L.) Extracted by Likens-Nickerson, Concentrated by Solvent Evaporation at Low Temperature (ca) and on Sep-Pak™ silica Cartridge (cb)*

	Component	% (ca)	% (cb)
1.	α-Pinene	3.8	0.7
2.	Camphene	0.2	0.5
3.	1-Octen-3-ol	2.2	4.0
4.	β-Pinene	1.7	0.5
5.	Limonene	1.4	0.6
6.	Phenylacetaldehyde	1.5	3.1
7.	Linalool	3.0	6.0
8.	α-Thujone	0.3	0.5
9.	4-Terpineol	0.1	0.2
10.	α-Terpineol	0.6	1.1
11.	Thymol	0.4	1.5
12.	Carvacrol	2.2	4.7
13.	[Not identified] (MW 150)	1.7	3.4
14.	Eugenol	1.8	3.0
15.	β-Caryophyllene	21.0	9.8
16.	α-Humulene	4.1	3.1
17.	Germacrene-D	11.5	5.7
18.	Caryophyllene oxide	13.0	24.3

Table 3, which refers to chiral analysis, compares the Likens-Nickerson extract concentrated by evaporation (3a) with the same extract analysed after two months (3b). The same table compares the plot relating to the Likens-Nickerson extract concentrated with the Sep-Pak (3c) and gives the enantiomeric distribution of all three samples.

Table 2	*Enantiomeric distribution of some monoterpenoid components of* Teucrium chamaedrys *(L.) extracted by Likens-Nickerson and analysed without concentration*

Table 2a	*Samples analysed immediately after extraction*

Compound	Enantiomers		Enantiomeric excess
1. α-Pinene	S(–) 93.1%	R(+) 6.9%	ee S(–) 86.2
3. β-Pinene	R(+) 99.9%	S(–) 0.1%	ee R(+) 99.8
4. Limonene	S(–) 39.6%	R(+) 60.4%	ee R(+) 20.8
5. 1-Octen-3-ol	S(+) 0.1%	R(–) 99.9%	ee R(–) 99.8
6. Linalool	R(–) 53.6%	S(+) 46.4%	ee R(–) 7.2
7. 4-Terpinenol	R(+) 66.8%	S(–) 33.2%	ee R(+) 33.6
8. α-Terpineol	R(+) 51.4%	S(–) 48.6%	ee R(+) 2.8

Table 2b	*Samples Analysed 2 Months after Extraction*

Compound	Enantiomers		Enantiomeric excess
1. α-Pinene	S(–) 93.0 %	R(+) 7.0%	ee S(–) 86.0
3. β-Pinene	R(+) 91.4%	S(–) 8.6%	ee R(+) 82.8
4. Limonene	S(–) 30.3%	R(+) 69.7%	ee R(+) 39.4
5. 1-Octen-3-ol	S(+) 0.1%	R(–) 99.9%	ee R(–) 99.8
6. Linalool	R(–) 52.9%	S(+) 47.1%	ee R(–) 5.8
7. 4-Terpinenol	R(+) 68.1%	S(–) 31.9%	ee R(+) 36.2
8. α-Terpineol	R(+) 48.8%	S(–) 51.2%	ee S(–) 2.4

4 DISCUSSION AND CONCLUSIONS

The first point to be made concerns the difference in composition resulting from the use of two different concentration methods on the Likens-Nickerson extract. This is shown clearly in Table 1.

As regards the identification of certain components of *Teucrium chamaedrys* not given in the references quoted in the introduction, it is worth considering the isomer of carvacrol and thymol, with a mass spectrum that can nevertheless be characterized (compound 13 in Table 1). Nor is the content of this compound negligible compared with that of its two known isomers. The occurrence of other isomers of thymol and carvacrol (with MW 150) in different natural herbs, such as *Thymus vulgaris* (L.) and *Origanum vulgare* (L.), has already been noted elsewhere.[3,4]

The other compounds identified, on which there was no previous information, are phenylacetaldehyde, eugenol, and germacrene-D.

Some of the enantiomeric distribution data are fairly typical. In particular, 1-octen-3-ol is, in practice, present only in the form of R(–), while the enantiomeric distribution of limonene seems to depend on whether the extract concentrated is examined immediately or two months after concentration. It is therefore sufficient to consider the data in Table 3. The other compounds considered from the point of view of enantiomeric distribution do not appear to undergo analogous changes. The substantial prevalence of the form S(–) for α-pinene, of the form R(+) for β-pinene, and the very low value for the enantiomeric excess of the form R(−) for linalool may be considered fairly typical elements, besides what has already been said about 1-octen-3-ol.

Table 3 *Enantiomeric Distribution of Some Monoterpenoid Components of* Teucrium chamaedrys *(L.) Extracted by Likens-Nickerson and Concentrated in Different Ways*

Table 3a *Low Temperature Evaporation; Immediate Analysis*

Compound	Enantiomers		Enantiomeric excess
1. α-Pinene	S(−) 95.1 %	R(+) 4.1%	ee S(−) 90.2
3. β-Pinene	R(+) 99.9%	S(−) 0.1%	ee R(+) 99.8
4. Limonene	S(−) 43.3%	R(+) 56.7%	ee R(+) 13.4
6. 1-Octen-3-ol	S(+) 0.1%	R(−) 99.9%	ee R(−) 99.8
8. Linalool	R(−) 53.3%	S(+) 46.7%	ee R(−) 6.6
9. 4-Terpinenol	R(+) 54.0%	S(−) 46.0%	ee R(+) 8.0
10. α-Terpineol	R(+) 50.1%	S(−) 49.9%	ee R(+) 0.2

Table 3b *Low Temperature Evaporation; Analysis after 2 Months*

Compound	Enantiomers		Enantiomeric excess
1. α-Pinene	S(−) 95.7 %	R(+) 4.3 %	ee S(−) 91.4
3. β-Pinene	R(+) 94.9%	S(−) 5.1%	ee R(+) 89.8
4. Limonene	S(−) 11.6%	R(+) 88.4%	ee R(+) 76.8
6. 1-Octen-3-ol	S(+) 0.1%	R(−) 99.9%	ee R(−) 99.8
8. Linalool	R(−) 52.2 %	S(+) 47.8%	ee R(−) 4.4
9. 4-Terpinenol	R(+) 66.7%	S(−) 33.3%	ee R(+) 33.4
10. α-Terpineol	R(+) n.c.d.	S(−) n.c.d	ee n.c.d

Table 3c *Sep-Pak Concentration; Analysis after 2 Months*

Compound	Enantiomers		Enantiomeric excess
1. α-Pinene	S(−) 95.7 %	R(+) 4.3 %	ee S(−) 91.4
3. β-Pinene	R(+) 95.8%	S(−) 4.2%	ee R(+) 91.6
4. Limonene	S(−) 13.5%	R(+) 86.5%	ee R(+) 73.0
6. 1-Octen-3-ol	S(+) 0.1%	R(−) 99.9%	ee R(−) 99.8
8. Linalool	R(−) 51.6%	S(+) 48.4%	ee R(−) 3.2
9. 4-Terpinenol	R(+) n.c.d	S(−) n.c.d	ee n.c.d
10. α-Terpineol	R(+) n.c.d	S(−) n.c.d	ee n.c.d

n.c.d.: not correctly detectable

ACKNOWLEDGEMENTS

We wish to thank F. Morelli for his collaboration.

REFERENCES

1. F. Chialva, G. Gabri, P.A.P. Liddle and F. Ulian, *Journal of HRC & CC*, 1982, **5**, 182.
2. F. Tateo, G. Cantele, M. Bononi, 'Sulla Determinazione Quantitativa dell' Etilvanillina, Sostanza Aromatizzante Artificiale ad Impiego Limitato nei Prodotti Dolciari'. Proceedings of XII° Congresso Nazionale di Chimica Analitica, Firenze, ed. Università degli Studi di Firenze, 1995.
3. F. Tateo, A. Ferrillo, G. Salvatore, in Proceedings of '11èmes Journées Internationales Huiles Essentielles', Digne-Les-Bains, *Rivista Italiana Heppos*, 1993 (February), 222.
4. F. Tateo, G. Salvatore, M. Nicoletti, 'Presenza di Diisopropilcresoli in Olii Essenziali di Timo', Proceedings of '6° Convegno Nazionale della Società Italiana di Fitochimica', Fiuggi, 1992, ed. Università d egli Studi di Roma 'La Sapienza'.

WINE LACTONE – A POTENT ODORANT IDENTIFIED FOR THE FIRST TIME IN DIFFERENT WINE VARIETIES

H. Guth

Deutsche Forschungsanstalt für Lebensmittelchemie, Lichtenbergstrasse 4, 85748 Garching, Germany

1 INTRODUCTION

Two white wine varieties (Gewürztraminer and Scheurebe) which differ strongly in flavour profiles were selected for the evaluation of the flavour differences by aroma extract dilution analysis (AEDA) and static headspace analysis (SHA).[1]

AEDA and SHA yielded 41 and 45 odour active compounds for Scheurebe and Gewürztraminer wines, respectively.[1] The most potent odorants, with flavour dilution factors (FD factor) greater than or equal to ten, are listed in Table 1. Unknown compound (**1**), ethyl isobutyrate, ethyl 2-methylbutyrate, 3-methylbutanol, 2-phenylethanol, 3-ethylphenol and 3-hydroxy-4,5-dimethyl-2(5*H*)-furanone showed high FD factors in both varieties. 4-Mercapto-4-methylpentan-2-one was detected only in the variety Scheurebe whereas *cis*-rose oxide was perceived only in Gewürztraminer (Table 1).

Among the perceived odorants there was one unknown compound which belongs to the most important flavour compounds in both varieties. This trace component in wine with coconut, woody and sweet odour was named 'wine lactone' (**1**). The details of the identification experiments, syntheses of stereoisomers, assignment of stereochemistry, analytical properties, and evaluation of odour threshold of the isomers are described in the present paper.

2 EXPERIMENTAL

Enrichment of wine lactone (**1**): Gewürztraminer wine (10 l) was extracted with pentane (2 l), the solvent extract was concentrated (1 ml), and purified by column chromatography on silica gel with pentane/diethylether (70+30, v/v) and, in addition, by HPLC (hypersil silica gel 60, 5μm, 500 × 4.6 mm; isocratic elution with pentane/diethylether (70+30, v/v); flow rate 2 ml min^{-1}; UV detection at 215 nm) and preparative GC on a stainless steel column (3 m × 2 mm) packed with FFAP (10%,w/w); with the following temperature program: after 1 min at 80 °C, the temperature was raised by 8 °C min^{-1} to 250°C and then kept at 250 °C for 10 min.

Syntheses of the reference compounds, and MS and NMR analyses were described in a recent publication.[2] Odour thresholds were determined by a gas chromatographic olfactometric method[3] using 2(*E*)-decenal (odour threshold 2.7 ng l^{-1} air) as the standard.

Table 1   *Results of Aroma Extract Dilution Analysis of Gewürztraminer and Scheurebe Wines*

Compound	FD-Factor	
	Scheurebe	*Gewürztraminer*
Unknown (**1**)	1000	1000
Ethyl isobutyrate	100	10
Ethyl butyrate	10	10
Ethyl 2-methylbutyrate	100	100
Ethyl 3-methylbutyrate	10	10
2-Methylpropanol	10	10
3-Methylbutanol	100	100
Ethyl hexanoate	10	10
cis-Rose oxide	<1	10
4-Mercapto-4-methylpentan-2-one	10	<1
Ethyl octanoate	10	10
Acetic acid	10	10
Linalol	10	10
Butanoic acid	10	10
2-/3-Methylbutanoic acid	10	10
2-Phenylethanol	100	100
5-Ethyl-4-hydroxy-2-methyl-3(2*H*)-furanone	10	10
Ethyl *trans*-cinnamate	10	10
3-Ethylphenol	100	100
3-Hydroxy-4, 5-dimethyl-2(5*H*)-furanone	100	100

3 RESULTS AND DISCUSSION

The unknown compound (wine lactone) with coconut, woody and sweet odour was identified by MS analysis after enrichment from 10 l Gewürztraminer wine (as described above) as 3a,4,5,7a-tetrahydro-3,6-dimethyl-2(3*H*)-benzofuranone (**1**) (see Figure 1).

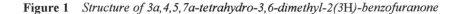

Figure 1   *Structure of 3a,4,5,7a-tetrahydro-3,6-dimethyl-2(3*H*)-benzofuranone*

Comparison of MS and chromatographic data of the isolated wine lactone with those of the synthesised reference compound (Figure 2) confirmed the proposed structure of compound (**1**).

To our knowledge, this odorant has not yet been detected in wine or food. Southwell,[4] who investigated the essential oil metabolism of koala animals after feeding on the leaf of *Eucalyptus punctata*, identified the terpene lactone tentatively by ^{1}H-NMR in the excreted urine, but without assignment of stereochemistry.

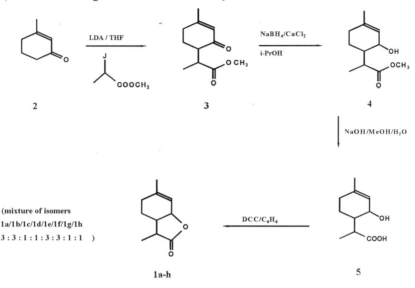

Figure 2 *Syntheses of lactones 1a/1b, 1c/1d, 1e/1f and 1g/1h*

Because of the three asymmetric centres in the molecule, there exist eight different stereoisomers. Since the concentration of (**1**) in wine was very low only MS and chromatographic data were available. To identify the stereochemistry of wine lactone, stereochemical controlled syntheses for the eight isomers were necessary.

Preparation of stereoisomer lactones (**1a/b–1g/h**) followed a route using 3-methylcyclohex-2-enone **2** as starting material (Figure 2). Separation of isomeric lactones (**1a/b–1g/h**) was performed on different stationary GC phases and by HPLC; RI values and retention times are summarized in Table 2. After separation of the mixture of lactones by HPLC (Table 2), NMR, high resolution MS and IR experiments were performed on the diastereomerically pure isomers[2] (**1a/1b**), (**1c/1d**), (**1e/1f**) and (**1g/1h**). The assignment of absolute configuration of 3a,4,5,7a-tetrahydro-3,6-dimethyl-2(3*H*)-benzofuranones (**1a–1h**) was achieved by syntheses[2] starting from (+)-(4*R*)-limonene and (-)-(4*S*)-limonene. The stereoisomers (**1a–1h**) were further characterized by circular dichroism measurements (CD)[2].

Comparison of MS and chromatographic data (Table 2 and Figure 3) of compound (**1**), isolated from different white wine varieties, with those of synthesized lactones (**1a–1h**) indicated that wine lactone (**1**) is identical with the (3S,3aS,7aR)-3a,4,5,7a-tetrahydro-3,6-dimethyl-2(3*H*)-benzofuranone (**1a**).

Table 2 *HRGC and HPLC Data of 3a,4,5,7a-Tetrahydro-3,6-dimethyl-2-(3H)-Benzofuranone Stereoisomers*

Stereoisomer	HRGC			HPLC
	FFAP (RI)	SE-54 (RI)	Chiral phase^a (t_R [min])	Silica gel (t_R [min])
Wine lactone (1)	2192	1455	8.4	7.7
1a (3S,3a S,7aR)	2192	1455	8.4	7.7
1b (3R,3aR,7a S)	2192	1455	8.0	7.7
1c (3R,3a S,7aR)	2314	1496	9.0	9.5
1d (3 S,3aR,7a S)	2314	1496	8.6	9.5
1e (3 S,3a S,7a S)	2129	1422	7.3	6.0
1f (3R,3aR,7aR)	2129	1422	7.7	6.0
1g (3R,3a S,7a S)	2206	1466	8.1	6.9
1h (3 S,3aR,7aR)	2206	1466	8.3	6.9

^{a} Separation of the enantiomers was performed on borosilicate glass capillary (20m × 0.25 mm) coated with a chiral stationary phase (octakis-(3-O-butyryl-2,6-di-O-pentyl)-γ-cyclodextrin); trade name Lipodex E; on column injection of the sample; temperature program: 70 °C for 1 min, increased at 40 °C min^{-1} to 170 °C and then at 8 °C min^{-1} to 200 °C, held there for 10 min.

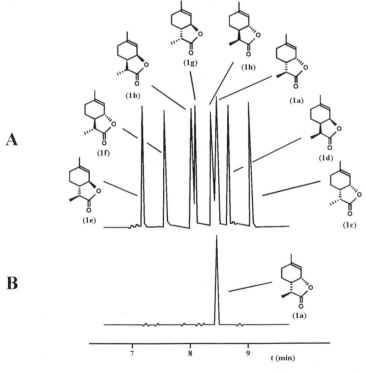

Figure 3 *A: separation of 3a,4,5,7a-tetrahydro-3,6-dimethyl-2(3H)-benzofuranone isomers by HRGC on a chiral stationary phase (Lipodex E); B: isolated lactone from wine*

The odour thresholds of the lactones (**1a–1h**) are compared in Table 3. Low values were found for (**1a**), (**1e**) and (**1h**). Comparison of the two compounds of each enantiomeric pair showed that the (3*S*)-configuration correlated to a lower threshold. The large differences of the odour threshold values observed for *e.g.* (**1a**) of 0.00001 ng l^{-1} air and (**1b**) of >1000 ng l^{-1} air, clearly demonstrate that the threshold was significantly influenced by the stereochemistry of the odorant.

Table 3   *Odour Threshold Values of 3a,4,5,7a-Tetrahydro-3,6-dimethyl-2-(3H)-Benzofuranone Stereoisomers*

Stereoisomer	Odour Threshold (ng l^{-1} in air)
1a (3*S*,3a*S*,7a*R*)	0.00001–0.00004
1b (3*R*,3a*R*,7a*S*)	>1000
1c (3*R*,3a *S*,7a*R*)	>1000
1d (3 *S*,3a*R*,7a *S*)	80–160
1e (3 *S*,3a *S*,7a *S*)	0.007–0.014
1f (3*R*,3a*R*,7a*R*)	14–28
1g (3*R*,3a *S*,7a*S*)	8–16
1h (3*S*,3a*R*,7a*R*)	0.05–0.2

The concentrations of wine lactone in different wine varieties were quantified by isotope dilution assays,[5] using (3*SR*, 3a*SR*, 7a*RS*)-3a,4,5,7a-tetrahydro-3[^{2}H$_3$],6-dimethyl-2(3*H*)-benzofuranone as the internal standard. The amounts varied between 0.07 and 0.2 μg l^{-1} in different wine varieties (Gewürztraminer, 1992 vintage: 0.10 μg l^{-1}; Gewürztraminer, 1995 vintage: 0.20 μg l^{-1}; Scheurebe, 1993 vintage: 0.10 μg l^{-1}; Spätburgunder red wine, 1992 vintage: 0.09 μg l^{-1}, all from the Ballrechten-Dottingen winery, Germany; Sauvignon Blanc, 1995 vintage, from the Morgenhof winery, Stellenbosch, South Africa: 0.07 μg l^{-1}).

REFERENCES

1. H. Guth , *J. Agric. Food Chem.*, in preparation.
2. H. Guth, *Helv. Chim. Acta*, 1996, **79**, in press.
3. F. Ullrich and W. Grosch, *Z. Lebensm. Unters. Forsch.*, 1987, **184**, 277.
4. I.A. Southwell, *Tetrahedron Lett.*, 1975, **24**, 1885.
5. H. Guth and W. Grosch, *J. Am. Oil Chem. Soc.*, 1993, **70**, 513.

PRODUCTION OF CHIRAL FLAVOUR COMPONENTS BY ENZYMIC METHODS

M. Nozaki, N. Suzuki and S. Oshikubo

Takasago Central Research Laboratory, Takasago International Corporation, 1-4-11, Nishi-Yawata, Hiratsuka, Kanagawa 254, Japan

1 INTRODUCTION

The character impact compound of mushroom is known to be 1-octen-3-ol which occurs predominantly in the (R) configuration[1] although the sensory properties of the enantiomers of 1-octen-3-ol[2] and other enantiomers[3] are known to differ. Optical resolution of racemic 1-octen-3-ol by porcine pancreatic lipase (PPL) has been reported[4] although the enantiomeric excess (ee) of the product was only 60%. This paper describes experiments using enzymes to resolve racemic mixtures of alcohols.

2 OPTICAL RESOLUTION OF ALCOHOLS

2.1 Resolution of 1-Octen-3-ol and Octan-3-ol

Racemic 1-octen-3-ol (2 g) and vinyl acetate (6.7 g) were mixed in hexane (100 ml) and *Candida antartica* lipase (CAL) (1 g) added. After stirring at 25 °C for three hours, the reaction was terminated by filtering off the enzyme and concentrating the hexane solution *in vacuo*, followed by silica gel chromatography. The (S)-1-octen-3-yl acetate was obtained at 45% yield (GC analysis) with an ee of 95% while the remaining (R)-1-octen-3-ol was present at 44.5% and 98% ee. The same system was used to resolve racemic octan-3-ol (racemic octan-3-ol, 10 g; vinyl acetate, 3.6 g; CAL, 0.5 g) over an incubation period of 18 hours and this gave a 51% conversion to (R)-3-octyl acetate with 93% ee.

2.2 Resolution of Alkan-2-ols

Optically active alkan-2-ols and their acetates are known as key components of some foods – for instance, blackberry, corn, coconut and banana.[5] The enantiomer distribution depends on their origin.[6] Optical resolution of alkan-2-ols by PPL has been reported in the ester hydrolysis mode[5] but the optical purity of the alcohols produced was not satisfactory. Using transesterification with vinyl acetate, CAL gave the best results of the lipases tested with the (R)-alkan-2-ol converted to the acetate while the (S)-form remained unconverted.

Racemic alkan-2-ol (5 g) was mixed with vinyl acetate in a 2:1 molar ratio (alcohol to vinyl acetate) with CAL (0.5 g) in hexane (100 ml). After 18 hours at 25 °C, GC analysis showed about 50% conversion of alcohol to acetate and after termination of the reaction and separation of the substrate and product (as described in 2.1), the optical purity of both compounds varied between 89% and 97% depending on the alkan-2-ol used (Table 1).

Table 1 *Optical Purity (ee %) of Alkan-2-ol Substrate and Acetate Product after Transesterification with Vinyl Acetate and CAL*

Substrate	(R)-Acetate	(S)-Alcohol	E
2-Pentanol	89	91	54
2-Hexanol	94	93	110
2-Heptanol	94	97	135
2-Octanol	93	96	108

3 STEREOSELECTIVE OXIDATION OF ALKAN-2-OLS

Various different types of microbial oxidation can be applied to alkan-2-ols but those that require regeneration of a co-factor (*e.g.* alcohol dehydrogenase and NAD^+) are difficult to apply because of the need to regenerate the co-factor and because of the expense of the co-factors themselves. However, the alcohol oxidase enzyme uses molecular oxygen to reoxidise the flavin co-factor associated with the enzyme and thus does not require complicated co-factor regeneration systems. Methylotrophic yeast such as *Candida boidinii*[7-9] *Pichia pastoris*[10] and *Candida maltosa*[11] contain alcohol oxidase activity and can convert alcohols to aldehydes. *C. boidinii* SA051[12] used in this study was derived from *C. boidinii* AOU-1 by ultraviolet radiation and had the highest alcohol oxidase activity among the mutants as assayed by formaldehyde production. Previously, we have shown the feasibility of using oxidation for the large scale preparation of aldehydes and shown some reaction selectivity (regio-, chemo- and stereo-).[13,14] In this paper, the stereoselectivity of microbial oxidation is studied.

The stereoselective oxidation of 1,3-butanediol by *C. boidinii* SA051 showed a preference for the (*R*) enantiomer which was converted to 1-hydroxy-3-butanone[13,14] while the remaining 1,3-butanediol showed some optical activity with the (*S*) configuration at 27% optical purity. The system was applied to the stereoselective oxidation of 2-alkanols (200 mg) in potassium phosphate buffer (20 ml; 0.1 M; pH 7.5) containing *C. boidinii* cells (600 mg dry weight) under an atmosphere of pure oxygen at 27 °C until the conversion was about 50% as measured by GC. This degree of conversion was achieved at different times depending on the substrate (2-pentanol, 6 hours; 2-hexanol, 5 hours; 2-heptanol, 47 hours). The reaction was halted by filtering off the biomass and the solution extracted with ethyl acetate, which was dried on $MgSO_4$ and concentrated *in vacuo*. After separation on a silica gel column, the optically active alcohol and the corresponding ketone were obtained.

Table 2 *Optical Purity of Alkan-2-ols after Treatment with* C. boidinii *(30 g l⁻¹) using a Substrate Concentration of 1%*

Substrate	Time (hours)	Conversion (%)	ee (%)	E
2-Butanol	3	61.1	<1	—
2-Pentanol	6	60.6	72.8	5.8
2-Hexanol	5	45.3	75.1	46
2-Heptanol	47	47.2	66.8	14
2-Octanol	48	<10	—	—

The stereochemistry of all of the alcohols obtained showed the (*R*) configuration (as compared with reference compounds); thus the enzyme oxidation prefers the (*S*) configuration for the alkan-2-ols. The optical purity of the alcohols was measured and varied depending on the substrate as shown in Table 2.

The *C. boidinii* system separates the 2-alkanol enantiomers by conversion of the (*S*) form to the alkanone but the reduction of this compound by baker's yeast is well documented (Prelog's rule). Therefore, treatment of the 2-alkanone by baker's yeast can regenerate the (*S*)-2-alkanol and this provides a way of separating the two enantiomers of the 2-alkanols. In fact, incubating the 2-hexanone (from the *C. boidinii* reaction) with baker's yeast gave (*S*)-2-hexanol in a 78% yield with an ee of 95%. This means that both enantiomers of 2-alkanol could be produced with high enantioselectivity.

4 SENSORY EVALUATION OF THE ENANTIOMERS

The optically active alkan-2-ol enantiomers were purified by forming the 3,5-dinitrobenzoyl esters and recrystallisation. The optical purity after this step was determined by GC of the camphanic acid ester derivatives and the values were in excess of 99%. The acetates were formed by acetylation of the corresponding purified alcohols. Enantiomers of 1-octen-3-ol were purified by the same recrystallisation method and optical purity determined by GC on a chiral column without derivatisation. The sensory properties of these compounds are listed in Table 3.

Table 3 *Sensory Properties of Enantiomers of 1-Octen-3-ol, Alkan-2-ols and Acetates*

Compound	(R) form	(S) form
1-Octen-3-ol	Intensive mushroom, fruity, soft odour	Herbaceous, musty, weak mushroom
2-Pentanol	Light, seedy, sharp	Heavy, wild berry, ripe, dusty, acerola, weak, astringent
2-Hexanol	Mushroom, dusty, oily	Mushroom, green, ripe, berry, astringent, metallic
2-Heptanol	Fruity, sweet, oily, fatty	Mushroom, oily, fatty, blue cheese, mouldy
2-Octanol	Creamy, cucumber, fatty, sour	Mushroom, oily, fatty, creamy, grape
2-Pentyl acetate	Fruity, muscat, green, metallic, chemical	Fruity, apple, plum, metallic
2-Hexyl acetate	Sour, fruity, cherry, plum. strawberry	Sweaty, sour, fruity, plum, nectarine
2-Heptyl acetate	Green, fatty, banana, methyl ketone	Mushroom, earthy, acerola, wild berry
2-Octyl acetate	Methyl ketone, fatty burnt, boiled vegetable	Methyl ketone, fruity, plum dusty

5 CONCLUSION

CAL showed an excellent ability for the optical resolution of racemic alcohols, especially alkan-2-ols and alkan-3-ols. Also *Candida boidinii* SA051 demonstrated their ability to oxidise alkan-2-ols stereoselectively. In the production of optically active alkan-2-ols, both lipase and microbial redox modes could be available. It was proved that these biocatalysts were suitable for the production of chiral flavour components. Thus, the given chiral flavour components possessed different sensory properties between enantiomers.

ACKNOWLEDGEMENTS

We thank Prof. Y. Tani and Prof. Y. Sakai for their generous donation of *Candida boidinii* SA051. We also thank Prof. K. Nakamura for fruitful discussions about the enzymic optical resolution of 1-octen-3-ol.

REFERENCES

1. M. Gessner, W. Deger and A. Mosandl, *Z. Lebensmitt. Unters. Forsch.*, 1988, **186**, 417.
2. A. Mosandl, G. Heussinger and M. Gessner, *J. Agric Food Chem.*, 1986, **34**, 119.
3. A. Mosandl, C. Gunther, M. Gessner, W. Deger, G. Singer and G. Heussinger, in 'Bioflavor 87', ed. P. Schreier, Walter de Gruyter, Berlin, 1988, p. 55.
4. P. Schreier, in 'Flavor Precursors', A.C.S. Symposium Series 490, ed. R. Teranishi, A.C.S., Washington, D.C., 1992, p. 32.
5. K.-H. Engel, in 'Bioflavor 87', ed. P. Schreier, Walter de Gruyter, Berlin, 1988, p. 75.
6. A. Mosandl *et al.*, *J Agric Food Chem.*, 1991, **39**, 1131.
7. Y. Sakai and Y. Tani, *Agric. Biol. Chem.*, 1987, **51**, 2617.
8. Y. Shachar-Nishri and A. Freeman, *Appl. Biochem. Biotechnol.*, 1993, **39/40**, 387.
9. D.S. Clark *et al.*, *Bioorganic and Medicinal Chem. Lett.*, 1994, **4**, 1745.
10. W.D. Murray, S.J.B. Duff and P.H. Lanthler, 'Production of natural flavor aldehydes from natural source primary alcohols C2–C7', U.S. Patent 4871669, 1989.
11. S. Mauersberger *et al.*, *Appl. Microbiol. Biotechnol.*, 1992, **37**, 66.
12. Y Sakai and Y Tani, *Agric. Biol. Chem.*, 1987, **51**, 2177.
13. M. Nozaki, Y. Washizu, N. Suzuki and T. Kanisawa, in 'Bioflavour 95', ed. P Etievant and P Schreier, INRA, Paris, 1995, p. 255.
14. M. Nozaki, N. Suzuki and Y. Washizu, in 'Biotechnology for Improved Foods and Flavors', A.C.S. Symposium Series 637, eds. G.R. Takeoka, R. Teranishi, P.J. Williams and A. Kobayashi, A.C.S., Washington, D.C., 1996, p. 188.

SECTION 4

THERMALLY GENERATED FLAVOUR

IDENTIFICATION OF THE KEY ODORANTS IN PROCESSED RIBOSE–CYSTEINE MAILLARD MIXTURES BY INSTRUMENTAL ANALYSIS AND SENSORY STUDIES

P. Schieberle[*] and T. Hofmann[†]

[*]Institute for Food Chemistry, Technical University of Munich and [†]Deutsche Forschungsanstalt für Lebensmittelchemie, Lichtenbergstraße 4, 85748 Garching, Germany

1 INTRODUCTION

The Maillard reaction between reducing carbohydrates and amino acids plays an important role in the generation of the characteristic flavours of certain processed foods, such as bread crust, roast meat, roast coffee and chocolate. Model studies performed some years ago[1] have already shown that depending on either the amino acid or carbohydrate used and, also, on the conditions of the thermal treatment, certain food flavours can be mimicked.[2] Today, such precursor systems are known as reaction or processed flavours and, in particular, cysteine is used to generate meat-like flavourings.

Numerous studies have been performed to identify the volatiles generated from cysteine in the presence of ribose especially, and in a series of studies,[3,4] more than 120 volatiles have been characterized. Although very odour-active compounds, such as 2-furfurylthiol and 2-methyl-3-furanthiol, have been found, correlations between the sensory potency of single odorants and the sensory impressions evoked by the reaction systems are rarely reported. However, such data would be very helpful, e.g., to optimize the odours of reaction flavours with respect to the original food flavour to be imitated. To characterize key odorants in food aromas, the following concepts have been developed in our group:[5,6]
- Screening of odour-active compounds by aroma extract dilution techniques;
- Quantification of key odorants by stable-isotope dilution assays;
- Calculation of odour activity values;
- Flavour simulation with reference odorants in the food matrix.

By using these concepts, the aim of the present study was (i) to characterize the key odorants in a thermally treated ribose–cysteine mixture and (ii) to gain preliminary insights into factors governing the yields of selected odorants.

2 EXPERIMENTAL

L-Cysteine (3.3 mmol) and ribose (10 mmol) were dissolved in phosphate buffer (100 ml; pH 5.0; 0.5 mol l^{-1}) and reacted for 20 min at 145 °C. Volatiles were then isolated by extraction with diethyl ether and sublimation *in vacuo*.[7] Odour-active compounds were detected in the extracts by aroma extract dilution analysis[5,6] and identified by using synthesised reference odorants. Quantifications were performed by means of stable-isotope dilution assays.[6]

Table 1   *Most Odour-active Compounds in the Thermally Treated Cysteine–Ribose Solution*[7]

Odorant	FD^a	Odorant	FD^a
2-Furfurylthiol	1024	Bis(2-methyl-3-furyl)disulfide	128
3-Mercapto-2-pentanone[b]	512	2-Acetyl-2-thiazoline	64
2-Methyl-3-furanthiol	256	4-Hydroxy-5-methyl-3(2H)-furanone	64
5-Acetyl-2,3-dihydro-1,4-thiazine	256	4-Hydroxy-2,5-dimethyl-3(2H)-furanone	32
3-Mercapto-2-butanone	128		

[a] Flavour dilution factor determined by aroma extract dilution analysis.[5,6]
[b] Containing smaller amounts of the isomeric 2-mercapto-3-pentanone.

3 RESULTS AND DISCUSSION

3.1 Identification of Odorants

Preliminary experiments, in which the ratio of cysteine and ribose and, also, the pH-value, had been varied succeeded in the production of a solution eliciting an intense roasty, meat-like odour after thermal treatment. In the volatile fraction, 29 odour-active compound were detected,[7] among which the eight compounds listed in Table 1 were identified with the highest flavour dilution (FD) factors. Among them, the 5-acetyl-2,3-dihydro-1,4-thiazine is reported for the first time among the aroma compounds of foods or reaction flavours.

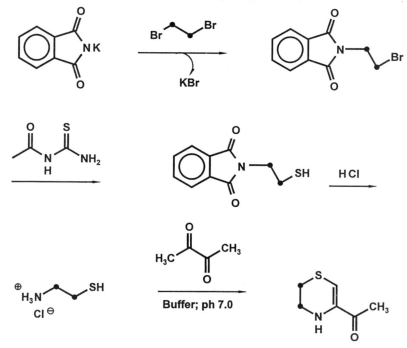

Figure 1   *Synthetic route used in the preparation of deuterium-labelled 5-acetyl-2,3-dihydro-1,4-thiazine*

3.2 Quantification of Odorants and Calculation of Odour Activity Values

Stable isotope dilution assays (SIDA) were developed to quantify the most important odorants identified in the mixture (T. Hofmann and P. Schieberle, *J. Agric. Food Chem.*, in preparation). In Figure 1, the synthesis of the deuterium labelled 5-acetyl-2,3-dihydro-1,4-thiazine is shown as an example. The target compound is obtained by reacting deuterium-labelled 1-amino-2-mercaptoethane with 2,3-butandione. The key intermediate $[^2H]_4$-1-amino-2-mercaptoethane is produced by treatment of deuterium-labelled 1,2-dibromoethane with potassium phtalimide followed by N-acetylthiourea and subsequent acid hydrolysis. The structures of the other labelled internal standards used are summarized in Figure 2.

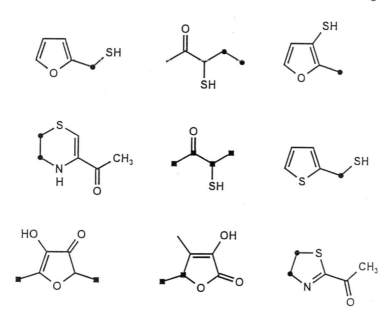

Figure 2   *Structures of isotopically labelled internal standards used to quantify selected key odorants:* ■ *carbon-13 label;* • *deuterium label*

Some of the odorants are extremely labile. However, because the internal standard will degrade to the same extent during work as the odorant, SIDA supplies exact data on the actual concentrations. The limit is set, of course, by a complete degradation during the enrichment procedure. Compared to other odorants containing a thiol group, 2-methyl-3-furanthiol (MFT) is rapidly degraded, even when stored at lower temperatures.[8] The degradation is significantly enhanced with increasing temperatures, and, therefore, especially if large solvent volumes containing low amounts of MFT have to be concentrated, the odorant and the labelled standard may be completely degraded.

To overcome this problem, two alternative enrichment procedures were developed: (a) the thiols were derivatised with 2-vinylpyridine prior to extraction (Method II; Table 2) or (b) enrichment was performed by the 'purge and trap' technique (Chrompack TCT system) using headspace extraction (Method III; Table 2).

Table 2   *Quantification of Three Selected Thiols in a Thermally Treated Cysteine–Glucose Solution – Comparison of Three Different Enrichment Procedures*

Thiol	Concentration ($\mu g\ l^{-1}$)		
	Method I[a]	Method II[b]	Method III[c]
2-Methyl-3-furanthiol	d.n.p.	17	21
2-Furfurylthiol	27	26	30
3-Mercapto-2-butanone	27.6	30.0	27.9

[a] Labelled internal standards were added to the thermally treated reaction mixture and, after extraction with diethyl ether and sublimation *in vacuo*, analysed in the acidic fraction (treatment with sodium hydroxide, 0.1 mol l^{-1}) by mass chromatography;[6]

[b] Before extraction, the labelled standards and the three flavour compounds were derivatised with 2-vinylpyridine;

[c] Isolation of the volatiles and internal standards was performed by 'purge and trap'; d.n.p.: determination not possible (*cf* text).

In Table 2, quantitative data obtained for MFT are compared with results obtained for the more stable 2-furfurylthiol and 3-mercapto-2-butanone. A cysteine–glucose mixture was chosen in the comparative study, because of the lower amounts of MFT formed from this carbohydrate. The results showed that 2-furfurylthiol, as well as 3-mercapto-2-butanone, can be determined very precisely by using each of the three methods. However, MFT, due to its extreme instability, has to be stabilized either by derivatization or, alternatively, has to be determined by the more careful headspace-enrichment technique.

Table 3   *Concentrations and Odour Activity Values (OAV) of Eleven Key Odorants in the Ribose–Cysteine Mixture*

Odorant	Conc. ($\mu g\ l^{-1}$)	OAV[a]
2-Methyl-3-furanthiol	198	28286
2-Furfurylthiol	121	12100
3-Mercapto-2-pentanone	599	856
2-Mercapto-3-pentanone	253	361
2-Methyl-3-thiophenethiol	7	350
5-Acetyl-2,3-dihydro-1,4-thiazine	424	340
2-Thenylthiol	5	119
3-Mercapto-2-butanone	342	114
4-Hydroxy-5-methyl-3(2H)-furanone (Norfuraneol)	545300	64
3-Hydroxy-4,5-dimethyl-2(5H)-furanone (Sotolon)	16	53
4-Hydroxy-2,5-dimethyl-3(2H)-furanone (Furaneol)	185	19
2-Acetyl-2-thiazoline	7	7

[a] Odour activity values were calculated by dividing the concentrations by the odour thresholds (orthonasal) in water.

In Table 3, the results of the quantitative determination of twelve selected odorants in the cysteine–ribose mixture are summarized. The amounts ranged between 5 $\mu g\ l^{-1}$ (2-

thenylthiol) and 545 mg l^{-1} (4-hydroxy-5-methyl-3(2*H*)-furanone; Norfuraneol). To reveal their flavour contribution, the odour activity values (OAV: ratio of concentration to odour threshold) were calculated on the basis of odour thresholds (orthonasal) in water. The results indicated 2-methyl-3-furanthiol and 2-furfurylthiol (FFT) to show by far the highest OAVs, as the character-impact odorants of the reaction mixture (Table 3). As further important odorants were established 3-mercapto-2-pentanone, 2-mercapto-3-pentanone, 2-methyl-3-thiophenethiol and 5-acetyl-2,3-dihydro-1,4-thiazine. It is interesting to note, that the first three compounds have also been reported as key odorants in beef and chicken broth,[8] whereas the last three have not.

3.3 Aroma Simulation

To simulate the overall aroma, the twelve key odorants were dissolved in phosphate buffer (pH 5.0) in the same concentrations as occur in the reaction mixture (Table 3), and evaluated by a sensory panel in comparison to the odour of the complete reaction mixture. Six assessors were asked to evaluate, on a seven point scale (0 to 3.0), the intensity of seven given odour qualities. Mean values were then calculated and are displayed as spider web diagrams in Figure 3.

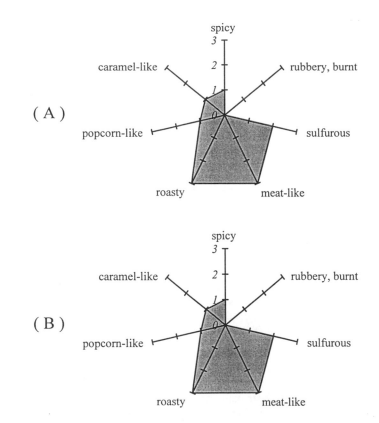

Figure 3 *Odour profiles of (A) the thermally treated mixture of cysteine–ribose; and (B) the reconstituted flavour mixture containing twelve key odorants (cf Table 3)*

In the complete ribose–cysteine mixture (Figure 3A), roasty, meaty odour qualities were scored highly. A comparison of Figure 3A and the data in Table 3 suggests that especially the roasty, coffee-like smelling 2-furfurylthiol and the meat-like, sulfury smelling 2-methyl-3-furanthiol predominantly generate this odour impression.

In the overall odour of the flavour model mixture, roasty, meat-like notes also predominated (Figure 3B). The overall odours of the reaction mixture and the reconstituted flavour mixture were nearly identical and were both described as beef broth-like. The data clearly indicate that these twelve aroma compounds are indeed responsible for the overall aroma of the reaction mixture.

Table 4   *Influence of the pH on the Odour Activity Values (OAV)[a] of Selected Key Odorants in the Ribose–Cysteine Mixture*

Odorant	OAV at pH		
	3.0	5.0	7.0
2-Methyl-3-furanthiol	79000	28286	3571
2-Furfurylthiol	22900	12100	1200
2-Methyl-3-thiophenethiol	900	350	50
2-Thenylthiol	191	19	<20
3-Mercapto-2-pentanone	361	856	424
2-Mercapto-3-pentanone	151	361	209
5-Acetyl-2,3-dihydro-1,4-thiazine	2	340	22
4-Hydroxy-2,5-dimethyl-3(2H)-furanone (Furaneol)	<1	19	208
3-Hydroxy-4,5-dimethyl-2(5H)-furanone (Sotolon)	40	53	83
2-Acetyl-2-thiazoline	<1	7	74

[a] See footnote 'a' in Table 3.

3.4 Flavour Changes Induced by Changing the pH Value

Compared with the mixture processed at pH 5.0, reacting the same amounts of cysteine and ribose at either pH 3.0 or pH 7.0, induced significant changes in the overall odours. Quantification of ten of the key odorants reflected these flavour differences by significant changes in the OAVs (Table 4). Compared with pH 5.0, pH 3.0 gave OAVs higher by factors of 3 and 2, respectively, for 2-methyl-3-furanthiol and 2-furfurylthiol and, also, for the analogous thiophene derivatives. On the other hand, the OAVs of the roasty smelling 5-acetyl-2,3-dihydro-1,4-thiazine and the caramel-like smelling 4-hydroxy-2,5-dimethyl-3(2H)-furanone were significantly decreased at pH 3.0. On the contrary, increasing the pH to 7.0 resulted in a decrease of MFT and FFT, whereas 4-hydroxy-2,5-dimethyl-3(2H)-furanone increased.

4 CONCLUSIONS

The results have shown that a systematic approach to elucidating key odorants by combining sensory and analytical techniques, provides a useful tool to characterize food ingredients, such as reaction flavours. The data are the basis for systematic studies on flavour formation, the results of which will then enable the manufacturer to optimise the yields of desired flavour compounds. Furthermore, helpful data are provided to mimic the original food flavour, *e.g.*, by selection of suitable precursor structures.

REFERENCES

1. K.O. Herz and S.S. Chang, *Adv. Food Research*, 1970, **18**, 1.
2. M.J. Lane and H.E. Nursten, in 'The Maillard Reaction in Food and Nutrition', A.C.S. Symposium Series 215, American Chemical Society, Washington, D.C., 1983, p. 141.
3. F.B. Whitfield, D.S. Mottram, S. Brock, D.J. Puckey and L. Salter, *J. Sci. Food Agric.*, 1988, **42**, 261.
4. L.J. Farmer, D.S. Mottram and F.B. Whitfield, *J. Sci. Food Agric.*, 1989, **49**, 347.
5. W. Grosch, *Trends in Food Science and Technology*, 1993, **4**, 68.
6. P. Schieberle, in 'Characterization of Foods – Emerging Methods', ed. A. Goankar, Elsevier, Amsterdam, 1995, p. 403.
7. T. Hofmann and P. Schieberle, *J. Agric. Food Chem.*, 1995, **43**, 2187.
8. T. Hofmann, P. Schieberle and W. Grosch, *J. Agric. Food Chem.*, 1996, **44**, 251.
9. U. Gasser and W. Grosch, *Z. Lebensm. Unters. Forsch.*, 1990, **190**, 1.

STUDIES ON INTERMEDIATES GENERATING THE FLAVOUR COMPOUNDS 2-METHYL-3-FURANTHIOL, 2-ACETYL-2-THIAZOLINE AND SOTOLON BY MAILLARD-TYPE REACTIONS

T. Hofmann and P. Schieberle

Deutsche Forschungsanstalt für Lebensmittelchemie, Lichtenbergstraße 4, D-85748 Garching, Germany

1 INTRODUCTION

The very potent odorants, 2-methyl-3-furanthiol (MFT; meat-like), 2-acetyl-2-thiazoline (AT; popcorn-like), and 3-hydroxy-4,5-dimethyl-2(5*H*)-furanone (Sotolon; SOT; hydrolysed protein-like) for example, have recently been established as important odorants in several thermally treated foods. MFT significantly contributes to the overall odour of beef and chicken broth,[1] roasted coffee[2] and tuna fish,[3] AT to the flavours of roast beef[4] and roasted white sesame seeds,[5] and SOT to the odours of stewed beef,[6] roasted coffee[7] and rye bread crust.[8]

Although also a biosynthetic formation pathway starting from 4-hydroxyisoleucine has as been established for SOT,[9] it is generally assumed in the literature that Maillard-type reactions are involved in the generation of the three odorants during thermal processing of foods. Furthermore, on the basis of quantitative experiments using stable-isotope dilution assays, thiamine was recently established to be an important precursor of MFT in beef and pork.[10]

By application of aroma-extract dilution techniques to the volatiles generated during heating of cysteine in the presence of carbohydrates[11] and quantitative measurements, we (Schieberle and Hofmann, this book) have recently established the three compounds among the key odorants of such model mixtures.

The following investigation was, therefore, undertaken to gain insights into the intermediates and formation pathways leading to MFT, AT and SOT from cysteine and carbohydrates.

2 EXPERIMENTAL

The three odorants were quantified by using the stable isotope-dilution assays recently reported (Schieberle and Hofmann, this book). Quantification was performed by mass chromatography, using an ion trap detector (ITD 800; Finnigan, Bremen, Germany) running in the chemical ionization mode with methanol as the reactant gas. The aroma compounds and internal standards were isolated by extraction with diethyl ether and sublimation *in vacuo*.[8]

The isolation of the two precursors of 2-acetyl-2-thiazoline formed in a mixture of cysteamine and 2-oxopropanal (20 mmol each) in phosphate buffer (2 l; 0.5 mol l^{-1}; pH 7.0) is described elsewhere.[12,13]

3 RESULTS AND DISCUSSION

3.1 2-Methyl-3-furanthiol

In the first experiment, MFT was quantified in aqueous solutions of cysteine reacted at 145 °C in the presence of either ribose, glucose or rhamnose. The results revealed ribose as the most effective precursor (Table 1). At pH 5.0, for instance, 10 and 25 times more MFT was generated from ribose than from glucose and rhamnose, respectively. The influence of the pH on the yields of MFT was, however, different for the three carbohydrates under investigation. While in the presence of glucose or rhamnose, the MFT generation went through a maximum at pH 5.0, ribose generated the highest amounts of MFT at pH 3.0. The last result compares well with the literature.[14]

Table 1   *Formation of 2-Methyl-3-furanthiol (MFT) from Carbohydrates and Cysteine[a]*

Carbohydrate	MFT [µg]		
	pH 3.0	*pH 5.0*	*pH 7.0*
Ribose	55.3	19.8	2.5
Glucose	0.3	1.9	0.4
Rhamnose	0.1	0.8	0.1

[a]   Cysteine (3.33 mmol) and the carbohydrate (10.0 mmol) were dissolved in phosphate buffer (100 ml; 0.5 mol l^{-1}) and heated for 20 min at 145 °C in an autoclave.

Table 2   *Amounts of 2-Methyl-3-furanthiol (MFT) Generated from Hydroxyacetaldehyde and Mercapto-2-propanone*

Model	MFT [µg]	mol %
A[a]	268.1	0.24
B[b]	1553.9	1.39

[a]   Hydroxyacetaldehyde (1 mmol) was reacted with mercapto-2-propanone (1 mmol) in phosphate buffer (50 ml; 0.5 mol l^{-1}; pH 5.0) in an autoclave for 20 min at 145 °C;
[b]   Hydroxyacetaldehyde (1 mmol) and mercapto-2-propanone (1 mmol) were intimately mixed with silica gel (3.0 g) containing 20.4 mg KH_2PO_4 and 0.3 ml water and were then dry-heated for 5 min at 180 °C.

The different influence of the pH on MFT formation suggests different pathways, from pentoses and hexoses. To generate a C-5 skeleton, for example, a retro-Aldol reaction of mercapto-2-propanone and hydroxyacetaldehyde might be proposed. Mercapto-2-propanone has been shown to be generated in significant amounts from 2-oxopropanal and H_2S[15] and hydroxyacetaldehyde is a well-known carbohydrate degradation product.[16]

To establish this hypothesis, mixtures of mercapto-2-propanone and hydroxy-acetaldehyde were thermally reacted either in aqueous solution or by dry-heating on silica gel. The last mentioned system was chosen to simulate roasting conditions. As shown in Table 2, high amounts of MFT are produced from both intermediates with dry-heating favouring the formation of the odorant. Under the latter conditions, nearly six times greater amounts of MFT were formed compared with the reaction in the presence of water.

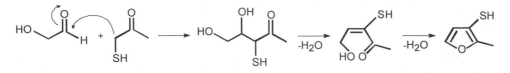

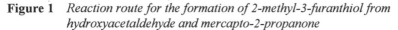

Figure 1   *Reaction route for the formation of 2-methyl-3-furanthiol from*
hydroxyacetaldehyde and mercapto-2-propanone

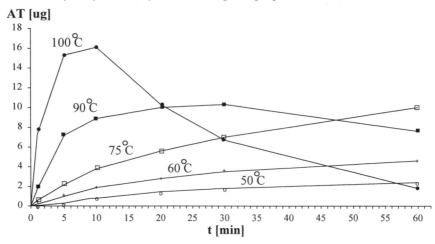

Figure 2   *Time course of the generation of 2-acetyl-2-thiazoline from 2-(1-hydroxyethyl)-*
4,5-dihydrothiazole (150 µg) during heating in aqueous solution

This result is well in line with the reaction route proposed in Figure 1, since two molecules of water have to be eliminated from the key intermediate, 3-mercapto-4,5-dihydroxypentane-2-one, which should be facilitated under dry-heating conditions.

3.2 2-Acetyl-2-thiazoline

In a previous paper,[12] we confirmed data[17] indicating that 2-acetyl-2-thiazoline is formed by reacting 2-oxopropanal with cysteamine. Furthermore, on the basis of NMR-measurements and synthesis the previously unknown 2-(1-hydroxyethyl)-4,5-dihydrothiazole (HDT) was characterized as an important intermediate in the generation of AT.[12] As shown in Figure 2, significant amounts of AT are formed simply by heating the HDT in water. The highest amounts were obtained by a thermal degradation of HDT for 10 min at 100 °C (Figure 2).

Although this corresponds to a nearly 10% yield of AT, the fate of the remaining amount of HDT in the reaction mixture was of interest. HDT (**II** in Figure 3) might tautomerize into 2-acetylthiazolidine (ATD; **I** in Figure 3). To elucidate the role of ATD in the formation of AT, synthetized ATD was refluxed for 20 min in tap water and the amounts of HDT, AT, and the unreacted ATD were determined by stable-isotope dilution assays.[13] The results were compared with an experiment, in which pure HDT was reacted. It was revealed (Table 3) that 2-acetyl-2-thiazoline is much more easily liberated from ATD than from HDT, because, after 20 min of boiling, five times more AT had been formed.

Further results indicated that, especially at lower temperatures, ATD is formed from 2-oxopropanal and cysteamine as the first reaction product (Figure 3), whereas a thermal treatment induces a tautomerization into HDT, undoubtedly via an intermediate enaminol (*cf* Figure 3).

ATD as well as HDT have to be oxidized to generate the AT. To gain an insight into the oxidation mechanism, the experiments summarized in Table 4 were performed. The presence of copper (II) ions significantly increased the yields of the AT from HDT, implying oxidation catalysed by a transiton metal. However, oxygen is also involved in this reaction, because elimination of oxygen from the reaction system drastically lowered the formation of AT (Experiment 4, Table 4).

Table 3 *Generation of 2-acetyl-2-thiazoline (AT) from the precursors 2-acetylthiazolidine (ATD) and 2-(1-hydroxyethyl)-4,5-dihydrothiazole*[a]

Precursor	Amounts present after 20 min at 100 °C		
	ATD [µg]	HDT [µg]	AT [µg]
ATD	18.1	76.4	51.5
HDT	22.6	120.1	10.9

[a] The precursors (150 µg each) were refluxed for 20 min in tap water (10 ml).

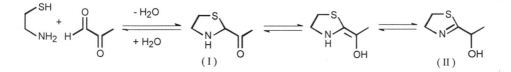

Figure 3 *2-Acetylthiazolidine and 2-(1-hydroxyethyl)-4,5-dihydrothiazole formed as reaction products of cysteamine and 2-oxopropanal*

Table 4 *Influence of copper ions on the formation of 2-acetyl-2-thiazoline (AT) from 2-(1-hydroxyethyl)-4,5-dihydrothiazole (HDT)*[a]

Expt.	Addition of (µmol)	AT [µg]
1	without (control experiment)	9.8
2	Cu^{2+} (0.01)	20.1
3	Cu^{2+} (0.01)	38.4
4	without (argon atmosphere)[b]	1.7

[a] HDT (190 µg) was heated for 20 min in water (10 ml) at 80 °C;
[b] The solvent water was degassed in an ultrasonic bath and subsequently flushed with argon.

On the basis of these results, a hypothetical formation route was devised (Figure 4) showing the role of oxygen and copper ions in the oxidation process. This reaction pathway, however, needs to be proved through further experiments.

3.3 3-Hydroxy-4,5-dimethyl-2(5*H*)-furanone (Sotolon)

The carbon backbone of Sotolon (SOT) is methyl branched and, therefore, direct formation from the intact C-6 skeleton of a hexose is not possible. We, therefore, assumed that SOT might originate from an Aldol-reaction of C-2 and C-4 compounds, previously formed by retro-Aldol cleavage of the carbohydrate skeleton. This prompted us to determine the amounts of SOT generated by heating a mixture of the possible precursors, hydroxyacetaldehyde and butan-2,3-dione (Experiment 1, Table 5). Significant amounts of SOT were formed at pH 5.0 (Table 5), while lowering as well as increasing the pH led to much lower yields. Dry-heating of the precursor mixture also generated SOT, but the amounts were nearly six times lower than the amounts generated in aqueous solution (*cf* Experiments 1 and 2, Table 5). In Figure 5, a reaction scheme illustrating the formation of SOT from both precursors is proposed.

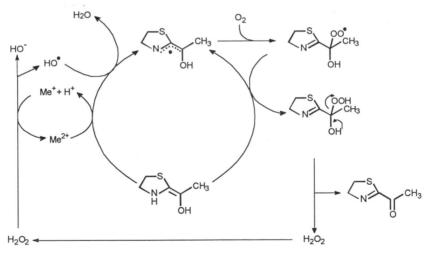

Figure 4   *Hypothesis for the formation of 2-acetyl-2-thiazoline from 2-acetylthiazolidine/ 2-(1-hydroxyethyl)-4,5-dihydrothiazole via a metal-catalysed oxidation of an intermediate enaminol*

Table 5   *Generation of Sotolon (SOT) from butan-2,3-dione and hydroxyacetaldehyde*

Experiment	pH 3.0	pH 5.0	pH 7.0
1[a]	13.5	764.7	273.1
2[b]	n.a.	131.5	n.a.

[a]   The reactants (1.0 mmol each) were dissolved in sodium/potassium phosphate buffer (50 ml; 0.5 mol l⁻¹) and heated for 20 min at 145 °C in a laboratory autoclave;

[b]   The reactants (1.0 mmol each) were intimately mixed with silica gel (3.0 g) containing 20.4 mg KH_2PO_4 and 0.3 ml water and were then dry-heated for 5 min at 180°C.
n.a.: Not analysed.

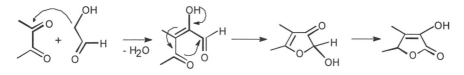

Figure 5 *Reaction pathway leading from butan-2,3-dione and hydroxyacetaldehyde to Sotolon*

4 CONCLUSION

The results clearly indicate that 2-oxopropanal, 2,3-butandione and hydroxyacetaldehyde are key intermediates in the generation of 2-methyl-3-furanthiol and 2-acetyl-2-thiazoline, as well as of 3-hydroxy-4,5-dimethyl-2(5H)-furanone. It is, therefore, very likely, that the three oxo-compounds are also involved in the generation of the three odorants during thermal processing of foods. The actual amount of a single odorant, however, will be significantly influenced by the presence or absence of further reactive intermediates, such as hydrogen sulfide or cysteamine, and, also, by the reaction conditions. To extend knowledge about how the formation of the above intermediates can be influenced is, therefore, an important task for the future.

REFERENCES

1. U. Gasser and W. Grosch, *Z. Lebensm. Unters. Forsch.*, 1990, **190**, 3.
2. P. Semmelroch and W. Grosch, *Lebensm. Wiss. Technol.*, 1995, **28**, 310.
3. D.A. Withycombe and C.J. Mussinan, *J. Food Sci.*, 1988, **63**, 658.
4. C. Cerny and W. Grosch, *Z. Lebensm. Unters. Forsch.*, 1993, **196**, 417.
5. P. Schieberle, in 'Progress in Flavour Precursor Studies', eds. P. Schreier and P. Winterhalter, Allured Publishing, Carol Stream, IL, 1993, p. 343.
6. H. Guth and W. Grosch, *J. Agric. Food Chem.*, 1994, **42**, 2862.
7. P. Semmelroch, G. Laskawy, I. Blank and W. Grosch, *Flavour and Fragrance J.*, 1995, **10**, 1.
8. P. Schieberle and W. Grosch, *Z. Lebensm. Unters. Forsch.*, 1994, **198**, 292.
9. Y. Sauvaire, P. Givardon, J.C. Baccou and A.M. Risteruca, *Phytochemistry*, 1984, **23**, 479.
10. W. Grosch and G. Zeiler-Hilgart, in 'Flavour Precursors – Thermal and Enzymatic Conversions', eds. R. Teranishi, G.R. Takeoka and M. Güntert, A.C.S. Symposium Series 490, American Chemical Society, Washington, 1992, p. 183.
11. T. Hofmann and P. Schieberle, *J. Agric. Food Chem.*, 1995, **43**, 2187.
12. T. Hofmann and P. Schieberle, *J. Agric. Food Chem.*, 1995, **43**, 2946.
13. T. Hofmann and P. Schieberle, *J. Agric. Food Chem.*, in press.
14. A. Meynier and D.S. Mottram, *Food Chem.*, 1995, **52**, 361.
15. T. Hofmann, Ph.D. Thesis, Technical University of München, 1995.
16. G. Bryne, D. Gardiner and F. Holmes, *J. App. Chem.*, 1966, **16**, 81.
17. T. Hayashi and T. Shibamoto, *J. Agric. Food Chem.*, 1985, **33**, 1090.

NEW ASPECTS OF THE FORMATION OF 3(2H)-FURANONES THROUGH THE MAILLARD REACTION

Imre Blank, Stéphanie Devaud and Laurent B. Fay

Nestec Ltd., Nestlé Research Centre, Vers-chez-les-Blanc, P.O. Box 44, 1000-Lausanne 26, Switzerland

1 INTRODUCTION

4-Hydroxy-2,5-dimethyl-3(2H)-furanone (**1**, furaneol, registered trademark of Firmenich) and 2-(or 5-)ethyl-4-hydroxy-5-(or 2-)methyl-3(2H)-furanone (**2**, homofuraneol) are potent flavour compounds contributing to the sensory properties of many natural products and thermally processed foods.[1] They give a caramel-like, sweet flavour and have low retronasal odour thresholds, *viz* 160 and 20 µg kg^{-1} water, respectively.[2] Homofuraneol exists in the tautomeric forms **2a** and **2b** in a ratio of 1:3 to 1:2.[2,3] However, only **2a** is odour-active.[4]

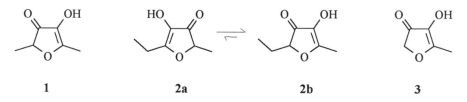

1	**2a**	**2b**	**3**

Furaneol can be formed by thermal degradation of rhamnose,[5] fructose[6,7] and hexose-phosphates.[8] In contrast, the formation of homofuraneol during heat-processing is not well understood. In general, the formation of furanones is thought to occur via the 2,3-enolisation leading to 1-deoxyosones as intermediates.[9,10] Analogously to the decomposition of hexoses, 4-hydroxy-5-methyl-3(2H)-furanone (**3**, norfuraneol) is formed from pentoses.[11] For this type of reaction, the carbon skeleton of the sugar determines the furanone formed, *i.e.* furaneol is produced from hexoses and norfuraneol from pentoses.

Recently, we described the formation of furaneol and homofuraneol from pentose sugars in Maillard systems containing glycine and alanine.[1] The key step of this reaction is chain elongation of the 1-deoxypentosone by the corresponding Strecker aldehydes (C$_5$ + C$_1$ and C$_5$ + C$_2$ reaction). In this paper, we support this mechanism with additional data and suggest sugar fragmentation as an alternative formation pathway of 3(2H)-furanones **1** and **2** from pentoses.

2 EXPERIMENTAL

D-Xylose, glycine and L-alanine (> 99%) were obtained from Fluka and [1-^{13}C]-D-Xylose (1-Xyl*), [2-^{13}C]-glycine (2-Gly*) and [3-^{13}C]-L-alanine (3-Ala*) from Cambridge Isotope Laboratories. The isotopic content of the labelled compounds was 99%. The reference compounds furaneol and homofuraneol were obtained from Aldrich and Givaudan-Roure, respectively.

Samples were prepared as described recently.[1] Xylose (5 mmol) and an amino acid (5 mmol; glycine or alanine) were dissolved in phosphate buffer (Na_2HPO_4, 5 ml, 0.2 mol l^{-1}, pH 7.0) and heated at 90 °C for one hour. Water (100 ml) was added to the reaction mixture, saturated with NaCl (40 g) and the pH adjusted to 4.0 (aqueous HCl, 2 mol l^{-1}). Neutral compounds were continuously extracted with Et_2O (50 ml) overnight using a rotation perforator (Normag). The organic phase was dried over Na_2SO_4 at 4 °C and concentrated to 0.5 ml.

GC-MS–MS was carried out as described recently:[1] MS Finnigan TSQ-700, GC HP-5890, autosampler HP-7673; ion source (150 °C); electron impact mode (70 eV); collision-induced dissociation (CID) of the molecular ions; collision energy: 10 eV; collision gas: argon (1.1 mTorr). Daughter spectra were recorded (20–200 Da). Chromatographic conditions were: carrier gas He (10 psi); cold 'on-column' injector; DB-FFAP fused silica capillary (30 m × 0.32 mm, film thickness 0.25 μm); temperature programme: 60 °C (2 min), 10 °C min^{-1} to 200 °C, 30 °C min^{-1} to 240 °C (10 min).

3 RESULTS AND DISCUSSION

The formation of furaneol and homofuraneol from pentose sugars has recently been described in phosphate-buffered (pH 6) Maillard reaction systems containing glycine and alanine.[1] We report here the formation of 3(2*H*)-furanones at pH 7 using [1-^{13}C]-D-xylose (1-Xyl*), [2-^{13}C]-glycine (2-Gly*), and [3-^{13}C]-L-alanine (3-Ala*) as precursors.

Table 1 *Relative Amounts (in %) of Norfuraneol, Furaneol and Homofuraneol Detected in Maillard Model Reactions Based on D-Xylose (Xyl), Glycine (Gly) and L-Alanine (Ala)*[a]

No.	Maillard System	Norfuraneol$_{tot}$	Furaneol$_{tot}$	Homofuraneol$_{tot}$
1	Xyl	93	3	4
2	Xyl–Gly	99.8	0.2	+[b]
3	Xyl–Ala	99.3	0.2	0.5
4	Xyl–2-Gly*	99.8	0.2	+
5	Xyl–3-Ala*	99.4	0.2	0.4
6	1-Xyl*–Gly	99.7	0.3	+
7	1-Xyl*–Ala	99.3	0.2	0.5
8	1-Xyl*–2-Gly*	99.7	0.3	+
9	1-Xyl*–3-Ala*	99.2	0.3	0.5

[a] Reaction conditions: phosphate buffer (pH 7, 0.2 mol l^{-1}), 90 °C, 1 hour;
[b] Compound marked with + was positively identified (relative amount less than 0.1%);
* Indicates labelling with ^{13}C.

3.1 Relative Amounts of 3(2*H*)-Furanones Formed

Norfuraneol was the major furanone compound in all samples analysed (Table 1), particularly in the presence of amino acids (systems 2–9), *i.e.* more than 99% of the total 3(2*H*)-furanones formed. In system 1, where only xylose was reacted, the total amount of

3(2*H*)-furanones was more than 100 times lower, thus indicating that the 1-deoxyosone of xylose was preferentially generated via the Amadori compound (Figure 1).

As shown in Table 1, both furaneol and homofuraneol were detected in all samples. However, the amount of these furanones corresponds to less than 1% of the total 3(2*H*)-furanones formed. Homofuraneol was preferentially generated in the presence of alanine (systems 3, 5, 7 and 9) compared to the reaction systems containing glycine. Conversely, the formation of furaneol was less dependent on the amino acid used. These data suggest different formation mechanisms of furaneol and homofuraneol from pentoses.

Even though the amount of norfuraneol was much higher than that of furaneol and homofuraneol, the last two dominated from the sensory point of view.[1] This is most likely due to the high retronasal odour threshold of norfuraneol (8300 µg kg^{-1} water).[2]

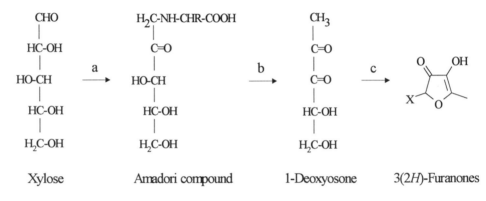

Xylose   Amadori compound   1-Deoxyosone   3(2*H*)-Furanones

Figure 1   *Schematic formation of 3(2H)-furanones detected in Maillard reaction systems based on D-xylose and glycine or L-alanine, i.e. norfuraneol (X = H), furaneol (X = CH₃) and homofuraneol (X = C₂H₅). Step **a** represents the early stage of the Maillard reaction including Amadori rearrangement; **b** the degradation of the Amadori compounds (R = H from glycine and R = CH₃ from alanine) via 2,3-enolisation; **c** cyclisation, giving rise to norfuraneol, or Strecker-assisted chain elongation (C₅ + C₁ and C₅ + C₂ reaction)[1] and recombination of sugar fragmentation products yielding furaneol and homofuraneol*

3.2 Formation of Labelled 3(2*H*)-Furanones

Formation and distribution of 3(2*H*)-furanones was investigated by GC-MS–MS in Maillard reaction systems containing the labelled precursors 1-Xyl*, 2-Gly* and 3-Ala*. Data were obtained by selecting molecular ions of unlabelled and labelled norfuraneol (*m/z* 114–117), furaneol (*m/z* 128–131) and homofuraneol (*m/z* 142–145) followed by monitoring the daughter ion spectra after CID.

3.2.1 Reaction of Xylose with Labelled Amino Acids (Table 2). Systems 4 and 5 resulted in unlabelled and singly labelled furaneol (Fur*) and homofuraneol (Hom*), respectively. Furaneol detected in the Maillard system Xyl–3-Ala* was unlabelled. These results indicate that furanones **1** and **2** are not exclusively formed by the reactions C₅ + C₁* and C₅ + C₂*, as recently described,[1] but also via recombination of sugar degradation products. This was particularly characteristic for furaneol of which about 55% was generated by sugar fragmentation. Norfuraneol was unlabelled, showing that Strecker aldehydes are not incorporated into the molecule.

Table 2 *Relative Amounts (in %)[a] of Norfuraneol (Nor), Furaneol (Fur) and Homofuraneol (Hom) formed in Maillard Model Reactions Based on Xylose (Xyl), Glycine (Gly) and Alanine (Ala)[b]*

No.	Maillard System	Nor	Nor*	Fur	Fur*	Fur**	Hom	Hom*	Hom**
4	Xyl–2-Gly*	100		55	45				
5	Xyl–3-Ala*	100		100			15	85	
6	1-Xyl*–Gly		100	10	65	25			
7	1-Xyl*–Ala		100	10	55	35	30	60	10
8	1-Xyl*–2-Gly*		100	5	45	50			
9	1-Xyl*–3-Ala*		100	5	60	35	15	10	60[c]

[a] Relative amounts of less than 1% of a compound are not reported in the table;
[b] Reaction conditions: phosphate buffer (pH 7, 0.2 mol l^{-1}), 90 °C, 1 hour;
[c] About 15% of the triply labelled homofuraneol was found in 1-Xyl*–3-Ala*;
*   Labelling with one ^{13}C atom;
** Labelling with two ^{13}C atoms.

3.2.2 Reaction of Labelled Xylose with Unlabelled Amino Acids (Table 2). In samples 6 and 7, norfuraneol was singly labelled. The mass spectrum with m/z 115 (M$^+$) and m/z 44 ($[^{13}CH_3CO]^+$) indicates that the methyl group is derived from C-1 of xylose.

The presence of unlabelled, singly and doubly labelled furaneol in both samples confirms its formation via a C_5^* + C_1 reaction and sugar fragmentation. The mass spectrum of 4-hydroxy-2-methyl-5-[^{13}C]methyl-3(2H)-furanone, formed as the major furaneol compound in the model reaction 1-Xyl*–Gly, is shown in Figure 2A. The fragmentation pattern is almost identical with that of the 2-[^{13}CH$_3$] isotopomer[1] detected in Xyl–2-Gly* (system 4). Most probably, this is due to the symmetry of the molecule.

Homofuraneol was mainly produced by a C_5^* + C_2 reaction in the Maillard system 1-Xyl*–Ala. The fragment m/z 44 in the mass spectrum of the singly labelled tautomer **2b** suggests the presence of a ^{13}CH$_3$CO substructure (Figure 2B), *i.e.* 2-ethyl-4-hydroxy-5-[^{13}C]methyl-3(2H)-furanone. This is in good agreement with the recently described formation mechanism[1] considering 1-Xyl* as sugar precursor (see also Figure 3).

3.2.3 Reaction of Labelled Xylose with Labelled Amino Acids (Table 2). Similar to samples 6 and 7, norfuraneol was singly labelled (Nor*), *i.e.* 4-hydroxy-5-[^{13}C]methyl-3(2H)-furanone (samples 8 and 9). It can be concluded that norfuraneol is directly formed from pentoses, without recombination of sugar fragmentation products.

Furaneol was generated by a C_5^* + C_1^* reaction (Figure 3), resulting in doubly labelled Fur**, and by sugar fragmentation forming singly labelled Fur*. The mass spectra of Fur* and Fur** found in systems 1-Xyl*–Ala and 1-Xyl*–3-Ala* were identical, thus suggesting the same formation mechanism. In agreement with this, similar distribution of furaneol compounds was found in samples 7 and 9. The relatively high level of Fur* in sample 8 (45%) is due to the pH (= 7). Under these conditions, sugar fragmentation via the 1-deoxyosone pathway is favoured.[10] At pH 6, the amount of Fur* was lower (12%).[1]

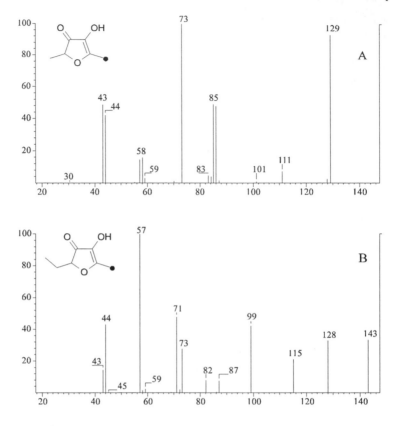

Figure 2   *GC-MS–MS spectra of 4-hydroxy-2-methyl-5-[^{13}C]methyl-3(2H)-furanone (A) and of the tautomer 2-ethyl-4-hydroxy-5-[^{13}C]methyl-3(2H)-furanone (B) obtained in Maillard systems 1-Xyl*–Gly and 1-Xyl*–Ala, respectively*

The predominant homofuraneol compound in system 1-Xyl*–3-Ala* (sample 9) was 2-(or 5-)[2-^{13}C]ethyl-4-hydroxy-5-(or 2-)[^{13}C]methyl-3(2*H*)-furanone (Hom**) formed by a C_5^* + C_2^* reaction (Figure 3). Formation involving sugar fragmentation is indicated by the presence of Hom, Hom* and Hom***.

4 CONCLUSION

The formation of norfuraneol, furaneol and homofuraneol was studied in Maillard systems by reacting D-xylose with glycine or L-alanine in a pH 7 phosphate-buffered aqueous solution at 90 °C for one hour. Norfuraneol was found to be the major component. The relative amounts of furaneol and homofuraneol were less than 1%. Experiments using the ^{13}C-labelled precursors suggest incorporation of the Strecker degradation products formaldehyde and acetaldehyde into the pentose moiety forming furaneol and homofuraneol, respectively (Figure 3). However, both furanones were partly generated by sugar fragmentation, particularly furaneol. On the contrary, homofuraneol was preferably formed by a C_5 + C_2 reaction in the presence of alanine.

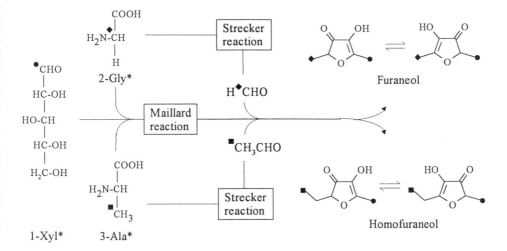

Figure 3 *Schematic formation of differently labelled furaneol and homofuraneol via a*
$C_5^{\bullet} + C_1^{\bullet}$ *and a* $C_5^{\bullet} + C_2^{\blacksquare}$ *reaction, respectively. The symbols indicate the*
origin of the carbon atoms

REFERENCES

1. I. Blank and L.B. Fay, *J. Agric. Food Chem.*, 1996, **44**, 531.
2. U.A. Huber, *Perfum. Flavorist*, 1992, **17** (4), 15.
3. L. Re, G. Maurer and G. Ohloff, *Helv. Chim. Acta*, 1973, **56**, 1882.
4. G. Bruche, A. Dietrich and A. Mosandl, *Z. Lebensm. Unters. Forsch.*, 1995, **201**, 249.
5. J.E. Hodge, B.E. Fisher and E.C. Nelson, *Am. Soc. Brew. Chem. Proc.*, 1963, 84.
6. F.D. Mills and J.E. Hodge, *Carbohydr. Res.*, 1976, **51**, 9.
7. P.E. Shaw, J.H. Tatum and R.E. Berry, *J. Agric. Food Chem.*, 1968, **16**, 979.
8. P. Schieberle, in 'Flavor Precursors – Thermal and Enzymatic Conversions', eds. R. Teranishi, G.R. Takeoka and M. Güntert, American Chemical Society, Washington, D.C., 1992, p. 164.
9. J.E. Hodge, F.D. Mills and B.E. Fisher, *Cereal Sci. Today*, 1972, **17**, 34.
10. F. Ledl and E. Schleicher, *Angew. Chem. (Int. Ed. Engl.)*, 1990, **29**, 565.
11. M.S. Feather, *Progr. Food Nutr. Sci.*, 1981, **5**, 37.

OFF-FLAVOUR FORMATION IN HEAT STERILIZED MEAT IN TRAYS

S. Langourieux[*] and F.E. Escher

Department of Food Science, Laboratory of Food Chemistry and Food Technology, Swiss Federal Institute of Technology, Zürich (ETH), CH-8092 Zürich, Switzerland

1 INTRODUCTION

Heat sterilization is one of the most important preservation techniques for producing microbiologically safe foods storable at room temperature. The increasing consumer demand for easy-to-use and smaller portions, and with a favourable geometry to reduce heat damage, has led to heat processed convenience foods packed in trays rather than in cans.[1]

Nevertheless, sterilization and storage still lead to a considerable loss of organoleptic and nutritional quality, due to oxidative and heat-induced changes,[2,3] in particular if these products contain meat preparations. Off-flavours are developed via the autoxidation of polyunsaturated fatty acids and oxidative effects on proteins, peptides and amino acids.[4-6] Oxidative stability of various heat-processed menu components has been studied in a comprehensive project at the ETH, Zürich, in collaboration with the food canning industry. The influence of the atmosphere above the product and the influence of a natural antioxidant were particularly investigated. For meat products in aluminium trays, removal of air by flushing with nitrogen during filling associated with a storage temperature of 20 °C significantly increased the storage stability.[1] But in this case, it also leads to the formation of a strong sulfurous off-flavour after the sterilization process. In turn, addition of a rosemary extract as an antioxidant resulted in a very effective stabilization of the product and no sulfurous off-flavour was detected sensorially when the trays were opened.[7]

Our objective in this work was first to investigate this off-flavour and then to determine the influence of the process parameters on its formation.

2 MATERIAL AND METHODS

2.1 Sample Preparation

The meat-fat mixture was prepared according to Güntensperger and Escher.[1] To this ground meat-fat mixture (90 g) was added 88 mg per kg fat of citric acid or 44 mg per kg fat of desaromatized rosemary extract (RE-S, FIS S.A., Switzerland) or samples were further processed without antioxidant. Deionized water (180 g) was added and the samples

[*]Present address: Laboratoire de Génie Biologique et Sciences des Aliments, Unité de Microbiologie et Biochimie industrielles, Université de Montpellier II, F-34095 Montpellier cedex 5, France

were put into aluminium trays (Type 73300, 138 × 138 × 28 mm, W. Wagner GmbH, Germany). The trays were sealed with a peelable aluminium lid foil (Flexalpeel 45/1000, Alusingen, Germany) with or without prior removal of air by evacuating to 200 mbar and flushing back with nitrogen to atmospheric pressure (CPM H175VG, Packaging Machinery, England). This operation was repeated three times before sealing. The trays were then sterilized in hot water in a batch retort (Pilot Rotor 400, Stock GmbH, Germany) without agitation at 121 °C or 131 °C to an F_0 value of 8–11 min. The trays were then stored in the dark at 20 °C for 14 hours or one month before analysis.

To investigate the effects of the different experimental parameters on the response variable (headspace concentration of volatile compounds), an ANOVA analysis was performed (SAS v.6.10, SAS Institute Inc., U.S.A.).

2.2 Analysis of the Volatile Compounds of the Headspace

2.2.1 Purge and Trap of the Headspace. The headspace from each tray was collected (without opening the tray) using two needles. The first one purged the tray with purified nitrogen (20 ml min^{-1}) and the second one drove the headspace gas into a PTFE column (14 cm length, 0.5 cm internal diameter) packed with Tenax GR, (Alltech). After an hour of purging, the Tenax trap was removed and replaced by another and this purge lasted a further three hours. Diethyl ether (20 ml) was used for the extraction of the traps with 1-butanethiol (Aldrich) and n-dodecane (Sigma) as internal standards. The total ether solution was further concentrated with a Kuderna-Danish concentrator to a volume of approximately 0.1 ml. No internal standard was added to the extracts for olfactometry.

2.2.2 Gas Chromatography (GC) and Gas Chromatography–Olfactometry (GC–O). A Hewlett-Packard model 5890 Series II gas chromatograph equipped with an HP-5 column (30 m length, 0.32 mm internal diameter, 0.25 mm film thickness) was used for the analysis of the extracts (1 µl). Nitrogen was used as the carrier gas (1.1 ml min^{-1}). The temperature programme started at 30 °C and increased at a rate of 1 °C min^{-1} to 40 °C, then increased at a rate of 10 °C min^{-1} to 250 °C and held for 10 min. Injector, FID and FPD (flame photometric detector) temperatures were set at 220, 250, and 150 °C respectively. The gas chromatograph was also equipped with an olfactometric detector (SGE, Australia). The data were analysed according to the CHARM analysis[11] with the help of software derived from Origin 3.5 (DMP Ltd., Switzerland).

2.2.3 Gas Chromatography–Mass Spectrometry Analyses (GC–MS). A Fisons GC 8000 series gas chromatograph equipped with the same column and operated at the same conditions as for the GC analysis and coupled to a Finnigan MAT SSQ710 mass spectrometer was used with helium as the carrier gas. The source, analyser and transfer line temperatures were 150, 70 and 220 °C respectively. The ionization voltage applied was 70 eV, the emission current was 0.2 mA and the electron multiplier voltage 1000 kV. Mass spectra obtained from a scan range of 33–250 *m/z* were compared with those of known compounds in the NIST library.

2.2.5 Analysis of Hydrogen Sulfide. Based on method of Jacobs *et al.*[9,10]

2.2.6 Ethane Determination. As described by Güntensperger and Escher[1]

3 RESULTS AND DISCUSSION

3.1 Determination of the Volatile Compounds Responsible of the Off-flavour

A 2 ml sample of the headspace of a non-flushed and a nitrogen-flushed tray were directly injected and a GC–O analysis carried out. The results are given in Figure 1. Short chain alkanes from C1 to C6 and hydrogen sulfide were identified. No odour was detected for the 'C-tray' (no flushing) but a strong spoiled egg like odour was detected for the 'F-trays' (flushing with nitrogen) at a retention of 1.8 min corresponding to hydrogen sulfide. The same odour was perceived when opening the 'F-trays'.

'C-' and 'F-trays' were also extracted using the purge and trap method and GC, GC–MS and GC–O analysis were performed. CHARM profiles of C and F headspace samples are shown in Figure 2. CHARM profiles revealed differences between the two samples. Dimethyl trisulfide, 2-pentyl furan and octanal are important contributors to the odour

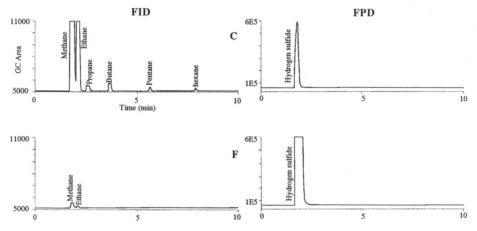

Figure 1   *FID and FPD chromatograms of 2 ml directly injected tray headspace (C: no removal of air before sealing; F: flushed with nitrogen)*

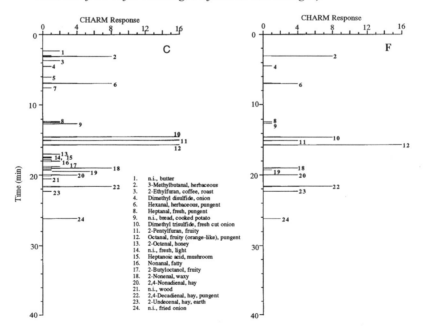

Figure 2   *CHARM profiles of headspace purge and trap extracts of two different trays (C: no removal of air before sealing; F: flushed with nitrogen)*

perceived in the 'C-trays' while only octanal persists in 'F-trays'. It can be concluded that hydrogen sulfide is mainly responsible for the off-flavour in the 'F-trays' but flushing with nitrogen also greatly modifies other odour active compounds.

No typical roasted or meaty notes were found in the purge and trap extracts. This is due to the fact that the samples were sterilized without prior pre-cooking and in the presence of a large proportion of water. All the identified odour active compounds were already described in the literature as meat aroma compounds.[11–15] They are formed by the autoxidation of lipids and by the degradation of sulfur-containing amino acids.

Hydrogen sulfide and dimethyl trisulfide were chosen as criteria for the off-flavour. To follow lipid oxidation, headspace concentration of ethane was also measured.

Headspace concentration of hydrogen sulfide, ethane and dimethyl trisulfide, analysed 14 hours and one month after the sterilization are summarized in Table 1. The most significant parameter in the ANOVA of the data was the atmosphere above the product. Flushing the trays with nitrogen significantly reduced the concentration of ethane and dimethyl trisulfide, whereas it increased hydrogen sulfide concentration.

Table 1 *Calculated Means of Headspace Concentrations of Hydrogen Sulfide, Ethane and Dimethyl Trisulfide from ANOVA*

	Temperature (°C)		Atmosphere		Antioxidant		
	121	*131*	*C*	*F*	*AO_0*	*AO_1*	*AO_2*
Hydrogen sulfide (mg l^{-1})							
T0	12.05	9.16	4.57	16.64	9.29	12.21	10.31
T4	7.79	5.37	0.09	13.06	6.39*	6.30*	7.04*
Ethane (mg l^{-1})							
T0	13.12	9.24	18.47	5.04	13.36	9.48	11.21
T4	93.97	87.59	175.02	13.30	84.22	63.38	123.52
Dimethyl trisulfide (mg l^{-1})							
T0	6.96	3.48	8.04	2.35	4.86*	3.05*	6.27*
T4	n.d.	n.d.	n.d.	n.d.	n.d.	n.d.	n.d.

T0, T4: 14 h and one month storage respectively;
*   Not significant at 0.05 level; n.d. Not determined.

3. 2 Influence of the Process Parameters on Lipid Oxidation

In previous work,[1,7,10] flushing with nitrogen and use of a rosemary extract were very effective in preventing lipid oxidation. Similar effects of vacuum and/or modified atmosphere have already been reported for beef patties, sliced cooked meat and whole milk powder.[6,16–18]

3. 3 Influence of the Process Parameters on Off-flavour

3.3.1 Influence of the Sterilization. For the two sterilization temperatures, the total sterilization effect was the same but the thermal treatments were very different with a cooking value of 80 min and 120 min for 131 °C and 121 °C respectively. A 10 °C increase of the sterilization temperature led to a decrease of about 25% in hydrogen sulfide and 30% in dimethyl trisulfide concentrations after 14 hours of storage. Hydrogen sulfide is formed by thermal denaturation of proteins and/or thermal degradation of sulfur-containing amino acids.[19-22] This can easily be explained since hydrogen sulfide is formed from the –SH groups of the proteins or the free amino acids.

Dimethyl trisulfide is formed indirectly in the Strecker degradation of methionine, under the effect of a thermal process. This produces methional, which is further degraded to methanethiol. Methanethiol can be transformed into dimethyl disulfide in the presence of oxygen.[23] Dimethyl trisulfide can be generated from dimethyl disulfide easily. The role of the temperature in these reactions explains the differences in the concentration of dimethyl trisulfide for the two sterilization temperatures.

3.3.2 Influence of Flushing with Nitrogen. Flushing with nitrogen has a very significant role on the hydrogen sulfide concentration in two ways. First, it increases the concentration initially formed (16.64 µg l^{-1} in 'F-trays' against 4.57 µg l^{-1} in 'C-trays'). The redox potential is strongly reduced in the absence of oxygen and a low redox potential favours the breakdown of the disulfide groups of the proteins and the formation of cysteine from free cystine.[24] As a consequence, formation of hydrogen sulfide is also favoured.

Secondly, the decrease in hydrogen sulfide concentration after one month's storage was higher for 'C-trays' (about 97%) than for 'F-trays' (about 22%). One can assume that either the reactivity of hydrogen sulfide is strongly reduced without oxygen or that the compounds with which it can react (*e.g.*, the secondary products of the lipid oxidation) were not present in 'F-trays'. It is known that protein oxidation can occur through radical reactions promoted by lipid hydroperoxides. A radical can be formed at the α-carbon of the amino acids, but the side chains are also susceptible to damage and cysteine/cystine are among the most labile amino acids. In the presence of oxygen, the radical initiation on sulfur substituents through the formation of thiyl radicals takes place very easily.[25-27] One can assume that hydrogen sulfide is also able to form thiyl radicals. This has already been described for high temperatures.[19] Thus the reaction between lipid hydroperoxides and thiyl radicals is highly probable. In the absence of oxygen, such as in the case of 'F-trays', hydrogen sulfide formed in excess due to the redox potential decrease, cannot react further because of the absence of the secondary products of the lipid oxidation. It is assumed that the decrease of hydrogen sulfide concentration after one month's storage was due only to its dissolution in water.

For the formation of dimethyl trisulfide, the role played by the oxygen explains why its concentration was smaller in 'F-trays' than in 'C-trays'.

3.3.3 Influence of Antioxidant. Rosemary extract is a primary antioxidant which can react with radicals to convert them into more stable products. Citric acid is a secondary antioxidant and has the ability to chelate metal ions which catalyse the lipid oxidation and so inhibits the formation of lipid peroxides. The significant role played by the antioxidant on the initial concentration of hydrogen sulfide supports the hypothesis that thiyl radicals are involved. With the more efficient antioxidant (rosemary extract), the concentration of hydrogen sulfide was higher. Consequently, the concentration of hydrogen sulfide was the smallest without antioxidant.

The antioxidants have no effect on the concentration of dimethyl trisulfide.

4 CONCLUSION

This study investigated the influence of different antioxidative treatments applied to sterilized meat in trays on the formation of off-flavour. The removal of oxygen by flushing with nitrogen prior to sterilization was very effective in inhibiting lipid oxidation but gave rise to off-flavour due to an accumulation of hydrogen sulfide and to a diminution of other odour active compounds such as dimethyl trisulfide. Both hydrogen sulfide and dimethyl trisulfide are characteristic compounds of meat aroma, but when the balance between the concentrations is modified, the flavour turns into off-flavour. A residual oxygen quantity must remain in the tray in order to permit formation of thiyl radicals and to prevent accumulation of hydrogen sulfide. In this perspective, use of rosemary extract as an antioxidant could be the optimal solution, both preventing lipid oxidation and formation of off-flavour.

REFERENCES

1. B. Güntensperger and F.E. Escher, *J. Food Science*, 1994, **59**, 689.
2. J. Kanner, in 'Lipid Oxidation in Food', ed. A.J. St. Angelo, American Chemical Society, NY, 1992, p. 55.
3. J.R. Vercellotti, A.J. St. Angelo and A.M. Spanier, in 'Lipid Oxidation in Food', ed. A.J. St. Angelo, American Chemical Society, NY, 1992, p. 1.
4. J.I. Gray and A.M. Pearson, 'Advances in Food Research', eds. A.M. Pearson and T.R. Dutson, Van Nostrand Reinhold, NY, 1987, Vol. 3, p. 221.
5. D. Ladikos and V. Lougovois, *Food Chem.*, 1990, **35**, 295.
6. A.M. Spanier, A.J. St. Angelo and G.P. Shaffer, *J. Agric. Fd. Chem.*, 1992, **40**, 1656.
7. B. Güntensperger and F.E. Escher, 86th American Oil Chem. Soc. Annual Meeting, San Antonio, Texas, 1995.
8. T.E. Acree, J. Barnard and D.G. Cunningham, *Food Chem.*, 1984, **14**, 273.
9. M.B. Jacobs, M.M. Braverman and S. Hochheiser, *Anal. Chem.*, 1957, **29**, 1349.
10. S. Langourieux and F.E. Escher, *J. Food Science*, submitted for publication.
11. A.M. Galt and G. MacLeod, *J. Agric. Fd. Chem.*, 1984, **32**, 59.
12. G. MacLeod and J.M. Ames, *Flavour and Fragrance J.*, 1986, **1**, 91.
13. D.W. Baloga, G.A. Reineccius and J.W. Miller, *J. Agric. Fd. Chem.*, 1990, **38**, 2021.
14. T.D. Drumm and A.M. Spanier, *J. Agric. Fd. Chem.*, 1991, **39**, 336.
15. N. Ramarathnam, L.J. Rubin and L.L. Diosady, *J. Agric. Fd. Chem.*, 1993, **41**, 933.
16. H. Stapelfeldt, H. Bjorn, L.H. Skibsted and G. Bertelsen, *Z. Lebensm. Unters. Forsch.*, 1993, **196**, 131.
17. G. Hall and H. Lingnert, *J. Food Quality*, 1984, **7**, 131.
18. G. Hall, J. Andersson, H. Lingnert and B. Olofsson, *J. Food Quality*, 1985, **7**, 153.
19. M. Fujimaki, S. Kato and T. Kurata, *Agr. Biol. Chem.*, 1969, **33**, 1144.
20. M. Boelens, L.M. Van der Linde, P.J. De Valois, H.M. Van Dort and H.J. Takken, *J. Agric. Fd. Chem.*, 1974, **22**, 1071.
21. C.-K. Shu, M.L. Hagedorn, B.D. Mookherjee and C.T. Ho, *J. Agric. Fd. Chem.*, 1985, **33**, 438.
22. C.-K. Shu, M.L. Hagedorn, B.D. Mookherjee and C.T. Ho, *J. Agric. Fd. Chem.*, 1985, **33**, 442.
23. L. Schutte, *CRC Critical Reviews in Food Technology*, 1974, **4**, 457.
24. K. Hofmann, *Die Fleischwirtschaft*, 1977, **10**, 1818.
25. H.W. Gardner, *J. Agric. Fd. Chem.*, 1979, **27**, 220.
26. H.W. Gardner, in 'Xenobiotics in Foods and Feeds', eds. J.W. Finley and D.E. Schwass, American Chemical Society, Washington, D.C., 1983.
27. D.A. Lillard, in 'Warmed-Over Flavor of Meat', eds. A.J. St. Angelo and M.E. Bailey, Academic Press, Inc., 1987, p. 63.

STUDIES ON THE AROMA OF ROASTED COFFEE

Werner Grosch, Michael Czerny, Robert Wagner and Florian Mayer

Deutsche Forschungsanstalt für Lebensmittelchemie, Lichtenbergstrassße 4, D-85748 Garching, Germany

1 INTRODUCTION

Potent odorants of ground roasted coffee and of coffee brews have been identified on the basis of aroma extract dilution analysis (AEDA) and gas chromatography–olfactometry of headspace samples (GC–O-H).[1-4] Twenty-two of them were quantified in brews prepared from *Arabica* (*Coffea arabica*) and *Robusta* coffees (*Coffea canephora* var. *Robusta*), and corresponding aroma models were made to verify to what extent these odorants contributed to the odour of coffee.[5,6] The overall odour of the models was characterised by a panel as clearly coffee-like. However, there were some differences in the intensities of the aroma notes, and, in particular, the earthy/musty note was weaker in the models than in the original brews.[6]

In continuation of this work, the screening for potent odorants was repeated to check the results mentioned above. The concentration of the extracts (containing the food volatiles) to a small volume before the start of a dilution experiment might be a shortcoming of AEDA. Therefore, aroma extract concentration analysis (AECA), which has been introduced as an alternative,[7] was performed in addition to AEDA. AECA starts with GC–O of the original extract which is then concentrated stepwise by distilling off the solvent and, after each step, an aliquot is analysed by GC–O.[7]

In agreement with the results mentioned above, two earthy smelling odorants of unknown structure were detected by AEDA and AECA. These compounds were identified, and their odour activities were compared to those of structure homologues. The new odorants and twenty-three further compounds causing sweetish/caramel, earthy, roasty/sulfurous, smoky, green and buttery odour notes were quantified in roasted ground coffees from Colombia and Kenya. The results are reported in the present paper.

2 SCREENING FOR POTENT ODORANTS

Ground coffee (medium roasted, 5 g) was extracted with a mixture of water, CH_2Cl_2 and methanol (4:5:10, v/v/v, 300 ml) and then with CH_2Cl_2 (300 ml). The volatile fraction including the solvents was distilled off from the non-volatile material.[6] The distillate (512 ml) was divided into two portions. One half was subjected to AEDA and the other to AECA.

In AEDA the distillate was concentrated to 2 ml and then diluted as a series of 1:1 dilutions, and each dilution was analysed by GC–O. 4-Vinylguaiacol (**16** in Table 1), 4-

ethylguaiacol (**17**), 4-hydroxy-2,5-dimethyl-3(2*H*)-furanone (**19**) and β-damascenone (**23**) were identified as the most potent odorants. They were still found by GC–O after dilution of the distillate to its initial volume of 256 ml.

Table 1 *Evaluation of Potent Odorants of Ground Roasted Arabica Coffee from Colombia by AEDA and AECA*

No	Odorant	RI^a	Volume^b (ml)	
			AEDA	ECA
1	2,3-Butanedione	585	32	32
2	2,3-Pentanedione	695	16	16
3	2-Methyl-3-furanthiol	870	16	32
4	2-Furfurylthiol	913	6	32
5	Methional	909	64	64
6	Unknown(sulfurous, burnt)	950	16	8
7	3-Methyl-2-buten-1-thiol	821	16	32
8	3-Mercapto-3-methyl-1-butanol	970	64	8
9	3-Mercapto-3-methylbutylformate	1023	64	32
10	2-Ethyl-3,5-dimethylpyrazine	1083	64	64
11	Unknown(earthy)	1103	64	16
12	2,3-Diethyl-5-methylpyrazine	1155	64	64
13	Unknown(earthy)	1182	64	64
14	3-Isobutyl-2-methoxpyrazine	1186	64	32
15	Guaiacol	1093	128	64
16	4-Vinylguaiacol	1323	256	256
17	4-Ethylguaiacol	1287	256	32
18	Vanillin	1410	32	16
19	4-Hydroxy-2,5-dimethyl-3(2*H*)-furanone	1065	256	256
20	2-Ethyl-4-hydroxy-5-methyl-3(2*H*)-furanone	1140	128	128
21	3-Hydroxy-4,5-dimethyl-2(5*H*)-furanone	1107	64	16
22	5-Ethyl-3-hydroxy-4-methyl-2(5*H*)-furanone	1193	32	32
23	β-Damascenone	1395	256	256

[a] Retention index (RI) on the capillary SE-54;
[b] Volume of the distillate at which the odorant was at first (AECA) or at last (AEDA) perceived by GC–O.

The odorants **16**, **19** and **23** were also detected in AECA in an aliquot of the original distillate of 256 ml (Table 1). After concentration of this sample to 128 ml, the odour of furanone **20** was perceived, and after concentration to 64 ml, five further odorants (**5**, **10**, **12**, **13**, **15**) were detected. In most cases, the volumes of the distillate, at which the odorant was detected in AEDA and AECA, differed by not more than one dilution or concentration step (Table 1). Exceptions were **8**, **11**, **17** and **21** which were all still perceived at higher dilutions in AEDA than in AECA. Possibly, progressing from a concentrated to a diluted sample, enhances the sensitivity of the panellist for the detection of odorants in AEDA. This might be in contrast to AECA in which, at the beginning of the procedure, the odorants have to be smelled during GC–O of highly diluted samples.

Nevertheless, all the potent odorants detected previously in AEDA[1,2] were also found in the experiments reported in Table 1. Only one unknown, **6**, smelling sulfurous, burnt was new. Further experiments showed that the retention indices of **6** on three capillaries of different polarity and the odour of **6** agreed with the corresponding properties of 2(1-mercaptoethyl)-furan. The earthy smelling, unknown compounds **11** and **13** had been previously perceived with relatively high FD-factors in AEDA[2] and GC–O-H[4] of coffee brews.

3 ADDITIONAL EARTHY ODORANTS

As the earthy odour quality of the unknown compounds **11** and **13** (Table 1) agreed with that of **10** and **12**, we suggest that they were also alkylpyrazines.

Shibamoto and Bernhard[8] have detected alkylpyrazines bearing an ethenyl group as Maillard reaction products of glucose-ammonia model systems. This result, in combination with the detection of **10** and **12** as potent odorants, led to the assumption that similarly structured ethenyl pyrazines might occur in roasted coffee, *e.g.* 2-ethenyl-3,5-dimethylpyrazine (**11**), 2-ethenyl-3-ethyl-5-methylpyrazine (**13a**) and 3-ethenyl-2-ethyl-5-methylpyrazine (**13b**).

To prove this hypothesis, the pyrazines **11**, **13a** and **13b** were synthesized, their chromatographic, spectroscopic and sensory properties were determined and finally compared to those of the unknown earthy smelling odorants of coffee.[9] At first, in a *Robusta* coffee sample, the two unknown earthy odorants were identified as 2-ethenyl-3,5-dimethylpyrazine (**11**) and 2-ethenyl-3-ethyl-5-methylpyrazine (**13a**) by comparison of GC and MS data with those of the corresponding reference substances.[9]

4 STRUCTURE-ODOUR ACTIVITY RELATIONSHIPS OF PYRAZINES

More than seventy alkylpyrazines are reported in the literature as components of the volatile fraction of coffee.[10] However, as shown above, only **10**, **11**, **12** and **13a** were identified as potent odorants. This discrepancy prompted us to compare the odour threshold values of alkylpyrazines in air. A total number of forty-five pyrazines was investigated; an extract of the results comparing eighteen compounds is shown in Figure 1.

Trimethylpyrazine had the lowest threshold in the line mono-, di-, tri- and tetramethylpyrazine (**P1** to **P6**). Substitution of the methyl group in position 2 by an ethyl group leads to pyrazine **10** having a 9600-fold lower odour threshold than **P5**. Ethyl groups in positions 3 (**P7**) or 5 (**P8**) of the pyrazine molecule were not so effective in lowering the odour threshold as those in position 2. A propyl instead of an ethyl group in position 2 was too bulky, the odour threshold increased from 0.01 ng l^{-1} (**10**) to 23 ng l^{-1} (**P9**). The thresholds of **11**, **12** and **13a** were as low as that of **10**. It was therefore not surprising that only these very odour-active pyrazines as well as **10** were perceived during GC–O of diluted aroma extracts of coffee.

The odour threshold of **13b** was 9500 times higher than that of **13a** (Figure 1) indicating that an ethenyl group was only tolerated in position 2, but not in 3. A further increase of the threshold was found, when the ethenyl group was in position 5 (**P12** and **P13**).

The great differences in the odour threshold values discussed here allows the conclusion that only a small number of alkylpyrazine molecules matches the geometry of the receptor for the earthy odour impression. Most alkylpyrazine molecules occurring in coffee are either too small or too large.

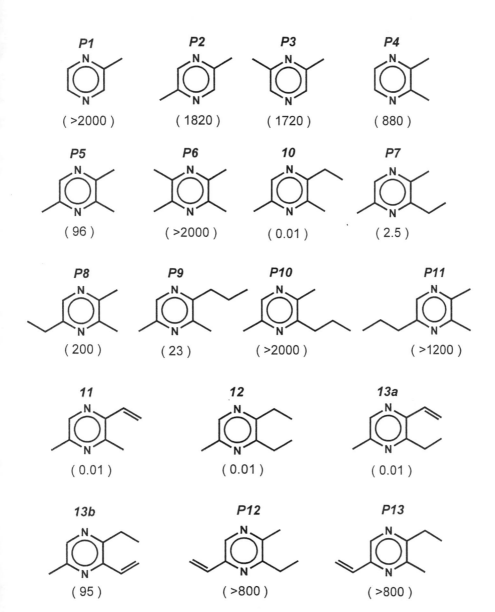

Figure 1 *Odour thresholds of alkylpyrazines in air. The values given in brackets and expressed as ng l⁻¹ were determined by AEDA using (E)-2-decenal as internal standard.² Numbering 10, 11, 12, 13a refers to Table 2*

5 CONCENTRATIONS OF KEY ODORANTS

The flavour of *Arabica* coffees depends on the provenance, *e.g.* ground roasted coffee from Colombia smells more intense smoky, roasty and earthy and less intense sweetish/caramel-like than coffee from Kenya. To get a first insight into the compounds causing these aroma differences, twenty-five potent odorants (Table 2), which had been screened by the dilution experiments reported here and in earlier papers,[2,4] were quantified in the two types of coffee using isotope dilution assays. Most of these analytical procedures were adapted from studies on coffee[5,6] and meat flavours.[11] However, assays for pyrazines 11 and 13a as well as for 2-methyl-3-furanthiol (3 in Table 2) were developed in the present study.

Table 2 *Concentrations (mg kg⁻¹) of Potent Odorants in Ground Roasted Arabica Coffee from Colombia and Kenya* [a]

No.[b]	Odorant	Colombia[c]		Kenya
		Sample A	Sample B	
1	2,3-Butanedione	48.4	50.8	58.7
2	2,3-Pentanedione	34.0	39.6	38.4
3	2-Methyl-3-furanthiol	0.148	—	0.195
4	2-Furfurylthiol	1.68	1.7	1.3
5	Methional	0.201	0.240	0.226
9	3-Mercapto-3-methylbutylformate	0.077	0.130	0.156
10	2-Ethyl-3,5-dimethylpyrazine	0.324	0.330	0.272
11	2-Ethenyl-3,5-dimethyl-pyrazine	0.053	—	0.037
12	2,3-Diethyl-5-methylpyrazine	0.069	0.095	0.061
13a	2-Ethenyl-3-ethyl-5-methylpyrazine	0.014	—	0.013
14	2-Isobutyl-3-methoxypyrazine	0.067	0.083	0.050
15	Guaiacol	3.0	4.2	1.91
16	4-Vinylguaiacol	70.0	64.8	55.4
18	Vanillin	3.41	4.8	3.9
19	4-Hydroxy-2,5-dimethyl-3(2*H*)-furanone	112	109	131
20	2-Ethyl-4-hydroxy-5-methyl-3(2*H*)-furanone	16.8	17.3	16.9
21	3-Hydroxy-4,5-dimethyl-2(5*H*)-furanone	1.36	1.47	1.71
22	5-Ethyl-3-hydroxy-4-methyl-2(5*H*)-furanone	0.104	0.16	0.159
23	β-Damascenone	0.222	0.195	0.236
24	Dimethyltrisulfide	0.028	—	0.029
25	Acetaldehyde	139	—	141
26	Propanal	17.4-	—	16.9
27	Methylpropanal	34.2	—	21.7
28	2-Methylbutanal	20.2	—	11.6
29	3-Methylbutanal	17.8	—	10.4

[a] The coffee beans were medium roasted. Data are means of at least two assays; maximum SD ± 10%;
[b] Numbering 1–5, 9–16, 18–23 refers to Table 1;
[c] Sample B was analysed by Semmelroch and co-workers.[5,6,13]

The rapid oxidation of **3**[12] and its labelled internal standard was prevented in the analytical procedure by a reaction with 1-buten-3-one. The unlabelled (resulting from the analyte) and the labelled 4-(2-methyl-3-furanthio)-2-butanone formed were then analysed by GC–MS.

The concentrations of the odorants found in coffees from Colombia (sample A) and Kenya are compared in Table 2. Data obtained in a preceding investigation of some potent odorants of Colombia coffee (sample B) were included in Table 2.

The concentrations of **1**, **2**, **4**, **5**, **10**, **14**, **16**, **21** and **23** agreed to a large extent in the two Colombian coffees, differing by no more than 25%. Only phenols **15** and **18** (40%), thiol **9** (70%), pyrazine **12** (38%) and furanone **22** (54%) were much higher in sample B. As 2-furfurylthiol (**4**) was not stable,[13] it decreased from 1.7 mg kg^{-1} to 1.08 mg kg^{-1}, when the *Arabica* coffee from Colombia was stored for one year at −35 °C.[6]

Thiol (**4**) and guaiacol (**15**) were lower in the Kenyan than in the two Colombian coffees. This difference might be the reason for the weaker smoky and roasty odour notes, respectively, in the former. The lower concentrations of pyrazines (**10**) and (**11**) might attenuate the earthy note in the overall odour of the Kenyan. In addition, the concentrations of the Strecker aldehydes **27**, **28** and **29** were 57% to 74% smaller in the Kenyan than in the Colombian (A). On the other hand, the concentrations of the furanones (**19**) and (**21**) were higher and this might enhance the caramel-like odour in the Kenyan.

In conclusion, the concentration differences of potent odorants were so small in *Arabica* coffees of different provenances that they can be ascertained only by sensitive and reliable analytical methods.

ACKNOWLEDGEMENT

This work was supported by the FEI (Forschungskreis der Erhrungsindustrie e.V., Bonn, Germany), the AIF and the Ministry of Economics. Project No. 10316N.

REFERENCES

1. W. Holscher, O.G. Vitzthum and H. Steinhart, *Café, Cacao, Thé*, 1990, **34**, 205.
2. I. Blank, A. Sen and W. Grosch, *Z. Lebensm. Unters. Forsch.*, 1992, **195**, 239.
3. W. Holscher and H. Steinhart, *Z. Lebensm. Unters. Forsch.*, 1992, **195**, 33.
4. P. Semmelroch and W. Grosch, *Lebensm. Wiss. Technol.*, 1995, **28**, 310.
5. P. Semmelroch, G. Laskawy, I. Blank and W. Grosch, *Flavour Fragrance J.*, 1995, **10**, 1.
6. P. Semmelroch and W. Grosch, *J. Agric. Food Chem.*, 1996, **44**, 537.
7. R. Kerscher and W. Grosch, *Z. Lebensm. Unters. Forsch.*, in press.
8. T. Shibamoto and R.A. Bernhard, *Agric. Biol. Chem.*, 1977, **41**, 143.
9. M. Czerny, R. Wagner and W. Grosch, *J. Agric. Food Chem.*, in press.
10. L.M. Nijssen, C.A. Visscher, H. Maarse, L.C. Willemsens and M.H. Boelens, 'Volatile Compounds in Food', 7th edition, TNO, Nutrition and Food Research Institute, Zeist, The Netherlands, 1996.
11. H. Guth and W. Grosch, *J. Agric. Food Chem.*, 1994, **42**, 2862.
12. T. Hofmann, P. Schieberle and W. Grosch, *J. Agric. Food Chem.*, 1996, **44**, 251.
13. W. Grosch, P. Semmelroch and C. Masanetz, 15th International Conference on Coffee Science (ASIC 15), Montpellier, France, Vol. II, p.545 (1994).

MECHANISM OF PYRIDINE FORMATION FROM TRIGONELLINE UNDER COFFEE ROASTING CONDITIONS

G.P. Rizzi and R.A. Sanders

The Procter and Gamble Company, Cincinnati, OH 45253-8707, U.S.A.

1 INTRODUCTION

Trigonelline (**1**) is, next to caffeine, the most prevalent non-protein nitrogen compound in raw coffee beans. During roasting, trigonelline is decomposed and its decomposition is associated with the formation of pyridine and nicotinic acid in roasted coffee.[1] The pyridine content of coffee increases with time and temperature of roasting, the highest value being *ca.* 200 ppm.

The mechanism of trigonelline degradation has not been extensively studied. Products isolated from trigonelline pyrolyzed at 180–230 °C (15–16 min.) include pyridine plus eleven monocyclic and nine bicyclic pyridine derivatives.[2]

The objective of this work was to examine the organoleptic role of pyridine as it affects the overall aroma acceptance of brewed coffee and to elucidate the formation mechanism of pyridine and its derivatives from trigonelline under coffee roasting conditions.

2 EXPERIMENTAL SECTION

2.1 Coffee Roasting Studies

Coffee beans were roasted over ranges of time and temperature to obtain a matrix of nine products representing a typical span of coffee roasting conditions. The products were ground and brewed and the coffee brews were analysed by a combination of sensory and instrumental techniques. For sensory measurement, the brews were sniffed by panellists and rated on eight (0–20 point) aroma attribute scales and a similar overall preference scale. For instrumental measurement, volatiles were isolated from brews by solid phase extraction, desorbed by solvent and analysed by gas chromatography. Numerical correlations among sensoric and objective data sets were obtained *via* the 'Correlation Tool' in Excel (Microsoft Co., version 4.0). Correlation data ranged from +1.00 (100% positive correlation) through 0.00 (no correlation) to −1.00 (100% negative correlation).

2.2 Trigonelline Pyrolysis Experiments

2.2.1 Materials. Trigonelline hydrate was prepared by passing the chloride salt through an anion-exchange resin (OH-form). Synthesis of methyl-d₃-labelled trigonelline began with methyl-d₃ alcohol. The sodium salt of methyl-d₃ alcohol reacted with *p*-toluenesulfonyl chloride to form methyl-d₃ *p*-toluenesulfonate. Reaction of the labelled methyl tosylate with nicotinic acid gave methyl-d_3 trigonelline *p*-toluenesulfonate which was passed through an anion-exchange resin to obtain methyl-d_3 trigonelline hydrate.

 2.2.2 Methods of Pyrolysis and Analysis. Trigonelline was flame-sealed into glass ampoules and heated in a thermostated oil bath at 200–205 °C for 30 min. Evidence of the total conversion of starting material was provided by the disappearance of the C-2 ring hydrogen NMR signal at 9.19 ppm. Solubility tests indicated pyrolysis products were mostly intractable polymeric material. For HPLC analysis, the pyrolysis product was slurried in methanol, filtered, diluted with an 80/20 MeOH–water mixture and analysed on a reversed phase [C-18] column coupled to a diode-array detector. For GC and GC–MS analysis, the pyrolysis product was added to water and subjected to atmospheric Likens-Nickerson (SDE) extraction using methylene chloride. SDE volatiles were quantitated with 4,4'-dipyridyl as an internal standard and the products were identified by comparing data with library spectra and authentic standards. In unsealed reactions, carbon dioxide was swept from the reaction zone with nitrogen into barium hydroxide solution and quantified as barium carbonate. Similarly, methylamine was trapped in dilute sulfuric acid, liberated with alkali and quantified as its N-benzoyl derivative.

Table 1   *Trigonelline Pyrolysis Products*

Product	$mg\ g^{-1}$ (1)	% yield	Identity
Carbon dioxide	250	88	a
Nicotinic acid	48	6.1	b, d, e
Pyridine	27	5.3	c, d, e
Methylamine	5.3	2.7	a, e
Methyl nicotinoate	5.3	0.61	b, d, e
4-Phenylpyridine	4.9	0.49	c, d
3-Methylpyridine	1.5	0.25	c, d, e
Ethyl tetrahydroquinolines	0.76	0.073	c
Methyl tetrahydroquinolines	0.42	0.044	c
4-n-Propylpyridine	0.22	0.028	c, d
Methyl pyridyl piperidine	0.18	0.032	c
4-Ethylpyridine	0.14	0.021	c, d
Dimethyl/ethylpropylpyridine	0.13	0.014	c
A tetrahydroquinoline	0.13	0.015	c
3-Methyl-5-propylpyridine	0.12	0.014	c
3-Ethyl-4-methylpyridine	0.11	0.014	c, d
Isoquinoline	0.03	0.004	
3-Aminobiphenyl	trace		c
N-methylnicotinamide	trace		b, d
N, N-dimethylnicotinamide	trace		b, d

a   Chemical derivative;
b   HPLC–DAD;
c   GC–MS + exact mass;
d   Known compound available;
e   Found in coffee.

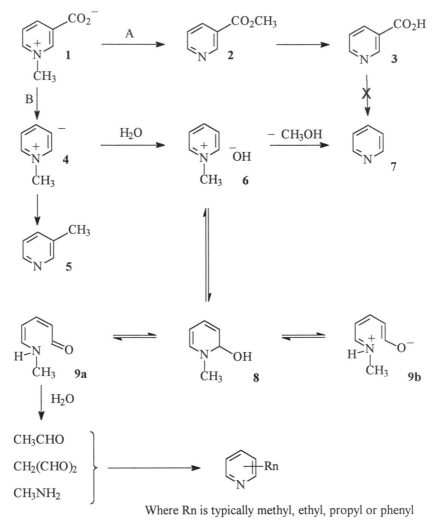

Where Rn is typically methyl, ethyl, propyl or phenyl

Figure 1   *Mechanism of Pyridine Formation in the Pyrolysis of Trigonelline*

3 RESULTS AND DISCUSSION

3.1 Coffee Roasting Studies

A coffee roasting study showed that high levels of pyridine in roasted coffee can adversely contribute to human acceptance of brewed coffee aroma. The level of brew pyridine gave strong positive correlations with panellists' perceptions of smoky (+0.813) and ashy (+0.750) aroma character. These data, plus the fact that smoky and ashy also gave negative correlations (−0.83 and −0.84 respectively) with overall aroma preference, suggested that a high level of pyridine in coffee brew could be a detrimental factor in consumer acceptance. The origin of brew pyridine was probed in model studies of trigonelline pyrolysis.

3.2 Pyrolysis of Trigonelline Hydrate

Pyrolysis of trigonelline hydrate (1) under mild coffee roasting conditions (200–205 °C for 30 min) produced a diverse mixture of volatile pyridine derivatives (Table 1). Trigonelline appears to degrade by two primary processes (Path A and Path B, Figure 1), one of which involves decarboxylation. The non-decarboxylative route (Path A) is a consequence of betaine chemistry that permits ester formation by the migration of an alkyl group. For trigonelline, the N-methyl group migrates to form methyl nicotinoate (2). Also, in the presence of water, some of the methyl ester is hydrolyzed to produce nicotinic acid (3). The chemical yields of (2) and (3) account for only *ca.* 7% of trigonelline reacted. Decarboxylation based on carbon dioxide yield (88%) is the major primary process. The source of carbon dioxide must be trigonelline itself since a separate experiment showed that nicotinic acid did not decarboxylate under the pyrolysis conditions. Trigonelline decarboxylation is favoured by an electronic inductive effect of the N-methyl iminium group.[3] We postulated that loss of carbon dioxide could lead to a zwitterion intermediate (4) which can rearrange by methyl migration to form 3-methylpyridine (5) or undergo hydrolysis to form 1-methylpyridinium hydroxide (6). Thermally induced methanol loss from (6) can lead to pyridine (7).[4]

To test the proposed mechanism, we pyrolyzed methyl-d_3-labelled trigonelline. Our expectation for mechanistic proof was to obtain significant quantities of 3-(methyl-d_3) pyridine and totally unlabelled pyridine.

3.3 Pyrolysis of Methyl-d_3 Trigonelline Hydrate

Pyrolysis of methyl-d_3 trigonelline gave essentially the same distribution of products as the unlabelled material. However, GC–MS analysis of reaction products revealed an unexpected widespread distribution of deuterium label (Table 2). The occurrence of predominately unlabelled pyridine (59% d_0) is consistent with the intermediacy of 1-methylpyridinium hydroxide (6). However, the relatively low yield of d_3-3-methylpyridine (11%) suggested that methyl migration alone cannot explain the formation of 3-methylpyridine (5).

Table 2 *Deuterium Content (% M^+) of Methyl-d_3 Trigonelline Pyrolysis Products*

Product	d_0	d_1	d_2	d_3	d_4	d_5	d_6
Pyridine	59	29	11	2	0.2		
3-Methylpyridine	10	37	39	11	2	0.3	
4-Ethylpyridine	3	18	34	30	13	3	0.5
4-Propylpyridine	4	16	35	30	13	3	0.4
3-Ethyl-4-methylpyridine	6	19	34	27	11	2	0.3
4-Phenylpyridine	12	25	30	21	11	2	0.4

Corrected for (M^+−1) and natural abundances.

To explain the formation of (5) and the extensive migration of deuterium label we postulate the intermediacy of 2-hydroxy-1-methyl-1,2-dihydropyridine (8), the pseudobase form of (6). Pseudobases of this type are well known and their ring-chain tautomerism is well documented.[5] We suggest that H–D exchange occurs in the open chain moieties of the pseudobase (9a and 9b) resulting in partial deuteration at C-3 and C-5 of the pyridine ring and in reduced labelling of the N-methyl group. Canonical charge separation in the open chain moiety (9b) can conceivably provide enough activation to allow intramolecular H–D exchange between anionic centres. The result of exchange is a modified form of (6) in which some N-methyl deuterium has exchanged with pyridine ring hydrogens. Thermal decomposition of partially ring-labelled (6) explains the observed formation of pyridine-d_1 (29%) and pyridine-d_2 (11%).

Besides undergoing thermal loss of methanol, the partially ring-labelled (6) may also serve as an alkylating agent to transfer its partially-labelled methyl group to the zwitterion (4). Products of this alkylation are the observed mono (37% d_1) and di (39% d_2)-deuterated 3-methylpyridines and partially deuterated pyridine.

In summary, the results of the deuterium labelling experiment support the mechanism for trigonelline decarboxylation in Figure 1 involving 1-methylpyridinium hydroxide and its pseudobase as key intermediates.

The formation of higher alkylpyridine homologues and bicyclic pyridines is a matter of greater speculation. The observation of methylamine as a trigonelline pyrolysis product suggests hydrolysis of the open-chain form of the pseudobase (9). Theoretically, hydrolysis of pseudobases derived from pyridine and 3-methylpyridine are also a source of acetaldehyde, propionaldehyde and malonaldehyde. Subsequent recombinations of these aldehydes with each other in aldol condensations and with methylamine in so-called Chichibabin reactions can explain the formation of many complex pyridines.[6] However, little evidence is currently available to support reasonable mechanisms thus leaving the subject open for further study.

ACKNOWLEDGEMENTS

The authors thank T. Morsch and D. Patton for performing numerous GC and GC–MS analyses and L. Boekley, G. Dria and D. Burgard for help with roasting experiments and with statistical analysis of data.

REFERENCES

1. E.B. Hughes and R.F. Smith, *J. Soc. Chem. Ind.*, 1949, **68**, 322.
2. R. Viani and I. Horman, *J. Food Sci.*, 1974, **39**, 1216.
3. P. Haake and J. Mantecon, *J. Amer. Chem. Soc.*, 1964, **86**, 5230.
4. K. Suyama and S. Adachi, in 'Amino-Carbonyl Reactions in Food and Biological Systems', eds. M. Fujimaki, M. Namaki and H. Kato, Elsevier, New York, 1986, p. 95.
5. J.W. Bunting, 'Advances in Heterocyclic Chemistry', Academic Press, New York, 1979, Vol. 25, p. 1.
6. E. Klingsberg, in 'The Chemistry of Heterocyclic Compounds', ed. A. Weissberger, Interscience, New York, 1960, p. 99.

CARBOHYDRATE CLEAVAGE IN THE MAILLARD REACTION

H. Weenen and W. Apeldoorn

Quest International, Postbus 2, 1400 CA Bussum, The Netherlands

1 INTRODUCTION

The Maillard reaction plays an important role in the formation of the aroma of all thermally treated foodstuffs. Aroma compounds which are formed by the Maillard reaction include carbohydrate derived substances, amino acid derived substances and compounds which originate from both carbohydrates and amino acids. Flavour substances which originate from carbohydrates, whether fully or partially, are generally formed by either of two major routes:

1. Isomerisation followed by cyclisation;
2. Fragmentation followed by intermolecular condensation.

Since the second route often gives relatively low yields (most probably because of the inefficiency of the fragmentation step), we chose to study carbohydrate fragmentation in more detail.

An interesting observation on carbohydrate cleavage was reported by Hayami.[1] Decomposition of specifically C-labelled pentoses and hexoses in concentrated aqueous phosphate buffer (40%, pH 6.7) revealed that hydroxyacetone was a major product, with the methyl group of hydroxyacetone originating from one of the terminal C-atoms. The mechanism proposed for these observations consists of isomerisation of 1-deoxyglycosone to a 1-deoxy-2,4-dioxo species, followed by hydrolytic cleavage (β-cleavage).

A study on base-catalysed fructose degradation (NaOH aq, pH 11.5) also pointed to hydroxyacetone as the main cleavage product.[2] Model studies indicated that hydroxyacetone and 1-hydroxy-2-butanone were the precursors of five carbocyclic compounds, which were observed in the fructose cleavage experiments.

In a series of very interesting papers, Hayashi and co-workers[3,4] convincingly demonstrated that glyoxal and pyruvaldehyde are formed when glucose is reacted with an amine such as β-alanine (aqueous solution, pH 9.3). They also measured browning activity, and found that glycolaldehyde, glyceraldehyde, pyruvaldehyde and glyoxal are very potent browning agents, with relative browning activities of 2109, 1967, 654 and 122, respectively when compared to glucose.

Severin et al.[5] determined sugar degradation products after heating fructose for 10 minutes at 160 °C in 10% aq. Na_2CO_3, using GC after reaction with O-(trimethylsilyl) hydroxylamine and subsequent acetylation. Glyceraldehyde, dihydroxyacetone, erythrose and pyruvaldehyde were detected.

Three carbohydrate cleavage mechanisms have been suggested as being responsible for carbohydrate fragmentation:[6]

1. Retroaldolisation;
2. β-Dicarbonyl cleavage; and
3. α-Dicarbonyl cleavage.

The first two mechanisms are well documented and generally accepted cleavage mechanisms in carbohydrate chemistry. Some evidence also appears to exist for α-dicarbonyl cleavage, which is supposed to involve immonium ion formation, followed by hydrolytic cleavage, with an immoniumbetaine as leaving species.[7]

In previous papers, we have reported on the mechanism of the formation of methylated pyrazines[6,8] and on carbonyl-containing carbohydrate cleavage products.[9] The main carbonyl-containing pyrazine precursor detected was hydroxyacetone, while only small amounts of pyruvaldehyde were observed, and no other α-dicarbonyl carbohydrate fragments. Since this might be due to the derivatisation method used, we have now used *o*-diaminobenzene as a derivatisation agent, which specifically derivatises α-dicarbonyl species to quinoxaline derivatives.

Carbohydrates which were investigated include glucose, fructose, xylose and 3-deoxyglucosone. These were heated in solution with cyclohexyl amine, alanine or without added amine. In addition the Amadori rearrangement product of glucose and alanine (1-fructosylalanine) was heated with and without added amine/amino acid. The results are discussed in relation to the formation of flavour substances in the Maillard reaction.

2 EXPERIMENTAL

2.1 Reaction

A mixture of an aqueous phosphate buffer solution (30 ml 0.33 M, pH 8) containing carbohydrate (0.55 M) with or without amine/amino acid (0.6 M) and n-butanol (15 ml) containing 4-methylquinoline as internal standard, was heated for an hour at 100 °C with magnetic stirring.

2.2 Derivatisation

An aliquot of the organic phase (2 ml) was added to 2 ml phosphate buffer (2 ml, 1 M, pH 4) containing 50 mg *o*-diaminobenzene. This mixture was reacted for two hours at 40 °C (while stirring) and analysed by GC. Every reaction was derivatised twice and each derivatisation reaction mixture analysed in duplicate.

3 RESULTS

3.1 Glucose

When glucose was reacted with alanine in phosphate buffer at pH 8 and 100 °C for one hour, glyoxal, pyruvaldehyde, diacetyl and 2,3-pentanedione are formed in about equal amounts. The formation of pyruvaldehyde can be explained as the product of retro-aldolisation of either 1- or 3-deoxyglucosone, as shown in Scheme 1.[8]

The formation of glyoxal can be explained as the product of oxidation of glycolaldehyde. In Scheme 2, two hypothetical oxidation mechanisms for glycolaldehyde are shown, which are characterized by a base-catalysed hydride shift to a hydride acceptor (electrophile) *e.g.*, a carbonyl moiety, or by oxidation with an α-dicarbonyl species, in a Strecker degradation type mechanism.

The formation of diacetyl can be understood as resulting mainly from 1-deoxyglucosone via isomerisation, H_2O elimination and β-cleavage, as shown in Scheme 3. 3-Deoxyglucosone also gives rise to diacetyl, but the yields are much lower. The formation of 2,3-pentanedione is more difficult to explain, but can possibly be understood as originating from the aldol condensation product of diacetyl and formaldehyde (Scheme 4). The resulting α,β-enedione would be a very good hydride acceptor, and would react swiftly with hydride donors such as glycolaldehyde (see Scheme 2).

Cyclohexylamine gave much higher yields, especially of pyruvaldehyde and glyoxal, but also of diacetyl. The main factor expected to cause these higher yields is the fact that alanine will react with α-dicarbonyls (Strecker degradation) to give non-α-dicarbonyl products, while imine formation in the reaction of α-dicarbonyls with cyclohexylamine is reversible.

Table 1 *Formation of α-Dicarbonyl Products*

Amine	Carbohydrate	α-Dicarbonyl Products (μg)			
		glyoxal	*pyruv-aldehyde*	*diacetyl*	*2,3-pent-anedione*
[None]	glucose	26	11	—	—
[None]	fructose	28	15	—	—
[None]	xylose	62	17	—	—
[None]	3-done	23	57	—	—
[None]	fru-ala	103	101	98	18
Alanine	glucose	58	43	41	38
Alanine	fructose	45	28	22	25
Alanine	xylose	27	81	28	42
Alanine	3-done	16	56	11	21
Alanine	fru-ala	81	67	81	22
Cyclohexylamine	glucose	618	865	227	39
Cyclohexylamine	fructose	691	1104	265	89
Cyclohexylamine	xylose	591	925	614	101
Cyclohexylamine	3-done	317	583	146	25
Cyclohexylamine	fru-ala	509	454	232	39

3-done = 3-deoxyglucosone; fru-ala = 1-fructosylalanine

3.2 Various Carbohydrates without Amine/Amino Acid

Interestingly, in the absence of amine or amino acid, no diacetyl or 2,3-pentanedione was formed (Table 1). Possibly β-cleavage (involved in diacetyl formation) requires an amine as nucleophile, or activation of the carbonyl as imine. Diacetyl and 2,3-pentanedione formation are apparently linked, in agreement with the proposed mechanism in Scheme 4. Thermal decomposition of the Amadori product ARP of glucose and alanine clearly gave the highest yields of α-dicarbonyl containing fragments. Obviously the presence of alanine, whether bound as ARP or free after elimination, was responsible for the observed high level of carbohydrate fragmentation.

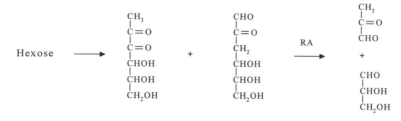

Scheme 1 *Formation of pyruvaldehyde from 1- and 3-deoxyglycosone (RA = retroaldolisation)*

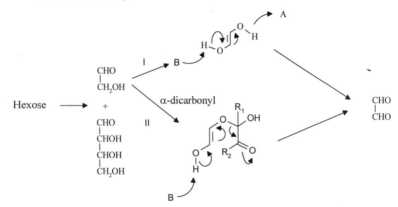

Scheme 2 *Formation of glyoxal via retroaldolisation and oxidation (B = base, A = electrophile)*

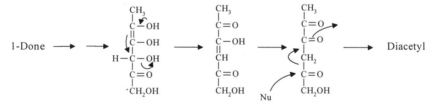

Scheme 3 *Formation of diacetyl from 1-deoxyglucosone (1-Done)*

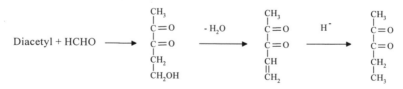

Scheme 4 *Formation of 2,3-pentanedione from diacetyl*

3.3 Various Carbohydrates with Alanine

With alanine as exogenously added base, the highest yields of glyoxal and diacetyl were again obtained from 1-fructosylalanine; however, the highest yields of pyruvaldehyde and 2,3-pentanedione were obtained from xylose (Table 1). The observation that 1-

fructosylalanine was only slightly better than glucose, suggests that the formation of cleavage products does not necessarily require the ARP as an intermediate.

Similarly, the relatively poor yield of fragmentation products (except pyruvaldehyde) from 3-deoxyglucosone suggests that 3-deoxyglucosone is not a necessary intermediate for all α-dicarbonyl fragments. The relatively low reactivity of 3-deoxyglucosone with regard to fragmentation is in agreement with the earlier observed relative stability of 3-deoxyglucosone,[9] and the observation that 3-deoxyglucosone is a relatively poor methylated pyrazine precursor.[9]

Interestingly, 2,3-pentanedione yields are relatively high in reactions with alanine. Apparently acetaldehyde, which is formed from alanine by Strecker degradation, enhances 2,3-pentanedione formation. This can be explained by a mechanism in which hydroxyacetone reacts with acetaldehyde as indicated in Scheme 5.[10]

3.4 Various Carbohydrates with Cyclohexylamine

With cyclohexylamine as the exogenously added base, the highest yields of carbohydrate fragments were obtained with the free carbohydrates and not the ARP (Table 1). Most probably, alanine, which can be regenerated from 1-fructosylalanine, reacts with the α-dicarbonyl species formed in Strecker degradation reactions, causing decomposition of these α-dicarbonyl species. This also explains the much higher yields (> 10×) obtained with cyclohexylamine, in comparison with alanine. Even 1-fructosylalanine gives higher yields of α-diones, when reacted in the presence of cyclohexyl amine. Two factors may play a role here: the basicity of cyclohexyl amine and possibly a protective effect of cyclohexyl imine formation.

3.5 3-Deoxyglucosone

3-Deoxyglucosone gave relatively high yields of pyruvaldehyde, in agreement with Scheme 1. Interestingly xylose also gave relatively high yields of pyruvaldehyde. We speculate that this may be due to the lower number of deoxyglycosone bicyclic structures[6,9] which are possible for deoxypentosones in comparison with deoxyhexosones, and therefore higher reactivity of 1- and 3-deoxyxylosone in comparison to 1- and 3-deoxyglucosone.

4. DISCUSSION AND CONCLUSIONS

Carbohydrate cleavage in the Maillard reaction of alanine with a reducing sugar results in the formation of the following α-dicarbonyl fragments: glyoxal, pyruvaldehyde, diacetyl and 2,3-pentanedione. The extent of formation of these α-dicarbonyl fragments depends on the sugar used. In a reaction with an amino acid such as alanine, xylose and 3-deoxyglucosone generate pyruvaldehyde as the main product, but for glucose and fructose glyoxal was observed as the main product.

The ARP of glucose and alanine was observed to generate glyoxal, pyruvaldehyde and diacetyl in about equal amounts (*ca.* 100 μg), but significantly less 2,3-pentanedione (*ca.* 20 μg). In this respect the ARP differed significantly from its precursors glucose and alanine, which were found to result in the formation of about 60 μg glyoxal, and methyl glyoxal, diacetyl and pentanedione each with a yield of about 40 μg.

Interestingly, without amine present, no detectable amounts of diacetyl and pentanedione were observed. Apparently the mechanisms responsible for diacetyl and pentanedione formation only take place in the Maillard reaction, and not in base-catalysed cleavage.

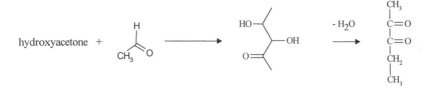

Scheme 5 *Formation of 2,3-pentanedione from acetaldehyde[10]*

The results observed in this study suggest that carbohydrate fragmentation occurs from deoxyglycosones, ARPs and the starting carbohydrates. It was observed that the ARP of glucose and alanine is a relatively efficient α-dicarbonyl precursor.

Cyclohexylamine clearly resulted in the highest levels of α-dicarbonyl species detected. This can be explained by a combination of two factors.

1. The amine functionality catalyses fragmentation;
2. The absence of an α-carboxylic acid group (as in alanine) excludes degradation of α-dicarbonyls by Strecker degradation.

The proposed mechanisms for the formation of glyoxal and pentanedione include redox steps. The mechanisms proposed could possibly be of general relevance for redox steps in the Maillard reaction, but are speculative and require further study.

ACKNOWLEDGEMENTS

We thank J. Eshuis for a generous gift of 1-fructosylalanine, and M. Roeten for her technical assistance. We acknowledge useful comments from J. Kerler and J. van der Ven.

REFERENCES

1.   J. Hayami, *Bull. Chem. Soc. Jap.*, 1961, **34**, 927.
2.   P.E. Shaw, J.H. Tatum and R.E. Berry, *J. Agr. Food Chem.*, 1968, **16** (6), 979.
3.   T. Hayashi, S. Mase and M. Namiki, *Agric. Biol. Chem.*, 1980, **50** (8), 1959.
4.   T. Hayashi and M. Namiki, *Agric. Biol. Chem.*, 1986, **50** (8), 1965.
5.   T. Severin, J. Hiebl and H. Popp-Ginsbach, *Z. Lebensm. Unters. Forsch.*, 1984, **78**, 284.
6.   H. Weenen, S.B. Tjan, P.J. de Valois, N. Bouter, A. Pos and H. Vonk, in 'Thermally Generated Flavors', eds. T. Parliment, M.J. Morello and R.J. McGorrin, American Chemical Society, Washington, D.C., 1993,142.
7.   F. Ledl and E. Schleicher, *Angew. Chem. Int. Ed.*, 1990, **29** (6), 565.
8.   H. Weenen and S.B. Tjan, in 'Flavor Precursors', eds. R. Teranishi; G.R. Takeoka and M. Güntert, American Chemical Society, Washington, D.C., 1992, 217.
9.   H. Weenen and S.B. Tjan, in 'Trends in Flavour Research', eds. H. Maarse and D.G. van der Heij, Elsevier Science B.V., Amsterdam, 1994, 327.
10. T.F. Hofmann, 'Charakterisierung intensiver Geruchstoffe in Kohlenhydrat/Cystein-Modellreaktionen und Klärung von Bildungswegen. Ein Beitrag zur Maillard-Reaktion', Dissertation, Technical University of Munich, 1995, p. 151.

EFFECTS OF FREE RADICALS ON PYRAZINE FORMATION IN THE MAILLARD REACTION

Anna Arnoldi and Emanuela Corain

Dipartimento di Scienze Molecolari Agroalimentari, Università di Milano, via Celoria 2, 20133 Milano, Italy

1 INTRODUCTION

The flavour of processed foods depends on the heat treatments necessary for their preparation. The reaction between amino acids and sugars (Maillard reaction) produces many volatile heterocyclic compounds, whose structures, odour thresholds and concentrations determine the aroma of a particular food. Generally foods also contain fats which can in part be oxidised to aldehydes and ketones by a radical chain mechanism[1] during storage. These lipid degradation products (aldehydes and ketones) can influence the flavour of food by their presence and through interaction with the Maillard reaction products.

Few studies have focused on the interaction of lipids with the Maillard reaction: of these, many are related to the formation of meat flavour. The subject was extensively reviewed by Whitfield,[2] and recently studies on low-moisture systems have indicated that the effects of phospholipids is increased in the presence of water.[3] Scattered information is available on how these interactions influence vegetable flavours. Defatting coconuts before roasting entirely modifies the pleasant aroma, which becomes similar to that of hazelnut.[4] The addition of corn oil to zein and corn starch enhances the formation of pyrazines;[5] some of them have long alkyl chains: their mechanism of formation has been explained recently.[6]

In order to get a better insight into the effect of vegetable oils on the Maillard reaction, we studied model systems containing glucose, an amino acid (cysteine or lysine), water and a lipid (corn oil or extra virgin olive oil); and compared them to systems containing methyl stearate or lacking any lipophilic phase.[7] Subsequently, we added to the model systems free radical initiators or scavengers, which directly influenced the formation of some Maillard reaction products.[8] The work described in this paper was carried out to gain a better comprehension of this phenomenon.

2 EXPERIMENTAL

Equimolar solutions of glucose–lysine and fructose–lysine (30 ml, 1.27 M) were heated for 3 hours at 100 and 120 °C with stirring in a closed tube in the presence of a variable amount of azobisisobutyrronitrile (AIBN), BHT or 2,2,5,5-tetramethyl-2,5-dihydro-3-pyrrolcarboxyamide (TMPC). The pH of the solutions was corrected to the desired values by the addition of NaOH or HCl. Groups of tubes were immersed in an oil bath at the desired temperature and each group contained a reference sample without any additive. At the end of the heating time, the pH of the mixtures was adjusted to pH 10 with 0.32 M

NaOH; 0.50 mg of quinoxaline (internal standard) was added, and the mixture was extracted for 8 hours with dichloromethane in an apparatus for continuous extraction. After drying and careful concentration of the solvent to 1 ml, the mixtures were analysed by GC and GC–MS. GC–MS analyses were performed on a Finnigan TSQ70 instrument. The column was a 30 m × 0.25 mm Supelcowax-10. The temperature programme was 70 °C for 8 min, then 5 °C min^{-1} to 180 °C. Ions were generated by EI at 150 °C. The compound identification was done by comparison with authentic samples whenever possible, by determination of Kovats indices, and by comparison with the NIST mass spectral library. Compound amounts were expressed in μg per model system and were calculated from the total ion chromatogram based on the amount of the internal standard added. A correction factor of 1 was used for all the compounds. The recovery of pyrazines and furans was estimated to be of the order of 80%. Initial pH values of the model systems were 5, 6, 7, 8, and 9, while final pHs were in the range 3.6–3.8 for all the systems.

3 RESULTS AND DISCUSSION

The most important classes of compounds formed by reacting lysine with glucose or fructose are pyrazines, furans and pyrroles. The last two classes of compounds are only slightly influenced by the presence of the three additives. Only 2-furanmethanol increases at all the pH values with both sugars by the addition of BHT, an effect less pronounced in the presence of TMPC. Conversely, interesting observations were made for the pyrazines.

The reaction of lysine with glucose (Figure 1) produces large amounts of these heterocycles: pyrazine is the most abundant, followed by 2-methylpyrazine, 2,5-dimethylpyrazine, 2,6-dimethylpyrazine and trimethylpyrazine. Their yields increased greatly as the pH changed from 5 to 8. The antioxidants BHT and TMPC increased the formation of unsubstituted pyrazine and 2-methylpyrazine. At pH 5–7, TMPC is more effective than BHT, while at pH 8 and 9 their efficiency is reversed. Besides experiments with 60 mg of additive at 120 °C, others were performed with 20, 40, 60, 80 mg at 100 and 120 °C, showing the same trend. The opposite occurs after the addition of the free radical initiator AIBN. With 60 mg of this additive, pyrazine was reduced to a tenth at pH 5, to a quarter at pH 7 and to a half at pH 8. A similar, but less pronounced effect, was observed for 2-methylpyrazine (which is reduced by a half at pH 5), while the other pyrazines remained unchanged. Also, in this case, the effect was proportional to the amount of additive added (results not shown).

The reaction of lysine with fructose (Figure 2) produces large amounts of pyrazines, but in the order 2,5-dimethylpyrazine > pyrazine > 2-methylpyrazine. Only at pH does 8 pyrazine exceed 2,5-dimethylpyrazine. The addition of the free radical scavengers has similar effects to those described for glucose: pyrazine, 2-methylpyrazine and 2,5-dimethylpyrazine increase especially at high pHs where BHT is more effective than TMPC. AIBN reduces pyrazine to a eighth at pH 5 and by 1.3 times at pH 8, the effect is smaller for 2-methylpyrazine, while 2,5-dimethylpyrazine seems to be unaffected.

The observation that the dimethylpyrazines and trimethylpyrazine are less affected by these additives although, for the presence of benzylic hydrogens, they are expected to be more susceptible to autoxidation, suggests they are not directly involved in this process. On the contrary, it seems to indicate that the alkylimine of 2-hydroxyacetaldehyde (glycolic aldehyde), considered to be the intermediate of the formation of pyrazine and 2-methylpyrazine, is particularly sensitive to autoxidation, while the alkylimine of 2-hydroxypropanal (2-methylglycolic aldehyde), producing 2,5-dimethylpyrazine, is less

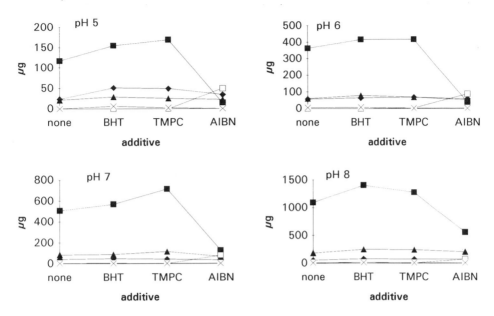

Figure 1 *Pyrazines obtained reacting glucose and lysine for 3 hours at 120 °C in the presence of additives (60 mg):* ■ *pyrazine,* Δ *2-methylpyrazine,* ♦ *2,5-dimethylpyrazine,* □ *2,3-dimethylpyrazine,* × *trimethylpyrazine. Each point is the average of two replicates*

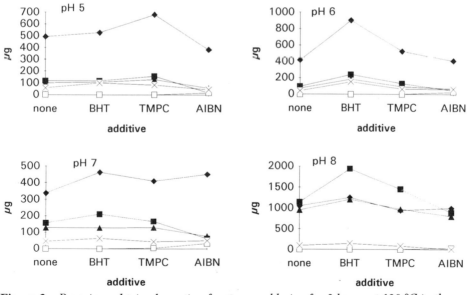

Figure 2 *Pyrazines obtained reacting fructose and lysine for 3 hours at 120 °C in the presence of additives (60 mg):* ■ *pyrazine,* Δ *2-methylpyrazine,* ♦ *2,5-dimethylpyrazine,* □ *2,3-dimethylpyrazine,* × *trimethylpyrazine. Each point is the average of two replicates*

sensitive. These results support the Hayashi and Namiki theory[9] according to which the alkylimine of glycolic aldehyde can be formed directly from the sugar fragmentation, without formation of an Amadori intermediate, which is indispensable for the formation of furans and pyrroles.

ACKNOWLEDGEMENT

This work was supported by the Italian National Research Council (CNR), Progetto Strategico 'Innovazione Produttiva nelle Piccole e Medie Imprese'. Sottoprogetto: 'Qualità Tipica degli Alimenti: Metodologie Innovative di Indagine'.

REFERENCES

1. H.W.-S Chan, in 'Autoxidation of Unsaturated Lipids', ed. H.W.-S. Chan, Academic Press, London, 1987, p. 1.
2. F.B. Whitfield, *Crit. Rev. Food Sci. Nutr.*, 1992, **31**, 1.
3. D.S. Mottram and F.B. Whitfield, *J. Agric. Food Chem.*, 1995, **43**, 984.
4. S. Saittagaroon, S. Kawakishi and M. Namiki, *Agric. Biol. Chem.*, 1984, **48**, 2301.
5. T.-C. Huang, L.J. Bruechert, T.G. Hartman, R.T. Rosen and C.-T. Ho, *J. Agric. Food Chem.*, 1987, **35**, 985.
6. E.-M. Chiu, M.-C. Kuo, L.J. Bruechert and C.-T. Ho, *J. Agric. Food Chem.*, 1990, **38**, 58.
7. A. Arnoldi, in 'Thermally Generated Flavors – Maillard, Microwave, Extrusion Process', eds. T.H. Parliment, M. J. Morello and R. J. McGorrin, A.C.S. Symposium Series 543, American Chemical Society, Washington, D.C., 1994, p. 240.
8. A. Arnoldi, in 'Maillard Reactions in Chemistry, Food, and Health', eds. T.P. Labuza, G.A. Reineccius, V.M. Monnier, J. O'Brien and J.W. Baynes, Royal Society of Chemistry, Cambridge, U.K., 1994, p. 410.
9. T. Hayashi and M. Namiki, *Agric. Biol. Chem.*, 1980, **44**, 2575.

AROMA VOLATILES FROM THE EXTRUSION COOKING OF MODEL SYSTEMS CONTAINING PROLINE AND ORNITHINE

W.L.P. Bredie,[*†] D.S. Mottram[*] and R.C.E. Guy[§]

[*]The University of Reading, Department of Food Science and Technology, Whiteknights, Reading, RG6 6AP, U.K.
[§]Campden and Chorleywood Food Research Association, Chipping Campden, GL55 6LD, U.K.

1 INTRODUCTION

Extrusion cooking is an energy-efficient process for manufacturing cereal products with a range of textural and functional properties. However, the desirable flavour associated with conventionally baked cereals does not develop to the same extent in extrusion cooking, resulting in products of inferior eating quality. In attempts to stimulate flavour generation during extrusion, the merits of modifying the process and design variables and using different cereal feeds have been reported.[1–3] Other studies have investigated the use of flavour precursors in cereal extrusion, such as the fortification of flour with thiamin.[4] Addition of carbonyl and nitrogen reactants in wheat flour[5] or autolysed yeast[6] have been shown to stimulate the formation of pyrazines.

In bread, 2-acetyl-1-pyrroline (AP) and 2-acetyltetrahydropyridine (ATHP) have been identified as important contributors to the typical crust aroma.[7,8] These compounds are also important in the flavour of other cereals.[9] Both AP and ATHP have roasted, biscuit-like aromas and their odour threshold values are lower than those reported for alkylpyrazines. In bread, ornithine produced by the yeast in the dough, is a likely precursor for AP. In the absence of ornithine, proline has been suggested to be the source for ATHP and AP in popcorn.[10,11] The proposed route to AP from proline or ornithine involves the intermediate formation of 1-pyrroline and its reaction of 2-oxopropanal.[12] The present study was part of investigations into alternative ways for improving *in situ* flavour generation in extrusion cooking of cereals. The aim of the work was to evaluate the potential of proline and ornithine for the generation of AP and ATHP in the extrusion cooking of starch-based model feedstocks.

2 EXPERIMENTAL

2.1 Extrusion Processing

Soft wheat starch with added precursors, L-proline (> 99%) or L-ornithine (99%), and D-glucose or 2-oxopropanal, was extruded on an APV Baker MPF 50D twin-screw extruder (product temperature: 165 °C; moisture: 18%; residence time: 35 s; feedrate: 800 g min^{-1}; and screw speed: 350 rpm). Extrusion feedstocks were prepared by blending

---

[†]Present address: The Royal Veterinary and Agricultural University, Department of Dairy and Food Science, 1958 Frederiksberg C, Denmark

combinations of the precursors (each 44 mmol kg^{-1} starch) with the starch *prior* to extrusion. 2-Oxopropanal (2.2 mmol kg^{-1} starch) was added with the water feed during extrusion.

2.2 Analysis of Extrudates

2.2.1 Dynamic Headspace Analysis. The extrudate (10 g) was placed in a conical flask fitted with a Dreschel head and wetted by adding 20 ml of distilled water. Volatiles were swept onto Tenax TA (nitrogen, 50 ml min^{-1}, 1 hour, 37 °C) contained in the CHIS headspace trapping system (SGE Ltd., Milton Keynes, U.K.). 1,2-Dichlorobenzene was used as the internal standard. The volatiles were analysed on an HP 5980/5972A GC–MS equipped with a CHIS injector (SGE). Volatiles were thermally desorbed at 250 °C on to the GC column (50 m × 0.32 mm internal diameter coated with BPX-5 (SGE)) while the oven was held at 0 °C for 5 min. The temperature was then raised to 60 °C over 1 min and held for 5 min, before programming to 250 °C at a rate of 4 °C min^{-1}. Mass spectra were recorded in the electron impact mode at ionisation voltage of 70 eV and source temperature of 280 °C. Four replicates of each extrudate were analysed and quantities of components calculated by comparison of peak areas with those of the internal standard in the GC–MS chromatograms.

2.2.2 Solvent Extraction Analysis. A 25 g portion of extrudate was mixed with 250 ml of distilled water and the suspension extracted with three 50 ml portions of pentane/diethyl ether (1:1). The combined solvent extract was frozen overnight and 1,2-dichlorobenzene as internal standard added. The extract was concentrated to 1 ml using a Kuderna-Danish apparatus and further by placing it in a gentle stream of nitrogen. The extract was analysed by GC–MS using a HP 7673 autosampler with automated split/splitless injection. The oven temperature was held at 50 °C for 2 min before programming at a rate of 4 °C min^{-1} to 250 °C and a hold of 10 min. Other GC–MS conditions were similar as in section 2.2.1. Extrudates were analysed in duplicate.

Table 1 *Approximate Quantities (μg kg^{-1} extrudate) of Selected Compounds Generated in Starch Extruded with Different Combinations of Flavour Precursors*

Amino acid/carbonyl	AP		ATHP		1-Pyrroline
source	DHA	SEA	DHA	SEA	DHA
Proline	tr	6.3	tr	170	2.4 (0.9)
Proline/glucose	0.3 (0.2)	22	1.7 (0.8)	1200	0.8 (0.4)
Proline/2-oxopropanal	0.1 (0.1)	27	2.1 (3.0)	4900	4.5 (1.5)
Ornithine/glucose	3.0 (1.5)	290	nd	100	9.2 (1.1)
Ornithine/2-oxopropanal	1.5 (0.8)	120	nd	120	18.0 (3.7)

tr: trace; nd: not detected; standard deviations shown in parentheses.

3 RESULTS AND DISCUSSION

Five combinations of feedstocks with the amino acids proline or ornithine were extruded using glucose or 2-oxopropanal as carbonyl sources, and starch as the base material (Table 1). The levels of AP and ATHP were determined by dynamic headspace analysis (DHA) and solvent extraction analysis (SEA). 2-Propionyl-1-pyrroline, another biscuit-like odorant recently described in popcorn,[11] could not be detected by either DHA or SEA of the extrudates. SEA gave a much higher recovery of AP and ATHP than DHA (Table 1) and unless indicated, the data from SEA are discussed.

AP was formed in the systems containing proline and glucose together with much larger amounts of ATHP. The small quantities of both AP and ATHP found when starch was extruded with proline alone, indicated that proline had reacted with traces of reducing sugars and/or free carbonyl groups in the starch polymers. When the glucose was replaced with 2-oxopropanal, higher amounts of ATHP were generated, but the level of AP was unaffected. Other studies on reaction systems containing proline in aqueous buffer solutions have shown an increased yield of AP relative to ATHP when glucose was replaced by 2-oxopropanal.[12] However, in the present work, the low moisture level in the starch–proline extrusions did not favour the production of AP over ATHP. Higher levels of AP in starch extrusions were achieved when proline was replaced by ornithine. Even though a relatively high level of 2-oxopropanal was added in the starch–ornithine extrusion, this carbonyl compound was less effective than glucose in producing AP. The most prominent volatiles in the starch–ornithine extrudates were alkylpyrazines (Table 2), which were present in only trace amounts in the starch–proline extrudates.

Table 2 *Approximate Quantities ($\mu g\ kg^{-1}$ extrudate) of Pyrazines in Headspace (DHA) from Starch Extruded with Ornithine and Carbonyl Precursors*

Pyrazine	Ornithine/glucose	Ornithine/2-oxopropanal
Pyrazine	23 (9)	5.3 (1.9)
Methylpyrazine	11 (5)	6.1 (0.7)
2,3-Dimethylpyrazine	0.7 (0.1)	8.0 (0.3)
2,5-Dimethylpyrazine	nd	2.2 (0.6)
Ethylpyrazine	1.8 (0.4)	nd
Vinylpyrazine	1.0 (0.4)	nd

nd: not detected; standard deviations shown in parentheses

Generation of pyrazines in the ornithine-containing extrudates may be explained by the Strecker degradation of ornithine (Figure 1) followed by condensation of the formed α-aminoketones. In seems likely that the essential step in producing AP from ornithine is the formation of 4-aminobutanal, which may rapidly condense to 1-pyrroline. In model systems, 1-pyrroline has been shown to be an important precursor for AP.[12] The higher level of 1-pyrroline in the ornithine extrudates, compared to those with proline, may therefore suggest that ornithine is a better precursor for 1-pyrroline and consequently AP. Relatively high amounts of ATHP were also identified in the ornithine extrudates. ATHP was not reported in the previous study on reactions between ornithine and 2-oxopropanal or fructose in aqueous buffers.[10] Formation of ATHP from ornithine may involve the reaction between hydroxypropanone and 4-aminobutanal (Figure 1).

In conclusion, significant amounts of AP and ATHP were generated from proline and ornithine precursors in starch-based model extrusions. Proline was an important source for ATHP, whereas ornithine gave more AP. 2-Oxopropanal was less effective than glucose in generating AP, suggesting either that the reaction between the oxoaldehyde and 1-pyrroline in the extrusion system is influenced by other factors, or the existence of an alternative mechanism.

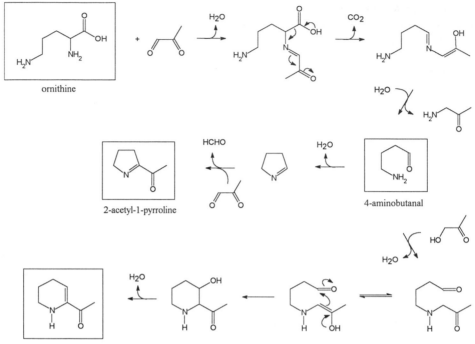

2-acetyltetrahydropyridine

Figure 1   *Scheme for the formation of AP and ATHP from ornithine and oxopropanal*

ACKNOWLEDGEMENTS

This study was financially supported by the MAFF/DTI Food Processing Sciences LINK Scheme (project CSA 2172). The authors thank Hazel Ling and Andrew Dodson for technical assistance, and Harry Nursten and Jennifer Ames for advice.

REFERENCES

1. J.A. Maga and C.E. Sizer, *Lebensm.-Wiss. u. -Technol.*, 1979, **12**, 15.
2. S.M. Fors and C.E. Eriksson, *J. Sci. Food Agric.*, 1986, **37**, 991.
3. H. Guth and W. Grosch, *Z. Lebensm. Unters. Forsch.*, 1993, **196**, 22.
4. C.T. Ho, L.J. Breuchert, M.C. Kuo and M.T. Izzo, in 'Flavor Chemistry – Trends and Developments', eds. R. Teranishi, R.G. Buttery and F. Shahidi, American Chemical Society, Washington, 1989, p. 504.
5. H.V. Izzo, T.G. Hartman and C.T. Ho, in 'Thermally Generated Flavors – Maillard, Microwave, and Extrusion Processes', eds. T.H. Parliment, M.J. Morello and R.J. McGorrin, American Chemical Society, Washington, D.C., 1994, p. 328.
6. H.V. Izzo and C.T. Ho, *J. Food Sci.*, 1992, **57**, 657.
7. P. Schieberle and W. Grosch, *Z. Lebensm. Unters. Forsch.*, 1985, **180**, 474.
8. P. Schieberle and W. Grosch, *Z. Lebensm. Unters. Forsch.*, 1987, **185**, 111.
9. R.G. Buttery and L.C. Ling, *J. Agric. Food Chem.*, 1995, **43**, 1878.
10. P. Schieberle, in 'Flavour Science and Technology', eds. Y. Bessière and A.F. Thomas, John Wiley and Sons, Chichester, 1990, p. 105.
11. P. Schieberle, *J. Agric. Food Chem.*, 1991, **39**, 1141.
12. P. Schieberle, *J. Agric. Food Chem.*, 1995, **43**, 2442.

A COMPARISON OF THREE SUGARS AND INOSINE MONOPHOSPHATE AS PRECURSORS OF MEAT AROMA

L.J. Farmer,[*†] T.D.J. Hagan[*] and O. Paraskevas[†]

[*]Department of Agriculture for Northern Ireland, Food Science Division and [†]The Queen's University of Belfast, Newforge Lane, Belfast BT9 5PX, U.K.

1 INTRODUCTION

While many hundreds of volatile compounds have been detected in cooked meats,[1] it is only during the past ten years that the important aroma compounds responsible for the characteristic flavour of cooked meat have been identified.[2-4] These compounds are formed by a variety of reactions which occur during the cooking of the meat. A number of furanthiols with 'meaty', 'roasted' odours and very low odour thresholds[2,3] may be formed either by the Maillard reaction between cysteine and reducing sugars[5] or by the thermal breakdown of thiamine.[6] The precursors for these odour compounds are all present in meat, but which of them are limiting for flavour formation is unknown.

Initial sensory screening experiments suggested that the addition of reducing sugars or IMP increased the 'meaty' odour of cooked meat samples to a greater extent than the addition of other potential precursors, such as cysteine or thiamine. Inosine monophosphate (IMP) has been shown to increase the formation of the furanthiols believed to be important for meat flavour when added to beef at ten times the natural concentration.[7] In this preliminary study, the relative importance of IMP and three reducing sugars was investigated in beef and pork by adding them to the raw homogenised meat at concentrations approximately twice and four times those reported in raw meat. The aromas of the cooked samples were assessed by sensory analysis. In addition, a comparison was conducted of these precursors added at approximately equimolar concentrations. The headspace volatiles from cooked beef with precursors added at four times the reported concentrations were examined by gas chromatography (GC)–odour assessment and GC–mass spectrometry (GC–MS).

2 MATERIALS AND METHODS

2.1 Materials

Minced lean beef (*M. semimembranosus*) was purchased from a local supplier while lean pork (*Longissimus dorsi*) was obtained from pigs produced at the Agricultural Research Institute for Northern Ireland, Hillsborough. The lean pork was separated from the fat and minced. Solutions of ribose, IMP, glucose-6-phosphate (G6P) and glucose (Sigma Chemical Co.) were prepared in distilled water such that, when added to meat at 10 ml per 100 g fresh minced meat, the added concentrations listed in Table 1 were obtained. Samples of each meat with added precursors were homogenised in a domestic food processor and the samples were held at 4 °C for 18 hours. A control sample was prepared using distilled water only. The pH of the meat samples was monitored after equilibration.

Table 1   *Concentrations of Precursors (mg per 100 g) Reported in the Literature[8] and added to Beef and Pork in 2× and 4× Treatments*

Precursor	Mean reported in literature	'2×' treatment	'4×' treatment	Equimolar comparison[a]
IMP	85	170	340	340
Ribose	300	600	1200	135
Glucose	70	140	280	140
G6P	150	300	600	300

[a] Molar concentrations in range 0.78–1.00 mmoles per 100g meat.

2.2 Sensory Evaluation

Samples (20 g) of meat were cooked in sealed aluminium foil containers for fifteen minutes at 120 °C. Treated and control samples were then presented to thirty panellists who assessed them using a paired comparison test. Panellists were asked to decide which samples had the "more meaty" or "more roasted" aroma.

2.3 Gas Chromatography–Odour Assessment

Homogenised beef (10 g, control or treated at four times the reported concentrations) was heated, in a glass vial covered with a lid containing a Teflon-coated septum, for 30 min in a water bath held at 100 °C. Using a preheated (60 °C) gas-tight syringe, 5 ml of headspace volatiles were injected on to a trap containing Tenax GC (SGE Ltd.), which was subjected to a partial vacuum by connection to a water pump. Moisture was removed from the trap by flushing with nitrogen (50 ml min^{-1}, for 5 min). The volatiles were thermally desorbed on to the front of a fused silica capillary column (CPWAX 52CB, 50 m × 0.32 mm internal diameter × 0.2 µm film thickness, Chrompak U.K. Ltd.) cooled in solid CO_2, using a 'Unijector' (SGE Ltd.) in a Hewlett Packard 5890 Series II gas chromatograph. The effluent was split between a flame ionisation detector and an odour port, which comprised a heated fused silica transfer line (220 °C) connected to a PTFE cup, flushed with an auxiliary flow of moist air. GC–odour assessment was conducted on cooked meat from each treatment by at least two assessors.

2.4 Gas Chromatography–Mass Spectrometry

Homogenised beef samples (50 g, control or treated at '4×' the reported concentrations) were cooked at 100 °C for 30 min in Erlenmeyer flasks, loosely covered with aluminium foil. The headspace volatiles from freshly cooked meat (held at 80 °C) were collected on to Tenax traps by dynamic headspace concentration using a stream of nitrogen (50 ml min^{-1}, for 1 hour). The volatiles were analysed using a Kratos MS25 RFA operating at 70 eV and scanning over the range 30–378 mass units. The capillary column and injection technique were as described for GC–odour assessment.

3 RESULTS AND DISCUSSION

From the results of the sensory comparisons (Table 2), it is evident that the addition of the potential precursors caused a significant increase in 'meaty' or 'roasted' aroma in many of the treated meat samples; in no case did these precursors cause a decrease in these odours compared with the control.

Table 2 *Paired Comparison of 'meaty' and 'roasted' aroma of Beef and Pork with and without Added Precursors at two and four times the Literature Concentrations and at approximately Equimolar Concentrations*

Added Precursor	'Meaty'		'Roasted'	
	Beef	*Pork*	*Beef*	*Pork*
'2x'				
IMP *vs* Control	NS[a]	+**	NS	NS
RIB *vs* Control	+**	+***	NS	+***
GLU *vs* Control	NS	NS	NS	NS
G6P *vs* Control	NS	NS	+*	NS
'4x'				
IMP *vs* Control	+***	+**	+**	+***
RIB *vs* Control	+**	NS	+**	+***
GLU *vs* Control	NS	+*	+**	NS
G6P *vs* Control	+*	NS	+***	+***
Equimolar				
IMP *vs* Control	+***	+**	+**	+***
RIB *vs* Control	na	NS	na	NS
GLU *vs* Control	NS	NS	NS	NS
G6P *vs* Control	NS	NS	NS	+*

[a] NS = no significant difference; + = increase in 'meaty' or 'roasted' aroma; * = $P<0.05$, ** = $P<0.01$; *** = $P<0.001$, na = not assessed.

At four times the reported concentration in meat, IMP caused increases in both 'meaty' and 'roasted' aroma in beef and pork; in most cases no effect was observed at twice the reported concentration. Ribose generally gave increases in both odours at both concentrations. Samples of meat treated with glucose-6-phosphate gave more 'roasted' aroma but showed much less effect on 'meaty' aroma. Glucose only affected 'meaty' or 'roasted' aroma at four times the literature concentration; no effects were observed at the lower concentration. The results obtained for ribose are broadly in agreement with those of Hudson and Loxley,[10] who reported that the addition of xylose to minced lamb caused a more 'mild', 'sweet', 'meaty' aroma and flavour. The threshold of detection for this effect of xylose was 500 mg 100 g^{-1} meat which is similar to the '2x' ribose treatment used in this study.

Decreases in pH as small as 0.5 pH unit can increase the formation of furanthiols and possibly, therefore, meaty flavour.[9] However, determination of the pH of the samples showed that the addition of precursors had little or no effect on pH and that any small pH changes did not explain the observed effects.

The effects of the four precursors were also compared at approximately equimolar concentrations (Table 2). At a given concentration, IMP gave a greater increase in both 'meaty' and 'roasted' aroma than any of the other precursors including ribose. This suggests that IMP is not acting solely as a source of ribose for the formation of meat flavour compounds by the Maillard reaction, but that some other mechanism is involved.

Table 3 shows the principal odours detected by GC–odour assessment of control and treated cooked beef, together with the tentative identities of some of the compounds responsible; GC–MS studies are continuing. A slightly different cooking method was used for the instrumental analyses to facilitate the sampling of volatiles and to enhance reproducibility. Although there was some variation in the odours detected, due to the small volumes of odour analysed, some effects were apparent which are consistent with the changes in cooked meat aroma, as judged by the sensory panel.

Ribose increased 'popcorn', 'sweet' aroma (LRI 1023) and decreased 'potato' odours (LRI 1440, 1510), while IMP increased a 'burning', 'roasted meat' aroma (LRI 1940). These changes could change the balance of the overall aroma such that an increase in 'meaty' and 'roasted' aromas is perceived. Glucose-6-phosphate increased 'popcorn' (LRI 1327) and 'meaty', 'burnt' (LRI 2032) aromas and decreased 'green', 'tomato' odour (LRI 1078); these changes are in accord with an increase in overall 'roasted' aroma. No major changes in individual odours were detected in beef treated with glucose; this compound also had the least effect on the sensory results.

The 'meaty' aroma at LRI 1657 was important in all these samples; the compound, 2-methyl-3-(methyldithio)furan was detected at this LRI, by GC–MS, in the headspace from treated beef samples. This compound has been reported recently to contribute to the flavour of cooked beef and pork.[11] Mottram and Madruga[7] observed an increase in many sulphur-containing furans, including 2-methyl-3-(methyldithio)furan, on addition of IMP to beef at ten times its natural concentration.

IMP and glucose-6-phosphate both appeared to suppress the 'green' odour at LRI 1078, probably caused by hexanal which was identified at this LRI. To test the effect of these precursors on the products of lipid oxidation pathways, the peak areas corresponding to the more abundant products of lipid oxidation, identified in the gas chromatography runs used for GC–odour assessment, were compared (Table 4). 2-Furfural was also monitored. These data confirm that IMP, and to a lesser extent, glucose-6-phosphate and ribose, reduce the formation of n-aldehydes and other lipid oxidation products. The suppression of certain lipid oxidation products has also been observed in heated model systems containing lipids and the Maillard reactants, cysteine and ribose.[12] In contrast, 2-furfural, a product of the Maillard reaction, is increased by both ribose and, to a lesser extent, glucose-6-phosphate. The effect of ribose would be expected from the known mechanism of formation of 2-furfural,[13] and also due to the high concentration of ribose added. The glucose showed little effect on these volatile products.

Which precursors are limiting for aroma formation in a piece of meat will be strongly influenced by the concentrations in the raw meat, which in turn will be affected by individual animal, conditioning time of the meat *etc*. This natural variation in the concentrations of precursors in meats mean that those which are limiting may differ between meat from different sources. Further studies are ongoing to investigate how the natural concentrations of these compounds influence aroma formation in meat.

Table 3 *Principal Odours in GC–Odour Assessments of Cooked Beef (with and without added precursors)*

LRI[a]	Odour Description	Odour score					Suggested Identities	Method of Identification
		CTL[b]	RIB	IMP	GLU	G6P		
<1000	Acrid, nasty	2[c]	(3)	3	3	3	(Not yet identified)	
1023	Popcorn, sweet	(2)	2	—	2	3	(Not yet identified)	
1078	Green, floral, tomatoes	3	2	—	3	—	Hexanal	MS + LRI
1280	Meaty, sweet	2	1	1	2	2	2-Methyl-3-furanthiol	LRI + odour
1294	Mushrooms	3	2	3	3	2	1-Octen-3-one	MS + LRI
1327	Popcorn	(4)	—	—	2	3	A dimethylpyrazine	MS + LRI
1369	Metallic, geranium	3	3	3	3	3	Dimethyltrisulphide	MS + LRI
1429	Potato, sweet, popcorn	2	2	2	3	3	An ethyldimethylpyrazine	MS + LRI
1440	Potatoes	4	1	3	3	3	Methional	MS + LRI
1510	Potatoes, stale	2	—	2	2	2	2-Nonenal	MS + LRI
1657	Meaty	3	3	3	3	3	2-Methyl-3-(methyldithio) furan	MS + LRI(lit)
1940	Burning paper, roasted	(1)	—	2	2	2	(Not yet identified)	
1957	Meaty, popcorn, sweet	2	1	3	3	3	(Not yet identified)	
2032	Meaty, burnt	(1)	—	—	—	2	(Not yet identified)	

[a] Linear retention indices (LRI) were calculated with respect to an external standard containing the n-alkanes (C_{10} to C_{28});

[b] CTL = control; RIB = with ribose; IMP = with IMP; GLU = with glucose; G6P = with glucose-6-phosphate;

[c] Odours scored on a scale of 1 = very weak to 5 = very strong; **2**, 2, (2) or — indicates whether odour detected in > **50%**, *25%–50%*, (≤25%) or none of the analyses.

Table 4   *Mean Peak Areas for Selected Volatile Components from Static Headspace Collections from '4×' Treatments*

Identity	Control	RIB	IMP	GLU	G6P
Hexanal	1217[a]	689[ab]	337[b]	939[ab]	326[b]
Heptanal	797[a]	190[b]	121[b]	391[ab]	94[b]
Octanal	222[a]	72[ab]	64[b]	156[ab]	77[ab]
Nonanal	149	68	48	123	49
1-Octen-3-one	21[a]	12[ab]	7[b]	14[ab]	12[ab]
1-Hexanol	67	52	18	61	18
1-Heptanol	38[a]	7[ab]	2[b]	27[ab]	14[ab]
2-Furfural	10[a]	160[c]	3[a]	6[a]	41[b]

[a,b,c] Values are means of at least three replicate analyses. For each compound, values which do not share a common superscript are significantly different ($P < 0.05$)

These preliminary experiments demonstrate that 'meaty' and 'roasted' aromas can be increased in cooked meats by the addition of sugar-related precursors, suggesting that such compounds may be a limiting factor for flavour formation. It appears that these precursors not only increase 'meaty' and 'roasted' type aromas, probably formed from the Maillard reaction, but also suppress the formation of odour compounds derived from lipid oxidation, thus changing the overall balance of the aroma.

REFERENCES

1. D.S. Mottram, in 'Volatile Compounds in Foods and Beverages', ed. H. Maarse, Marcel Dekker, New York, 1991, p. 107.
2. U. Gasser and W. Grosch, *Z. Lebens.-Unters. Forsch.*, 1988, **186**, 489.
3. U. Gasser and W. Grosch, *Z. Lebens.-Unters. Forsch.*, 1990, **190**, 3.
4. W. Grosch, G. Zeiler-Hilgart, C. Cerny and H. Guth, in 'Progress in Flavour Precursor Studies', eds. P. Schreier and P. Winterhalter, Allured Publishing Company, Carol Stream, IL, 1993, p. 329.
5. L.J. Farmer, D.S. Mottram and F.B. Whitfield, *J. Sci. Food Agric.*, 1989, **49**, 347.
6. P. Werkhoff, J. Bruning, R. Emberger, M. Guntert, W. Kopsel, M. Kuhn and H. Surburg, *J. Agric. Food Chem.*, 1990, **38**, 777.
7. D.S. Mottram and M. S. Madruga, in 'Trends in Flavour Research', eds. H. Maarse and D.G. Van der Heij, Elsevier, Amsterdam, 1993, p. 339.
8. O. Paraskevas, M.Phil. thesis, The Queen's University of Belfast, in preparation.
9. L.J. Farmer and D.S. Mottram, in 'Flavour Science and Technology', eds. Y. Bessiere and A.F. Thomas, John Wiley and Sons, Chichester, 1990, p. 113.
10. J.E. Hudson and R.A. Loxley, *Food Technology in Australia*, 1983, **35**, 174.
11. P. Werkhoff, J. Bruning, R. Emberger, M. Guntert and R. Hopp, in 'Recent Developments in Flavor and Fragrance Chemistry', eds. R. Hopp, and K. Mori, VCH, Weinheim, 1994, p. 183.
12. L.J. Farmer and D.S. Mottram, *J. Sci. Food Agric.*, 1992, **60**, 489.
13. G. Vernin and C. Parkanyi, in 'The Chemistry of Heterocyclic Flavouring and Aroma Compounds', ed. G. Vernin, Ellis Horwood Ltd, Chichester, 1982, p. 151.

A GAS CHROMATOGRAPHY–OLFACTOMETRIC STUDY OF COOKED CURED HAM – IMPACT OF SODIUM NITRITE

A.S. Guillard, J.L. Le Quéré and J.L. Vendeuvre[*]

Laboratoire de Recherches sur les Arômes INRA, 17 rue Sully 21034 Dijon, France
[*]CTSCCV, 7 av. du Général de Gaulle, 94700 Maisons-Alfort, France

1 INTRODUCTION

Curing of meat before cooking imparts a characteristic flavour to the product. Among the ingredients added with brine, sodium nitrite is thought to be a major contributor to this flavour. For instance, it has been shown that addition of sodium nitrite changes the profile of volatile compounds of cooked cured meat qualitatively (*i.e.* formation of nitrogen compounds) as well as quantitatively (*i.e.* decrease of volatile lipid oxidation products).[1,2]

Several compounds have been identified[3,4] but, in spite of a number of studies, no single compound or class of compounds has been found to be responsible for the characteristic flavour of cooked cured meat products, nor have the involved mechanisms been elucidated.[5]

In an attempt to have a sensorial approach to the aroma of cooked cured ham, we have performed analysis by gas chromatography–olfactometry (GC–O) on representative extracts of salted cooked pork muscle, cured and not cured, by means of an aroma extract dilution analysis (AEDA) procedure.[6]

The extraction method was first evaluated for its sensorial representativeness.[7] The extracts obtained were tested for their similarity with the product by a trained panel. The most representative extract was then assessed in GC–O by three trained individuals who provided descriptions of the odours detected. In order to estimate the quantitative impact of the odours described, serial dilutions were evaluated, until no odour was detectable.

2 EXPERIMENTAL

1.1 Cooked Pork Samples

10 kg of pork *semi membranosus* muscle from commercial sources were selected for similar visual appearance and mean pH value 5.7 (standard deviation 0.1) and divided randomly in two lots of 5 kg each. Intramuscular brine injection (10 % w/w) was performed with a pumping needle. For one portion of the meat, the level of sodium nitrite was adjusted to inject 100 mg kg^{-1} muscle (cured pork). The other portion was treated with brine containing no sodium nitrite (uncured pork). No spices were added to the brine in order to avoid interference from their volatile components.

The two products (cured and uncured pork) were put in two 5 kg sealed cook-in-bag pouches and cooked to a core temperature of 65 °C. After cooling to a core temperature of 3 °C and a further 24 hours temperature stabilization, 100 g slices were cut, wrapped in an aluminium foil, placed individually in a polyethylene bag sealed under vacuum, and frozen to

–20°C to minimise oxidation. The aluminium foil was used to protect the sample from migration of oligomers from the polyethylene bag.

The experimental products were compared with a commercial product by an expert sensory panel.

1.2 Isolation of Volatiles

Cooked cured and uncured pork meat (250 g) were ground frozen. The volatile constituents of each sample were extracted by vacuum hydrodistillation (meat mixed with ultrapure water) and collected in glass traps cooled with liquid nitrogen.[8,9] The contents of the traps were adjusted to pH 10 with 1 N sodium hydroxide. The neutral volatiles were then extracted with twice distilled dichloromethane (1 vol. CH_2Cl_2 : 4 vol. extract) and dried over anhydrous sodium sulphate. The extract was concentrated to a final volume of 500 µl using a Kuderna-Danish concentrator fitted with a Snyder column.

1.3 Gas Chromatography–Olfactometry

Analyses were performed using a Hewlett Packard HP5890 series II gas chromatograph equipped with a J&W scientific on-column injector. A 2 m. uncoated deactivated fused silica pre-column (0.32 mm internal diameter) was connected to a DB-FFAP J&W Scientific fused-silica capillary column (30 m × 0.32 mm internal diameter, film thickness 0.25 µm), using a press-fit glass connector. A splitting system was installed at the end of the column to divide the effluent with a 1:1 split ratio between the flame ionization detector (FID) and the sniffing port. H_2 carrier gas was used at a velocity of 37 cm s^{-1} at 143 °C. The temperature was programmed from 34 °C to 44 °C at a rate of 10 °C min^{-1}, held for 2.5 min. at 44 °C and raised to 220 °C at a rate of 3 °C min^{-1}. FID detection of compounds and olfactory results were recorded simultaneously, using hardware and software devices developed in the laboratory (P. Mielle and R. Almanza, Coconut® INRA 1987–1993).

Three selected persons were trained for GC–olfactometry analysis. To facilitate correlation between odours and FID signals, retention indices for each compound were calculated using a solution of C_{10} to C_{24} n-alkanes, which was chromatographed prior to each extract analysed.

1.4 Gas Chromatography–Mass Spectrometry

For GC–MS studies, the HP5890 was coupled with a Nermag R10-10C quadrupole mass spectrometer. Chromatographic conditions were the same as described above but the carrier gas was helium (velocity 35 cm s^{-1} at 220 °C) instead of hydrogen and injection was splitless–split. The column was connected directly to the ion source via a heated transfer line (260 °C). Mass spectra were obtained by electron impact (ei) at 70 eV (scanning from 25 to 300 a.m.u.). Positive chemical ionization (ci) mass spectra were generated at 70 eV, using methane as the reagent gas (scanning from 60 to 300 a.m.u.).

3 RESULTS AND DISCUSSION

Fewer volatile compounds were extracted from cured pork, as compared to uncured pork (Figure 1). This effect of nitrite has already been observed. One explanation is the decrease of lipid oxidation due to the action of sodium nitrite.[5] For example, hexanal (peak 1) and nonanal (peak 12), typical products of lipid oxidation, were observed in higher amounts in the uncured pork extract. Chromatographic profiles obtained from both extracts might look relatively poor in total amount of compounds but overall odour was judged significantly representative of the product.

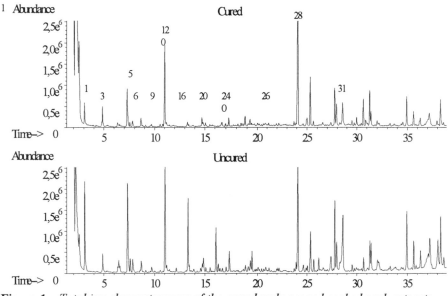

Figure 1 *Total ion chromatograms of the cured and uncured cooked pork extracts*

Differences in odour profile between the two extracts were observed in quality and dilution value (Table 1). For most of the odours detected, at least two sniffers out of three agreed in the detection, and often in the dilution value. For example, in the cured pork, odours 9 (meaty, onion like descriptors), 16 (cooked potatoes descriptor) and 28 (cooked rice, pop corn like descriptors) had the highest dilution value for the three sniffers.

Some odours were product specific, such as odours 13 (bread like descriptors) and 19 (iodine like descriptors) in cured pork and odours 11 (herbaceous, ivy like descriptors), 21 (carrot, vegetable like descriptors) and 25 (excrement, faeces descriptors) in uncured pork. Differences in dilution values were also perceived between the products.

When a peak was detected at the same retention time as an odour, identification of the compound was performed. Some aldehydes, ketones and N- and/or S-containing compounds were identified. Definitive identification will be confirmed by GC–O analysis of the pure compound. In some cases, no compound was detected at the retention time of an odour. Extraction of an higher amount of product will be undertaken in order to isolate sufficient material for identification.

REFERENCES

1.   S. Erduran and J.H. Hotchkiss, *J. Food Sci.*, 1995, **60** (5), 946.
2.   D.S. Mottram, S.E. Croft, R.L.S. Patterson, *J. Sci. Food Agric.*, 1984, **35**, 233.
3.   D.S. Mottram, *J. Agric. Food Chem.*, 1984, **32**, 343.
4.   F. Shahidi, L.J. Rubin, L.A. D'Souza, *CRC Crit. Rev. Food Sci. Nutr.*, 1986, **24**, 141.
5.   N. Ramarathnam and L.J. Rubin, in 'Flavour of Meat and Meat Products', ed. F. Shahidi, Blackie Academic & Professional, 1994, p. 174.

6. W. Grosch, *Trends in Food Sci. & Tech.*, 1993, **4**, 68.
7. P.X. Etievant, L. Moio, E. Guichard, D. Langlois, I. Leschaeve, P. Schlich, E. Chambellant, in 'Trends in Flavour Research', eds. H. Maarse and D.G. van der Heij, 1994, p. 179.
8. D.A. Forss, V.M. Jacobsen and E.M. Ramshaw, *J. Agric. Food Chem.*, 1967, **15**, 104.
9. J.P. Dumont, G. Pradel, S. Roger and J. Adda, *Le Lait*, 1976, **56**, 18.

Table 1   *Odorant Compounds Detected in Cooked Cured and Uncured Pork Meat Extracts: Dilution Value and Identification*

Odour No	RI[1] (FFAP)	Dilution Value for Sniffer 1, 2, 3		Tentative Identification[2]
		Cured	*Uncured*	
1	1070	3, 3, 0	5, 5, 3	hexanal (ei, ci, ri)
2	1144	3, 0, 0	2, 2, 0	3-heptanone (ei, ci)
3	1173	0, 3, 0	2, 2, 1	heptanal (ei, ci, ri)
4	1233	0, 3, 0	2, 4, 2	2,6-dimethylpyridine (ei, ci)
5	1273	2, 3, 4	2, 4, 3	octanal (ei, ci, ri)
6	1286	4, 2, 2	5, 4, 1	1-octen-3-one (ei)
7	1310	4, 3, 0	1, 3, 0	ni[3]
8	1317	3, 0, 2	2, 0, 0	mixture (alcohol+ketone+?)
9	1338	7, 5, 5	2, 2, 1	heteroatomic (N,S) (ei)
10	1350	2, 4, 1	0, 2, 0	a pyridine compound (ei)
11	1356	0, 0, 0	2, 1, 0	2-nonanone (ei, ci)
12	1377	3, 3, 4	1, 4, 2	nonanal (ei, ci, ri)
13	1408	5, 4, 3	0, 0, 0	ni
14	1414	4, 3, 2	4, 2, 2	ni
15	1425	5, 5, 0	4, 6, 0	mixture of compounds
16	1435	7, 7, 6	6, 7, 6	methional (ei)
17	1442	0, 2, 2	0, 2, 0	ni
18	1463	4, 4, 0	2, 0, 0	ni
19	1470	2, 0, 2	0, 0, 0	ni
20	1478	2, 0, 0	2, 2, 0	decanal (ei, ci, ri)
21	1485	0, 0, 0	3, 2, 3	ni
22	1501	0, 3, 2	4, 5, 0	an aldehyde (ei)
23	1519	4, 2, 0	2, 4, 0	ni
24	1534	3, 2, 3	3, 4, 1	linalool (ei, ri)
25	1614	0, 0, 0	1, 2, 0	ni
26	1618	6, 2, 4	3, 4, 0	a phenyl ketone (ei)
27	1655	0, 0, 4	1, 2, 4	ni
28	1722	7, 7, 7	5, 6, 6	piperidine aldehyde(ei, ci)
29	1748	2, 0, 2	3, 5, 4	acetyl piperidine (ei, ci)
30	1789	4, 3, 0	2, 3, 0	ni
31	1793	4, 3, 0	3, 3, 0	ni

[1] Retention index; [2] Identification by RI and/or mass spectrometry (ei, ci); [3] Not identified.

INFLUENCE OF HIGH HYDROSTATIC PRESSURE ON THE GENERATION OF VOLATILES IN A HEATED GLUCOSE–LYSINE MODEL SYSTEM

Vanessa M. Hill, Jennifer M. Ames and David A. Ledward

Department of Food Science and Technology, University of Reading, Whiteknights, Reading, RG6 6AP, U.K.

1 INTRODUCTION

The potential of high pressures for food preservation and processing was first recognised at the beginning of the twentieth century, but it has only recently become commercially feasible.[1] The application of high pressure as a preservation process has achieved recognition due to the superior colour, flavour and nutrient retention in the products, compared to those produced by thermal processing.[2]

Aqueous sugar–amino acid solutions, when incubated, will develop flavour and brown colour, due to the Maillard reaction.[3] It is known that the rate of colour development is enhanced by such factors as heat,[4] and increasing alkalinity over a pH range of 6–10.[5] However, little is known about the effect of high pressures, in the range 100–800 MPa, on Maillard colour and flavour development. Hill et al.[6] report that in a glucose–lysine model system, with an initial pH of above ca. 7, the rate of colour development, at a pressure of 600 MPa, is enhanced compared to atmospheric pressure, whilst the reverse is seen at lower pH values.

This study looked at flavour development of a model Maillard system, which was subjected to a combination of high pressure and moderate temperature. The results were compared with those of the same system reacted at atmospheric pressure.

2 EXPERIMENTAL

One molal aqueous glucose–lysine solutions (initial pH 10.1) were prepared and 100 ml aliquots were incubated, in sealed polyethylene bags, either at atmospheric pressure and a temperature of 60±0.1 °C, or in a high pressure rig at 600 MPa and 60±2 °C. Samples were incubated in the pressurised and unpressurised systems to the same degree of browning. Further details are reported in Hill et al.[6] After incubation, the solutions were cooled and the internal standard (1,2-dichlorobenzene) was added for semi-quantification. Volatiles were extracted with ether:n-pentane (80:20, v/v). The solvent extract was washed with distilled water, dried over anhydrous sodium sulfate and concentrated to a final volume of 0.1 ml.

Replicate extracts were analysed by gas chromatography–mass spectrometry (GC–MS), using splitless injection onto a BPX5 column.

3 RESULTS AND DISCUSSION

The solutions were incubated for approximately 8 hours at atmospheric pressure and 5 hours at high pressure to achieve the same absorbance at 420 nm (0.44–0.48 absorbance units for a 250-fold dilution). The resultant solutions were dark brown, very slightly viscous, and had a sweet, 'digestive biscuit' odour. The solution subjected to high pressure seemed to have a slightly fainter aroma than that incubated at atmospheric pressure.

Table 1   *Identity of Fifteen Compounds Isolated from a Glucose–Lysine System, pH 10.1, Incubated at Atmospheric Pressure and 600 MPa*

Identity	Ref. LRI[a]	Expt. LRI		Amount (μg)		RPY[e]
		Atmos. press.	600 MPa	Atmos. press.[c]	600 MPa[d]	
2-Methyl-3-(2H)-furanone	820	821	820	14	4	29
4-Hydroxy-2,5-dimethyl-3-(2H)-furanone		1078	1074	108	17	16
Methylpyrazine	833	835	836	257	3	1
2,5- and/or 2,6-Dimethylpyrazine	925	927	923	3951	110	3
2,3-Dimethylpyrazine	930	930	928	127	6	5
Trimethylpyrazine	1006	1014	1012	758	20	3
3-Ethyl-2,5-dimethylpyrazine	1095	1094	NA	28	ND	NA
3-Methyl-1,2-cyclopentanedione	1043	1043	1043	25	4	16
2-Acetyl-1,4,5,6-tetrahydro-pyridine	1062	1062	1060	30	9	30
2-Acetylpyrrole	1087	1085	NA	38	ND	NA
2,5-Dimethyl- 2,5-cyclo-hexadiene-1,4-dione	1129	1130	NA	7	ND	NA
2,3-Dihydro-5-hydroxy-6-methyl-(4H)-pyran-4-one		1104	1104	51	4	8
2,3-Dihydro-3,5-dihydroxy-(4H)-pyran-4-one		1165	1169	442	5	1
5-Formyl-6-methyl-2,3-dihydro-(1H)-pyrrolizine	1475[b]	1484	1484	17	5	29
7-Acetyl-5,6-dimethyl-2,3-dihydro-(1H)-pyrrolizine		1779	1778	78	12	15

[a] LRI quoted are of authentic samples run on a BPX5 column;
[b] LRI from a BPX5 equivalent column;
[c] Average of 4 runs. Average coefficient of variation ~ 15%;
[d] Average of 5 runs. Average coefficient of variation ~ 35%;
[e] RPY = Relative percentage yield. Amount formed at 600 MPa compared to amount formed at 1 atmosphere, expressed as a percentage;
ND = not detected;
NA = not applicable.

Table 1 lists the identities of fifteen of the larger peaks obtained by GC–MS analysis of the solvent extracts, together with the quantitative data. It is clear that the application of high pressure caused a substantial decrease in the yields of all the compounds listed in Table 1, despite the system having attained the same degree of browning. Three of the compounds were not detected in the pressurised system, probably because they were produced at levels below the limit of detection. The yields of the twelve remaining compounds were 30% or less of those obtained at atmospheric pressure.

Pyrazines dominated the reaction products in this model system, at both atmospheric and high pressure. 2,5- and/or 2,6-Dimethylpyrazine was the most abundant reaction product in both systems. Pyrazine formation occurs via Strecker degradation,[3] and is generally associated with roasting temperatures. In this 60 °C model system, pyrazines dominated due to the high initial pH (10.1). Under such alkaline conditions, pyrazine production is enhanced due to both the increased degradation of the sugar molecule to compounds including dicarbonyls, and to the increased reactivity of the amino group of the amino acid, towards these dicarbonyls.[7] The application of high pressure caused a massive decrease in pyrazine concentrations, with yields in the range of 1%–5% of those at atmospheric pressure. Of all the compounds listed in Table 1, yields of pyrazines were suppressed most by pressure.

Two 2,3-dihydro-(1*H*)-pyrrolizines were tentatively identified in this study, *i.e.*, the 5-formyl-6-methyl- and 7-acetyl-5,6-dimethyl- derivatives. Tressl *et al.*[8] characterised twenty-two 2,3-dihydro-(1*H*)-pyrrolizines, and claimed them to be proline-specific Maillard reaction products. However, Apriyantono and Ames[9] tentatively identified three 2,3-dihydro-(1*H*)-pyrrolizines in an aqueous xylose–lysine model system, which was refluxed at pH 5. They proposed a mechanism of formation involving pyrrolidine as an intermediate, with subsequent steps based on the work of Tressl *et al.*[8]

2-Acetyl-1,4,5,6-tetrahydropyridine, with a cracker-like odour and low odour threshold, was identified in this study. It is another compound that Tressl *et al.*[11] suggested to be a proline-specific product.

Several furanones and pyranones are associated with caramel aromas. Two members of each class were identified but the yields of each were greatly reduced at high pressure.

Of the remaining compounds listed in Table 1, levels of both 2-acetylpyrrole and 2,5-dimethyl-2,5-cyclohexadiene-1,4-dione, were suppressed below the limit of detection in the pressurised system. However, both these compounds, as well as 3-methyl-1,2-cyclopentanedione, were only formed in relatively small quantities in the atmospheric pressure system.

The model system used in this study was at a relatively high initial pH. Hence, the reaction pathways principally favoured by the Amadori rearrangement product will be those involving the 1-deoxyosone and sugar fragmentation.[3,7]

It is apparent that application of pressure during the Maillard reaction caused a dramatic reduction in yields of low molecular weight products at the same level of colour formation.[6]

It is well understood that reactions with negative volumes of activation will be accelerated by pressure. Thus, the types of reactions expected to be accelerated by pressure include those in which the number of molecules decrease, as in condensation reactions.[12] Therefore, it is possible that the intermediates formed during the Maillard reaction under pressure react quickly in 'condensation-like' reactions to give favoured decreases in volume, with the coloured melanoidins as the end products. The greater rate of colour formation in the pressurised system[6] supports this. Ion formation should also involve volume decreases, due to hydration of the charges produced, and such processes are favoured by pressure.[12] Huang *et al.*[13] reacted 3-hydroxy-2-butanone and ammonium acetate under weakly acidic conditions and saw an increase in tetramethylpyrazine formation at high pressure. They attributed this to formation of an ionic intermediate. Therefore, certain Maillard reaction pathways leading to volatile compounds are favoured by high pressure, under certain conditions and for certain reaction precursors.

In conclusion, for the same degree of colour development, systems reacted under high pressure had far lower levels of low molecular weight compounds (volatiles) than the same systems reacted at atmospheric pressure. Results so far have not indicated the formation of any different Maillard products with the application of high pressure.

ACKNOWLEDGEMENT

This work was funded by the Ministry of Agriculture, Fisheries and Food (MAFF), U.K., via a studentship awarded to V.M.H.

REFERENCES

1. V.B. Galazka and D.A. Ledward, *Food Technol. Int. Europe*, 1995, 123.
2. J.C. Cheftel, in 'High Pressure and Biotechnology', eds. C. Balny, R. Hayashi, K. Heremans, P. Masson and J. Libbey, Eurotext Ltd, London, 1992, p. 195.
3. F. Ledl and E. Schleicher, *Angew. Chem. Int. Ed. Engl.*, 1990, **29**, 565.
4. J.M. Ames, *Trends in Food Sci. Technol.*, 1990, **1**, 150.
5. S.H. Ashoor and J.B. Zent, *J. Food Sci.*, 1984, **49**, 1206.
6. V.M. Hill, D.A. Ledward and J.M. Ames, *J. Agric. Food Chem.*, 1996, **44**, 594.
7. P.E. Koehler and G.V. Odell, *J. Agric. Food Chem.*, 1970, **18**, 895.
8. R. Tressl, D. Rewicki, B. Helak, H. Kamperschroer and N. Martin, *J. Agric. Food Chem.*, 1985, **33**, 919.
9. A. Apriyantono and J.M. Ames, *J. Sci. Food Agric.*, 1993, **61**, 477.
10. S. Fors, in 'The Maillard Reaction in Food and Nutrition', eds. G.R. Waller, M.S. Feather, A.C.S. Symposium Series 215, American Chemical Society, Washington, D.C., 1983, p. 185.
11. R. Tressl, B. Helak, N. Martin and D. Rewicki, in 'Amino–Carbonyl Reactions in Food and Biological Systems', eds. M. Fujimaki, M. Namiki and H. Kato, Elsevier, Amsterdam, 1986, p. 235.
12. K. Matsumoto, A. Sera and T. Uchida, *Synthesis*, 1985, **1**, 1.
13. T.-C. Huang, H.-Y. Fu and C.-T. Ho, *J. Agric. Food Chem.*, 1996, **44**, 240.

COMPARISON OF SOME AROMA IMPACT COMPOUNDS IN ROASTED COFFEE AND COFFEE SURROGATES

W. Holscher

Kraft Jacobs Suchard, Coffee Research & Development, Weser-Ems-Str. 3-5, 28309 Bremen, Germany

1 INTRODUCTION

Over the last decade, the application of combined gas chromatography and olfactometry (*e.g.* aroma extract dilution analyses) has proven to be extremely useful for the identification of the character impact compounds of various foodstuffs. Substantial progress has been made in the area of coffee aroma research. It was found that coffee aroma does not consist of a single key coffee compound but is a mixture of many odorants from several chemical classes and with different aroma nuances.[1-3] Many of these compounds occur in other foodstuffs as well. Therefore, the question why well balanced real coffee aroma is generated only during roasting of green coffee and not from other sources still has not yet been answered satisfactorily. Two classes of compounds have attracted special interest since their presence in roasted coffee can be related to the composition of the corresponding green material: methoxyisoalkylpyrazines and thiols, such as 3-mercapto-3-methylbutyl formate, 3-mercapto-3-methyl-butanol and 3-methyl-2-buten-1-thiol. Methoxyisoalkylpyrazines are native green coffee constituents. The same holds true for the most likely precursor of the thiols, *i.e.*, prenol (3-methyl-2-buten-1-ol).

The objective of this study was to further clarify the role of these odorants in well-balanced and natural coffee aroma in coffee related foodstuffs. Their presence was screened in coffee surrogates by means of GC–MS and GC–olfactometry. Coffee surrogates are made from roasted chicory, malt, rye or barley or blends of same. They possess some coffee-like flavour characteristics but lack well balanced roasted coffee aroma and exhibit more *grainy* or *malty* notes.

2 EXPERIMENTAL

2.1 Sample Material

Colombian Arabica and Indonesian Robusta coffees were roasted to a medium degree. Coffee surrogates were either from pure roasted chicory or commercially available brands. They consisted of roasted malt, two brands containing blends of roasted malt, barley, rye and chicory and a spray dried extract made from a similar blend. For raw material analyses the corresponding green coffees were used as well as unroasted barley and rye.

2.2 Isolation of Volatiles

Roasted and ground coffees (50 g), as well as the coffee surrogates (in the case of green coffee and cereals, 100 g coarsely ground material was used), were placed in a 2 l round bottom flask of a simultaneous distillation–extraction device[4] together with 1 l of distilled water, an anti-foam and the internal standard (2,3-dimethoxytoluene). Simultaneous distillation–extraction was carried out for two hours using 50 ml of a mixture of n-pentane–diethyl-ether (1:1, v/v) as solvent. The raw extract was dried over anhydrous sodium sulfate, filtered off into a 50 ml finger flask and concentrated to about 0.5 ml by means of a Vigreux column (250 × 10 mm) at a water bath temperature of 45 °C.

2.3 GC–MS and GC–Olfactometry

GC–MS was performed on a Hewlett Packard 5890 series II gas chromatograph combined with a MSD 5872 and equipped with a DB-WAX capillary column (60 m × 0.25 mm, film thickness 0.25 mm; J&W Scientific). The extracts were injected by a programmable variable temperature injector system (Gerstel, Mülheim, Germany). The temperature profile ranged from 60 °C to 200 °C at a heating rate of 12 °C s^{-1}. The oven temperature profile was 35 °C held for 1 min, then 40 °C min^{-1} to 60 °C and 3 °C min^{-1} to 220 °C, held for 10 min. All quantitative data were calculated as values relative to the data of Colombian coffee that were set to 1 for normalisation.

GC–olfactometry was conducted on a Carlo Erba MEGA 5300 gas chromatograph using the same capillary column mentioned above but with an inner diameter of 0.32 mm. The GC-effluent was split 1:1 by means of a glass-cap-cross splitter[5] for flame ionisation detection as well as for simultaneous olfactometric evaluation by the human nose. The oven temperature profile was 45 °C held for 1 min, then 40 °C min^{-1} to 60 °C and 3 °C min^{-1} to 220 °C. Sniffing was performed by two trained assessors. Identification of compounds was carried out on the basis of retention- and MS-data as well as sensory properties in comparison with authentic reference chemicals.

3 RESULTS AND DISCUSSION

In the early 1990s, two important coffee odorants were reported in food for the first time: 3-mercapto-3-methylbutyl formate and 3-mercapto-3-methylbutanol.[6,7] In the meantime, the formate has been shown to contribute to beer staleness.[8] Compounds with tertiary mercapto-function show interesting structure–odour relationships as depicted in Figure 1. The basic *catty*, *sweaty* odour is obviously related to a tertiary amyl mercaptan structure (**I**) which also occurs in compound (**II**). The methoxy group and the keto-function of compounds (**III**) and (**IV**) do not influence the overall aroma perception dramatically, whereas, the ester functions of compounds (**V**) and (**VI**) appear to introduce a distinct *fruity* character to the basic *catty* smell. However, desterification to the free 3-mercapto-3-methyl-butanol was linked to an odour conversion from *sweaty*, *catty* to a *sulfurous*, *boiled meat-like* character in combination with a steep drop of the odour theshold value by three orders of magnitude. It is interesting to note that compounds (**V**), (**VI**) and (**VII**) have been reported as artificial flavouring substances in the patent literature.[9] These compounds were suggested for aromatisation of blackcurrant or coffee products. The authors found that these kinds of compound possess extremely unpleasant smells in the concentrated state. However, at high dilution they exhibit or enhance the natural character of *fruity*, *meat-like* or *roasty* aroma nuances. Another derivative, 3-mercapto-3-methylbutylacetate, was synthesised and found to have an even more *fruity* than *sweaty* character than the

Table 1 *Volatiles in Roasted Coffee and Coffee Surrogates*

	Colombian Arabica	Indonesian Robusta	Roasted Chicory	Roasted malt	Roasted malt, barley, rye, chicory	Roasted malt, barley, rye, chicory (spray dried extract)
Hexanal	1	2.3	0.5	0.9	4.4	11
3-Methyl-2-buten-1-thiol	1	1.2	—	—	—	—
Prenol	1	0.6	—	—	—	—
2-Ethylpyrazine	1	1.8	<0.1	0.5	0.4	0.1
2,3,5-Trimethylpyrazine	1	2	<0.1	0.7	0.8	0.2
2-Furanmethanethiol	1	1.7	0.1	0.1	0.2	0.1
2-Methoxy-3-isopropylpyrazine	1	1.3	—	—	—	—
Methional	1	0.8	<0.1	0.3	0.3	0.6
2-Ethyl-3,5-dimethylpyrazine	1	2.7	<0.1	0.6	0.8	0.2
3-Mercapto-3-methylbutylformate	1	0.7	—	—	—	—
2-Methoxy-3-isobutylpyrazine	1	0.4	—	—	—	—
2-Acetylpyridine	1	1.3	<0.1	0.6	0.5	0.3
5-Methyl-5H-6,7-dihydro-cyclopentapyrazine	1	2.4	<0.1	1	0.8	0.3
2-Phenylacetaldehyde	1	1	0.1	0.2	0.2	0.4
3-Mercapto-3-methylbutanol	1	0.8	—	—	—	—
3-Methylbutyric acid	1	0.5	<0.1	0.1	<0.1	0.1
β-Damascenone	1	1.2	<0.1	<0.1	0.1	0.2
Cyclotene	1	0.9	0.3	0.4	0.2	0.4
Guaiacol	1	6	<0.1	0.7	0.5	0.1
4-Ethylguaiacol	1	1	11	<0.1	0.8	0.9
4-Vinylguaiacol	1	2.4	<0.1	0.1	0.1	0.1

Semi-quantitative data given as values relative to levels found in Colombian coffee which were set to 1 for normalisation; —: not detectable at a detection limit of about 1 μg kg^{-1}; compounds listed in the order of their retention on DB-WAX.

corresponding formate, however, no evidence was found in previous work that this compound occurs naturally in coffee.

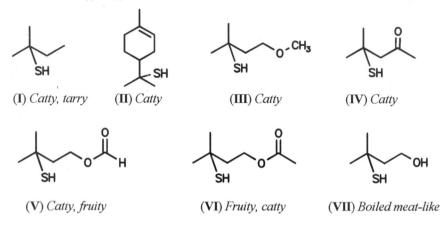

(I) *Catty, tarry*   (II) *Catty*   (III) *Catty*   (IV) *Catty*

(V) *Catty, fruity*   (VI) *Fruity, catty*   (VII) *Boiled meat-like*

Figure 1   *Structure–odour relationships of various odorants with a tertiary mercapto-function*

In earlier communications, it was suggested that compounds such as 3-methyl-3-methyl-butylformate, the corresponding free alcohol and a third compound, 3-methyl-2-buten-1-thiol, were obviously related to each other by common precursors. The precursors are very probably intermediates of the isoprenoid biosynthesis cascade, *e.g.*, isomers of active isoprenes, such as prenylpyrophosphate or the corresponding free alcohol prenol.[7,10] This assumption was supported by roast model reactions using mixtures of prenol and hydrogen sulfide-donating amino acids. It could be demonstrated that at least 3-mercapto-3-methylbutanol and 3-methyl-2-buten-1-thiol were formed in significant amounts. As expected, the three thiols of interest were detected in the Arabica as well as in the Robusta coffee variety. No evidence, however, was found that the coffee surrogates under investigation contained any of these compounds (Table 1). It can definitely be assumed that the absence of these thiols accounts for the less coffee-like character of coffee surrogates to some extent. With regard to the precursor chemistry, all roasted samples were screened for the presence and absence, respectively, of the likely precursor compound prenol. Prenol could be detected in both the Arabica and the Robusta coffee variant. As Table 1 shows, no traces of prenol were detectable in the coffee surrogates under investigation. The same exercise was executed for raw materials such as the corresponding green coffees and raw barley and rye. In agreement with previous findings, green Arabica and Robusta coffees were found to contain prenol in the low ppm level[11] (Table 2). The unroasted barley and rye, respectively, did not show detectable amounts of prenol. This finding may be considered as further support for its role as possible aroma precursor.

Methoxyisoalkylpyrazines such as 2-methoxy-3-isobutylpyrazine and the corresponding isopropyl derivative were present in both botanical varieties of roasted coffee. Similar to the prenol, no traces of the pyrazines could be detected in the coffee surrogates. The amounts of 2-methoxy-3-isobutylpyrazine, which possesses a strong *vegetable-like*, *bellpepper-like* or *green* odour, occurred in significantly lower amounts in Robusta coffee. This is in accordance with recent communications dealing with quantitative differences of the aroma impact compounds in Arabica and Robusta coffee.[3] Methoxyisoalkylpyrazines are naturally occurring compounds that are widely spread among various plants such as bell pepper or peas. These important odorants were first reported by Vitzthum *et al.*[12] as powerful odorants that determine the characteristic smell of green coffee. Present findings suggest that there is a strong link between raw material composition and final flavour profile of the roasted material. These methoxyisoalkylpyrazines will not be formed during roasting. They are sufficiently stable to survive normal roasting conditions. It can be assumed that methoxyisoalkylpyrazines do not occur in the raw materials used as coffee surrogates in contrast to green coffee. As Table 2 shows, actually no methoxyisoalkylpyrazines were found in raw barley and rye after steam distillation.

Table 2 *Detection of Prenol and Methoxyisoalkylpyrazines in Green Coffee and Cereals*

	Green Arabica	Green Robusta	Raw Barley	Raw Rye
Prenol	+	+	—	—
3-Methoxy-3-isopropylpyrazine	+	+	—	—
3-Methoxy-3-isobutylpyrazine	+	+	—	—

+ Identified on the basis of retention data, GC–MS and GC–olfactometry;
— Not detectable at a detection limit of about 1 µg kg^{-1}.

In contrast to the previously mentioned compounds, some known aroma impact compounds of coffee, *e.g.*, 2-furanmethanethiol, were identified in the coffee surrogates. 2-Furanmethanethiol is one of the few odorants that actually exhibits true coffee-like aroma characteristics. Literature screening on the basis of comprehensive volatiles data[13] indicated that its presence in roasted chicory, rye, barley or malt has not yet been reported. Although the levels are significantly lower compared with roasted coffee, a coffee-like aroma impact is still very likely at these lower concentrations.

Coffee surrogates were screened for other compounds that have been reported as constituents of coffee surrogates[13] and which are known impact compounds of roasted coffee as well. Among those were various alkylpyrazines, aldehydes and guaiacols (Table 1). All compounds occurred in lower amounts than in roasted coffee. Traces of most of the compounds were found in roasted chicory.

It may be stated, in summary, that important coffee odorants such as methoxyisoalkyl-pyrazines and thiols derived from prenol are specifically absent in coffee-related beverages such as roasted chicory, malt, rye or barley. The findings throw some light onto how coffee aroma is formed and may offer an explanation as to why natural roasted coffee aroma can only be generated from green coffee.

ACKNOWLEDGEMENT

I thank Dr. H. Bade-Wegner for skilful mass spectral analyses as well as I. Bendig and P. Gießmann for substantial technical assistance.

REFERENCES

1. W. Holscher, O.G. Vitzthum and H. Steinhart, *Café, Cacao, Thé*, 1990, **34**, 205.

2. I. Blank, A. Sen and W. Grosch, '14ᵉ Coll. Scient. Int. sur le Café, San Francisco, 1991', ASIC, Paris, 1992, p. 117.
3. P. Semmelroch and W. Grosch, *J. Agric. Food Chem.*, 1996, **44**, 537.
4. T.H. Schultz, R.A. Flath, T.R. Mon, S.B. Eggling and R. Teranishi, *J. Agric. Food Chem.*, 1977, **25**, 446.
5. W. Bretschneider and P. Werkhoff, *J. High Res. Chromatogr. & Chromatogr. Comm.*, 1988, **11**, 543.
6. W. Holscher and H. Steinhart, '14ᵉ Coll. Scient. Int. sur le Café, San Francisco, 1991,' ASIC, Paris, 1992, p. 130.
7. W. Holscher, O.G. Vitzthum and H. Steinhart, *J. Agric. Food Chem.*, 1992, **40**, 655.
8. P. Schieberle, *Z. Lebensm. Unters. Forsch.*, 1991, **193**, 558.
9. J. Stoffelsma and J. Pijpker, German Patent DE 2316456 C2, 1973.
10. R. Tressl, M. Holzer and H. Kamperschroer, '10ᵉ Coll. Scient. Int. sur le Café, Salvador, 1982', ASIC, Paris, 1983, p. 279.
11. W. Holscher and H. Steinhart, in 'Food Flavors: Generation, Analysis and Process Influence', ed. G. Charalambous, Elsevier Science, Amsterdam, 1995, p. 785.
12. O.G. Vitzthum, P. Werkhoff and E. Ablanque, '7ᵉ Coll. Scient. Int. sur le Café, Hamburg, 1975', ASIC, Paris, 1976, p. 115.
13. L.M. Nijssen, C.A. Visscher, H. Maarse, L.C. Willemsens and M.H. Boelens, 'Volatile Compounds in Food', TNO, 7th edition, Zeist, The Netherlands, 1996.

INFLUENCE OF BLANCHING ON AROMA COMPOUNDS IN LEEKS DURING FROZEN STORAGE

M.A. Petersen, L. Poll, M. Lewis* and K. Holm

The Royal Veterinary and Agricultural University, Department of Dairy and Food Science, Rolighedsvej 30, 1958 Frederiksberg C, Denmark

*Long Ashton Research Station, Department of Agricultural Sciences, Long Ashton, Bristol, BS18 9AF, U.K.

1 INTRODUCTION

In Denmark, there has been an increase in consumption of frozen vegetables. Most often, vegetables are blanched before freezing to avoid formation of off-flavours during storage. However, it is not common practice to blanch leeks since blanching causes undesirable textural changes. Furthermore the strong sulfurous odour of leeks is believed to cover up minor amounts of developing off-flavours. However, such oxidizing enzymes as lipoxygenase and peroxidase are present in leeks, and significant amounts of off-flavours may therefore be formed during longer storage if blanching is omitted.

The aim of this study was, therefore, to investigate the changes in aroma composition in raw and blanched leeks during frozen storage.

2 EXPERIMENTAL

Leeks (*Allium ampeloprasum* var. Bulgarsk Kæmpe) grown to maturity in fields on Funen, Denmark, were cleaned, and the light green stem was cut into 1 cm slices. The slices were steam-blanched at 95 °C for 0 or 180 s, cooled in ice water for 5 min, drained for 5 min, and finally frozen under nitrogen at –50 °C for 10 min. The leeks were then packed in plastic bags that were sealed and stored at –20° C. Samples were analysed after 0, 26, 54, 83, 112 and 140 days. All analyses were carried out in duplicate.

Volatiles were collected from an aqueous suspension of homogenized leek, containing 2 ml 4-methyl-1-pentanol (50 ppm) as internal standard, using a dynamic headspace technique with a Porapak Q trap. The trap was eluted with 200 mg diethylether that was concentrated to 50 mg by gently blowing nitrogen over the surface.

Separation and quantification was carried out on a Hewlett-Packard 5890A gas chromatograph equipped with a DB-Wax column. Concentrations of aroma compounds are given as 'relative peak areas', *i.e.*, area of peak representing the given compound divided by the area of the internal standard peak. Identification was done using a Kratos MS 80 RFA GC–MS equipped with a Superox II column. Sniffing (GC–O) was done by one of the authors using the same equipment as for GC–MS, except that the end of the column was attached to a small glass funnel. The odour type was described and marks were given for intensity on a scale from 0 to 5.

3 RESULTS AND DISCUSSION

In a typical leek sample, 103 volatile compounds were identified by GC–MS with varying certainty. In the odour run, 71 odours were detected and, of these, 22 could be related to identified compounds with reasonable certainty. Table 1 lists these data. It is seen that odours characterized as 'leek', 'onion' or 'sulfury' primarily stem from sulfides and thiols, while 'green', 'dusty' and 'dry' odours were mostly caused by aldehydes. However, this division was not always found since some sulfides were described as 'green' or 'dry' and one aldehyde as 'oniony'. 2-Propen-1-thiol and acetic acid have not been reported in leeks before, but they have been found in onions.[1] Our findings are in good agreement with reports that methyl propyl disulfide and dipropyl trisulfide are very characteristic aroma compounds in leeks.[2,3]

Table 1   *GC–sniffing of a leek sample (raw, stored for 82 days at –20 °C)*

Time	Intensity	Odour	Identification	Code in Figure 2	Pattern
6.01	1	Faint onion	1-Propanethiol[*]	s1	A
8.00	2.5	Sulfury, onion	2-Propene-1-thiol[**]	s2	A
10.07	1	Smoky, choking	Pentanal[*]	a1	D
13.25	4	Green leaves	Hexanal[**]	a2	E
14.09	1.5	Green, vegetable, dry	2-Methyl-2-butenal[*]	a3	B
17.06	2	Sweetish, dry dusty	2-Methyl-2-pentenal[**]	a4	C
20.47	4.5	Strong, green, dry	Methyl propyl disulfide[**]	s3	A
23.47	1.5	Green onion	(E)-Propenyl methyl disulfide[**]	s4	A
28.15	4	Dusty, dry, crusts	Dimethyl trisulfide[***] +		
28.22	4	Metallic, dry	Dipropyl disulfide[***] (not separated)	s5	A
29.50	1.5	Green, bit meaty	(Z)-Propenyl propyl disulfide[**]	s6	A
30.37	3	Dry, woody	Allyl propyl disulfide[**]	s7	A
30.54	3.5	Green, sweetish	(E)-Propenyl propyl disulfide[**]	s8	A
32.47	1.5	Urinary	Acetic acid[**]	ac	A
33.55	1.5	Oniony, dry	2,4-Heptadienal[**]	a5	E
35.12	2.5	Dry, oniony	Methyl propyl trisulfide[*]	s9	A
35.19	4	Strong, dry, aldehyde	2-Nonenal[** a]		
41.17	2	Acidic, dry, sulfury	Dipropyl trisulfide[***]	s10	A
44.16	1.5	Smoky, green, dry	3,5-Diethyl-1,2,4-trithiolane (t)	s11	A
45.44	2	Green, sulfury, bit fruity	(Z)-Propenyl propyl trisulfide[*]	s12	A
46.30	4	Cooked, green, bit sulfury	(E)-Propenyl propyl trisulfide[*]	s13	C
46.54	2	Dry, sweetish	2,4-Decadienal[**]	a6	A
60.44	4	Strong dry, bit sulfury	(Z)-Propenyl propyl tetrasulfide[*]	s14	C

[***], [**], [*], (t): certainty of identification (ranging from 'certain' to 'tentative');
[a] Not found by GC, only by GC–MS and GC–O.

Different patterns of changes in concentration of aroma compounds were seen during frozen storage. For all aroma compounds in Table 1, except pentanal, hexanal and 2,4-heptadienal, concentrations were low and relatively constant in blanched samples, while concentrations were higher and more fluctuating in raw samples. An example of this is (Z)-

propenyl propyl disulfide (Figure 1A) which represents the most common pattern (compounds marked with 'A' in Table 1). Twelve of 14 sulfur containing compounds follow pattern 'A'. Figure 1B shows the pattern for 2-methyl-2-butenal (marked 'B' in Table 1). Figure 1C shows the pattern for 2-methyl-2-pentenal (pattern 'C'). For pentanal the concentration was highest in blanched samples (pattern 'D'), while concentrations of hexanal and 2,4-heptadienal were almost equal in raw and blanched samples but increasing with time (pattern 'E').

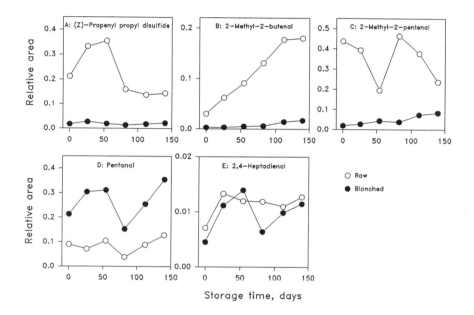

Figure 1   *Changes in concentration of (Z)-propenyl propyl disulfide, 2-methyl-2-butenal, 2-methyl-2-pentenal, pentanal and 2,4-heptadienal during frozen storage of leeks*

The low and constant concentration of aroma compounds in blanched leeks is most likely due to destruction of the enzymes normally responsible for the production of the aroma compounds themselves or precursors, eventually combined with some leaching out of aroma during blanching.

The increasing amount of sulfides in unblanched samples is probably due to alliinase activity in the disrupted cells. This enzyme starts a chain of reactions by catalyzing the cleavage of S-methyl-, S-propyl and S-propenyl-cysteine to the corresponding sulfenic acids. Also aldehydes are produced from these reactions.[4-6] Aldehydes, especially hexanal and unsaturated aldehydes, can, however, also be produced by a chain of reactions starting with the oxidation of unsaturated fatty acids, either catalyzed by lipoxygenase or due to autoxidation.[7] The production of pentanal, hexanal and 2,4-heptadienal in blanched samples is probably due to non-enzymatic reactions, even though precursors for these aldehydes may be produced enzymatically in the time from cutting the leeks to blanching.

To get an overview of the changes desribed above, a principal component analysis (PCA) was carried out. Two principal components accounted for 71% of the total variation. Loadings and scores plot are seen in Figure 2.

From the scores plot it is seen that the samples are clearly divided into three groups: blanched samples, raw samples stored for 54 days or less, and raw samples stored for 82 days or more. Blanched samples exhibit no systematic changes in the concentrations of aroma compounds during storage, though some fluctuations are seen. In contrast to this, raw samples move from one part of the score plot to another, indicating systematical changes in aroma during storage.

When loadings and scores plot are compared, it is seen that blanched samples are characterized by having high concentration of pentanal and to a lesser degree of hexanal, and having low concentration of sulfides and other aldehydes.

During storage, raw samples move from a region totally dominated by sulfides to a region where the aldehydes, hexanal, 2-methyl-2-butenal, 2,4-heptadienal and 2,4-decadienal, are prominent together with two thiols.

These results indicate that blanching has had the desired effect of stabilizing the aroma of the leeks, but this is achieved at the expense of the aroma that characterizes the freshly frozen leeks. The aroma of the unblanched samples is relatively constant for approximately two months, but they then change so that the aroma profile is influenced more by aldehydes and less by sulfides.

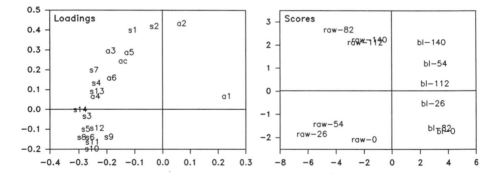

Figure 2   *Loadings and scores plot from PCA of concentrations of aroma compounds from Table 1 during frozen storage of raw and blanched leeks. Codes from Table 1 are used in the loadings plot while synbols in the scores plot indicate treatment (raw or blanched) and days of storage. x-axis is PC1 (55%) and y-axis is PC2 (16%)*

From Table 1, it is seen that odour of sulfides very often is described as 'leek', 'onion' or 'sulfury', while aldehydes are given descriptions like 'green' and 'dry'. In the literature other descriptions can also be seen, *e.g.*, the odour of 2-methyl-2-butenal is also described as 'paint-like'. It would therefore be expected that the balance between leek-flavour and off-flavour is worsened during long time storage of unblanched leek, while blanched leek is more stable, but with a less intense leek flavour. However, whether this is of practical significance has to be confirmed by sensory analyses.

REFERENCES

1. K.O. Abraham, M.L. Shankaranarayana, B. Raghavan and C.P. Natarajan, *Lebensm.-Wiss. u. -Technol.*, 1976, **9**, 193.
2. L. Schreyen, P. Dirinck, F.V. Wassenhove and N. Schamp, *J. Agric. Food Chem.*, 1976, **24**, 336.
3. L. Schreyen, P. Dirinck, F.V. Wassenhove and N. Schamp, *J. Agric. Food Chem.*, 1976, **24**, 1147.
4. S. Schwimmer and M. Mazelis, *Biophysics*, 1963, **100**, 66.
5. J.R. Whitaker, *Adv. Food Res.*, 1976, **22**, 73.
6. T.-H. Yu and C.-T. Ho, in 'Shelf Life Studies of Foods and Beverages', ed. G. Charalambous, Elsevier Science Publishers B.V., 1993, p. 501.
7. T. Galliard, in 'Recent Advances in the Chemistry and Biochemistry of Plant Lipids', eds. T. Galliard and E.I. Mercer, Academic Press, 1975, p. 319.

AROMA COMPOUNDS FORMATION IN THE INTERACTION OF L-ASCORBIC ACID WITH α-AMINO ACIDS

S.M. Rogacheva, R. Verhe[*] and T.D. Obretenov

Department of Organic Chemistry, Higher Institute of Food and Flavour Industries, Plovdiv 4000, Bulgaria

[*]Department of Organic Chemistry, Faculty of Agricultural and Applied Biological Sciences, The University, Gent 9000, Belgium

1 INTRODUCTION

After reducing sugars, L-ascorbic acid appears to be the most widely studied carbonyl component in the processes of non-enzymic browning. This is due, on one hand, to its significant presence in food products and its biological activity, and, on the other, to the interesting chemical transformations it undergoes in the course of these processes. A series of intermediates arising from such interactions has been known. In the course of our own investigations into the processes of melanoidin formation with the participation of L-ascorbic acid, our attention has been drawn to the nature of volatiles obtained.[1,2]

2 EXPERIMENTAL

The interaction between L-ascorbic acid and α-amino acids – lysine, glycine and glutamic acid – was carried out in a round-bottomed flask at molar ratio 1:1, water content 20%, temperature 100 °C and reaction time up to 100 hours. The heating was performed in a glycerol bath with refluxing and intense stirring. The reaction mixture was purged with an inert gas (at about 1 l per hour). The volatile components collected were trapped in diethyl ether in three collectors at a temperature of less then –20 °C. After the reaction was ended, the ether solutions with volatiles were dried with anhydrous $MgSO_4$. The reaction mixture was diluted three-fold with distilled water, saturated with NaCl and extracted four-fold with diethyl ether. The ether extracts collected were washed twice with water and treated the same way as above. These volatile and non-volatile fractions were separated and identified by gas chromatography–mass spectrometry. The analyses were performed using GC–MS with the following conditions: column: DB-5 (30 m × 0.25 mm internal diameter); detector: MSD; carrier gas: helium; column temperature: 220 °C. The mass spectra of the separated compounds obtained were identified using the NIST–EPA–MSDB Mass spectral database, version 3.01, mass spectral fragmentation schemes, monographs and articles.

3 RESULTS AND DISCUSSION

A total of 187 compounds were identified by GC–MS in the volatile and non-volatile fractions. A significant difference in their quantitative and qualitative composition, dependent on the nature of the α-amino acid, reaction time and collection conditions was observed. Summarized data is presented in Table 1.

Table 1 *Number of the Components from* L-*Ascorbic acid (AA)–α-Amino Acid Interactions*

Compounds	Model system		
	AA–Gly	AA–Lys	AA–Glu
Aliphatic hydrocarbons	15	6	2
Aromatic hydrocarbons	9	9	4
Aliphatic O-containing compounds	17	11	6
Carbocyclic O-containing compounds	4	4	3
Furans	33	5	19
N-containing heterocyclic compounds	6	25	3
Phenols	2	3	
Pyrans	2	2	
Miscellaneous		2	2
Unidentified-from degradation of AA[*]	8	12	9
Other unidentified	17	24	16

[*] Compounds also found in GC–MS analysis of thermal degradation of AA alone.

The volatile intermediates are of particular interest because of their importance in the aroma of foods. The components, detected in the high concentrations, are considered to be the main participants in the aroma formation processes under model conditions. Oxygen- and nitrogen-containing aliphatic and heterocyclic compounds were predominantly found (Table 2). 2-Furfural and 2-acetylfuran were among the major components in the model systems AA–Gly and AA–Glu. The accumulation of 2-furancarboxylic acid, among the non-volatiles in the extracted fractions, paralleled the decrease of the former compounds. Cyclohexanone was found in the model systems at high pH (Gly pH$_i$ 6.20) while diacetyl only at very acid pH (Glu pH$_i$ 3.10). The compounds referred to seem to be produced from L-ascorbic acid via its transformations. L-Ascorbic acid, normally a stable compound, particularly in acidic solution in the absence of oxygen, is oxidized in the presence of oxygen to dehydroascorbic acid, which undergoes hydrolysis of the lactone ring to 2,3-diketo-L-gulonic acid. In food preparation in the presence of oxygen, the degradation of L-ascorbic acid is a complex reaction and a large number of low molecular weight compounds may be expected to be present. Our findings suggest that L-ascorbic acid undergoes intensively the process of retroaldolisation. The compound, with main fragments m/z 45(100), 43, 73, 61, was also found in the reaction of L-ascorbic acid alone and is the major volatile component in all model systems. Presumably, this is the main degradation product of L-ascorbic acid with the predominant participation in the further processes.

It is noteworthy that, in AA–Lys model, pyrazines and pyrimidines were significant products from the interaction of L-ascorbic acid degradation products with a highly reactive amino acid L-lysine. Undoubtedly, L-lysine, a rich source of nitrogen, and the alkali pH of the reaction media (pH$_i$ 9.74), create favourable conditions for cyclisation to form N-containing heterocyclic compounds. Pyrazines can be obtained by condensation of dicarbonyl compounds derived from L-ascorbic acid with ammonia or by dimerisation of L-lysine followed by various transformations. Some pyrrolidines, indols and quinolines were found in the non-volatile fractions. Such reactions appeared to dominate and no other heterocyclic compounds (furans) can be obtained.

Table 2   *Relative Amounts (%) of the Major Identified Volatiles in L-Ascorbic Acid–α-Amino Acid Model Systems Varying the Reaction Time*

Compound	T_R (min)	AA–Gly			AA–Lys			AA–Glu		
		1 h	7 h	50 h	1 h	7 h	50 h	1 h	7 h	50 h
1-hydroxy-2-propanone	4.320	4			5	4				
diacetyl	4.563							18	3	
2-ethoxy-1-propanol	4.637							5		
diethoxyethane	4.670	tr*				3				8
toluene	5.297	tr	6			2		9		tr
unidentified – M⁺ 74(12) 45(100) 43(40) 44(6) 42(5) 59(4) 46(4) 61(2)	6.081	tr			4					tr
unidentified – 45(100) 43(98) 73(92) 61(71) 63(15) 44(15) 42(8) 74(5)	6.404	42	38	50	36	73	18	50	59	58
methylpyrazine	6.455	tr					23			
2-furfural	6.556	tr						7	31	
unidentified – M⁺ 96(2), 45(100) 43(74) 73(34) 61(27) 44(15) 41(12) 42(11)	6.723						4			tr
2-methylpyrimidine	6.753						13			
unidentified – M⁺ 128(5) 43(100) 77(77) 45(71) 49(52) 42(35) 79(23) 59(22)	7.585							8		tr
cyclohexanone	7.788	5	35	8	4	5	3			
alkyl benzene MW 120	8.057		5							
2-acetylfuran	8.143	31	3	29					tr	tr
alkyl benzene MW 120	8.300		2							
2,6-dimethylpyrazine	8.302						12			
alkyl benzene MW 120	8.681		5				tr			
trimethylpyrazine	10.156						2			
unidentified – M⁺ 150(11) 146(100) 148(70) 111(50) 75(37) 50(26) 74(24) 113(13)	10.491							2	tr	tr
3-ethyl-2,5-dimethylpyrazine	11.839						1			
2,6-bis(1,1-dimethylethyl)-4-methylphenol	20.240		tr	3	3					
4-tert-butylbenzoic acid	20.405									5

*   tr – Traces.

Strecker-type aldehydes such as lysylaldehyde, formaldehyde were not detected. The most probable explanation is that they were consumed in various reactions due to their high reactivity.

4 CONCLUSION

In heated model systems, L-ascorbic acid is a precursor for the production of a variety of flavour and aroma constituents, as well as participants in Strecker degradation reactions with free amino acids. It is also noteworthy that the nature of the α-amino acid plays an important role in the generation of aroma compounds. An acidic amino acid Glu causes mainly L-ascorbic acid decomposition. Under basic conditions the condensation reactions of L-ascorbic acid with Lys dominate and a high number of N-containing heterocyclic compounds are accumulated.

REFERENCES

1. S.M. Rogacheva, M.J. Kuntcheva, I.N. Panchev and T.D. Obretenov, *Z. Lebensm. Unters. Forsch.*, 1995, **200**, 52.
2. G. Vernin, S. Chakib, S. Rogacheva, T. Obretenov, and C. Parkanyi, 'Thermal decomposition of ascorbic acid', A.C.S. 211th National Meeting, 1996, New Orleans, LA, U.S.A.

THE ROLE OF FORMALDEHYDE IN THE MAILLARD BROWNING OF GLUCOSE–GLYCINE REACTION

R. Vasiliauskaitë,* B.L. Wedzicha† and P.R. Venskutonis*

*Department of Food Technology, Kaunas University of Technology, Radvilënø pl. 19, Kaunas 3028, Lithuania

†Procter Department of Food Science, University of Leeds, Leeds, LS2 9JT, U.K.

1 INTRODUCTION

The Maillard reaction is very common in many thermally processed foods. Its mechanism is very extensive but poorly understood. The Maillard reaction is responsible for discoloration, flavour formation, and reduction in the nutritional value of foods. The most abundant final products of this reaction are high molecular weight polymeric brown compounds, so-called melanoidins. The significance of these in relation to the nutritional, psychological, toxicological properties of food, has been considered by many scientists. However, very little is known about the structure of Maillard polymers.

One of the numerous reactions taking place during the Maillard browning is the Strecker degradation of amino acids, which leads to the formation of Strecker aldehydes.[1] The further fate of these aldehydes is not fully understood. They can be incorporated into melanoidins or they can participate in the formation of volatiles[2,3] or both.

The aim of this work is to investigate the incorporation of formaldehyde into melanoidins and its influence on the volatiles formed during the Maillard browning of glucose and glycine.

2 MATERIALS AND METHODS

All reaction mixtures were prepared by dissolving glucose and glycine to prepare 1 M and 0.5 M solutions, respectively, and adding formaldehyde to give final concentrations in the range of 0–50 mM. The pH was adjusted to 5.5 using sodium hydroxide (1 M) and hydrochloric acid (1 M) before making up to the final volume. All kinetic experiments were carried out at 55 °C. To determine the extent of browning and volatiles produced during the browning, samples were withdrawn at timed intervals and absorbances were measured in the range 400–500 nm. For the radiochemical investigation, samples were prepared as described above. To aliquots (10 ml) of each mixture, approximately 4.6 MBq ^{14}C-glucose (labelled uniformly) of negligible mass was added and the volume made up to 20 ml. Similarly, the two reaction mixtures (10 ml) were labelled with 4.6 MBq ^{14}C$_1$-glycine and 4.6 MBq ^{14}C$_U$-glycine. An aliquot (25 ml) of the solution containing formaldehyde was mixed with 18.5 MBq ^{14}C-formaldehyde and made up to 50 ml with water. For the separation of melanoidins (M_r > 12000) the samples were dialysed against water by placing 1 ml of each reaction mixture into the dialysis tubes which were placed in beakers containing continuously stirred water (1 l). The water was changed three times every 12 hours. The retentates were made up to 50 ml, 1 ml of this stock solution was mixed with 10 ml of scintillation fluid, shaken and counted in the counting chamber for 10 min.

For the investigation of volatiles, reaction mixtures (prepared as described above) were purged with purified nitrogen (flow rate 60 ml min^{-1}) for two hours with subsequent adsorption on Tenax.[4] The volatiles collected were desorbed with ether, concentrated with nitrogen and analysed by means of capillary GC and GC–MS. The approximate concentrations of detected compounds were calculated (ng g^{-1} glucose) based on the peak area of an internal standard (n-octane) added to the reaction mixture before volatile collection.

3 RESULTS AND DISCUSSION

There is no single wavelength at which it is best to measure browning and therefore to investigate the effect of formaldehyde on the glucose–glycine reaction, absorbances of reaction mixtures were measured in the range of 400–500 nm. The results obtained at 450 nm are illustrated in Figure 1.

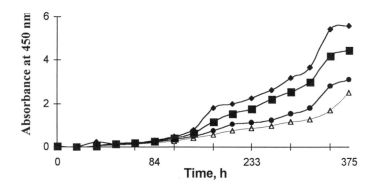

Figure 1 *Graph of absorbance at 450 nm, A_{450}, against time for the glucose–glycine reaction with added formaldehyde: -◆- 0; -■- 1 mM; -●- 10 mM; -Δ- 30 mM*

It is evident that there seemed to be a reduction in browning of samples containing formaldehyde. The sensitivity of the reaction to low concentrations of formaldehyde was particularly striking; a concentration of 1 mM was sufficient to reduce the absorbance markedly.

One of the most important aims in this work was to investigate whether or not formaldehyde becomes incorporated into melanoidin. It is possible to calculate the amount of formaldehyde incorporated into melanoidin using the number of counts obtained for samples with $^{14}C_1$-glycine and $^{14}C_U$-glycine. The procedure is illustrated by considering the reaction analysed after 23 days for $^{14}C_1$-glycine. When $^{14}C_1$-glycine is incorporated into the melanoidin the product may or may not become labelled with ^{14}C, depending on whether the binding of glycine involved the whole molecule (including C_1) or was via the Strecker aldehyde (after losing C_1 as CO_2). One can, therefore, write a general stoichiometric equation:

$$a\ NH_2CH_2^*COOH \rightarrow b\ MelCH_2^*COOH + c\ MelCH_2 + {}^*CO_2$$

Where b and c represent the relative amounts of non-decarboxylated and decarboxylated glycine incorporated and Mel denotes the rest of the melanoidin structure. The amount of ^{14}C incorporated into the melanoidin after 23 days can be used as an example, and this was equivalent to a concentration of glycine of 2.55 mmol dm^{-3}. Consider now the situation when $^{14}C_U$-glycine was used. In this case the stoichiometric equation is:

$$a\ NH_2^*CH_2^*COOH \rightarrow b\ Mel^*CH_2^*COOH + c\ Mel^*CH_2 + {}^*CO_2$$

The concentration of the doubly labelled melanoidin species is 2.55 mmol dm^{-3} because this must be the same as the concentration of singly labelled species when $^{14}C_1$-glycine was used. However, the apparent concentration of $^{14}C_U$-glycine in the melanoidin is 3.80 mmol dm^{-3}. It is assumed that the difference is because additional glycine was incorporated after it became decarboxylated, *i.e.*, converted to formaldehyde. By difference, the amount of formaldehyde incorporated was $(3.77 - 2.55) \times 2$, where the factor 2 allows for the fact that the decarboxylated glycine has a specific activity which is half that of $^{14}C_U$-glycine.

Hence, it is concluded that based on the present model, the melanoidin obtained after 23 days contained both complete and decarboxylated glycine. In this experiment there was approximately one molecule of formaldehyde for each molecule of whole glycine present. This calculation was repeated for all the reaction times and for the reaction with and without added formaldehyde (50 mM). The data are summarised in Table 1.

Table 1 *Concentrations (mmol dm^{-3}) of Radio-labelled Whole Glycine Molecules and Formaldehyde Residues Incorporated into Melanoidins Prepared with and without added Formaldehyde*

Days	No formaldehyde added		50 mM formaldehyde added	
	Glycine	Formaldehyde	Glycine	Formaldehyde
10	0.31	0.23	0.95	1.34
12	0.82	1.30	1.51	0.94
15	1.00	1.26	1.81	0.81
18	1.53	1.42	2.82	0.21
20	2.00	2.44	2.99	0.50
23	2.55	2.42	3.80	0.26
25	2.92	4.24	4.20	0.7

The result listed above suggested that formaldehyde became incorporated into the melanoidins. The data obtained for samples containing formaldehyde also confirm this conclusion, but in the presence of formaldehyde, the added formaldehyde took the place of formaldehyde produced during the Strecker degradation.

Further investigation showed that added formaldehyde also changed the profile of the volatiles compounds. In total, over 30 volatile compounds were tentatively identified and screened quantitatively. The sensory assessment showed that samples without formaldehyde had a strong caramel like odour, samples containing 5 mM of formaldehyde had a medium strong aroma, whilst samples with 30 mM had a weak aroma. The changes in total amount of volatile compounds are given in Figure 2. The graphs presented indicate that, in the samples containing formaldehyde, the total amount of volatile compounds were lower than in the sample without added formaldehyde.

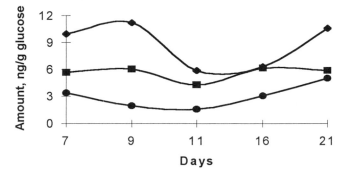

Figure 2 *Changes in total amount of volatile compounds produced during the Maillard browning of glucose and glycine with different concentrations of formaldehyde: -♦- 0; -■- 5 mmol; -●- 30 mmol*

It can be concluded that formaldehyde inhibits the browning of the glucose–glycine reaction. The radiochemical investigation showed that formaldehyde becomes incorporated into melanoidins. Formaldehyde also inhibits the formation of caramel flavour and causes changes in the composition and amounts of volatiles produced during the Maillard browning of glucose and glycine.

REFERENCES

1. J.E. Hodge, *Agric. Food Chem.*, 1953, **1**, 928.
2. J.P. Danehy, *Adv. Food Res.*, 1986, **30**, 77.
3. S. Fors, in 'The Maillard Reaction in Foods and Nutrition', eds. G.R. Waller and M.S. Feather, A.C.S. Symposium Series 215, American Chemical Society, Washington, D.C., 1983, p. 185.
4. Ch. Chen, *J. Agric. Food Chem.*, 1982, **30**, 1211.

VOLATILE FLAVOUR-ACTIVE DEGRADATION PRODUCTS ARISING FROM ANALOGUES OF DEOXYALLIIN AND ALLIIN

J. Velíšek, R. Kubec and M. Dolezal

Department of Food Chemistry and Analysis, Institute of Chemical Technology, Faculty of Food and Biochemical Technology, Technicka 1905, 166 28 Prague, Czech Republic

1 INTRODUCTION

Important natural constituents of many plants belonging to the genus *Allium* and *Brassica* are non-protein amino acids S-alk(en)yl-L-cysteines and their sulfoxides. The characteristic flavour arises in part from the enzymatic degradation of these amino acids when cellular tissue of the vegetables is disrupted by cutting, slicing or chopping. The enzyme alliinase (E.C. 4.4.1.4) converts S-allyl-L-cysteine sulfoxide (alliin) and its analogues into corresponding alk(en)yl thiosulfinates, the pungent principles of raw garlic and onion. These labile intermediates are subsequently transformed to a number of secondary products. Culinary processing can cause thermal denaturation of alliinase and thus a considerable amount of aroma precursors remains unchanged and could participate in the developing of characteristic flavour of processed food.

The volatile components of raw, dried, baked and boiled garlic and onion have been well documented and the contributions of alk(en)yl thiosulfinates and their transformation products to the flavour of processed vegetables have been studied.[1] A few papers report the contribution of alliin and its analogues to the flavour of thermally processed food.[2-4]

This paper reports on the major volatile compounds generated by thermal degradation of the amino acid alliin and its methyl- and propyl- analogues in aqueous solution.

2 EXPERIMENTAL

2.1 Synthesis of S-Alk(en)yl-L-Cysteine Sulfoxides

Alliin was synthesised according to the procedure of Yu *et al.*[2] and its (+)-stereoisomer was isolated by repeated crystallisation. S-methyl- and S-propyl-L-cysteine sulfoxides were synthesised in the same way. The structure of these compounds was confirmed by ^{1}H- and ^{13}C-NMR and by IR spectroscopy. Purity (> 99%) was checked by HPLC after derivatisation with *o*-phthaldialdehyde.[5]

2.2 Thermal Decomposition of S-Alk(en)yl-L-Cysteine Sulfoxides

Amino acid (50 mg) was placed in a 5 ml ampoule, water was added and the ampoule was sealed. After equilibration for 24 hours, the ampoule was heated in an oven at the given temperature, then cooled to −18 °C and crushed under water (total amount 2.45 ml). The solution was extracted with 5 ml of diethyl ether. The extract obtained was dried with anhydrous Na_2SO_4 and analysed by GC.

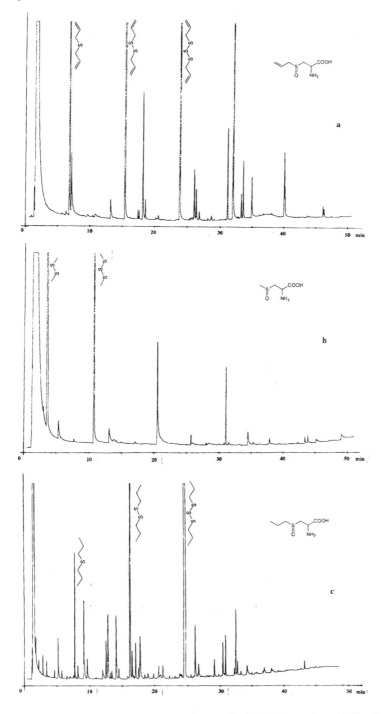

Figure 1 *Separation of amino acid pyrolysates by GC (HP-5 column), (a) alliin, (b) S-methylcysteine sulfoxide,(c) S-propylcysteine sulfoxide*

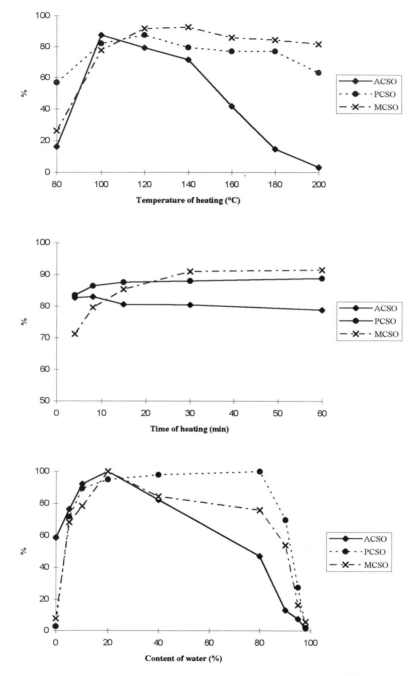

Figure 2   *Changes in the content of total dialk(en)yl sulfides (measured as % total volatiles) as influenced by time of heating (a), temperature (b), and water content (c). ACSO = alliin; MCSO = S-methylcysteine sulfoxide; PCSO = S-propylcysteine sulfoxide*

2.3 Gas Chromatographic Analyses and Gas Chromatography–Mass Spectrometry

A Hewlett-Packard 5890 chromatograph fitted with a flame ionisation detector and an HP-5 or HP-INNOWax fused silica capillary column (30 m × 0.25 mm internal diameter, Hewlett-Packard) was used. The operating conditions were as follows: injector temperature, 220 °C; detector temperature 250 °C; nitrogen carrier gas flow rate of 2 ml min^{-1}; and temperature program for HP-5 column: 40 °C(3 min), increased at 4 °C min^{-1} to 240 °C (10 min) and for the HP-INNOWax column: 40°C (3 min), increased at 4 °C min^{-1} to 190 °C (10 min). One µl of sample was injected, using a split of 5:1.

GC–MS analysis was carried out using a Hewlett-Packard G1800A chromatograph. The operating conditions were the same as described above. Mass spectra were obtained by EI ionisation at 70 eV over the range of 15–400 mass units.

3 RESULTS AND DISCUSSION

Figures 1a, 1b and 1c show typical chromatograms obtained from analysing volatiles arising from alliin and its methyl- and propyl analogues pyrolysed in the presence of 10% (w/w) of water at 120 °C for 60 min, respectively. As can be seen, the major volatiles of all the three amino acids were the corresponding dialk(en)yl sulfides, dialk(en)yl disulfides and dialk(en)yl trisulfides. The minor degradation products were mainly alkyl substituted pyridines.

Figure 2a shows that the highest content of the major three sulfides (expressed in % of total volatile material) arose at temperatures from 120 to 140 °C. Alliin decomposed during the first 15 min of heating and the amount of volatiles remained constant on prolonged heating. The methyl analogue of alliin produced sulfides more slowly than alliin itself. Sulfides, formed from the propyl analogue of alliin, decomposed on heating for longer than 10 min (Figure 2b). Addition of water had a marked influence on the amount of volatiles produced by amino acid pyrolysis. A small amount of volatiles arose from dry amino acids and at water contents greater than 80% (w/w).

The compounds produced from alliin resembled cooked onion and garlic, those originating from S-methylcysteine sulfoxide possessed typical flavour of cooked cabbage and volatiles arising from S-propylcysteine sulfoxide had cooked onion-like flavour. At higher temperatures, sulfurous and burned notes prevailed.

ACKNOWLEDGEMENT

Supported by Grant Agency of the Czech Republic, grant number 509/94/0318.

REFERENCES

1. E. Block, *Angew. Chem. Int. Ed. Engl.*, 1992, **31**, 1135.
2. T.-H. Yu, C.-M. Wu, R.T. Rosen and T.G. Hartman and C.-T. Ho, *J. Agric. Food Chem.*, 1994, **42**, 146.
3. T.-H. Yu, C.-M. Wu and C.-T. Ho, *J. Agric. Food Chem.*, 1994, **42**, 1005.
4. T.-H. Yu, C.-M. Wu and C.-T. Ho, *Food Chem.*, 1994, **51**, 281.
5. J. Velíšek, B.H. de Vos and A. Schouten, *Potrav. Vedy*, 1993, **11**, 445.

NOVEL METHODS OF FLAVOUR ANALYSIS

MONO-DIMENSIONAL CHIRAL GAS CHROMATOGRAPHY WITH DYNAMIC HEADSPACE CONCENTRATION: APPLICATION TO THE AROMAS OF MUSHROOM

S. Breheret, K. Feuillassier and T. Talou

National Polytechnic Institute of Toulouse, Agro-Industrial Chemistry Laboratory, ENSCT
118 route de Narbonne, F-31077 Toulouse Cedex, France

1 INTRODUCTION

Aroma compositions have been investigated extensively. Most of the identifications were achieved by GC–MS measurements which do not allow stereochemical assignment of chiral constituents.

Recently, the importance of enantiomeric separations to aroma and fragrance compounds has attracted increasing attention.[1,2] Indeed, enantioselectivity during biosynthesis may serve as an endogenic parameter in the authenticity and provenance control of natural flavour and fragrance compounds. It could also be indicative for some biogenetic pathways, and for establishing phenotype variations.

Derivatization of enantiomers using chiral derivatization agents is a method of separating chiral flavour compounds on achiral stationary phases.[3-6] In the last decade, new chiral stationary phases have been developed, i.e., modified α-, β- and γ-cyclodextrins which have been successful in the GC separation of several enantiomeric compounds.[7-15] The majority of the published results was obtained by combining conventional capillary columns with such cyclodextrin chiral capillary columns in a double-oven gas chromatograph (two-dimensional gas chromatograph) with split-injection.[16-23] This system has been demonstrated as a powerful method for the direct stereoanalysis of chiral volatiles whose performance could be increased by acetylation of samples allowing a better separation of chiral compounds. However the two-dimensional GC requires a high financial and technical investment. On the other hand, dynamic headspace concentration is well established as a rapid and suitable method for direct gas chromatographic analysis of aroma compounds.[24,25] A preliminary study carried out on industrial mushroom flavourings by combining headspace concentration with chiral gas chromatographic separation was very promising in terms of 1-octen-3-ol ratio titration and its relative natural *versus* nature-identical flavourings differentiation.[26]

The method used in this study was based on mono-dimensional chiral chromatography with α- and β-cyclodextrin capillary columns and dynamic headspace concentration without preliminary chemical derivatization of the isolate. The purpose of this study was to examine the effect of different industrial processes, *viz* canning and deep-freezing, on the natural enantiomeric excess of major chiral compounds of two mushrooms of commercial significance: the cultivated mushroom or the 'Paris mushroom' (*Agaricus bisporus*) and the wood blewitt (*Lepista nuda*).

In this study we have examined the natural enantiomeric excess of 1-octen-3-ol, the major chiral compound in mushroom aroma and especially in the flavour of *Agaricus bisporus*,[27] and of linalool which is a key compound of the aroma of *Lepista nuda*.

2 MATERIALS AND METHODS

2.1 Mushroom Samples

Commercial strains of two cultivated mushrooms, the 'Paris mushroom' (*Agaricus bisporus* (Lange) Imbach) and the wood blewitt (*Lepista nuda* (Bull.: Fr) Cooke) (Basidiomycetes), were kindly supplied by the Mushroom Research Station of INRA (Bordeaux, France), the Mushroom Museum (Maine and Loire, France). A commercial strain was bought from a market in Toulouse.

Table 1   *Odour Description of Mushrooms Studied and their Major Chiral Compounds*

Mushroom	Odour description of the basidiocarp[29]	Major chiral compound containing	Relative amount of chiral compound (%) [27,28]
A. bisporus	Typical mushroom-like odour	1-octen-3-ol	10 to 80
L. nuda	Fresh, flowery, vitamin C-like	linalool	20

Basidiocarps were sampled with different processes:

2.1.1 Fresh Mushroom. Fresh cubes of basidiocarps were directly analysed three or four days after collecting.

2.1.2 Freezing Process. Fresh entire basidiocarps were frozen in plastic bags at −20 °C three or four days after collecting. The frozen cubes were used directly for analyses.

2.1.3 Blanching and Canning Process. Fresh entire basidiocarps were blanched in water for three minutes, three or four days after collecting. They were then canned and sterilized for one hour. Cubes of canned mushrooms were used directly for analyses.

2.2 Reference Chiral Compounds

The order of elution was assigned by using racemates and enantiomerically pure references.

Racemates: 1-octen-3-ol (Acros), Linalool (Sigma).

Optically active references: $R(-)$-1-octen-3-ol (Acros), $R(-)$-linalool (Fluka).

2.3 Monodimensional Chiral Gas Chromatography with Dynamic Headspace Concentration

Analyses were carried out using a dynamic headspace injector, DCI device (DELSI Instruments). This headspace injector was connected to a gas chromatograph DN 200 (DELSI Instruments) fitted with chiral gas capillary columns. The selection of the columns used in this study was based both on analytical criteria (direct detection) and on economical ones (similar price to a conventional GC column).

α-**DEX 120** (SUPELCO, 30 m, internal diameter 0.25 mm, 0.25 μm coated with 20% permethylated α-cyclodextrin in SPB-35 poly (35% phenyl / 65% dimethyl siloxane));

β-**DEX 110** (SUPELCO, 30 m, internal diameter 0.25 mm, 0.25 μm coated with 10% permethylated β-cyclodextrin in SPP-35 poly (35% diphenyl / 65% dimethyl siloxane)).

The pressure of the carrier gas (helium) was fixed at 1.2 bar and the split leak at 30 ml min^{-1}. The detector (FID) temperature was 230 °C. The dynamic headspace injector was directly connected to a specially designed cell (0.25 l capacity) in which samples were placed. Volatile compounds were concentrated on Tenax traps with a scavenger gas (helium) flow rate of 30 ml min^{-1} at room temperature. The odour profile description of chiral compounds was obtained using a sniffing-port with a split ratio FID–Sniffing of 1:1, and performed by the reference to the olfactory descriptors in 'Le Champ des Odeurs'[30].

3 RESULTS AND DISCUSSION

3.1 Enantioselective Analysis of Reference Chiral Compounds

Reference compounds (1 μl) were placed in the headspace cell and concentrated on Tenax for 30 seconds. Each analysis was run in triplicate.

3.1.1 Enantiomeric Separation of 1-Octen-3-ol. The separation was carried out on the α-DEX 120 capillary column with an isothermal temperature of 72 °C. The (*R*)-octen-3-ol enantiomer eluted in the second position (Table 2).

According to Sandra *et al.* (1989) the resolution (R_s) is defined as the separation of two peaks in terms of their average peak width at half height, $R_s = 1.177 \times (\Delta t_R / Wh_1 + Wh_2)$. By comparing our experimental R_s value (1.27) with the reference table, the enantiomer separation of 1-octen-3-ol was found to be approaching 99%.

Table 2   *Elution and Sensory Evaluation of (S)- and (R)-1-Octen-3-ol Enantiomers*

Enantiomer	Retention time (min)	Sensory evaluation
(*S*)-1-Octen-3-ol	21.08 ± 0.07	Herbaceous, musty
(*R*)-1-Octen-3-ol	21.76 ± 0.05	Intensive mushroom-like odour, musty-earthy

3.1.2 Enantiomeric Separation of Linalool. The separation was carried out on the β-DEX 110 capillary column with an isothermal temperature of 95 °C. The (*R*)-linalool enantiomer eluted in first position (Table 3). In the way described previously, the enantiomer separation of linalool was approaching 98% ($R_s = 1.13$).[31]

Table 3   *Elution Position and Sensory Evaluation of (R)- and (S)-Linalool Enantiomers*

Enantiomer	Retention time (mins)	Sensory evaluation
(*R*)-Linalool	16.15 ± 0.05	Flowery, fresh
(*S*)-Linalool	16.41 ± 0.06	Differs slightly in odour

3.2 Enantioselective Analysis of 1-Octen-3-ol in Headspace of *A. bisporus* on α-DEX 120

Basidiocarp cubes (30 g to 60 g according to the availability of the mushrooms) were placed in the headspace glass cell. Volatile compounds were collected and concentrated on Tenax for 30 minutes. Each analysis was run in triplicate. The mean percentage deviation of each enantiomer was around 5%. Two commercial strains of *A. bisporus* from different sources were studied (Tables 4 and 5).

The natural enantiomeric excess of 1-octen-3-ol in the headspace of both commercial strains of fresh *A. bisporus* was around 60%–70 % for the (*R*) enantiomer. The freezing process did not modify this excess for either strain. On the other hand, the canning process reversed the enantiomeric excess in favour of the (*S*) enantiomer to around 10%–15% for both commercial strains. In all canned products, 3-octanone, a major non-chiral compound in the aroma of fresh *A. bisporus* (30%–60 %), was absent, without it could be established a correlation with the racemization of 1-octen-3-ol.

Table 4 *Enantiomeric Excess of 1-Octen-3-ol in the Aroma of* A. bisporus *(Commercial Strain 1 from INRA)*

Sample	% (S)-1-octen-3-ol	% (R)-1-octen-3-ol	Enantiomeric excess (%)
Fresh basidiocarps	20	80	(R)-60
Frozen basidiocarps	14	86	(R)-72
Canned basidiocarps	57	43	(S)-14

Table 5 *Enantiomeric excess of 1-octen-3-ol in the aroma of* A. bisporus *(commercial strain 2 from vegetable market)*

Sample	% (S)-1-octen-3-ol	% (R)-1-octen-3-ol	Enantiomeric excess (%)
Fresh basidiocarps	19	81	(R)-62
Frozen basidiocarps	12	88	(R)-76
Canned basidiocarps	55	45	(S)-10

3.3 Enantioselective Analysis of Linalool in Headspace of *L. nuda* on β-DEX 110

Basidiocarp cubes (20 g to 50 g according to the availability of the mushrooms) were placed in the headspace glass cell. Volatile compounds were concentrated on Tenax for 30 minutes. Each analysis was run in triplicate. The mean percentage deviation of each enantiomer was around 5%. Two commercial strains of *L. nuda* from different sources were studied (Tables 6 and 7).

Table 6 *Enantiomeric Excess of Linalool in the Headspace of* L. nuda *(Commercial Strain 1 from Mushroom Museum)*

Sample	% (R)-linalool	% (S)-linalool	Enantiomeric excess (%)
Fresh basidiocarp	83	17	(R)-66
Frozen basidiocarps	82	18	(R)-64
Canned basidiocarps	86.5	13.5	(R)-68

Table 7 *Enantiomeric Excess of Linalool in the Headspace of* L. nuda *(Commercial Strain 2 from INRA)*

Sample	% (R)-linalool	% (S)-linalool	Enantiomeric excess (%)
Frozen basidiocarps	56	44	(R)-12

The natural enantiomeric excess of linalool for fresh *L. nuda* for the first commercial strain was around 66% for the (R) enantiomer. The freezing and the canning processes did not modify this excess. The enantiomeric excess of linalool in the aroma of the second strain remained a little in favour of the (S) enantiomer, but close to the racemate value.

While the source, the phenotype or the maturation–degradation stage of the strain could have changed the natural enantiomeric excess in the headspace of *L. nuda* (as reported by Casabianca and Graff for different essential oils)[23] the different processes applied in this study seemed to have little effect on it. According to the previous authors, pH is the major parameter influencing the racemization of linalool during hydrodistillation. In our study, the pH of the medium remained at neutral values after processing (heat or freeze treatments), which could explain the stable enantiomeric excess observed.

On the other hand, according to our previous work,[28] *L. nuda* also contained 4% of 1-octen-3-ol, but chiral analysis of this compound could not be achieved with our method due to the low amounts in the aroma.

4 CONCLUSION

In this study, mono-dimensional chiral gas chromatography with dynamic headspace concentration allowed:
  (i) A fast and satisfactory separation of the enantiomers of 1-octen-3-ol and linalool;
  (ii) Analysis of these chiral compounds in headspaces of mushroom.
This novel perspective on the chiral chromatography combining headspace concentration without any derivatization of the sample was very promising in terms of rapid and easy authenticity control of natural flavour and fragrance, in a product which contained a major chiral compound.

REFERENCES

1. A. Mosandl, *J. Chromatogr.*, 1992, **624**, 267.
2. A. Mosandl, *Food Rev. Int.*, 1995, **11**, 597.
3. R. Tressl and K.H. Engel, 'Analysis of volatiles', Walter de Gruyter & Co., Berlin, 1984, p. 323.
4. E.M. Gaydou, R.P. Randriamiharisoa, *J. Chromatogr.*, 1987, **396**, 378.
5. A.A. Rudmann and J.R. Aldrich, *J. Chromatogr.*, 1987, **407**, 324.
6. A.M. Stalcup, K.H. Ekborg, M.P. Gasper and D.W. Armstrong, *J. Agric. Food Chem.*, 1993, **41**, 1684.
7. G. Takeoka, R.A. Flath, T.R. Mon, R.G. Buttery, R. Teranishi, M. Güntert, R. Lautamo and J. Szejtli, *J. High Resolut. Chromatogr.*, 1990, **13**, 202.
8. W. Keim, A. Köhnes, W. Meltzow and H. Römer, *J. High Resolut. Chromatogr.*, 1991, **14**, 507.
9. A. Dietrich, B. Maas, V. Karl, P. Kreis, D. Lehmann, B. Weber and A. Mosandl, *J. High Resolut. Chromatogr.*, 1992, **15**, 176.

10. M. Eghbaldar, R. Fellous, G. George, L. Lizzani-Cuvelier, A.M. Loiseau and C. Schippa, *Parfums, Cosmétiques, Arômes*, 1992, **104**, 71.
11. C. Saturnin, R. Tabbachi and A. Saxer, *Chimia*, 1993, **47**, 221.
12. T.J. Betts, *J. Chromatogr. A*, 1994, **678**, 370.
13. C. Bicchi, A. D'Amato, V. Manzin, A. Galli and M. Galli, *J. Chromatogr.*, 1994, **666**, 137.
14. B. Koppenhoefer, R. Behnisch, U. Epperlein, H. Holzschuh, A. Bernreuther, P. Piras and C. Roussel, *Perfum. Flavor.*, 1994, **19**, 1.
15. C. Bicchi, V. Manzin, A. D'Amato and P. Rubiolo, *Flav. Fragr. J.*, 1995, **10**, 127.
16. C. Bicchi and A. Pisciotta, *J. Chromatogr.*, 1990, **508**, 341.
17. P. Kreis and A. Mosandl, *Flav. Fragr. J.*, 1992, **7**, 187.
18. P. Kreis and A. Mosandl, *Flav. Fragr. J.*, 1992, **7**, 199.
19. H. Casabianca and J.F. Graff, *J. High. Resolut. Chromatogr.*, 1994, **17**, 184.
20. P. Kreis and A. Mosandl, *Flav. Fragr. J.*, 1994, **9**, 249.
21. P. Kreis and A. Mosandl, *Flav. Fragr. J.*, 1994, **9**, 257.
22. H. Casabianca, *Rivista Italiana Eppos*, 1996, Gennaio'96, 205.
23. H. Casabianca and J.B. Graff, *Rivista Italiana Eppos*, 1996, Gennaio '96, 227.
24. T. Talou, M. Delmas and A. Gaset, *J. Agric. Food Chem.*, 1987, **35**, 774.
25. S. Breheret, T. Talou, B. Bourrounet and A. Gaset, 'Bioflavour 95', ed. INRA, Paris, 1995, p. 103.
26. T. Talou, K. Roule and A. Gaset, *Rivista Italiana Eppos*, 1996, Gennaio '96, 269.
27. J. Mau, R.B. Beelman and R.G. Ziegler, *Dev. Food Sci.*, 1994, **34**, 657.
28. B. Breheret, T. Talou, S. Rapior and J.M. Bessière, *J. Agric Food Chem.*, in press.
29. R. Courtecuisse and B. Duhem, 'Guide des Champignons de France et d'Europe', Delachaux et Niestlé, Lausanne, 1994.
30. J.N. Jaubert, G. Gordon and J.C. Doré. *Parfums, Cosmétiques, Arômes*, 1987, **78**, 71.
31. P. Sandra, *J. High Resolut. Chromatogr.*, 1989, **12**, 82.

TANDEM MASS SPECTROMETRY IN FLAVOUR RESEARCH

L.B. Fay,* I. Blank* and C. Cerny†

*Nestec Ltd., Nestlé Research Centre, Vers-chez-les-Blanc, P.O. Box 44, CH-1000 Lausanne 26, Switzerland

†Nestec Ltd., Nestlé R&D Center Kemptthal, CH-8310 Kemptthal, Switzerland

1 INTRODUCTION

During the last twenty years tandem mass spectrometry has become a major tool for analysis of complex mixtures[1,2] and is widely employed for forensic chemistry,[3] and environmental[4] and biological[5,6] applications. However, despite its great analytical potential, MS–MS is less frequently used for flavour research, and mass spectrometry coupled to gas chromatography remains the more popular spectroscopic technique for aroma analysis. This can be explained by the longer commercial availability of GC–MS instruments as opposed to MS–MS instruments, by the success of the GC–MS method itself and more recently of the HPLC–MS combination,[7] and by the high cost of the MS–MS apparatus. Nevertheless, because of its sensitivity and selectivity, MS–MS offers features allowing structural identification[8] as well as quantitative determination.

In this paper, examples of the use of MS–MS for flavour applications obtained with a quadrupole tandem mass spectrometer will be presented. These examples will highlight the use of MS–MS used alone and in conjunction with GC or HPLC, for identification and quantification of flavour compounds.

2 EXPERIMENTAL

All experiments were carried out using a Finnigan TSQ-700 mass spectrometer (Finnigan MAT, Bremen, Germany) working with argon as the collision gas and set to between 1 and 2 mTorr. A collision energy from 5 to 30 eV in the laboratory frame was used.

All GC separations were done using an HP-5890 gas chromatograph equipped with an HP-7673 autosampler. The column was a J&W Sci DB-Wax capillary column 30 m × 0.32 mm internal diameter with a 0.25 μm film thickness. Helium was used as carrier gas at a pressure of 10 psi. Furanone samples were injected in a splitless injector (280 °C) whereas pyrazine samples were injected with a cold on-column injector. The ionisation was either electron impact at 70 eV (furanones) or positive chemical ionisation with ammonia as reagent gas (pyrazines).

HPLC separations were made using a Waters 600-MS pump and a Waters 717 autosampler. The column was an HP Lichrosper 100 RP-18 (125 × 4 mm, 5 μm) operated at room temperature at a flow rate of 1 ml min^{-1}. The mobile phase was as follows: solvent A: 50 mM ammonium acetate, pH 5.5; solvent B: methanol. Flow was initially 100% A for

10 min, then going to 100% B over 15 min, and returning to 100% A over 5 min. The HPLC was connected to the MS via a thermospray interface.

Fast atom bombardment ionisation was generated with an Ion Tech saddle-field atom gun operated at 10 kV and 0.2 mA using xenon as the gas. All samples were dissolved in thioglycerol used as a matrix.

3 RESULTS AND DISCUSSION

3.1 Application of MS–MS for Qualitative Experiments

The use of capillary GC columns with high resolving power allows the analysis of complex flavour profiles. However, it is sometimes still very difficult to get enough resolution to resolve all the constituents of a complex flavour mixture. In such cases, contaminated GC peaks are obtained and the corresponding mass spectra are difficult to interpret. Tandem mass spectrometry offers a third dimension to GC–MS by adding the electronically-based selectivity of the collision induced dissociation (CID) process to the chemically-based selectivity of the chromatographic separation. As an example, Figure 1A shows the contaminated mass spectrum of a compound tentatively identified as 4-hydroxy-2,5-dimethyl-3(2*H*)-furanone. This contaminated spectrum obtained after GC–MS does not allow an unambiguous characterisation of the molecule. On the other hand, the daughter mass spectra of the same compound (Figure 1B) obtained in the same GC conditions, is free of interference.

At the present time MS–MS spectral libraries are not commercially available. In principle standard CID MS–MS spectra could be collected and stored in reference libraries. Nevertheless, there are several instrument parameters that can cause significant differences in the observed MS–MS spectra for a given molecule. However, a protocol to overcome this problem has recently been proposed.[9]

The Maillard reaction that occurs between sugars and amino acids on heating generates very complex mixtures. GC–MS–MS was employed to identify the caramel-like smelling compounds 4-hydroxy-2,5-dimethyl-3(2*H*)-furanone (HDMF) and 4-hydroxy-2(or 5)-ethyl-5(or 2)-methyl-3(2*H*)-furanone (HEMF) in Maillard reaction systems based on pentoses.[10]

Figure 2 shows the mass spectra of ^{13}C-labelled HEMF and HDMF obtained in the reaction systems D-xylose–[3-^{13}C]-L-alanine and D-xylose–[2-^{13}C]-glycine respectively. These spectra were obtained by GC–MS–MS and CID of the parent molecular ions at *m/z* 143 and 129 respectively. In the case of the ^{13}C-HEMF, the daughter ions at *m/z* 127 [M-^{13}CH$_3$]$^+$ and at *m/z* 114 [M-CH$_2$-^{13}CH$_3$]$^+$ revealed the presence of the ^{13}C atom in the ethyl group. The incorporation of the ^{13}C atom in the molecule was also indicated by the fragments at *m/z* 58, 72, 87 and 100, which correspond to the ions 57, 71, 86 and 99, respectively in the mass spectrum of the unlabelled HEMF. As indicated by the fragment [M-CO-CH$_3$]$^+$ at *m/z* 100 for the labelled HEMF and at *m/z* 99 for the unlabelled HEMF, as well as by the fragment [CO-CH$_3$]$^+$ at *m/z* 43 for both compounds, the methyl group of the labelled HEMF does not bear the ^{13}C atom. The ^{13}C atom is more difficult to locate in the spectra of the mass spectrum of ^{13}C-HDMF. Indeed, this molecule may occur as a 1:1 mixture of two isotopomers and the ion pairs 43/44, 57/58 and 85/86 of nearly equal intensity correspond to its symmetric structure.

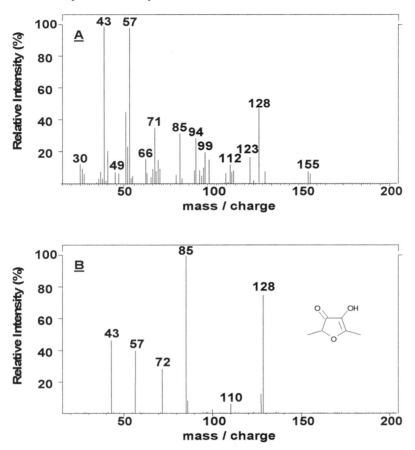

Figure 1   *Mass spectra of 4-hydroxy-2,5-dimethyl-3(2H)-furanone identified in a*
Maillard reaction system based on pentose and glycine. A: mass spectrum
obtained by GC–MS after electron impact ionisation at 70 eV; B: product ions
after collision induced dissociation (10 eV) of the molecular ion at m/z *128*
formed by electron impact ionisation

3.3 Application of MS–MS for Quantitative Measurements

Tandem mass spectrometry has been shown to be useful for quantification of compounds in complex biological matrices because of its selectivity and sensitivity.[11] The two examples that follow highlight the use of this technique for flavour applications. These are the quantification by fast atom bombardment MS–MS (FAB–MS–MS) of the Amadori compound N-(1-deoxy-D-fructos-1-yl)-glycine (DFG) and the determination of 8-oxocaffeine in coffee by HPLC–MS–MS.

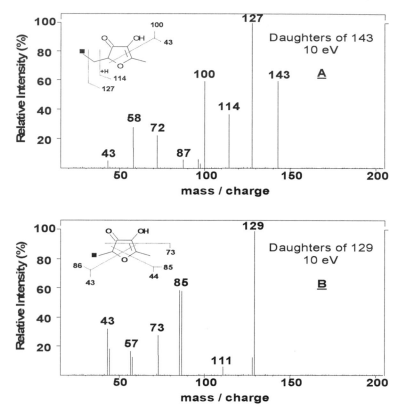

Figure 2 *GC–MS–MS spectra of one tautomer 2-([2-¹³C]ethyl)-4-hydroxy-5-methyl-3(2H)-furanone (A) and of 4-hydroxy-2-[¹³C]methyl-5-methyl-3(2H)-furanone (B) obtained in the reaction systems D-xylose/[3-¹³C]-L-alanine and D-xylose/[2-¹³C]-glycine respectively (the labelled positions are marked with ∎)*

To study the decomposition of DFG in model systems we developed a quantification method based on isotope dilution fast atom bombardment MS–MS after synthesis of the ¹³C-labelled DFG used as internal standard.[12] Isotope dilution with FAB–MS is known to show a high dynamic range[13] and therefore to be applicable for quantitative purposes.[14] From DFG, FAB–MS produces the commonly observed protonated molecular ion at m/z 238 with little fragmentation. The CID spectrum of the protonated molecular ions leads to spectra showing intense daughter ions at m/z 220 ([M+H-H$_2$O]$^+$) and at m/z 202 ([M+H-2H$_2$O]$^+$), and an additional fragment corresponding to the glycine residue at m/z 76. The daughter ion at m/z 220 was selectively recorded (selective reaction monitoring) to quantify DFG. The low mass fragment ion at m/z 76 was of too low an intensity to be used for quantification. Similar daughter ions at m/z 221 and 77 were obtained from ¹³C-labelled DFG used as an internal standard. A second order calibration curve was used to quantify DFG in model reactions and the repeatability of the method was found to be below 2%. It should be pointed out that this technique requires neither a purification nor a chromatographic step and is, therefore, very fast and straightforward. However, high salt concentrations or strong buffer solutions should be avoided. Buffers can lead to a complete suppression of all the signals.

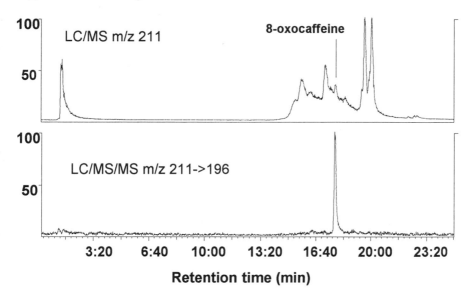

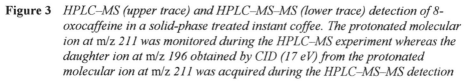

Figure 3 *HPLC–MS (upper trace) and HPLC–MS–MS (lower trace) detection of 8-oxocaffeine in a solid-phase treated instant coffee. The protonated molecular ion at m/z 211 was monitored during the HPLC–MS experiment whereas the daughter ion at m/z 196 obtained by CID (17 eV) from the protonated molecular ion at m/z 211 was acquired during the HPLC–MS–MS detection*

The second example of quantification by MS–MS is the determination of 8-oxocaffeine (1,3,7-trimethyluric acid) in coffee subjected to oxidative stress. It was demonstrated that caffeine in coffee is an efficient scavenger of the deleterious hydroxyl radical *in situ*, forming the major reaction product 8-oxocaffeine.[15] This compound was monitored by HPLC–MS–MS in the coffee extracts obtained by solid phase extraction on C_{18} cartridges. 8-Oxocaffeine was ionised using a thermospray interface and the protonated molecular ion at m/z 211 was submitted to CID, producing fragment ions at m/z 196, 153, 139, 111, 83 and 42. The daughter ion at m/z 196 was selectively recorded and the resulting chromatogram is presented in Figure 3 in comparison with the chromatogram of the $[M+H]^+$ ion obtained by HPLC–MS only. These data clearly demonstrate the gain in selectivity obtained by HPLC–MS–MS in comparison with HPLC–MS. To quantify 8-oxocaffeine, the coffee solutions were spiked with different known amounts of a standard compound and the amount in the non-spiked sample extrapolated from a linear regression equation. For all samples, there was a good linear relationship between oxo-caffeine concentration in the coffee and peak area, giving a correlation coefficient (*r*) of better than 0.996. This technique was applied successfully to instant, roasted and ground and green coffees.

4 CONCLUSION

As for many other analytical techniques, the growth of tandem mass spectrometry is linked to instrumental development. The commercial availability of triple quadrupole instruments and, more recently, of quadrupole ion trap mass spectrometers at affordable prices has allowed this tool to enter flavour laboratories. Moreover, improvement in computer control and instrument design has offered users easy-to-operate machines on a routine basis. Flavour analysis deals with complex matrices in which traces of key components are sometimes expected. Therefore, because of its features of structural identification as well as quantitative determinations, we should see an expansion in the use of tandem mass spectrometry (used or not used in conjunction with chromatographic separations) in our research area.

ACKNOWLEDGEMENTS

The authors are grateful to R. Fumeaux and S. Marti for expert technical assistance and to Dr. I. Horman for helpful discussions.

REFERENCES

1. K.L. Busch, G.L. Glish and S.A. McLuckey, 'Mass Spectrometry / Mass Spectrometry: Techniques and Applications of Tandem Mass Spectrometry', VCH Publishers Inc., Weinheim, Germany, 1988, p. 173.
2. J.V. Johnson and R.A. Yost, *Anal. Chem.*, 1985, **57**, 758A.
3. J. Yinon, *Mass Spectrom. Rev.*, 1991, **10**, 179.
4. D.F. Hunt, J. Shabanowitz, T.M. Harvey and M.L. Coates, *Anal. Chem.*, 1985, **57**, 525.
5. D. Favretto and P. Traldi, *Mass Spectrom. Rev.*, 1993, **12**, 313.
6. I.A. Blair, *Chem. Res. Toxicol.*, 1993, **6**, 741.
7. C. Salles, J.C. Jallageas, F. Fornier, J.C. Tabet and J.C. Crouzet, *J. Agric. Food Chem.*, 1991, **39**, 1979.
8. K.L. Busch and K. Kroha, 'Characterization and Measurement of Flavor Compounds', A.C.S. Symposium Series 289, Washington, D.C., 1985, p. 121.
9. K.R. Mohan, M.G. Barlett, K.L. Busch, A.E. Schoen and N. Gore, *J. Am. Soc. Mass Spectrom.*, 1994, **5**, 576.
10. I. Blank and L.B. Fay, *J. Agric. Food Chem.*, 1996, **44**, 531.
11. G.C. Thorne and S.J. Gaskell, *Biomed. Mass Spectrom.*, 1985, **12**, 19.
12. A.A. Staempfli, I. Blank, R. Fumeaux and L.B. Fay, *Biological Mass Spectrom.*, 1994, **23**, 642.
13. C.F. Beckner and R.M. Caprioli, *Biomed. Mass Spectrom.*, 1984, **11**, 60.
14. A.A. Staempfli, O. Ballèvre and L.B. Fay, *Rapid Comm. Mass Spectrom.*, 1992, **6**, 547.
15. R.H. Stadler and L.B. Fay, *J. Agric. Food Chem.*, 1995, **43**, 1332.

NEW TRENDS IN ELECTRONIC NOSES FOR FLAVOUR CHEMISTRY

T. Talou, J.M. Sanchez and B. Bourrounet

National Polytechnic Institute of Toulouse, Agro-Industrial Chemistry Laboratory, ENSCT
118 route de Narbonne, F-31077 Toulouse Cedex, France

1 INTRODUCTION

Qualification and quantification of aroma volatiles emitted by flavourings, plant extracts (essential oils, oleoresins, ...) or flavoured industrial products are two important factors of the quality control methodology used by the aromatic and food industries. Physico-chemical techniques (GC, GC–MS, HPLC) and sensory analysis (dynamic olfactometry, evaluation by a 'Nose' or by panellists) are the two classical methods used for this purpose, but they are time consuming, expensive and do not allow an immediate decision (off-line method).

Considerable interest has arisen in the past ten years in the use of arrays of gas sensors together with an associated pattern recognition technique to quantify, differentiate and identify complex mixtures of volatile compounds.[1] The principle of detection in such apparatus, labelled 'electronic noses', is based on the reversible electrical resistance changes of the sensing elements (metal oxides or conducting polymers) in the presence of volatiles combined with on-line computerized statistical processing of the data (FDA, PCA, ANN, fuzzy logic, ...).[2–4]

Many publications report the application of different prototypes or commercial devices for odour differentiation of industrial products (raw, extracts, processed, packagings, ...),[5–15] for process monitoring or for a particular application.[16] None of them compares effectively the efficiency of the new method to classical ones, especially GC and sensory analysis.

In continuation of our comparative work on mushrooms[17] and essential oils[18,19] differentiation, and in order to test the potential of an electronic nose as a tool for perfume and flavouring differentiation, the present paper reports the results of a study, using an electronic nose with conducting polymer gas sensors compared with headspace GC and with sensory analysis, for the differentiation of five different descriptors of three major odorous notes.

2 MATERIALS AND METHODS

2.1 Odorous Notes

Three odorous notes commonly used in perfumes and flavouring formulations were selected from our collection of natural extracts and synthetic substances used as aromatic bases by flavourists, assembled according both to the professional olfactory reference[20] 'Le Champ des Odeurs' and to the recommendations of the famous perfumers Carles[21] and

Roudnitska.[22,23] These notes (rustic, minty and citrus) are described by 15 different chemotyped essential oils and represented by 45 samples corresponding to three replicates of each product, provided by Berdoues SA (Toulouse, France) in glass screw-top containers and stored at room temperature (25 °C).

2.1.1 Rustic Notes. Spike lavender (*Lavandula latifolia*), lavender (*Lavandula angustifolia*), lavandin (*Lavandula hybrida*), thyme (*Thymus vulgaris*), rosemary (*Rosemarinus officinalis*).

2.1.2 Minty Notes. Peppermint (*Mentha piperita*), spearmint (*Mentha spicata*), sweet mint (*Mentha suavolens*), cornmint (*Mentha arvensis*), pennyroyal (*Mentha pulegium*).

2.1.3 Citrus notes. Mandarin (*Citrus reticulata*), bergamot (*Citrus bergamia*), lemon (*Citrus lemon*), lime (*Citrus limetta*), sweet orange (*Citrus auranthium*).

2.2 Electronic Nose Apparatus

The analyses were conducted with an AromaScanner A32/50S (AromaScan plc, Crewe, U.K.) combining a multisampler based on a static headspace autosampler unit (Tekmar, U.S.A.) with the analyser system using an array of 32 conducting polymer gas sensors,[24] the complete device being monitored by dedicated software including data processing. In summary, the sample headspace was generated at a set temperature to reach equilibrium in a heated inner carousel, a so-called platen. After equilibration, the vial was raised into a fixed needle and 1 ml sample loop of headspace was extracted. The loop was purged with carrier gas into a heated transfer line and delivered to the sensor array according to a 'stop flow' technique. The headspace was then held over the sensors in order to generate a stable response. By using a pneumatic valve, the sensors were then cleaned and the wash was removed by carrier gas allowed to the sensors to be ready for the next sample.

The experimental conditions were as follows:

2.2.1 Sample Preparation. Essential oil (1.5µl) was deposited with a microsyringe in a 22 ml glass vial which was sealed with a septum and capped. The vials were loaded into the multisampler upper carousel in a random order. The settings for the multisampler method were: platen temperature: 50 °C; platen equilibration time: 5 min; sample equilibration time: 15 min; carrier gas: nitrogen; static vial pressure: 1.5 psi; vial pressure: 4.5 psi; vial pressure time: 1 min; pressure equilibration time: 15 s; loop fill time: 15 s; loop equilibration time: 12 s; inject time: 1 min; sample loop temperature: 50 °C; transfer line temperature: 50 °C.

2.2.2 Sample Analysis. The array of 32 activated sensors was thermostatted at 35 °C. The acquisition parameters, *i.e.*, the detection threshold and the sampling interval, were fixed at 0.2 s and 1 s respectively. The settings of the valve sequence were: reference gas: 30 s; sample: 60 s; wash (2% butanol and 98% water): 30 s; reference: 160 s; carrier gas: nitrogen.

2.2.3 Data Analysis. The sampling time from which the database files were created was between T = 70s and T = 90s. These databases, based on the sensors' normalised response profiles, were averaged. A cluster analysis called Sammon mapping[25] was used for the statistical data processing. The axis of each map obtained was in the units of Euclidean Distance (ED) which measured the difference between the odours.

2.3 Capillary Gas Chromatography with Dynamic Headspace Concentration

Analyses were performed with a dynamic headspace injector (DRI) (Perichrom, Saulx-les-Chartreux, France) coupled to a gas chromatograph DN 200 (Delsi Instruments, Paris, France).

2.3.1 Sample Preparation. The DRI device was directly connected to a specially designed glass cell (0.25 l capacity) in which 5 mg of essential oil sample were placed. After a period of static equilibration (15 min), volatile compounds were concentrated on a Tenax TA trap cooled at –20 °C by liquid nitrogen with a scavenger gas (helium) flow rate of 30 ml min^{-1} at room temperature (25 °C) for 2 min. The trap was then heated at 250 °C to desorb volatiles onto the capillary GC column.

2.3.2 Sample Analysis. The GC separation was performed on a capillary wall coated column (SUPELCO SPB-1, 50 m, internal diameter 0.32 mm, 1μm film thickness). The oven temperature was programmed from 50 to 220 °C at 5 °C min^{-1}. Column inlet pressure of carrier gas (helium) and split leaks were respectively fixed at 17 psi and at 30 ml min^{-1}. The FID temperature was 230 °C.

2.3.3 Data Analysis: The recorded GC profiles were used as 'fingerprints' for direct comparison and differentiation of the essential oils.

2.4 Sensory Analysis

The sensory analysis was carried out by a panel of four non-specialist judges, trained for six months with the olfactory reference[20] 'Le Champ des Odeurs' from which the three odorous notes selected, *i.e.*, rustic, minty and citrus, were represented schematically by the pure chemicals d-camphor, l-menthol and d-limonene. An aliquot of each essential oil, *i.e.*, 45 samples (diluted in ethanol (90%) and applied to a testing strip) was presented at random to the judges in three distinct sessions corresponding to the different notes.

The responses measured for each note were: (i) differentiation of the samples corresponding to the same descriptor; (ii) differentiation and identification of the different descriptors, in order to evaluate the capacity of the panellists to differentiate between the five descriptors of the three odorous notes.

3 RESULTS AND DISCUSSION

3.1 Rustic Note

The sensory panel both successfully identified the replicates of the five different essential oils and differentiated these five descriptors. Their identification was reported as difficult due to the top camphoreous note in all the samples, only thyme being correctly identified (due to the particularly well known thymol odour).

Conversely, the differentiation (and therefore their identification) of the essential oils, was quite easy both by GC and by the electronic nose as shown by fingerprints and sensor odour mapping with five distinct population clusters (Figure 1). Therefore, the chemical composition of rosemary and spike lavender, which both contained, as main constituents, camphor, eucalyptol and linalool, generated similar GC profiles and two close cluster groups in the sensor odour mapping. It seemed that the differentiation by electronic nose was set up according to a diagonal axis corresponding to increasing content of camphor and of eucalyptol from lavandin to rosemary.

3.2 Minty Note

Although the panellists successfully identified the replicates, they could not differentiate peppermint from corn mint and sweet mint, only pennyroyal and spearmint with their respectively pungent and chewing gum like odours being identified without problem.

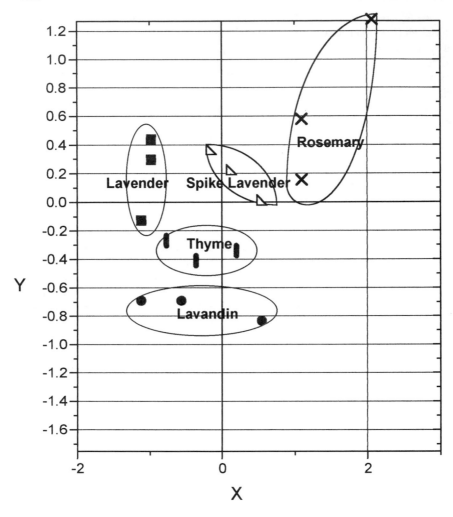

Figure 1 *Differentiation of rustic note descriptors by electronic nose*

In the case of GC, only peppermint and sweet mint presented large similarities in their fingerprints, preventing their accurate differentiation. The sensor odour mapping showed four distinct population clusters in which the pennyroyal and peppermint ones were very close while corn mint plots were more dispersed (Figure 2). No differentiation axis based on a key compound could be found although similarities in the chemical skeleton of pulegone, menthol, menthone and carvone, major compounds respectively of pennyroyal, peppermint and spearmint are well known.

3.3 Citrus Note

In this case, the high amount of limonene in all these citrus oils did not allow the judges to differentiate lemon from lime and bergamot from orange, only mandarin being accurately identified while all the replicates were satisfactorily checked. On the other hand, the similar

GC profiles due to very close chemical compositions of these essential oils did not permit the differentiation of them with this technique. Similar results were obtained with an electronic nose for which the odour mapping presented no clustered populations.

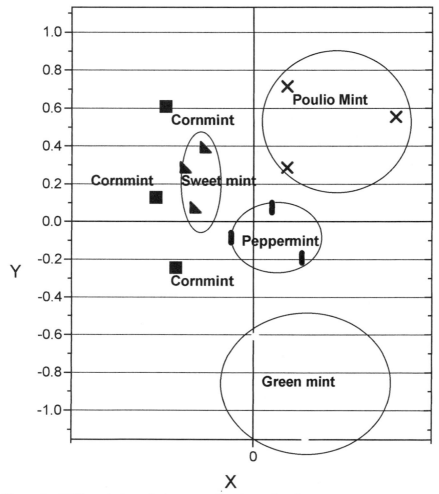

Figure 2   *Differentiation of minty note descriptors by electronic nose*

4 CONCLUSION

In this preliminary study, the use of an electronic nose for the differentiation and the identification of five descriptors of three major odorous notes of the olfactory reference 'Le Champ des Odeurs' allowed: (i) confirmation of the strong relationship between the results obtained by headspace GC and an electronic nose; (ii) demonstration that this kind of apparatus could perform as well as the human nose in differentiation of odorous mixtures without particular training. If these interesting results can be confirmed with the other 38 reference notes, this novel view on electronic nose technology could be very promising in terms of an alternative tool for the formulation of flavours and fragrances.

ACKNOWLEDGEMENTS

This work is a part of the program 'FLAVOUR 2000' carried out by the 'Electronic Nose Department' of the CATAR-CRITT Agrore sources (ones R&D associated structures of INPT-ENSCT) and sponsored by the Midi-Pyrénées County Council.

REFERENCES

1. T. Talou, *Odours and VOC's J.*, 1996, **H.S.**, 11.
2. J.W. Gardner and P.N. Bartlett, 'Sensors and Sensory Systems for an Electronic Nose', NATO ASI Series, Kluwer Acad. Pub., London, 1992, p. 317.
3. K. Persaud, *Analytical Proceedings*, 1991, **28**, 339.
4. K. Persaud and P. Pelosi, 'Sensors and Sensory Systems for an Electronic Nose', NATO ASI Series, Kluwer Acad. Pub., London, 1992, p. 237.
5. T. Aishima, *J. Agric. Food Chem.*, 1991, **39**, 752.
6. J.L. Berdagué and T. Talou, *Sciences des Aliments*, 1993, **13**, 141.
7. M. Egashira, Y. Shimizu and Y. Takao, *Sensors and Actuators B*, 1990, **1**, 108.
8. J.W. Gardner, H.V. Shurmer and T.T. Tan, *Sensors and Actuators B*, 1992, **6**, 71.
9. J.W. Gardner and P.N. Bartlett, 'Olfaction and Taste XI', Springer-Verlag, Tokyo, 1994, p. 690.
10. R. Olafsson, E. Martinsdottir, G. Olafsdottir, P.I. Sigfusson and J.W. Gardner, 'Sensors and Sensory Systems for an Electronic Nose', NATO ASI Series, Kluwer Acad. Pub., London, 1992, p. 257.
11. F. Windquist, E.G. Hörnsten, H. Sungren and I. Lundström, *Meas. Sci. Technol.*, 1993, **4**, 1493.
12. B. Bourrounet, T. Talou and A. Gaset, *Sensors and Actuators B*, 1995, **26–27**, 250.
13. C. Di Natale, F. Davide, A. d'Amico, G. Sberveglieri, P. Nelli and G. Faglia, 'Current Status and Future Trends', Proceedings EURO FOOD CHEM VIII, 1995, GOCh, Vienna, p. 131.
14. B. Bourrounet, T. Talou and A. Gaset, *Odours and VOC's J.*, 1996, **1**, 334.
15. J.F. Clapperton, *Odours and VOC's J.*, 1996, **H.S.**, 22.
16. T. Talou, K. Persaud and A. Gaset, 1992, Patents Fr 2 296 236, SP P9301997, I M193A0011972.
17. S. Breheret, T. Talou and A. Gaset, 'Bioflavour '95', ed. INRA, Paris, 1995, p. 103.
18. B. Bourrounet, M. Cazagou and T. Talou, *Rivista Italiana EPPOS*, 1995, Gennaio '96, 566.
19. B. Bourrounet, T. Talou and A. Gaset, *Odours and VOC's J.*, 1996, **H.S.**, 34.
20. J.N. Jaubert, G. Gordon and J.C. Doré. *Parfums, Cosmétiques, Arômes*, 1987, **78**, 71.
21. J. Carles, *Recherches*, 1961, **11**, 8.
22. E. Roudnitska, *Parfums, Cosmétiques, Arômes*, 1994, **115**, 47.
23. E. Roudnitska, *Parfums, Cosmétiques, Arômes*, 1994, **116**, 45.
24. K. Persaud and P. Pelosi, 1986, WO Patent 0 1599.
25. J.W. Sammon, *IEE Trans. Comp.*, 1969, **18**, 4.

PRODUCTION OF REPRESENTATIVE CHAMPAGNE EXTRACTS FOR OLFACTORY ANALYSIS

C. Priser,[*] P.X. Etievant[*] and S. Nicklaus[†]

[*]INRA, Laboratoire de Recherches sur les Arômes, 17 rue Sully, 21034 Dijon CEDEX, France

[†]Mumm–Perrier-Jouët, Vignobles et Recherches, 11 avenue de Champagne, 51206 Epernay CEDEX, France

1 INTRODUCTION

Gas chromatography–olfactometry (GC–O) analysis is a powerful way of determining key compounds of food aroma. However, the results of a GC–O analysis are only considered valid if the odour of the extracts is close to that of the original food prior to extraction.[1] This similarity in odour is not obvious since, according to the method chosen, different classes of compounds are preferentially extracted. A preamble to GC–O analysis should therefore be a systematic sensory comparison of the food to the extract.

Abbott et al.[2] and Moio et al.[3] used this type of approach on beer and wine respectively. Both used sensory analysis and GC–O detection to confirm that the selected extracts were good olfactory representatives. Abbott et al.[2] finally chose a method involving adsorption on a mixture of XAD resins and Moio et al.[3] demonstrated that an extraction with dichloromethane without concentration was more similar to the odour of the original wine. Because of the undesirable odour of dichloromethane and of its anaesthetic effects on olfactory receptors, the similarity between the extracts and the wine was established by GC–O detection.

Our aim was to compare the olfactory quality of different types of extracts obtained from champagnes and to choose the best one for GC–O analysis. Three types of extracts were selected. These were the solvent and resin extracts (already tested on beer and wine) and a demixtion extract. This latter type of extraction was described by Singleton.[4]

2 MATERIALS AND METHODS

2.1 Champagne Samples

Three commercial champagne samples, one vintage from 1985 (A), one from 1989 (B) and one non-vintage (C), were used. All three were obtained from important champagne houses, and were stored in a cellar for a maximum of six months before use.

2.2 Analytical Reagents

The different XAD resins were purchased from Fluka AG, Switzerland. A mixture containing equal weights of three different wet resins (XAD_2, XAD_7 and XAD_{16}) was washed in a Soxhlet apparatus with methanol for 12 hours. The resins were then rinsed before use with water (10×50 ml). All reagents used were of HPLC grade, and water was purified by a milli-Q system (Millipore. S.A Saint-Quentin, France). Smelling strips were purchased from GRANGER-VEYRON (Privas, France).

2.3 Methods

2.3.1 Solvent Extraction. The volatile constituents of three champagnes were isolated as described by Moio *et al.*[2] Champagne (200 ml), CH_2Cl_2 (20 ml) and sodium chloride (50 g) were poured into a flask (600 ml), cooled with melting crushed ice and magnetically stirred at 200 rpm for two hours. The champagne–CH_2Cl_2 emulsion formed during stirring was separated from the aqueous layer and frozen at –20 °C. The flask was then allowed to reach room temperature and the CH_2Cl_2 solution, progressively separated from the remaining champagne, was transferred without concentration into a small vial stored at –20 °C.

2.3.2 Resin Extraction. The volatile compounds from the three champagnes were extracted using the following procedure. Ten grams of the clean mixture of resins were placed in a flask (650 ml) with champagne (200 ml), sodium chloride (50 g) and diluted hydrochloric acid (3.2% w/v, 4 ml) and magnetically stirred at 200 rpm for two hours. The mixture was then poured into a glass column (internal diameter = 11 mm) stoppered with glass wool. Complete transfer of the resins was achieved by rinsing the flask with water (3 × 10 ml). Residual water was removed from the column with nitrogen and the volatile compounds eluted stepwise with absolute ethanol (10 × 2 ml, with 5 min wait between each addition) in a cooled flask. The final aliquot of ethanol was eluted under a flow of nitrogen. The ethanolic solution was stored at –20 °C until analysis. A blank sample was prepared by substituting champagne by water.

2.3.3 Wine Demixtion. Champagne (200 ml), ammonium sulfate (100 g) and absolute ethanol (11 ml) were added to a flask (600 ml), magnetically stirred for 30 min at 25 °C. The solution was then decanted into a separating funnel and the ethanolic phase transferred into a small vial stored at –20 °C. The percentage of ethanol was measured with a Salleron-Dujardin ebulliometer and the dry matter was estimated from the dry residue of 1 g of the extract after 15 hours in an oven at 115 °C.

2.3.4 Distillation. The champagne demixtion extract (10 ml) was first submitted to a vacuum distillation for an hour and a half at 5 Pa and the residue was further extracted on a cold finger cooled with liquid nitrogen for three and a half hours at $5×10^{-2}$ Pa. After the last distillation, the compounds on the cold finger were rinsed with the distillate obtained from the first distillation, and stored at –20 °C until used.

2.4 Sensory Analysis

The panel was composed of 17 subjects (13 men and 4 women) from the expert panel of Mumm–Perrier-Jouët company. All the panellists were wine professionals.

Champagne samples (10 °C, 70 ml) were presented to the panel in coded glasses. For the presentation of champagne extracts, smelling strips were used. The strips were dipped in the extracts and then introduced after 6 min (time necessary for solvent evaporation) in a flask hermetically closed with a cap. For the sensory evaluation, panellists had first to open the coded flasks. Champagne samples were assessed for odour and taste, and extracts were assessed for odour only, in isolated booths and during separate sessions.

A different panel (INRA) analysed the two extracts obtained from the demixtion. It was composed of 20 subjects (5 men and 15 women) from the laboratory. The champagne extracts were presented as described previously, using smelling strips, and were assessed in similar conditions.

2.5 Quantitative Descriptive Analysis of Champagne Samples and Extracts

Panellists were trained for 18 months to perform descriptive analysis on champagnes. A list of consensus descriptors previously established by this panel was used to describe the odours of the three wine samples. In separate sessions, champagne samples and extracts were presented to the panel, randomised over all subjects and all samples. Panellists were asked to take one glass at a time, to assess the aroma of the sample, and to rate the intensity of each descriptor on an unstructured scale of 100 mm marked with 'Least intense' at the left-hand end and with 'Most intense' at the right-hand end. All the data were analysed by FIZZ software (Biosystem, Dijon, France).

2.6 Two out of Five Test[5]

The subjects of the INRA panel (flavour chemists) were not specifically trained for this test. Five coded flasks containing the extracts, randomised over all subjects and all samples, were presented to the panellists, who were asked to find the two samples different from the three others. All the data were analysed by BINRISKS SAS® macro.[6]

2.7 Gas Chromatography–Olfactometry (GC–O)

The analysis was carried out using a Hewlett-Packard 5890 Plus chromatograph equipped with an on-column injector (J&W Scientific Inc), a flame ionisation detector, a sniffing port and a DB-Wax fused silica capillary column (30 m × 0.32 mm internal diameter; film thickness 0.5 μm) (J&W Scientific Inc.). The hydrogen carrier gas velocity was 50 cm s^{-1}, and the temperature of the injector and detector 250 °C. The oven temperature was programmed from 40 to 240 °C at 5 °C min^{-1}.

3 RESULTS AND DISCUSSION

Except for the distillate obtained after demixtion, all the champagnes and extracts obtained were presented to the 17 panellists for quantitative analysis description with two repetitions. A comparison of the profile of one of the three champagnes with the profiles of its corresponding extracts is illustrated in Figure 1.

It shows that the scores obtained by sniffing the extracts on strips after evaporation of the solvent were not very different from the scores obtained on the wine itself, thus validating the protocol chosen to prepare and to assess the samples.

In order to evaluate the global similarity of the extracts with the wines, three two-factor variance tests with interaction (product (Pr) + panellist (Pa) + product*panellist (Pr*Pa)) were performed on each of the 16 descriptors. The results, reported in Table 1, indicate that only three to five descriptors were scored differently between the samples. The odour profiles of the wine and the extracts were therefore judged similar for 11 to 13 descriptors out of 16. As often in sensory analysis, the panellist effect was significant for all the discriminant descriptors, indicating a different use of the scoring scale. From the interaction, we can observe that the panellists agreed when they scored 9 of the 13 descriptors. As regards the four remaining descriptors, for which the panellists disagreed, we again tested the significance of the product effect with a new Fisher value calculated using the interaction Pr*Pa instead of the residue.[7] This test confirmed the product effect only for the rubber descriptor in champagne B.

In order to evaluate the similarity of each of the three extracts with the wines, a Duncan test was carried out on the 10 remaining discriminant descriptors. These results also are given in Table 1.

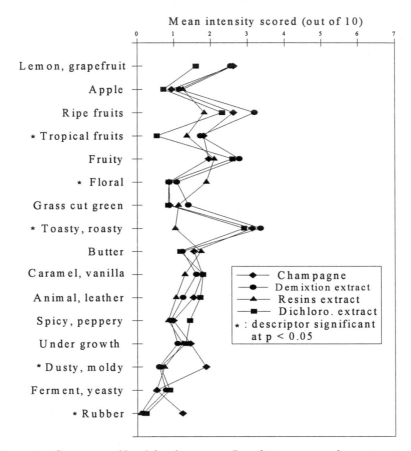

Figure 1   *Sensory profile of the champagne B and its corresponding extracts*

Table 1 *Evaluation of the Difference Between Wines and Extracts*

Type of Champagne	Significant descriptors	Rank[a]	Results of the two-way ANOVA			Duncan test: type of extract scored similarly to the wine
			Pr	Pa	Pr*Pa	
A	Toasty, roasty	1	*	***	NS	Demixtion and solvent
	Ripe fruits	2	*	***	*	
	Floral	7	*	**	NS	Resins
	Apple	12	**	***	**	
	Dusty, mouldy	14	**	**	NS	Demixtion and resins
B	Toasty, roasty	1	***	***	NS	Demixtion and solvent
	Dusty, mouldy	5	**	**	NS	No extract
	Tropical fruits	6	*	***	NS	Demixtion and resins
	Rubber	11	***	*	**	No extract
	Floral	15	*	***	NS	Demixtion and solvent
C	Lemon, grapefruit	1	**	***	NS	Demixtion
	Fruity	3	**	***	***	
	Apple	4	*	***	NS	Demixtion and resins

[a]   Rank of the average intensity of the descriptor in the wine profile;
NS   Not significant;
*   Significance: $p < 0.05$;
**   Significance: $p < 0.01$;
***   Significance: $p < 0.001$.

According to this table, the Duncan test showed that the demixtion extract scored similarly to the wine (on the selected variables) 7 times out of 10 whereas, for the resin extract and the solvent extract, the similarity occurred 4 and 3 times out of 10 respectively. According to the wine evaluation, the profile of the extract obtained by demixtion was therefore judged identical to the wine for 13 to 15 descriptors out of 16. However, the similarity was not total, as seen for example in the case of champagne B for which the dusty and the rubber notes were not well recovered in any of our extracts. Trying to find correlations between the concentration of the compounds responsible for these specific odours and the flavour profile of the wine will probably be a waste of time.

It is also interesting to note that the odour characteristic scored as the most intense in each of the three champagnes was also scored higher in the demixtion extracts (Figure 1 and Table 1), *i.e.*, the toasty and the ripe fruit notes in wine A, the toasty odour in wine B and the citrus odour in wine C.

We concluded from this experiment that demixtion was the best method among those tested to extract flavour active compounds in the three champagnes we studied. However, this extract contains high amounts of water (44%) which may cause problems in the subsequent GC analysis either in the injector or during chromatography (excess water affects the performance of some GC phases). In order to evaluate the possible formation of artefacts in the demixtion sample, the three types of extracts were injected and the odours of compounds eluted were detected by GC–O.

As expected, most descriptors were found common to the three extracts. However, the presence of many burnt descriptors in the profile of the demixtion extract seemed suspicious. After the injection of 2 µl of demixtion extract, dark spots, probably due to non-volatile products, were observed on the first 60 cm of the pre-column and on the glass liner.

In order to solve the problem of non-volatile products, a vacuum distillation was applied to the demixtion extract of the champagne C. A sensory analysis was then carried out to test the effect of the distillation on the sensory characteristics of the extract. A two out of five test was applied to an extract obtained from demixtion against the same extract submitted to distillation. Five correct responses among 20 were obtained. The data analysis allowed consideration of the type 1 (conclusion of a difference for similar samples) and type 2 risks (conclusion of a similarity for different samples). Since we cannot take the risk of using a distillate in which the aroma is different from that of the demixtion extract, we evaluated, using a p_c value[6] of 0.4, the type 2 risk which was found to 4.6%. We therefore concluded that the distillation did not perceptibly change the flavour characteristics of the extract as assessed by difference testing.

The method using a champagne distillation demixtion extract was chosen as its sensory profile was closest to the wine compared to a dichloromethane extract or a resins extract, and because it is compatible with a GC analysis using on-column injection. Further work will involve a GC–O analysis on distilled demixtion extracts in order to identify the flavour active constituents of champagne, and those responsible for flavour differences between the champagnes studied. These aromagrams will be compared to the GC–O profiles obtained from a nosespace extract obtained according to Taylor's method[8] in order to decide if the latter could replace the more traditional type of extracts in the GC–O analysis.

ACKNOWLEDGEMENT

Special thanks are expressed to the Company Mumm–Perrier-Jouët and to the Conseil Régional de la Bourgogne for financial support.

REFERENCES

1. P. Etiévant, L. Moio, E. Guichard, D. Langlois, I. Lesschaeve, P. Schlich and E. Chambellant, 'Trends in Flavour Research', Elsevier Science B.V., Amsterdam, 1994, p. 179.
2. N. Abbott, P. Etiévant, D. Langlois, I. Lesschaeve and S. Issanchou, *J. Agric. Food Chem.*, 1993, **41**, 777.
3. L. Moio, E. Chambellant, I. Lesschaeve, S. Issanchou, P. Schlich and P.X. Etiévant, *Ital. J. Food Sci.*, 1995, **3**, 265.
4. V.L. Singleton, *Am. J. Enol. Vitic.*, 1961, **12**, 1.
5. AFNOR, 'Analyse sensorielle', 1983.
6. P. Schlich, *Food Qual. Pref.*, 1993, **4**, 141.
7. D.S. Lundhal and M.R. McDaniel, *J. Sens. Studies*, **3**, 113.
8. R.S. Linforth and A.J. Taylor, *Food Chem.*, 1993, **48**, 115.

COMPARISON OF THREE SAMPLE PREPARATION TECHNIQUES FOR THE DETERMINATION OF FRESH TOMATO AROMA VOLATILES

A. Krumbein and D. Ulrich[*]

Institute for Vegetable and Ornamental Crops, Theodor-Echtermeyer-Weg 1, 14979 Großbeeren, Germany

[*]Federal Center of Breeding, Research on Cultivated Plant, Institute for Quality Analysis, Neuer Weg 22/23, 06484 Quedlinburg, Germany

1 INTRODUCTION

The results of aroma analysis are significantly influenced by the isolation method of flavour volatiles. To establish effective methods for the determination of fresh tomato aroma volatiles, which can be used for quality analysis, three different sample preparation techniques were tested: dynamic headspace analysis on a Tenax trap with solvent desorption of the volatiles; headspace solid phase microextraction with thermal desorption of volatiles (SPME); and liquid–liquid extraction with Freon. Compared to dynamic headspace and liquid extraction, the SPME is a relatively new technique. Response of detection, precision and overall analysis were compared. The methods were used to analyse the fruits of two tomato cultivars.

2 MATERIALS AND METHODS

2.1 Materials

The tomato cultivars 'DRW3126F1' (truss type) and 'Gourmet' (round type) were grown in soilless culture in a greenhouse. Fruits were harvested at the same stage of ripeness, marked by the same colour.

2.2 Sample Preparation Techniques

2.2.1 Dynamic Headspace Method on Tenax TA. A modified dynamic headspace method of Buttery *et al.*[1] with a small Tenax trap was used. After blending 500 g of tomatoes for 30 s and holding the mixture for 180 s, 500 ml of saturated calcium chloride solution were added and the mixture blended for 10 s. An internal standard (2-octanone) was then added and the mixture blended for another 10 s. The mixture was placed in a 3 l flask containing a magnetic stirrer, and purified air (150 ml min^{-1}) passed through the mixture and out of the flask through a Tenax trap (200 mg). The isolation was carried out for 150 min, then the trap was removed and volatiles extracted with 3 ml acetone. Samples were concentrated under nitrogen to a volume of 50 µl.

2.2.2 Headspace SPME. After blending 500 g of tomatoes for 30 s and holding the mixture for 180 s, 500 ml of saturated sodium chloride solution were added and the mixture blended for 10 s. Samples were then centrifuged at 5000 rpm for 30 minutes at 4 °C. An internal standard (2-octanone) was added to 12 ml of the supernatant and the solution placed into 20 ml headspace vials. Volatiles from the headspace were adsorbed for 10 min on a 100 µm polydimethylsiloxane fibre (Supelco 5-7300) at 30 °C, then desorbed for 2 min into the injection port of the GC.

2.2.3 Liquid–Liquid Extraction. Tomato fruits were blended and centrifuged with a procedure similar to that used with the SPME technique. Volatiles in the supernatant were extracted with Freon for 20 hours at room temperature,[2] and the extract concentrated to 10 µl in a water bath at 23 °C using a Vigreux distillation column.

2.3 GC Analysis

2.3.1 GC–FID. Instruments: HP5890A with FID; split 40 ml; injector temperature 250 °C; HP-Inno Wax column 60 m × 0.32 mm internal diameter; film 0.5 µm; 2 ml hydrogen min^{-1}; temperature programme: 3 min at 40 °C; from 40 °C at 1 °C min^{-1} to 60 °C, 2 min at 60 °C; from 60 °C at 5 °C min^{-1} to 180 °C, 10 min at 180 °C. The relative peak areas were normalized with the peak area of the internal standard.

2.3.2 GC–MS. Instruments: HP5890 Series II plus MSD 5972A; injector temperature 250 °C; Supelcowax 10 column 30 m × 0.25 mm internal diameter; film 0.25 mm; 1 ml helium min^{-1}; temperature programme: the same as GC–FID.

2.4 Statistics

All measurements were performed in triplicate. Tukey's test at the 5% level was used for mean comparison.

Table 1   *Tomato Volatiles Monitored*

Peak Number	Component
1	3 Methylbutanal
2	1-Penten-3-one
3	Hexanal
4	(Z)-3-Hexenal
5	3-Methyl-butanol
6	(E)-2-Hexenal
7	(E)-2-Heptenal
8	6-Methyl-5-hepten-2-one
9	(Z)-3 Hexenol
10	2-Isobutylthiazole
11	Phenylacetaldehyde
12	Methyl salicylate
13	Geranylacetone
14	2-Phenylethanol
15	β-Ionone

3 RESULTS AND DISCUSSION

In agreement with published data,[3,4] 54 aroma volatiles were identified by GC–MS in the dynamic headspace extract. Only a limited number, however, are essential to tomato flavour. Therefore, comparison of the isolation methods concentrated on fifteen volatile flavour components, which best characterized fresh tomato flavour based on their odour unit values.[5] Six aldehydes, three alcohols, four ketones, one ester and one sulfur-containing compound were included in the comparison (Table 1).

The chromatograms, which were obtained with the three different isolation methods, are presented in Figure 1. The higher volatile compounds, phenylacetaldehyde and 2-phenylethanol, could not be measured quantitatively with either the dynamic headspace or headspace–SPME method. Poor recovery of 2-phenylethanol may be the reason.[1] Surprisingly, the headspace–SPME was not able to recover the highly volatile flavour components 1-penten-3-one and 3 methyl-butanol. All important aroma volatiles were determined by liquid–liquid extraction.

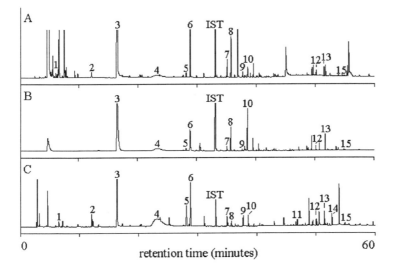

Figure 1 *Gas chromatograms of tomato cultivar 'DRW 3126 F1' obtained by different sample preparation techniques – A: dynamic headspace on Tenax; B: headspace SPME and D: Freon liquid–liquid extraction. Peak numbering is as shown in Table 1*

The precision data for all three methods were comparable. The relative standard deviation for all components was less than 15%.

With respect to isolation time, the liquid extraction method was much more time-consuming, with a sample preparation time of one hour and an extraction time of 20 hours. The headspace analysis on a Tenax trap required a sample preparation time of one hour and an adsorption time of 2.5 hours. The most rapid method was the headspace solid phase microextraction, with a sample preparation time of ten minutes and an adsorption time of ten minutes.

In principle, all three sample preparation techniques were suitable for characterizing the two cultivars. With all three isolation methods, the relative peak areas of the aroma volatiles hexanal and 2-isobutylthiazole differed significantly for the two cultivars. An example is given in Figure 2 for the SPME.

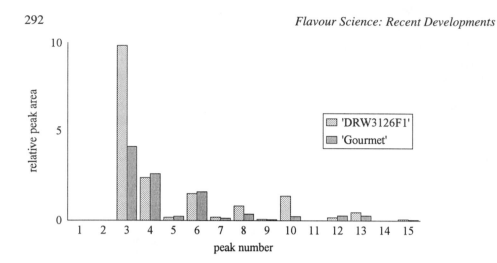

Figure 2   *Relative peak areas of aroma volatiles obtained by SPME. Peak numbering is as shown in Table 1*

4 CONCLUSIONS

In principle, all three sample preparation techniques were suitable for characterizing the two cultivars. The liquid–liquid extraction method recorded volatiles with a wide range of boiling point and polarity, but the amount of time required for the isolation procedure was a disadvantage. The dynamic headspace method on a Tenax trap was more effective, although the volatiles phenylacetaldehyde and 2-phenylethanol were not quantitatively measured. The most effective method was the headspace SPME. However, very volatile components 3-methylbutanal and 1-penten-3-one, as well as the volatiles phenylacetaldehyde and 2-phenylethanol were not quantitatively detected. Nevertheless, the new sample preparation technique SPME is a suitable screening method for tomato flavour.

REFERENCES

1.   R.G. Buttery, R. Teranishi and L.C. Ling, *J. Agric. Food Chem.*, 1987, **35**, 540.
2.   A. Rapp, H. Hastrich and L. Engel, *Vitis*, 1976, **15**, 29.
3.   R.G. Buttery, R. Teranishi, L.C. Ling, R.A. Flath, D.J. Stern, *J. Agric. Food Chem.*, 1988, **36**, 1247.
4.   Petro-Turza, *Food Rev. Int.*, 1986–1987, **2**, 309.
5.   R.G. Buttery, R. Teranishi, R.A. Flath and L.C. Ling, 'Flavor Chemistry: Trends and Development', eds. Teranishi, Buttery and Shahidi, Washington, D.C., 1989, p. 213.

DETERMINATION OF VOLATILE COMPONENTS IN CHEESE USING DYNAMIC HEADSPACE TECHNIQUES

R. Neeter, C. de Jong, H.G.J. Teisman and G. Ellen

Department of Analytical Chemistry, Netherlands Institute for Dairy Research (NIZO), P.O. Box 20, 6710 BA Ede, The Netherlands

1 INTRODUCTION

A large number of volatile compounds can be detected in cheese, belonging to chemical classes like alcohols, aldehydes and ketones, esters, fatty acids and sulfur compounds. It is the delicate balance among the numerous aroma compounds, many present only at the ppm or even ppb level, that determines the typical flavour of a particular food product, *e.g.*, a special type of cheese. In order to unravel the contribution of various volatile compounds to the flavour of a food product or to the difference in flavour between similar products such as different types of cheese, their identities and concentrations have to be established. The usual analytical techniques for this purpose are isolation of the volatile compounds from the food matrix,[1-5] followed by gas chromatography, preferably combined with mass spectrometry (GC–MS).

In this paper a dynamic headspace purge-and-trap technique, followed by GC and GC–MS, was used to investigate the formation of aroma components during ripening of Gouda cheese, and to study the difference in aroma profile between six-months-old Gouda and Proosdij cheeses.

2 EXPERIMENTAL

Gouda and Proosdij type cheeses were produced and stored during ripening in the experimental factory of NIZO according to standardized methods and circumstances. Prior to analysis for volatile compounds, the cheeses were tested by a panel of trained assessors for organoleptic quality. Gouda cheese was analysed immediately after production (pressed curd) and after six weeks' and six months' ripening. Proosdij cheese was only analysed after six months' ripening. A detailed description of the analytical procedure will be published elsewhere.[6] In brief, the procedure was as follows: 30 g of cheese was homogenized with 70 ml of distilled water in a stomacher. A 20 ml aliquot of the slurry was purged for 30 min at 40 °C with helium (flow rate: 150 ml min^{-1}). The volatile compounds were collected in a trap filled with Carbosieve SIII and Carbotrap. The trapped compounds were transferred to the TCT-injector of a gas chromatograph fitted with a capillary column and an FID, or coupled with a mass spectrometer. Structures of compounds were assigned by spectrum interpretation and comparison of spectra with bibliographic data.

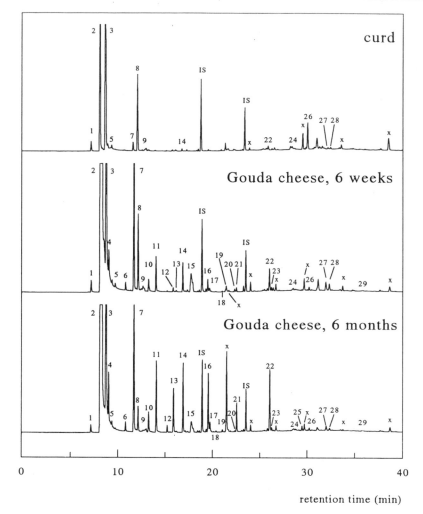

Figure 1　*Gas chromatograms of curd for Gouda cheese and of Gouda cheeses after 6 weeks' and 6 months' ripening. IS = internal standard, x = contaminant. Peak intensities have been normalized upon the first IS. See Table 1 for peak identities*

3 RESULTS AND DISCUSSION

The results are presented mainly in Figures 1 and 2 and in Table 1. In Gouda cheese curd only a few aroma components are present, predominantly ethanol, acetone and 2-butanone – compounds which are present in milk; this agrees with the finding that curd has almost no taste. After six weeks' ripening, the same compounds are found in Gouda cheese as in the curd and several other aroma compounds have been formed, *e.g.*, linear and branched alcohols and aldehydes, ketones, diacetyl and ethylbutyrate. After six months' ripening, essentially the same compounds are found as after six weeks, but most of them in higher concentrations, especially 2-methyl-propanol, 1-butanol, 2-pentanone, 3-methyl-butanol, ethyl butyrate, 2-heptanone and ethyl caproate.

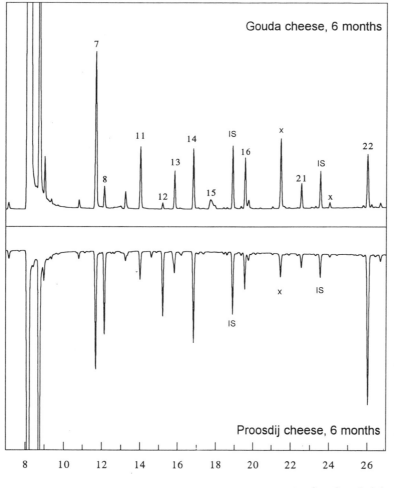

Figure 2 *Gas chromatograms (relevant parts) of Gouda cheese (positive) and Proosdij cheese (negative), both after 6 months' ripening. IS = internal standard, x = contaminant. Intensities of peaks have been normalized upon the first IS. See Figure 1 and Table 1 for peak identities*

Figure 2 shows that in six-months-old Proosdij cheese, almost the same aroma compounds are present as in Gouda cheese of the same age, but the concentrations are different. In particular the content of 3-methylbutanal is much higher in Proosdij cheese. Also higher amounts of 2-butanone, 2-pentanone and 2-heptanone are found in Proosdij cheese. In compensation, the levels of diacetyl, 2-methyl-propanol, 1-butanol, acetoin, 3-methyl-butanol and ethyl butyrate are somewhat higher in Gouda cheese than in Proosdij cheese. Results of sensory tests have shown that Proosdij cheese differs in taste from Gouda cheese, with extra attributes like sweet, caramel, bouillon and fruity. Sweet and caramel are possibly related to the presence of 3-methylbutanal, and bouillon and sweet may be due to the combined presence of 2-heptanone and sulfur compounds. To test this hypothesis, small amounts of 3-methylbutanal, 2-butanone, 2-pentanone and 2-heptanone were added to fat isolated from Gouda cheese. Experienced assessors indeed recognized the taste of this fat as Proosdij cheese-like.

Table 1   *Identities of Compounds (GC–MS) Corresponding with the Peaks in Figures 1 and 2*

No.	Compound	No.	Compound	No.	Compound
1	Acetaldehyde	11	2-Methylpropanol	21	Ethyl butyrate
2	Ethanol	12	3-Methylbutanal	22	2-Heptanone
3	Acetone	13	1-Butanol	23	Heptanal
4	Isopropyl alcohol	14	2-Pentanone	24	Hexanoic acid
5	Dimethylsulfide	15	Acetoin	25	Ethyl caproate
6	1-Propanol	16	3-Methylbutanol	26	Limonene
7	Diacetyl	17	2-Methylbutanal	27	2-Nonanone
8	Butanone	18	Dimethyldisulfide	28	Nonanal
9	Acetic acid	19	1-Pentanol	29	2-Decanone
10	Hexane	20	Hexanal		

A limitation of (dynamic) headspace techniques is that very polar and high-boiling compounds are not isolated or only isolated at a low yield from an aqueous liquid phase. Furthermore the amounts of trapped individual compounds are not directly proportional to their concentrations, but depend also upon their vapour pressure. Nevertheless, the release of aroma compounds from food in the human mouth is governed by comparable limitations. To discover if the components isolated by the dynamic headspace technique are representative of the aroma of Gouda cheese, the purged and trapped compounds from a slurry of Gouda cheese were, after thermal desorption from the trap, directly trapped in distilled water and in de-odorized butter oil, respectively. The smell and taste of the obtained solutions were strongly associated with Gouda cheese by trained assessors. We conclude that the dynamic headspace technique is a quick and easy technique to follow the ripening process of cheese and to differentiate between different types of cheese.

REFERENCES

1. R. Neeter and H.T. Badings, *Voedingsmiddelentechnologie*, 1989, **22** (5), 21.
2. R. Neeter and C. de Jong, *Voedingsmiddelentechnologie*, 1992, **25** (11), 11.
3. J.O. Bosset and R. Gauch, *Int. Dairy J.*, 1993, **3**, 359.
4. A.F. Wood, J.W. Aston and G.K. Douglas, *Austr. J. Dairy Technol.*, 1994, **49**, 42.
5. W.T. Yang and D.B. Min, *J. Food Sci.*, 1994, **59**, 1309.
6. W.J.M. Engels, R. Dekker, C. de Jong, R. Neeter and S. Visser, in preparation.

THE ISOLATION OF FLAVOUR COMPOUNDS FROM FOODS BY ENHANCED SOLVENT EXTRACTION METHODS

Karl Ropkins and Andrew J. Taylor

Department of Applied Biochemistry and Food Science, University of Nottingham, Sutton Bonington Campus, Loughborough, Leicestershire, LE12 5RD, U.K.

1 INTRODUCTION

A comparison of headspace volatiles and the total volatile content of a food can be used to produce a simple volatile generation–release model. While headspace is a relatively convenient technique, total volatile content requires time-consuming methods like Soxhlet, uses large volumes of organic solvents and produces complex extracts which require careful concentration before analysis. Supercritical fluid extraction (SFE) has already been shown to be a viable alternative to more conventional solvent extraction for a wide range of sample types.[1] The solvating power of the supercritical fluid can also be 'fine tuned', by optimising extraction temperature, pressure, time and flow rate, and by the addition of organic modifiers, to selectively fractionate a given sample producing cleaner, simpler extracts.[2] The objective of this study was to determine whether SFE could be used as an alternative to Soxhlet for the isolation of flavour compounds from a breakfast cereal.

2 METHODS

2.1 Sample Preparation

A model breakfast cereal formulation was prepared as outlined by Fast.[3] Wheat, malt extract, sucrose and salt (NaCl) were blended together and the water content of the resulting mixture adjusted to 30% (w/w). Samples were then transferred to Schott bottles, sealed and cooked at 121 °C and 105 kPa for 40 min. Determination of dry weight after cooking indicated that the water content remained constant during the sample processing. All samples were stored at −18 °C prior to analysis.

2.2 Headspace

Sub-samples (35 g) were thawed at room temperature overnight before analysis. The headspace above each sample was swept with dry nitrogen gas (50 ml min^{-1}) and volatiles concentrated onto Tenax TA traps (SGE, U.K.) for analysis.

2.3 Soxhlet

Samples (10 g) were ground with excess anhydrous Na_2HCO_3 to remove water and increase surface area. These samples were packed into 25 ml cellulose thimbles and extracted with 350 ml dichloromethane (Sigma, U.K.) in a Soxhlet apparatus for 12 hours (*ca.* 3-minute cycles). Samples were then concentrated by rotary evaporation under vacuum (> 30 °C) and careful nitrogen blow-down to about 500 µl before analysis.

2.4 SFE

Sub-samples (2.5 g) were ground and packed into 10 ml extract vessels with excess hydromatrix (Anachem, U.K.) for supercritical CO_2 (Air Products, U.K.) extraction using a Suprex AutoPrep 44 Supercritical Fluid Extractor. Details of the conditions used for the optimisation of the SFE method are given in Section 3 of this paper. The optimised volatile extraction was carried out at 150 atm and 45 °C for 45 min. The resulting extracts were depressurised and the gaseous CO_2 was vented off while the residues were cold trapped on a bed of silanised silica beads at −35 °C. At the end of the extraction period, residues were desorbed from the trap at 20 °C and flushed into collection vials with 500 µl of dichloromethane for analysis.

2.5 Gas Chromatography–Mass Spectrometry (GC–MS)

In all cases, volatile separation and analysis was by GC–MS (MD800, Fisons Instruments, U.K.), after the addition of an external standard (2-octanone). Headspace samples (on Tenax traps) were thermally desorbed at 250 °C and cryofocused onto the first coil of a GC column (BP-1; 25 m × 0.22 mm internal diameter; SGE, U.K.) using a liquid nitrogen cold trap. For Soxhlet and SFE extracts 1 µl aliquots were injected via a splitless injector at 210 °C. For both headspace and liquid injections, the temperature programme was the same (40 °C for 1 min, 10 °C min^{-1} to 280 °C and then held at final temperature for 10 min).

2.6 Data Analysis

For each method, samples were prepared and analysed in triplicate. Identification of the analytes was made by comparing their retention times and mass spectra to those of authentic reference compounds. Quantification was by peak area ratio calculation, after relative response factors for the analytes and the external standard had been taken into account. These factors were determined by comparing the peak areas of the external standard and authentic compounds in a reference mixture.

3 RESULTS

3.1 Headspace and Soxhlet

During headspace sampling, volatiles are swept from the air above a food and concentrated (usually onto a porous polymer) for analysis. Volatiles have to be already present in the headspace or released to the headspace during sampling to be isolated. Therefore, compared with other flavour isolation techniques, headspace extracts tend to be relatively free of impurities which would otherwise hinder analysis. As a result, headspace is widely used, particularly in the quality assurance of finished products, such as packaged food. Unfortunately non-volatiles and tightly bound and encapsulated volatiles are not amenable to headspace analysis.

To remove these 'bound' aroma compounds requires more aggressive extraction techniques such as Soxhlet. During Soxhlet extraction, the sample is repetitively extracted with large volumes of an organic solvent which must penetrate the sample matrix to bring out the analytes. The process is time-consuming and often extracts not only flavour compounds but also a wide range of other unwanted material such as lipids, free fatty acids and non-aroma hydrocarbons. The resulting extract is probably as close as we can presently get to a total volatile fraction. However, if care is not taken during the extraction and subsequent concentration steps, the very volatile aroma compounds can be lost.

Figure 1 shows the headspace and Soxhlet GC–MS profiles of the model breakfast cereal. Although the two profiles appear very different, many of the compounds in the headspace are also present in the Soxhlet extract. Differences in the distributions of these compounds probably reflect differences in volatility and sample–headspace partitioning, which affects the headspace more significantly than the Soxhlet extract.

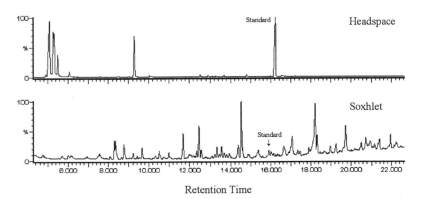

Retention Time

Figure 1 *The headspace and Soxhlet GC–MS profile of the model breakfast cereal*

3.2 SFE

Although primarily used for the analysis of environmental samples, one particularly promising alternative to Soxhlet is SFE and some work has already been carried out on extracting aromas from food samples.[4,5] Because supercritical fluids not only have solvating abilities similar to liquids but also diffusivities and mass transfer properties approaching those of gases, they should, in theory, offer improved extraction efficiencies and reduced extraction times compared with traditional solvents. Supercritical CO_2 is most commonly used because it is inert, non-toxic and relatively inexpensive.

As with all extraction techniques, SFE requires optimisation before efficient recoveries can be obtained. A wide range of parameters can affect the kinetics of SFE, such as extraction time, pressure, temperature and flow rate, as well as the presence of organic modifiers and the nature of both the matrix and the target analytes. Because the solvating power of supercritical CO_2 is dependent upon its density, a sample can be selectively extracted (fractionated) using temperature, pressure and/or organic modifiers. Extraction recoveries were compared over a wide range of pressures (150, 200, 250, 300, 350, 400 and 450 atm) and temperatures (45, 65 and 85 °C). At lower densities (low temperature and/or pressure), supercritical CO_2 had a solvating power comparable to pentane, and extraction appears to be preferentially of volatile and non-polar compounds, while at higher densities, comparable to polar solvents such as dichloromethane or toluene, extracts appear to contain mainly non-volatile and polar compounds (such as the fats and lipid material). As a result, 150 atm and 45 °C were selected as optimum extraction conditions for the recovery of volatiles from the breakfast cereal formulation. Dichloromethane and methanol were considered as organic modifiers. Although they improved volatile recoveries, the increased polarities of the resulting supercritical CO_2–organic solvent mixtures significantly increased fat breakthrough.

Because CO_2 is only a liquid under supercritical conditions, *i.e.*, at increased temperatures and pressures, it can easily be evaporated away by depressurising the solvent–extract mixture back to 1 atm. As this can be achieved at very low temperatures, volatile losses during extract concentration can be minimised. The SFE system used during this study cryotrapped the extract residues onto a bed of silanised silica beads, using a CO_2 cold trap. Volatile recoveries from the cold trap were compared over the range –5 to –45 °C. An optimum cold trap temperature of –35 °C was selected because although the best recoveries were observed below –40 °C, the trap tended to be unreliable at these temperatures and regularly blocked. Trapping onto solid absorbents, such as Tenax and C_{18}, improved volatile recovery at higher cold-trap temperatures but larger volumes of eluting solvent were required to recover the extracts (reducing the concentration factor). During the extraction process, analytes are being continually extracted from the sample matrix and then deposited in the trap site as the CO_2 is vented off. As volatiles already held within the trap can be stripped away by this venting process, good recoveries are achieved by optimising the extraction time (a good recovery without significant volatile losses). For the model breakfast cereal examined during this study, several extraction times were compared (15, 30, 45, 60 and 120 min) and optimum volatile recovery was estimated at 45 minutes.

Quantification of volatiles by optimised SFE and Soxhlet from a single batch of the model breakfast cereal are presented in Table 1. Volatile recoveries by SFE were found to be comparable to those of Soxhlet. However reduced co-extraction of fats and lipid-based material simplified analysis and prolonged the GC capillary column life-time.

Table 1   *A Comparison of Volatile Recoveries using Soxhlet and SFE Extractions*

Compound	Soxhlet Extraction		SFE Extraction		% recovery
	Concentration (ng g^{-1})	rsd (%)	concentration (ng g^{-1})	rsd (%)	
3-Methyl butanal	92	22	128	17	140
Dimethyl disulfide	60	20	0	0	0
Hexanal	412	8	378	12	92
2-Methyl pyrazine	14	12	10	17	70
2-Furancarboxaldehyde	232	10	241	19	104
2-Furanmethanol	1128	11	1479	18	131
C$_2$-Pyrazine	22	13	14	20	64

REFERENCES

1   R.M. Smith, *LC-GC International*, 1996, **9** (1), 8.
2   D.R. Gere and E.M. Derrico, *LC-GC International*, 1994, 7 (7), 370.
3   R.B. Fast, in 'Breakfast Cereals and How They are Made', American Association of Cereal Chemists, Minnesota, 1989, 15.
4   D.L. Taylor and D.K. Larick, *J. Food Sci.*, 1995, **60** (6), 1197.
5   M.J. Morello, in ACS Symposium Series, 1994, **543**, 95.

THE RECOVERY OF SULFUR AROMA COMPOUNDS OF BIOLOGICAL ORIGIN BY PERVAPORATION

A. Baudot, M. Marin and H.E. Spinnler

Institut National Agronomique Paris-Grignon, Laboratoire de Génie et Microbiologie des Procédés Alimentaires (INRA), C.B.A.I., F-78850 Thiverval-Grignon, France

1 INTRODUCTION

Sulfur aroma compounds are top notes of the flavour of several food products, such as bacteria surface ripened cheeses,[1] meat, vegetables and fruits. Such molecules are nowadays used in gravies, sauces, soups and frozen dishes. These aroma compounds can be extracted from the original products; however, low concentrations may hinder the technological and economical feasibility. A promising alternative technology concerns the production of concentrated natural labelled sulfur compounds by bioconversion.[2] In this case, the continuous and selective extraction of the aroma molecules from the biological medium is necessary in order to recover the bioproduct, detoxify the bioreactor and could also be useful in understanding the bioconversion pathways.[3]

The performances of the pervaporation process for the recovery of S-methyl thiobutyrate have been studied. This thioester, which displays a strong cauliflower fragrance, is a sulfur compound that can be produced by *Coryneform* bacteria. The effects of the culture medium constituents on the yield of the extraction have also been investigated.

2 THEORY

Pervaporation is a separation process based on a selective transport through a dense membrane driven by an evaporation of the permeate on the downstream side of the membrane. This separation process is particularly designed for coupling with a fermenter for a continuous extraction of volatile organic compounds: no fouling due to micro-organisms, non-stressful feed operating conditions (low temperature and atmospheric pressure in the feed). Pervaporation has already been used for the continuous removal of volatile compounds from fermentation broth, so as to increase the productivity of the bioreactor.[3]

Membrane processes can be characterized by performance parameters such as flux and selectivity which are indicators of the ability of the process to extract a chosen component.

Selectivity of pervaporation is usually defined as an enrichment factor β_i^{PVP} defined as the ratio between the permeate weight fraction ($w^i_{permeate}$) and the feed weight fraction (w^i_{feed}) of the compound i to be extracted:

$$\beta_i^{PVP} = \frac{w^i_{permeate}}{w^i_{feed}} \tag{1}$$

As a reference, this parameter can be compared with the performances of vacuum distillation with the same feed operating conditions. Making the analogy with β_i^{PVP}, the enrichment factor of the vapour–liquid equilibrium β_i^{VLE}, characteristic of distillation, can be defined as:

$$\beta_i^{VLE} = \frac{w_{vapour}^i}{w_{feed}^i} \quad \text{with} \quad w_{vapour}^i = \frac{x_i \, \gamma_i \, P_i^0(T_{feed}) \, M_i}{x_i \, \gamma_i \, P_i^0(T_{feed}) \, M_i + \sum_j x_j \, \gamma_j \, P_j^0(T_{feed}) \, M_j} \quad (2)$$

The subscript i refers to the aroma compound i to be extracted in multicomponent (Σ_j) mixtures. $P^0(T)$ refers to the saturated vapour pressure (Pa) at feed temperature T_{feed} (K). M is the molar weight (g mol^{-1}), x is the molar fraction (mol mol^{-1}) and γ, the activity coefficient of each compound in the feed. The activity coefficient of methylthiobutyrate was obtained by the mutual solubility method.[4] The saturated vapour pressures were estimated with the Lee–Kesler method[5] except for water (Antoine relation).

3 MATERIALS AND METHODS

Experiments were carried out with GKSS PEBA 40 homogeneous membranes (70 μm thick) in a plate and frame module. High flow rate combined with an optimized injection system of the liquid feed made polarization phenomena at the upstream face of the membrane negligible. The effective membrane area was 0.1 m^2. Permeate pressure was regulated with a dry air inlet near the vacuum pump. Permeate was condensed with a liquid nitrogen cold trap. More details about the pervaporation pilot plant are given elsewhere.[4]

Model solutions were prepared with methylthiobutyrate (Aldrich, 97% purity) in double-osmosed water. Several multicomponent mixtures, close to real fermentation media (dedicated to *Coryneform* bacteria culture), were prepared with the following compounds (1 l water qsp):

(1) Lactose	6 g	(4) Casaminoacids (Difco)	1 g
(2) Na$_2$HPO$_4$	4.7 g	(4) Biotrypcase (Bio Merieux)	1 g
(2) KH$_2$PO$_4$	2.7 g	(4) L-Methionine (Sigma)	2 g
(2) NaCl	10 g	(4) L-Leucine (Sigma)	2 g
(3) Sodium lactate (60%)	25 ml	(4) L-Cysteine (Sigma)	0.015 g
		(4) L-Tryptophan (Sigma)	0.0675 g

Analyses were performed with a FID–GC (4.5% mean standard error).

4 RESULTS AND DISCUSSION

All the pervaporation experiments were carried out with a constant methylthiobutyrate concentration in the feed of 50 ppm (w/w). The temperature in the feed was equal to 30 °C, which is suitable for a coupling of pervaporation with a bioreactor. Permeate pressure was varied from 260 Pa to 2200 Pa.

4.1 Water–Methylthiobutyrate Binary Mixture

Experiments with water–methylthiobutyrate binary mixtures were first performed in order to evaluate pervaporation performances with regard to the thioester extraction.

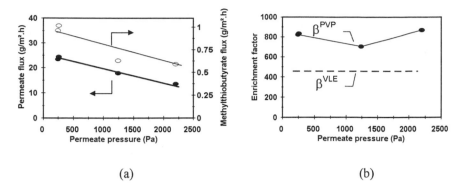

Figure 1   *(a) Permeate total flux and methylthiobutyrate flux (PEBA 40, 30 °C); (b) Enrichment factor of methyl thiobutyrate (PEBA 40, 30 °C)*

Permeate flux as well as methylthiobutyrate flux decreased with increasing permeate pressure (Figure 1(a)). The thioester partial flux reached values up to $1 \text{ g m}^{-2} \text{ h}^{-1}$ for a permeate pressure equal to 260 Pa. Enrichment factors for methylthiobutyrate were unaffected by permeate pressure variation (Figure 1(b)). The enrichment factors displayed by PEBA membranes were almost twice as high as the selectivity of vacuum distillation for the same 30 °C feed temperature ($\beta^{PVP}= 807 > \beta^{VLE}= 491$). The good performances offered by pervaporation membranes for the recovery of sulfur compounds can be related, as a common rule, to the high hydrophobicity of these molecules.[4,6]

4.2 Multicomponent Mixtures

The influence of the compounds of real fermentation media on pervaporation performances was studied. Indeed, it has been previously shown that organic compounds can lower the partial pressure of aroma compounds in the feed and thus alter the partial fluxes of these components through the pervaporation membrane.[7]

The feed mixture was rendered more and more complex (eventually close to a real culture medium) by successively adding culture medium compounds to the model binary mixture (described in Section 4.1): lactose (1), salts (2), sodium lactate (3) and amino-acids (4). Each series of experiments was performed for two or three days before each addition (conditioning the membrane). The enrichment factors remained constant, close to the binary mixture ones, whatever the tested permeate pressures (Table 1).

Whatever the complexity of the medium, the constituents had no significant influence on the performances of the pervaporation process (fluxes and enrichment factor). Indeed, no salting-out of the thioester from the media could occur, as the ionized molecules (salts, organic acids or amino-acids) and sugars (lactose) were highly diluted in the feed.

In all cases, the thioester was so concentrated in the condensate that a phase separation was observed when thawing the permeate: an organic phase (almost pure thioester) was always floating over an aqueous saturated phase in the cold trap.

Table 1   *Partial Flux and Mean Enrichment Factors (over 4–6 samples) in Methyl Thiobutyrate for Complex Mixtures*

Mixtures (PEBA 40 membrane, 30°C)	MTB flux / g m⁻² h⁻¹		Mean enrichment factor
	260 Pa	2200 Pa	
H₂O +MTB	1.01	0.63	807
H₂O +MTB + (1)	0.92	0.57	783
H₂O +MTB + (1) + (2)	0.96	0.53	785
H₂O +MTB + (1) + (2) + (3)	1.01	0.61	797
H₂O +MTB + (1) + (2) + (3) + (4)	0.98	0.53	720

MTB refers to methylthiobutyrate (50 ppm) and the numbers to the culture medium compounds described in Section 3.

5 CONCLUSION

Pervaporation with organophilic membrane (PEBA) proved to be an efficient process for the recovery of methylthiobutyrate (more selective than vacuum distillation). The constituents of a culture medium dedicated to thioester-producing *Coryneform* bacteria did not alter the performances of the pervaporation process (no polarization in the liquid feed). Pervaporation is well-suited for a continuous extraction of the aroma compounds when coupled with a thioester-producing reactor.

ACKNOWLEDGEMENTS

The authors would like to thank Electricité de France (EDF-DER, ADEI, Les Renardières, BP 1, route de Sens, F-77250 Moret-s-Loing, France) for its financial support.

REFERENCES

1. A. Cuer, G. Dauphin, A. Kergomard, S. Roger, J.P. Dumont and J. Adda, *Lebens.-Wissen. und -Technol.*, 1979, **12**, 258.
2. G. Lamberet, B. Auberger and J.L. Bergère, in 'Bioflavour '95', eds. P. Etievant and P. Schreier, INRA, Paris, 1995, p. 265.
3. T. Lamer, H.E. Spinnler, I. Souchon and A. Voilley, *Process Biochem.*, 1996, in press.
4. A. Baudot and M. Marin, *J. Membrane Sci.*, 1996, in press.
5. R.C. Reid, J.M. Prausnitz and B.E. Poling, 'Properties of Gases and Liquids', McGraw-Hill, New York, 1987.
6. J.T.M. Sluys, F.G.C.T. Sommerdijk and J.H. Hanemaaijer, in 'Récents progrès en Génie des Procédés', eds. P. Aimar and P. Aptel, **21**, Lavoisier, Paris, 1992, p. 401.
7. E. Favre, Q.T. Nguyen and S. Bruneau, *J. Chem. Technol. Biotechnol.*, 1996, **65**, 221.

EXTRACTION OF AROMA COMPOUNDS BY PERVAPORATION

I. Souchon,[*] C. Fontanini and A. Voilley

Laboratoire de Génie des Procédés Alimentaires et Biotechnologiques, ENSBANA, Université de Bourgogne, 1 Esplanade Erasme F-21000 Dijon, France

1 INTRODUCTION

The use of pervaporation in the food industry has been considered for several years.[1] This membrane process has many potential applications such as wine or beer dealcoholization,[2] juice concentration[3] and extraction of aroma compounds from various dilute media such as effluents from the food industry[4] or fermentation broth.[5]

Pervaporation consists of a partial vaporization of a liquid through a dense membrane. Solutes are dissolved in the polymer material at the upstream face of the membrane, diffuse into the polymeric network and are finally evaporated in the downstream gas phase. This evaporation occurs because of a continuous pumping which produces a low partial pressure in the downstream compartment.

Our study deals with the extraction of five aroma compounds which are often found in foodstuffs and which can be produced by micro-organisms. Their mass transfer has been studied in relation to their thermodynamic and physico-chemical properties.

2 MATERIALS AND METHODS

Five aroma compounds with different chemical functions and molecular weights have been used. Their physico-chemical and thermodynamic properties are given in Table 1.

Pervaporation experiments were performed using an experimental design described previously by Lamer et al.[7] All the experiments were carried out with a model solution (one aroma compound + water) at a constant feed concentration of 0.1 g l^{-1}. A dense homogeneous polydimethylsiloxane (PDMS) plane membrane, 150 µm thick (Dow Corning), was used for the extraction. This membrane has hydrophobic properties in order to increase the aroma compounds' affinity for the membrane.

A turbulent flow (Re ≈ 10^5) was maintained at the membrane surface in order to limit the boundary layer effect. The pervaporation experiments were carried out with a low downstream pressure (60 Pa) to avoid evaporation problems.

Each pervaporation experiment allowed the determination of two parameters:
1. The flux (Ji, g m^{-2} h^{-1}) of the component at steady state and defined as the mass of organic compound pervaporating through a membrane per unit area per unit time;

---

[*]Present address: Institut National Agronomique Paris - Grignon, LGMPA, CBAI, F78850 Thiverval Grignon, France

2. The selectivity, β, or enrichment factor defined as the ratio of the mass fraction of compounds in the pervaporate to that in the feed.

Solubility coefficients (Si), *i.e.* the partition coefficient of the aroma compound between the aqueous solution and the polymer, and diffusion coefficients (Dim) have been deduced from the kinetics of sorption of the aroma on PDMS.[7]

Table 1   *Physico-chemical and Thermodynamic Properties of the Aroma Compounds (25 °C)*

Compound	Odour	MW[a]	Volatility (Pa)		Solubility in water (g l^{-1}) T=25°C	log P [(d)]
			Pure[b]	In solution ×10^4 [c]		
Benzaldehyde	Bitter almond	106	70	5.8	7.0	1.5
2-Phenylethanol	Rose	122	1	0.04	16.0	1.4
2-Heptanone	Banana, fruity	114	373	110.9	4.3	1.8
2-Octanone	Floral, acid	128	107	141.0	1.5	2.3
2-Nonanone	Rose, tea	142	58	336.9	0.4	2.9

[a] MW = molecular weight (g mol^{-1});
[b] Vapour pressure of pure compound P_i^s at 25 °C (Pa);
[c] Henry's constant, He = $\gamma_i^\infty \times P_i^s$ with γ_i^∞, activity coefficient of aroma compound in water at infinite dilution;
[d] log P: hydrophobicity constant (Rekker[6]).

Table 2   *Fluxes and Selectivities of Pervaporation (T = 25 °C, Concentration = 0.1 g l^{-1}, PDMS membrane (150 µm), Re ≈ 10^5)*

Compound	Flux (g m^{-2} h^{-1})	Selectivity β
Benzaldehyde BZA	0.93	970
2-Phenylethanol PHE	0.04	37
2-Heptanone HEP	1.92	2500
2-Octanone OCT	2.1	2830
2-Nonanone NON	2.4	3200

3 RESULTS AND DISCUSSION

3.1 Pervaporation Performances – Flux of Aroma Compounds and Selectivities

The characteristics of mass transfer of aroma compounds, fluxes and selectivities are presented in Table 2.

The lowest flux is obtained with 2-phenylethanol which is the least hydrophobic compound (log P = 1.4), the most soluble in water (16 g l^{-1} at 25 °C) and the least volatile (Henry's constant = 0.04×10^4 Pa). At the other extreme, the highest flux is observed for 2-nonanone, which is the most hydrophobic compound (2.9), the least soluble in water (0.4 g l^{-1}) and the most volatile (336.9×10^4 Pa). The selectivity for 2-phenylethanol is not very high, but is of the same order of magnitude as the generally accepted value for many organic compounds such as alcohols. The other compounds studied present higher selectivities. For instance, a selectivity equal to 3200 in the case of 2-nonanone means that for a 0.1 g l^{-1} aroma compound solution, the 2-nonanone concentration in the pervaporate is 320 g l^{-1}. In fact the water solubility of 2-nonanone is 0.4 g l^{-1} and two phases were obtained in the condensate.

From Henry's constant and the thermodynamic laws of vapour–liquid equilibria, it is possible to determine the selectivity of the vapour–liquid equilibrium ($\beta_{L/V}$):

$$\beta_{L/V} = \frac{Y_i}{X_i}$$

where X_i is the mass fraction of the aroma compound in the aqueous phase and Y_i the mass fraction of the aroma compound from the condensables (water + aroma → compound) of the vapour phase. The selectivity of pervaporation is compared with that of the vapour–liquid equilibria in Figure 1.

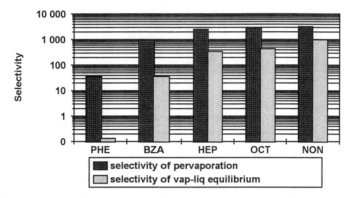

Figure 1 *Comparison of the selectivity of the pervaporation and the selectivity of the vapour–liquid equilibrium of aroma compounds (T = 25 °C)*

For all the compounds studied and under the operational conditions used, pervaporation is a process more selective than a process based only on the vapour–liquid equilibrium such as distillation. The extraction of the aroma compounds studied is at least three times more selective than the vapour–liquid equilibrium.

3.2 Importance of Sorption and Diffusion in the Pervaporation

To understand the mass transfer of aroma compounds better, two major parameters, solubility and diffusion coefficients in the membrane, have been determined and are presented in Figure 2 in relation to the flux of pervaporation. These results show that solubility and diffusion coefficients evolve in opposite directions. Little work has been done on the physico-chemical explanations for the influence of interactions on diffusivity of the aroma compounds in the polymer. However, when interactions are strong, *i.e.* when the solubility coefficient is high, diffusing molecules are retained by the polymer segments and the diffusion is lower.[7,8]

The organic fluxes increase with increasing sorption whereas they diminish when membrane diffusivity of the compounds increase. For instance, 2-nonanone has the highest flux and solubility coefficient, whereas its diffusion coefficient in PDMS is the lowest. Under the present experimental conditions, consequently, sorption is the principal step determining mass transfer in the process.

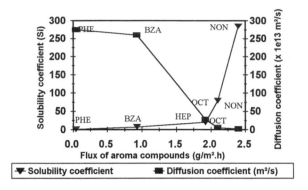

Figure 2   *Relation between diffusivity, sorption and pervaporation performances*

4 CONCLUSION

Pervaporation is a very useful process for the extraction of organic compounds from dilute solutions. Indeed, the selectivity is at least three times higher than the selectivity of the vapour–liquid equilibrium. The fluxes of the aroma compounds are relatively high, especially for the series of methylketones, and depend strongly on the sorption of the compounds in the membrane. At present, pervaporation has found some industrial applications in the chemistry industry (in particular for dehydrating organic solvents) but the extraction of aroma compounds is a very attractive potential application. A selective extraction–concentration could be performed with pervaporation at low temperature and without the use of organic solvent. For example, a continuous extraction integrated with a bioreactor could be envisaged with this technique, since a continuous recovery under non-stressful conditions is possible. Generally, pervaporation is an efficient process for the recovery of high value organic compounds such as aroma compounds from dilute aqueous solution.

REFERENCES

1. A. Voilley and T. Lamer, *I.A.A.*, 1990, **10**, 881.
2. J.L. Escudier, M. Le Bouar, M. Moutounet, C. Jouet and J.M. Barillere, 'Proceedings of the 3rd International Conference on Pervaporation Processes in the Chemical Industry', ed. R. Bakish, Englewood, 1988, p. 387.
3. F. Lutin, E. Deneve, J. Brunea, L. Idier and R. Buvet, 'Proceedings of the 5th World Filtration Congress', Nice, 1990 , **1**, 28.
4. I. Souchon, P. Godiard and A. Voilley, *Odours & VOCs J.*, 1995, **1** (2), 124.
5. T. Lamer, H.E. Spinnler, I. Souchon and A. Voilley, *Process Biochem.*, 1996, in press.
6. R.F. Rekker, in 'The Hydrophobic Fragmental Constant', eds. W.Th. Nauta and R.F. Rekker, Elsevier, Amsterdam, 1977, p. 389.
7. T. Lamer, M.S. Rohart, A. Voilley, H. Baussart, *J. Membrane Sci.*, 1994, **90**, 251.
8. J.M. Watson, P.A. Payne, *J. Membrane Sci.*, 1990, **49**, 171.

SENSORY METHODS IN FLAVOUR

FLAVOUR CHEMISTRY AND HUMAN CHEMICAL ECOLOGY

Terry E. Acree and Johanna M. Bloss[*]

Department of Food Science and Technology, Cornell University – Geneva, Geneva, New York 14456, U.S.A.
[*]Department of Biology, Boston University, Boston, MA, 02215, U.S.A.

1 INTRODUCTION

Odour, aroma, redolence and perfume describe '… the property of a substance that affects the sense of smell'.[1] This property is manifested in chemicals called odorants, and distinguishes them from other chemicals by the behaviour they stimulate in an organism and by the mode of action that mediates this behaviour. In simpler terms, one or more chemicals stimulate a pattern of responses in an olfactory nerve, and this pattern of neural responses causes a range of behaviours that benefit the survival of an organism. The book, 'Sensory Ecology: How Organisms Acquire and Respond to Information' by Dunsenbery[2] organizes observations on the sense of smell of many different organisms into a single paradigm: information molecules or *semiochemicals*. Table 1 shows the relationship between the source of an information molecule (the transmitter) and the organism that receives it. The central idea is that smell fits into a global concept based on the communication within organisms, and between organisms and their environment.

Table 1   *Information Molecules and their Modes of Action*

Mode of Action – Relationship of Transmitter and Receiver	Type of Information Molecule
Interorganismal	Semiochemicals (scents)
Intraspecific	Pheromones (sex attractants)
Interspecific	Allelochemicals
Benefit to Receiver	Kairomones (food scents)
Benefit to Sender	Allomones (lures)
Benefit to Both	Synomones (floral scents)
Intra-organismal	Hormones (*e.g.* insulin)

Adapted from Dunsenbery[2].

   When applied to humans, the flavour of fruits and vegetables is in general synomones since they induce people to both eat a nutrient-rich crop and to propagate the crop well beyond its potential in a non-agrarian ecology. Both benefit. However, the flavour of a roasting passenger pigeon or giant Moa may have contributed to their extinction – perhaps benefiting, although briefly, only the receiver. Are these flavours kairomones? Is the smell of burning cannabis an allomone, benefiting only the sender? Does cannabis compel the receiver to propagate it without any benefit to the receiver? Although answers to these and similar questions are not always simple, it is clear that odorants are semiochemicals and that the human perception of flavour and fragrance are elements of a *chemical ecology*. As defined on the web site of the Cornell Institute for Research in Chemical Ecology (CIRCE), chemical ecology '... deals with the chemical interactions of organisms, interactions that are pervasive at all levels of biological organization, from microbes to humans, and they operate in the most diverse biological contexts. Organisms find food and seek out mates on the basis of chemicals, ...' [3]. Understanding flavour as a process generally equivalent to the chemical ecology of all biology may help illuminate the physiology of olfaction, its evolution, and its psychological and social function.

2 THE G-PROTEIN RECEPTOR

As our knowledge of the comparative anatomy, physiology and biochemistry of olfaction developed, strong similarities between human olfaction and the semiochemistry of many other organisms emerged. Two features stand out as common in vertebrate and in at least some non-vertebrate systems:

1. Individual neurons in the olfactory system are broadly tuned to many different odorants[4,5], *i.e.* one odour-active chemical will react with several different receptors simultaneously; and
2. The olfactory epithelium is characterized by a zonal topography of odorant receptor expression,[6] *i.e.* different areas of the olfactory epithelium will respond differently to a particular odorant mixture.

In addition, the present belief is that all olfactory transduction systems, *i.e.* those parts of olfaction that detect the presence of an odorant and initiate neuronal signals, are characterized by the presence of a G-protein coupled receptor with minimal intracellular and extracellular loop structures. As shown in Figure 1, seven stitches (with short N- and C-termini segments) are sewn in and out of the receptor cell with the odorant receptor proteins lipophilic transmembrane regions.

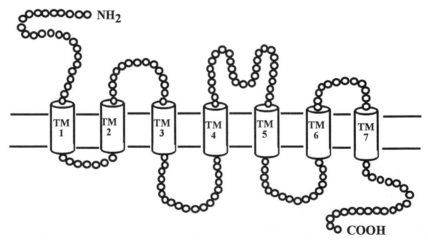

Figure 1 *Seven trans-membrane G-protein coupled receptor thought to be the ligand binding site for odorants (from Krieger 1994[7])*

Variation in the structure of the olfactory protein is thought to be responsible for the range of chemical structures detected by humans. The presence of small loops and segments have been used to argue that the olfactory system is one of the oldest chemical information systems in biology.[7] If so, then a high degree of homology should exist across diverse biological systems and along a major part of biological evolution in the structure and functions of semiochemicals. This paper will indicate some of the specific chemicals that have been found to function as odorants perceived by humans and as semiochemicals in other chemical ecologies and speculate on their implications.

3 INSECT SEMIOCHEMICALS

Insects use chemicals to orient their behaviour as much as, if not more than, any other form of information. They produce chemicals to communicate with individuals and they detect chemicals to learn about their environment. Probably the most common airborne semiochemicals produced by insects are *female* sex pheromones. An extensive list of sex pheromones of lepidoptera has been prepared[8] and displayed by H. Arn in an elegant format on the world wide web (http://www.nysaes.cornell.edu/pheronet). It describes almost 1,000 compounds that range from 7 to 22 carbon structures, and it includes alcohols, esters and even halogenated compounds. It may not be coincidental, but very few have odour for humans at the level at which they are active for the insects. It seems that the G-protein coupled receptors used by male lepidoptera to detect these compounds must be very different from the 1,000 or so used by humans to detect volatiles. However, other semiochemicals detected by insects are easily detected by humans. An example is the commercially important aroma chemical methyl jasmonate.

In an attempt to isolate the compounds released from a specialized organ on the male oriental fruit moth (*Grapholitha molesta* Busck.) which induced 'wing-raising' behaviour in females, Nishida[9] observed the smell of old lemon peels. Using gas chromatography–olfactometry, an odorant was detected in extracts of peels of lemon that smelled exactly like the lemon smelling compound found in the hair pencils of the insect, and it eluted from the GC at the same retention time. Identified as (–)-epi-methyl jasmonate, the compound was subsequently shown to be the only stereoisomer of methyl jasmonate with significant odour for humans as well as an essential part of the oriental fruit moth hair pencil pheromone. It is also the only stereoisomer of methyl jasmonate found in the peels of lemons. The striking parallel in selectivity and sensitivity of the olfactory receptors in humans and female oriental fruit moths may indicate a similar olfactory receptor protein in both organisms. While compounds reported in hair pencil glands of male lepidoptera often have recognizable odours for humans, those produced by females are usually not detected by human subjects.

4 MAMMALIAN SEMIOCHEMICALS

In 1982, while walking through a laboratory at Kyoto University, Nishida perceived an odour similar to that of Concord grapes. Investigating, he found that a colleague was dissecting the anal sac of the Japanese weasel, *Itatsi mustella*. The strong scent of the anal sac was due to the presence of large amounts of *o*-aminoacetophenone, a chemical also identified in scent glands of other carnivores.

Concord grapes and some similar smelling cultivars of the Labruscana type are characterized by the presence of methyl anthranilate. It has also been established that most Labruscana and all the similar smelling Muscadine grapes (collectively called fox grapes) have only sub-threshold levels of methyl anthranilate. However, when these grapes were examined for *o*-aminoacetophenone content, it became clear that *o*-aminoacetophenone was responsible for the 'foxy' character of many non-vinifera grapes. Perhaps this explains why the earliest European settlers to North America described their introduction to fox grapes, with a descriptor that accurately identified a stimulant important to the chemical ecology of a canine that they had experienced.

The high degree of similarity observed between the olfactory G-protein coupled receptors in rats and humans may be repeated across most of the biological spectrum from insects to humans. In every case, precepts in the brain, or some equivalent multivariate transformations thereof, are created from patterns of neuronal signals generated by multiple G-protein coupled responses, *i.e.* one stimulant chemical will fire more than one type of receptor and mixtures of stimulant chemicals make up a single semiochemical message. The central idea is that G-protein homology across the biological spectrum implies that the array of odorants acting as semiochemicals is limited and shared by many organisms. If this is true then many of the semiochemicals found in biology should also function as odorants in human foods.

5 GAS CHROMATOGRAPHY–OLFACTOMETRY: ODOUR SPECTRUM

In order to begin a study of the similarity between human food flavorants and semiochemicals in other animal systems, we need a list of important odorants in human foods. But what does important mean? Over the last thirty years flavour chemists have developed a system to rank the 'important' or at least the most potent odorants in a food. Called gas chromatography–olfactometry (GC–O), this technique involves formalized sensory testing of gas chromatographic effluents, and it results in graphs like the one below for green tea. The graphs are lists of dilution values, or titer from threshold, *vs.* gas chromatographic retention indices. Figure 2 shows a typical GC–O chromatogram.[10]

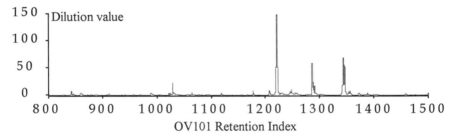

Figure 2   *The GC–O chromatogram prepared by the technique called Charm Analysis of extracts of brewed green tea*[10]

As shown in Figure 3, the usual GC–O data can be further transformed into an odour spectrum in which the fragments are the retention indices of the component odorants, and their response is relative potency normalized to the most potent odorant. In the odour spectra, charm values (or dilution values from AEDA) are modified by Steven's Psychophysical Law using the equation:

$$\Psi = k\Phi^n$$

where Ψ is perceived intensity of a stimulant, k is a constant, Φ stimulus level and n is Steven's exponent.[11,12] Clearly the compressibility of human olfaction is more visible in the spectrum than in the GC–O chromatogram.[10]

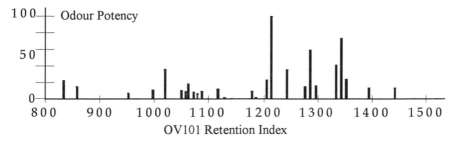

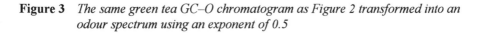

OV101 Retention Index

Figure 3 *The same green tea GC–O chromatogram as Figure 2 transformed into an odour spectrum using an exponent of 0.5*

Table 2 *A List of Odorant Flavours found in Apple, Dill Seed, Musk Melon, Cucumbers and Strawberries*[13-19]

Human Odorant	Descriptor	Other Semiochemistry	Ref.
Acetic acid	Vinegar	Vaginal secretions of cows	20
Butanoic acid	Rancid butter	Scent gland of garter snakes	21
2-Methylpropanoic acid	Sweaty	Scent gland of garter snakes	21
cis-3-Hexen-1-ol	Cut grass	Wasp orientation	22
(E)-2-Hexenal	Apple juice	Male insect pheromone, insect scent gland, wasp kairomone	22, 23, 24
(Z)-3-Hexenal	Grass	Host location cues, male insect pheromone, insect scent gland, wasp kairomone	22, 23, 24, 25
2-Heptanone	Crayon	Moth oviposition	26
1-Octen-3-ol	Mushroom	Insect food cue	27
(Z)-2-Nonenal	Cucumber	Insect host location cue	28
2,4-(E,Z)- decadienoate	Watermelon	Insect pheromone	23
1,3,5-(E,Z)-undecatriene.	Musk melon	Male insect aggregation pheromone	29
1,8-Cineole	Coke	Host location cues, mosquito oviposition repellent, fruit fly food cue	25, 30, 31
(R)(+)-Limonene	Citrus	Ant kairomone, larval secretion, army ant secretions	26, 32, 33, 34
Ethyl hexanoat	Apple	Moth kairomone	31
Dimethyl trisulfide	Cabbage	Parasitoid frass volatiles	25
Linalool	Orange oil	Spider mite kairomone, larvae defence, beaver scent markings	35, 36, 37
2-Phenylethanol	Beer, rose	Onion fly food cue, aggregation pheromone, insect orientation cue	29, 38, 39
Phenylacetaldehyde	Tea	Insect orientation cue, pitcher plant insect attractant	39, 40

Using published GC–O data[13–19] for four natural products apples, musk melon, dill seed, cucumber and strawberry, Table 2 was prepared. A data base of 284 citations and abstracts of papers in chemical ecology selected from Biosis and Chemical Abstract Service were searched for references to any of the 84 odorants found in the apples, musk melon, dill seed and other GC–O data. The compounds listed in Table 2 are some examples of these human food odorants and their function in other chemical ecologies. Only examples in which the authors used bioassays to show that the chemicals did indeed mediate behaviour were included. The broad range of animal systems reported for these semiochemicals is supported by the notion that the olfactory transduction systems of many organisms are very related. Although chemical ecologies are functionally divergent, their chemical vocabulary is not.

There is a practical consequence of this shared semiochemistry. It is entirely logical and possibly productive to study one chemical ecology using another as a probe. The discovery of epi-methyl jasmonate through the GC–O analysis of lemon peels may be more than a serendipitous event. It may be an example of an effective experimental approach to the study of some problems in chemical ecology.

6 HUMAN OLFACTION AS A SEMIOCHEMICAL PROBE

An example of just such an effort is shown in Figure 4, the odour spectrum of an air sample from a bat maternity cave. Millions of female Mexican free-tailed bats *Tadarida brasiliensis* rear their young in densely packed caves in Texas. These bats leave every night to feed and return to find and nurse their own young even at cluster densities as high as 5000 pups m^{-2}. Recent studies suggest a combination of special, acoustic and olfactory cues are used for pup identification.[41–44]

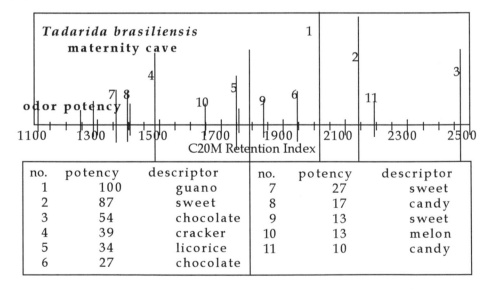

no.	potency	descriptor	no.	potency	descriptor
1	100	guano	7	27	sweet
2	87	sweet	8	17	candy
3	54	chocolate	9	13	sweet
4	39	cracker	10	13	melon
5	34	licorice	11	10	candy
6	27	chocolate			

Figure 4   *The odour spectrum of an air sample from the maternity cave of a bat and a list of the 11 most potent odorants and their associated odour descriptions (Bloss, 1996 unpublished)*

The special identity of the chemicals that act as individual recognition cues and the role of the background odour of the cave remain unknown. In an attempt to address some of these questions, one of us (Bloss) will identify the chemicals detected in odour spectra of the cave air and test their effects on bat behaviour. The underlying assumption is that bats and humans share some of the same G-protein mediated olfactory sensitivities and selectivities. Additionally, the semiochemicals used by bats to function in their environment are detectable by humans. In addition to the potential practical implication of a common semiochemical palette across many biological systems, there is an intriguing intellectual consequence. As humans experience their olfactory environment created by foods and fragrances, they are also experiencing the evolutionary history of biology. Among the precepts humans form from the odours in wines, for example, are the territorial markings of ancestral canines or the aphrodisial tools of an ancient moth. These conclusions are romantic thoughts for those who find comfort in being a part of the arc of life.

ACKNOWLEDGEMENT

This study was funded by grants to Thomas H. Kunz from the National Science Foundation, Texas Parks and Wildlife, and the Texas Nature Conservancy.

REFERENCES

1. J. Stein: Random House, Inc., 1975.
2. D.B. Dunsenbery, 'Sensory Ecology: How Organisms Acquire and Respond to Information', W.H. Freeman and Company, New York, 1992, Vol. 11.
3. Anonymous, Cornell Institute for Research in Chemical Ecology, CIRCE, 1995.
4. D. Lancet., *Ann. Rev. Neurosci.*, 1986, **3**, 668.
5. G.M. Shepherd, *Cell*, 1991, **67**, 845.
6. J. Strotmann, J. Hrieger, A. Beck, I. Wanner, T. Helfrich and H. Breer, in 'Advances in Biosciences', eds. R. Apfelbach, D. Müller-Schwarze, K. Reutter and E. Weiler, Elsevier Science Inc., Oxford, 1994, Vol. 93, 53.
7. J. Krieger, K. Raming, S. Kubick, P. de Geus and H. Breer, in 'Advances in Biosciences', eds. R. Apfelbach, D. Müller-Schwarze, K. Reutter and E. Weiler, Elsevier Science Inc., Oxford, 1994, Vol. 93
8. H. Arn, M. Tóth and E. Priesner, 'List of Sex Pheromones of Lepidoptera and Related Attractants', Eidgenössische Forschungsanstalt für Obst-, Wein- und Gartenbau, Wädenswil (Switzerland), 1992.
9. R. Nishida and T.E. Acree, *J. Agric. Food Chem.*, 1984, **32**, 1001.
10. K.M. King, M.S. Thesis, Cornell University, 1996.
11. S.S. Stevens, *American Scientist*, 1960, **48**, 226.
12. D.D. Roberts and T.E. Acree, *J. Agric. Food Chem.*, 1996, in press.
13. H. Kollmannsberger and R.G. Berger, *Chem. Mikrobiol. Technol. Lebensm.*, 1992, **14**, 81.
14. I. Blank and W. Grosch, *J. Food Sci.*, 1991, **56**, 63.
15. I. Blank, A. Sen and W. Grosch, *Food Chem.*, 1992, **43**, 337.
16. P. Schieberle, S. Ofner and W. Grosch, *J. Food Sci.*, 1990, **55**, 193.
17. P. Schieberle, *Dev. Food Sci.*, 1994, **35**, 345.
18. S.G. Wyllie, D.N. Leach, Y. Wang and R.L. Shewfelt, Abstract, 206th A.C.S. National Meeting, Vol. AGFD 82, Chicago, Illinois, 1993, 1.
19. D. Ulrich, A. Rapp and E. Hoberg, *Z. Lebensm.-Unters. Forsch.*, 1995, **200**, 217.

20. P. Hradecky, *J. Chem. Ecol.*, 1986, **12**, 187.
21. W.F. Wood, J.M. Parker and P.J. Weldon, *J. Chem. Ecol.*, 1995, **21**, 213.
22. D.W. Whitman and F.J. Eller, *J. Chem. Ecol.*, 1992, **18**, 1743.
23. J.R. Aldrich, G.K. Waite, C. Moore, J.A. Payne, W.R. Lusby, J. Kochansky and P., *J. Chem. Ecol.*, 1993, **19**, 2767.
24. P. Nagnan, P. Cassier, M. Andre, J.F. Llosa and D. Guillaumin., *International Journal of Insect Morphology and Embryology*, 1994, **23**, 355.
25. N.G. Agelopoulos and M.A. Keller, *J. Chem. Ecol.*, 1994, **20**, 1955.
26. J.K. Peterson, R.J. Horvat and K.D. Elsey, *J. Chem. Ecol.*, 1994, **20**, 2099.
27. G. Bengtsson, K. Hedlund and S. Rundgren, *J. Chem. Ecol.*, 1991, **17**, 2113.
28. R. Nishida, H. Fukami, T.C. Baker, W.L. Roelofs and T.E. Acree, *Semiochem: Flavors Pheromones, Proc. Am. Chem. Soc. Symp., Meeting Date*, 1983.
29. R.J. Bartelt, D.G. Carlson, R.S. Vetter and T.C. Baker, *J. Chem. Ecol.*, 1993, **19**, 107.
30. J.A. Klocke, M.V. Darlington and M.F. Balandrin, *J. Chem. Ecol.*, 1987, **13**, 2131.
31. D.C. Robacker, A.M.T. Moreno, J.A. Garcia and R.A. Flath., *J. Chem. Ecol.*, 1990, **16**, 2799.
32. D.W. Davidson, J.L. Seidel and W.W. Epstein, *J. Chem. Ecol.*, 1990, **16**, 2993.
33. K. Honda and N. Hayashi, *J. Chem. Ecol.*, 1995, **21**, 859.
34. N.J. Oldham, E.D. Morgan, B. Gobin, E. Schoeters and J. Billen, *J. Chem. Ecol.*, 1994, **20**, 3297.
35. M. Dicke, T.A. Van Beek, M.A. Posthumus, N. Ben Dom, H. Van Bokhoven and A. E. De Groot, *J. Chem. Ecol.*, 1990, **16**, 381.
36. S. Jonsson, G. Bergstrom, B.S. Lanne and U. Stensdotter, *J. Chem. Ecol.*, 1988, **14**, 713.
37. R. Tang, F.X. Webster and S.-D. Mueller, *J. Chem. Ecol.*, 1995, **21**, 1745.
38. S.M. Hausmann and J.R. Miller, *J. Chem. Ecol.*, 1989, **15**, 905.
39. K.F. Haynes, J.Z. Zhao and A. Latif, *J. Chem. Ecol.*, 1991, **17**, 637.
40. K. Jaffe, M.S. Blum, H.M. Fales, R.T. Mason and A. Cabrera, *J. Chem. Ecol.*, 1995, **21**, 379.
41. G.F. McCracken, *Science*, 1984, **223**, 1090.
42. M.K. Gustin and G. McCracken, *Animal Behaviour*, 1987, **35**, 13.
43. W.J. Loughry and G.F. McCracken, *Journal of Mammalogy*, 1991, **72**, 624.
44. G.F. McCracken and M.K. Gustin, *Ethology*, 1991, **85**, 305.

EFFECT OF CHILL FILTRATION ON WHISKY COMPOSITION AND HEADSPACE

J.R. Piggott, M.A. González Viñas, J.M. Conner, S.J. Withers and A. Paterson

Centre for Food Quality, University of Strathclyde, Department of Bioscience and Biotechnology, Glasgow, G1 1XW, U.K.

1 INTRODUCTION

Scotch whiskies, in common with many other distilled beverages, are matured at 50%–70% ethanol by volume, typically 63.4%[1,2] but are bottled at 40%–45%.[3] For heavier-bodied whiskies and those matured at higher strength, this can result in haze formation caused by high molecular-weight lipids and esters, and ethanol-soluble lignins, being less soluble in water than in ethanol. Most whisky is filtered prior to bottling to reduce the risk of haze formation. Haze formation may be slow, and so it may develop in the finished product in-bottle if it is not encouraged to form and then removed. Whisky is cooled to around 0 °C, held at this temperature for a time, and the problem compounds are removed by physical separation and adsorption by a filter.[4] The objective is to remove particulate material in order to present a clear, bright product to the consumer. Whisky chilling, however, requires a great deal of care at the production level. Filtration may remove some valuable flavour compounds and, if removed in excess, can subtly change the character of the whisky.[4] When diluted further to a typical ethanol concentration for consumption, the 'haze' in whisky reappears in the form of agglomerates predominantly of fatty acid esters. The agglomerates modify the solution concentration (and hence the headspace concentration) of many volatile components of the whisky. Thus the odour of a whisky is not a simple function of the total concentration of flavour compounds, but of the concentration of free flavour compounds in the solution, excluding the proportion in the agglomerates.[5-7] This work was carried out to test the hypothesis that the agglomerates could form a reservoir of flavour compounds and could thus permit the removal of potential flavour compounds by filtration without affecting the flavour at consumption.

2 EXPERIMENTAL

2.1 Materials

Samples of ten batches of matured Scotch malt whisky from a single Highland distillery were supplied before reduction (61% ethanol by volume) and after reduction with water to 43% and filtration at +2 °C and –2 °C. Samples were stored in full glass bottles at ambient temperature (approximately 20 °C) before analysis.

2.2 Sensory Analysis

A panel of 20 trained and experienced assessors described the aroma of the samples, using a vocabulary of 24 terms[8] on an intensity scale from 0 to 5. The samples were presented to the panel at 23% v/v ethanol in tulip-shaped nosing glasses similar to standard wine tasting glasses (BS5586:1978) but of approximately 150 ml capacity, covered with watchglasses and assessed in individual sensory booths under red lighting to minimise colour differences. Each sample was assessed independently in duplicate, eight per session, with presentation order being approximately balanced over assessors within sessions. Data were collected using the PSA-System.

2.3 Chemical Analysis

Dichloromethane (2 ml) was added to spirit (20 ml) in duplicate, diluted to 23% ethanol with water, shaken at approximately 20 °C and the dichloromethane separated and analysed immediately by gas chromatography–mass spectrometry (Finnegan-MAT ITS40) using a septum programmable injector held at 230 °C and a CP Wax 52 CB (30 m × 0.25 mm) column (Chrompack), with an initial temperature of 60 °C increasing after 5 min to 240 °C at 6 °C min^{-1}. Comparison of mass spectra and retention times with authentic compounds identified 33 compounds. Peak areas were standardised with 3,4-dimethylphenol as internal standard.

Headspace volatiles were analysed in glass vials (20 ml nominal) fitted with PTFE lined silicone-rubber septa in plastic screw caps. Samples were diluted to 23% v/v ethanol with water at 10 ml, equilibrated in a water bath at 25 °C for at least 30 min, 2.5 ml of headspace gas withdrawn using a 5 ml gas-tight syringe heated to 50 °C, and analysed using gas chromatography–mass spectrometry as above. Only one injection was made per vial, via a Chrompack thermal desorber in purge mode with cryofocussing at −75 °C. The initial oven temperature was 50 °C increasing after 4 min to 210 °C at 10 °C min^{-1}, and 12 compounds were quantified and identified by comparison of mass spectra and retention times with authentic compounds.

Wood-derived materials (aromatic acids, aldehydes and tannins) were measured by HPLC (Kratos 400 solvent pump, 430 low pressure gradient former and 470 autosampler with 20 μl loop) after dilution to 23% ethanol on a Spherisorb S5 ODS2 (250 mm × 4.6 mm) column using 5% formic acid:80% methanol gradient[9] with an LC–UV detector at 300 nm and integration. Peak areas were standardised on 2-naphthol.

2.4 Statistical Analysis

Chemical analytical data and aroma description data were separately analysed by principal components analysis[10] using the Unscrambler. Analysis of variance was carried out with Minitab.

3 RESULTS

3.1 Chemical Analysis

Peak areas from analysis of extracts (corrected to 43% ethanol) showed significant effects of filtration for all compounds determined (p < 0.001) except furan, furfural, phenylethyl alcohol, ethyl lactate, hexanol, benzaldehyde, ethyl nonanoate, isoamyl octanoate, vanillin, norisoprenoid and syringaldehyde. PCA showed that these compounds had lower loadings on the first component (54% variance) and that sample scores (Figure 1) showed differences due to process (p < 0.001). There were significant effects of product batch for dodecanol, tetradecanol, hexadecanol, hexadecanoic acid, γ-nonalactone, a phenylethyl ester and a further unidentified compound (p < 0.01 to p < 0.05), and for scores (p < 0.05) on the first component. Subsequent components showed no significant differences. In all cases the filtration temperature had little effect.

The headspace results did not show significant differences due to filtration for any compound analysed, but there were significant differences between product batches for all compounds (p < 0.001 to p < 0.01) except ethyl tetradecanoate. PCA showed that all these compounds had positive loadings on the first component (77% variance) and that sample scores (Figure 2) showed differences due to product batch (p < 0.001). Sample scores on the second component (11%) similarly showed an effect of product batch (p < 0.001).

Liquid chromatographic analysis for wood-derived materials showed significant differences between product batches (p < 0.001 to p < 0.05) for the seven compounds analysed, and significant effects of filtration (p < 0.05) for gallic acid and a late-eluting unidentified peak. PCA showed that all compounds measured except one unidentified peak had positive loadings on the first component (45% variance), and that sample scores (Figure 3) showed differences due to filtration (p < 0.05) and product batch (p < 0.001). Sample scores on the third component (18%) similarly showed an effect of product batch (p < 0.001).

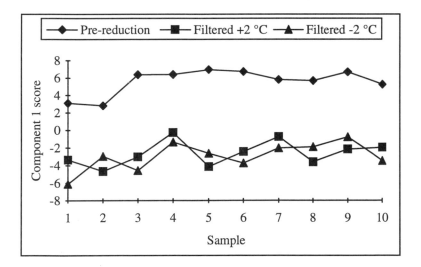

Figure 1 *Sample scores for whiskies plotted on the first principal component from PCA of GC analysis of solvent extracts*

3.2 Sensory Analysis

Analysis of variance did not show significant differences between the pre-reduction and reduced and filtered whiskies, but there were minor differences between batches in scores for phenolic (p < 0.05) and catty (p < 0.01). Similarly, there were no significant effects of batch or process on the sample scores from PCA of sensory data on the first two components, but the product batch had a small effect on the third component (p < 0.05).

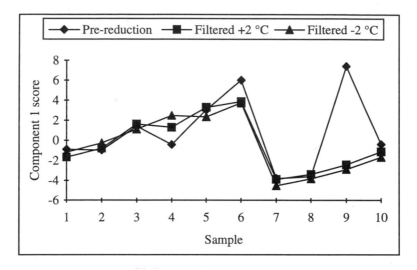

Figure 2   *Sample scores for whiskies plotted on the first principal component from PCA of GC analysis of headspace vapour*

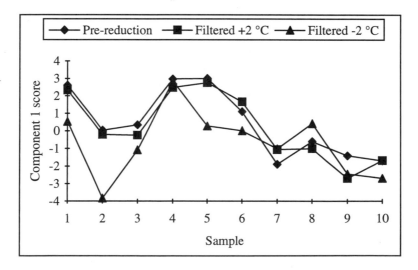

Figure 3   *Sample scores for whiskies plotted on the first principal component from PCA of HPLC analysis of solvent extracts*

4 DISCUSSION

Analysis of dichloromethane extracts provides as nearly as possible an accurate indication of the composition of the spirit.[11] As would be expected, the filtration step caused large differences between pre- and post-reduction samples (Figure 1), but there were no significant differences between the samples filtered at –2 °C and +2 °C. The differences between the product batches were consistent with differences in distillate caused by slight variations in congener production during fermentation.

In contrast with the extracts, the composition of the headspace is determined both by the concentration of volatile compounds and by the extent to which the volatiles are free in solution. Under the conditions of this experiment, agglomerates form predominantly from fatty acid ethyl esters,[5,6] and reduce the effective volatility of many congeners. The data presented here show differences between product batches, consistent with minor variations in the distillate composition, but no differences due to the filtration step (Figure 2). Thus it is apparent that, despite the large change in composition due to filtration, there was still sufficient of the required esters and wood components present to form the agglomerates and exert control over the composition of the headspace.

The liquid chromatographic analysis was predominantly for wood-derived materials, and found differences between product batches. There may be considerable natural variation in the process of whisky production, but the aim of blending is to stabilise the sensory characteristics of the product.[4] Differences in composition which are not reflected in sensory difference are therefore not relevant. Small effects of filtration were found for two compounds, and thus a small amount of these materials does appear to have been removed by filtration.

No significant differences were expected in the sensory data between samples taken before and after filtration, because no significant differences were found in the headspace analysis. The small differences found between product batches in the headspace data were not clearly reflected in the differences in sensory data, and were presumably too small to be reliably detected. As noted above, Scotch whisky is typically blended in batches from groups of casks to a predetermined sensory standard,[4] and thus small differences in headspace composition would be irrelevant. The minor sensory differences found here are consistent with slight variations in the extent of peating of the malt, which would result in small differences in the volatile phenols content of the whiskies. These were not analysed and so this cannot be confirmed.

Chill filtration of the samples analysed affected the composition of volatile compounds in solution but liberation of volatiles to the headspace is controlled by the extent of participation in agglomerates. Chill filtration in this case removed a proportion of the agglomerate-forming compounds but sufficient remained to stabilise the headspace composition and thus the sensory characteristics of the samples. The filtration temperatures used in the present work were not sufficiently different to significantly affect composition or aroma.

ACKNOWLEDGEMENTS

The authors wish to thank the U.K. Biotechnology and Biological Sciences Research Council and the Chivas and Glenlivet Group for financial support and technical assistance.

REFERENCES

1. J.M. Philp, in 'The Science and Technology of Whiskies', eds. J.R. Piggott, R. Sharp and R.E.B. Duncan, Longman, Harlow, U.K., 1989, p. 264.

2. R. Cantagrel, L. Lurton, J.P. Vidal and B. Galy, in 'Fermented Beverage Production', eds. A.G.H. Lea and J.R. Piggott, Blackie, Glasgow, 1995, p. 208.

3. Council Regulation (EEC) No 1576/89, *Official Journal of the European Communities* No L 160/1, 12.6.89.

4. M. Booth, W. Shaw and L. Morhalo, in 'The Science and Technology of Whiskies', eds. J.R. Piggott, R. Sharp and R.E.B. Duncan, Longman, Harlow, U.K., 1989, p. 295.

5. J.M. Conner, A. Paterson and J.R. Piggott, *J. Science Food Agric.*, 1994, **66**, 45.

6. J.M. Conner, A. Paterson and J.R. Piggott, *J. Agric. Food Chem.*, 1994, **42**, 2231.

7. J.R. Piggott, J.M. Conner, A. Paterson and M.S. Perez-Coello, in 'Bioflavour '95', eds. P. Etievant and P. Schreier, INRA Editions, Paris, 1995, p. 39.

8. K.V. Casteele, H. Geiger and C. van Sumere, *J. Chromatography*, 1983, **258**, 111.

9. J.R. Piggott and P.R. Canaway, in 'Flavour 81', ed. P. Schreier, Walter de Gruyter, Berlin, 1981, p. 33.

10. J.R. Piggott and K. Sharman, in 'Statistical Procedures in Food Research', ed. J.R. Piggott, Elsevier Applied Science, London, 1986, p. 181.

11. J.R. Piggott, J.M. Conner, J. Clyne and A. Paterson, *J. Science Food Agric.*, 1992, **59**, 477.

IDENTIFICATION AND SENSORY EVALUATION OF THE CHARACTER-IMPACT COMPOUNDS OF GOAT CHEESE FLAVOUR

J.-L. Le Quéré, C. Septier, D. Demaizières and C. Salles

INRA, Laboratoire de Recherches sur les Arômes, 17, rue Sully, 21034 Dijon, France

1 SUMMARY

The volatile compounds of various goat cheeses have been isolated in order to identify the character-impact odorants by a combination of instrumental analyses and sensory studies. Different extraction procedures have been studied in order to obtain a volatile fraction representative of the cheeses. The most representative extract, as determined by sensory evaluation, has been submitted to GC–MS and to GC–olfactometry, using the aroma extract dilution analysis (AEDA) method. The volatile fatty acids have been found to be the most important compounds for the characteristic goat flavour. Among them, branched-chain fatty acids (e.g., 4-methyloctanoic and 4-ethyloctanoic) have been found to be particularly important. Finally, incorporation of volatile extracts and individual synthetic components into a cheese model allowed the flavour impact-compounds identified to be evaluated sensorially.

2 INTRODUCTION

Cheeses made from goats milk are generally characterized by a strong typical flavour. The intensity of this 'goaty' flavour seems to be greatly dependent upon various factors, e.g., season, lactation period, feeding, milk yield, milk fat content and composition.[1] Recently, it has been argued that the sensory properties of goat cheeses may be influenced by the genetic polymorphism of the caprine α_{s1}-casein.[2,3] However, it has been assumed for a long time that the volatile free fatty acids, including n-chain, branched-chain and other minor fatty acids hydrolyzed from milk fats by lipases,[4] or present in milk as metabolic conjugates,[5] provide characteristic flavours to many dairy foods, especially cheese. It was also postulated than the typical goaty flavour originates from the lipid fraction. Particularly branched-chain fatty acids having 8–10 carbon atoms, some of which are specific to goats and sheep, were considered as important contributors to the flavour of goat cheeses.[6] Fatty acids exhibiting branching at the 4-position were found to have goaty–muttony–sheepy aroma notes,[6] and among them 4-ethyloctanoic acid exhibited an intense goat-like aroma with the lowest reported threshold (1.8 to 6 ppb) for any fatty acid.[6]

However, to our knowledge no clear demonstration of the role of potential key-aroma compounds on the sensory characteristics of goat cheese have appeared in the literature, and the possible relationship between goat genetic type for the α_{s1}-casein and flavour of the finished product has yet to be clarified.

In consequence, the volatile compounds of various goat cheeses, one type with the guarantee of protected origin ('Sainte-Maure') and one traditional local cheese ('Bouton de Culotte' de Saône et Loire), have been isolated in order to identify the character-impact odorants by a combination of instrumental analyses and sensory studies.

Our strategy includes a study of the representativeness of different cheese extracts using sensory analysis, *i.e.*, similarity tests (*versus* control) and descriptor citations, screening for potent odorants via dilution experiments using a gas chromatography–olfactometry method, identification and quantification of the odorants, and sensory evaluation of isolated fractions and of the identified key-flavour compounds in a matrix closely resembling the food product.[7]

3 METHODS AND MATERIALS

3.1 Recovery of the Volatiles

In order to determine the representativeness of the extracts, various extraction procedures have been tested to isolate the volatiles of one cheese sample ('Bouton de Culotte'). The vacuum distillation methods (vacuum hydrodistillation under reflux, primary vacuum distillation and secondary vacuum molecular distillation) have been conducted as previously described[8] with 100 g of cheese. A water extract was also obtained – 100 g of frozen cheese were homogenized with 600 ml of pure water. The mixture was stirred at 40 °C for one hour and then centrifuged at 3000g at 4 °C for 30 min. The pellet was then submitted to an identical extraction. Both supernatants were pooled and centrifuged at 13000g at 4 °C for 30 min and the fat layer was discarded. The supernatant and the aqueous distillates obtained with the distillation methods were extracted with CH_2Cl_2 after the pH was adjusted to 2 with 2 M HCl. The organic extracts were dried over anhydrous Na_2SO_4 and concentrated to 0.2 ml. For quantification, a known amount of an internal standard (pentyl pentanoate) was added prior to the concentration step.

3.2 Sensory Evaluation

3.2.1 Representativeness of the Extracts. The concentrated extracts were diluted in 400 ml of a citrate buffer (pH 5). Selected and trained panellists (19) evaluated the different extracts by smelling 20 ml of the buffer solutions contained in opaque glasses and compared their similarities with the odour of a control sample (5 g of cheese homogenized with 20 ml of water). The results were scored on an unstructured 100 cm scale where the control was arbitrarily scored 0, as previously described.[9]

3.2.2 Flavoured Cheese Model. The most representative extract, besides n-chain fatty acids (C6, C8 and C10) and branched-chain fatty acids (4-methyloctanoic and 4-ethyloctanoic), was incorporated in a flavourless cheese model.[7] The added amounts were calculated from the quantities measured in the cheese (related to the respective dry matter contents). These flavoured samples and the unflavoured cheese model were evaluated by a panel of eight trained assessors, who were asked to score the overall flavour and to describe the samples.

3.3 Separation, Evaluation and Identification of the Volatiles

3.3.1 Gas Chromatography (GC). The GC analyses were performed using a Hewlett Packard HP5890 Series II gas chromatograph fitted with a split–splitless injector, a flame ionization detector and a DB-FFAP fused silica capillary column (30 m × 0.32 mm internal diameter, 0.25 mm film thickness, J&W Scientific). Hydrogen carrier gas was used at a velocity of 37 cm s^{-1} at 143 °C. The injector and detector temperatures were 240 and 250 °C respectively. The oven temperature was raised from 40 °C to 220 °C at 3 °C min^{-1}. Quantification of peaks was expressed in mg kg^{-1} of cheese, relative to the internal standard.

3.3.2 Gas Chromatography–Olfactometry (GC–O). The GC–O analyses were performed according to a published procedure[10,11] using serial dilutions (1:2) of the cheese extracts. The sniffing panel consisted of three people chosen and trained as previously described.[10] Chromatographic conditions were as above, except the oven temperature, which was increased from 50 °C to 130 °C at 6 °C min^{-1} and then to 230 °C at 12 °C min^{-1} in order to allow a sniffing session to last only 20 min.

3.3.3 Gas Chromatography–Mass Spectrometry (GC–MS). The GC–MS analyses were carried out with a Nermag R 10-10 C quadrupole mass spectrometer coupled to a Hewlett Packard HP5890 series II gas chromatograph, fitted with a split–splitless injector and the same column as described above. The carrier gas (He) velocity was 35 cm s^{-1} and the column was directly connected to the ion source of the spectrometer through a heated transfer line maintained at 260 °C. The injection port was maintained at 240 °C, and oven temperature programmed in the same way as described above for GC–O. Electron-impact (EI) mass spectra were recorded at 70 eV on a 0.8 s cycle, the instrument scanning from 25 to 300 amu, with a source temperature of 150 °C.

Table 1 *Similarity Scaling of the Odours of Different Types of Goat Cheese Extracts in Five Independent Sessions[1]*

Extraction Method	Mean Values (19 Panellists)					Goat Descriptor Citation
	Session 1	Session 2	Session 3	Session 4	Session 5	
Water extraction	47(a)	44(a)	50(a)	47(a)	68(a)	+++
100 Pa hydrodistillation	60(a)	59(a)				++
10 Pa vacuum-distillation (A)			73(b)	79(b)		(+)
10 mPa vacuum-distillation (B)			90(b)	86(b)		+
(A) + (B)	85(b)	74(b)			95(b)	−
Hidden control					19(c)	

[1] Odour assessment, unstructured 100 cm scale, same letter (a, b or c) means results not significantly different at the 5% level within a session.

4 RESULTS AND DISCUSSION

Various extraction procedures (Table 1) have been tested in order to isolate a flavour extract as representative as possible of the flavour of a local goat cheese ('Bouton de Culotte' de Saône et Loire). A similarity test[9] was performed, where the panellists were asked to score the similarity of the odour of the extracts to the reference odour of a control cheese on an unstructured 100 cm scale (the control was scored 0). The trained panellists were also asked to quote descriptors, among which the 'goat' descriptor was particularly tracked. The results, summarized in Table 1, indicate that the best result, within each independent session, was obtained with a direct water extraction of the cheese volatiles. The higher score obtained for this water extract in the last session could easily be explained by the presence of a hidden control (scored 19) in the test. Therefore, this extraction procedure (water extraction of the ground cheese, followed by several steps of centrifugation and a final CH_2Cl_2 extraction) was chosen for the following steps of the study.

This most representative extract (total extract, *i.e.*, neutrals and acids) was submitted to GC–MS and to GC–olfactometry, using aroma extract dilution analysis (AEDA).[12,13] The diluted extracts were sequentially evaluated at the sniffing port of the gas chromatograph by a panel of three trained assessors. The most potent odorants were determined using the CHARM procedure,[10,12] described (descriptor citations during the GC–O experiment) and identified unambiguously by GC–MS and retention indices.

Figure 1 shows the CHARM-aromagram obtained by one assessor with a 'Bouton de Culotte' extract and the corresponding FID-chromatogram. Only the odorants detected after three serial dilutions of the extract have been retained for clarity. The most potent odorants (peaks 1 and 10) were detected until dilution 11 (CHARM value: 2048). Peaks 3, 5, 8 and 9 had CHARM values of 1024.

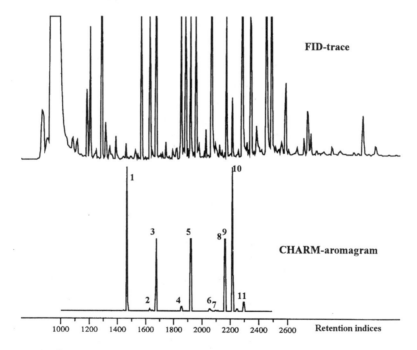

Figure 1 *FID-chromatogram and CHARM-aromagram obtained by one panellist of a representative 'Bouton de Culotte' cheese extract*

The complete results obtained by this assessor with the 'Bouton de Culotte' representative extract are summarized in Table 2 as an example. Among the most potent odorants, 3-(methylthio)propanal, 3-methylbutanoic acid, phenylethanol, δ-octalactone and nonanoic acid have been found to be particularly important. Among these, only nonanoic acid has been associated with a characteristic 'goat' odour. The other important compounds having a characteristic goat odour have been found in the straight-chain and branched-chain fatty acids. Beside the traditional 'goat' fatty acids (hexanoic and decanoic acid), the specific[4] branched-chain fatty acids, 4-methyloctanoic and 4-ethyloctanoic, have been described as typical of the goat odour. Surprisingly, octanoic acid has not been described as goatish by any of the three assessors, but as essentially rancid and pungent.

The amounts of 4-methyloctanoic acid, nonanoic acid and 4-ethyloctanoic acid in the extract were 0.26, 0.81 and 0.03 mg kg^{-1}, respectively, whereas the amounts of hexanoic, octanoic and decanoic acids reached 70, 113 and 28 mg kg^{-1}, respectively. This result confirms the importance of the fatty acids branched at the 4-position for the typical goat flavour, and particularly highlights the low detection threshold of 4-ethyloctanoic acid. However, among the acid fraction, another minor n-chain fatty acid, nonanoic acid, was revealed to be important.

Table 2 *Potent Odorants with their Associated Descriptors Determined by GC-O in a Representative 'Bouton de Culotte' Extract, and Identified by GC–MS[1]*

Peak number (Figure 1)	Retention Indices	Descriptors	Compounds	Potency
	1037	Fruity	Ethyl butanoate	
	1236	Mouldy	Ethyl hexanoate	
	1311	Mushroom	Oct-1-en-3-ol	
1	1463	Potato	3-Methylthiopropanal	+++
2	1629	Cheese	Butanoic acid	+
3	1678	Cheese	3-Methylbutanoic acid	++
4	1854	Goat	Hexanoic acid	+
5	1923	Rose	Phenylethanol	++
	1960	Rancid	Heptanoic acid	
6	2072	Rancid, pungent	Octanoic acid	+
7	2096	Goat	4-Methyloctanoic acid	+
8	2162	Fruity	δ-Octalactone	++
9	2165	Goat	Nonanoic acid	++
10	2224	Goat	4-Ethyloctanoic acid	+++
11	2239	Goat	Decanoic acid	+

[1] GC column: DB-FFAP (J & W Scientific), 30 m × 0.32 mm

Finally, in a last step, incorporation of volatile extracts and individual synthetic components in a flavourless and tasteless model cheese[7] allowed a sensory evaluation of the identified flavour impact compounds to be performed. The overall flavour of the representative volatile extract has been explained completely with the major fatty acids (Figure 2), but the typical 'goat' flavour has been associated more specifically with the branched-chain fatty acids, 4-methyloctanoic and 4-ethyloctanoic acids. For these acids, each of the eight trained panellists participated in the sensory evaluation of the model cheese has cited the 'goat' descriptor as the main attribute of the flavoured model cheese (Figure 2).

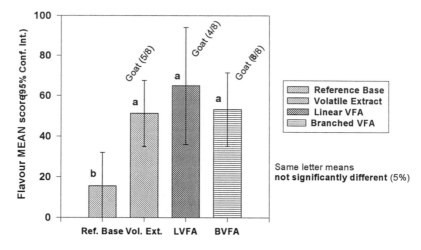

Figure 2   *Overall flavour scores (on a linear scale) of the cheese models: Ref. base –*
model cheese; Vol. Ext. – Ref. base plus volatile extract; LVFA – Ref. base plus
C₆, C₈ and C₁₀ linear fatty acids; BVFA – Ref. base plus 4-methyloctanoic and
4-ethyloctanoic acids. Duplicate sessions, eight trained panellists, 'goat'
descriptor citation by the panel

REFERENCES

1. K. Ronningen, *Acta Agric. Scand.*, 1965, **20**, 301.
2. F. Heil and J.P. Dumont, *Lait*, 1993, **73**, 559.
3. L. Vassal, A. Delacroix-Buchet and J. Bouillon, *Lait*, 1994, **74**, 89.
4. J.K. Ha and R.C. Lindsay, *J. Dairy Sci.*, 1993, **76**, 677.
5. V. Lopez and R.C. Lindsay, *J. Agric. Food Chem.*, 1993, **41**, 446.
6. C.P. Brennand, J.K. Ha and R.C. Lindsay, *J. Sensory Studies*, 1989, **4**, 105.
7. C. Salles, S. Dalmas, C. Septier, S. Issanchou, Y. Noël, P. Etiévant and J.L. Le
    Quéré, *Lait*, 1995, **75**, 309.
8. J.P. Dumont and J. Adda, *Lait*, 1972, **52**, 311.
9. P.X. Etiévant, L. Moio, E. Guichard, D. Langlois, I. Lesschaeve, P. Schlich and E.
    Chambellant, in 'Trends in Flavour Research', eds. H. Maarse and D.G. Van der Heij,
    Elsevier, Amsterdam, 1994, p. 179.
10. N. Abbott, P. Etiévant, S. Issanchou and D. Langlois, *J. Agric. Food Chem.*, 1993,
    41, 1698.
11. L. Moio, E. Chambellant, I. Lesschaeve, S. Issanchou, P. Schlich and P.X. Etiévant,
    Italian J. Food Sci., 1995, **7**, 265.
12. T.E. Acree, J. Barnard and D.G. Cunningham, *Food Chem.*, 1984, **14**, 273.
13. W. Grosch, *Trends Food Sci. Technol.*, 1993, **4**, 68.

QUANTITATIVE ANALYSIS OF NEW POTENT FLAVOUR COMPOUNDS IN BURGUNDY PINOT NOIR WINES

Victoire Aubry,* Christian Giniès,* Robert Henry† and Patrick Etiévant*

*Laboratoire de Recherches sur les Arômes, I.N.R.A., F-21034 Dijon CEDEX, France
†Laboratoire de Chimie Organique, I.N.S.A., F-69621 Villeurbanne CEDEX, France

1 INTRODUCTION

Four esters, ethyl dihydrocinnamate (A), ethyl cinnamate (B), methyl anthranilate (C) and ethyl anthranilate (D), identified recently in a Burgundy wine of *Vitis Vinifera* cv. Pinot Noir, were suspected of contributing to the typical aroma of Pinot Noir wines, according to the results of a gas chromatography–olfactometry analysis.[1] These compounds are, moreover, interesting since, for the first time, ethyl dihydrocinnamate has been identified in the aroma of wine and ethyl cinnamate reported as a key aroma compound of red wine, whereas ethyl anthranilate and methyl anthranilate, previously found in American wines (*Vitis Labrusca*), have now been found in *Vitis Vinifera* wines (Pinot Noir). In order to evaluate the importance of these potent odorants and to confirm their occurrence in other Pinot Noir wines, we report here a simplified extraction method and a gas chromatography–mass spectrometry (GC–MS) stable isotope dilution assay suitable for the reproducible quantification of the four esters at concentrations naturally found in wine.

2 EXPERIMENTAL SECTION

2.1 Wines and Chemicals

Thirty-three Pinot Noir wines were analysed, all of them sold as Burgundy wines. The four target esters were obtained from reliable retailers. The deuterated standards were synthesised.

2.2 Preliminary Experiments: Analysis of Model Mixtures

The stability of the esters was first studied in a buffered ethanolic model mixture (12%). Sealed tubes containing 10 ml of the hydroalcoholic solution were immersed in a bath regulated at 30 °C. At different periods, from 3 hours to 6 days, a 10 ml sample was removed and extracted with 5 ml dichloromethane. The concentrations of the esters were then determined.

In a second step, the efficiency of the extraction of the four esters with two solvents, pentane and dichloromethane, was evaluated in the same model solution. Ten ml of the hydroalcoholic solution were extracted once by 10 ml of solvent. The amounts of the esters extracted by each solvent were determined by GC. The concentration of the organic phase in a Kuderna-Danish apparatus at 55 °C was also tested with pentane as the solvent.

2.3 Preparation of Wine Extracts

The wine sample, previously mixed with a known amount of standards (the four selected deuterated esters), was divided into three 200 ml samples, each of them being extracted with 2 × 20 ml of pentane. The organic phase was dried over Na_2SO_4 and concentrated as above. A further concentration under a nitrogen stream was made down to 100 µl. The final concentration factor was 2000.

3 RESULTS AND DISCUSSION

3.1 Extraction and Concentration of Model Mixtures

As seen from Table 1, no hydrolysis nor degradation of the four unlabelled esters could be observed in the model solution after 140 hours at 30 °C, since the percentage of each ester extracted by dichloromethane was fairly constant.

Table 1   *Estimation of the Extraction Recovery of the Four Esters (A–D) at 30 °C with Time*

Sampling time (hours)	A	B	C	D
0	95%	97%	95%	94%
26	97%	99%	99%	98%
41	96%	99%	99%	98%
140	95%	98%	100%	98%

Table 2 shows that dichloromethane leads to better recovery of the esters than pentane. However, pentane was chosen because it is more selective for apolar substances like esters, with an acceptable yield better than 65% after a single extraction, and because it allows their quantitation without interference. It can be noticed in the same table that the phase equilibrium during the extraction is reached after only two minutes agitation. In practice, a time of 15 minutes was chosen.

The losses by evaporation or degradation during the concentration step of a pentane model solution were estimated to be 5% for A, 15% for B, 12% for C and 10% for D.

Table 2   *Extraction Recovery of the Esters in Different Solvents*

Type of solvent	Time for agitation (min)	A	B	C	D
Dichloromethane	20	95%	85%	90%	95%
	2	93%	85%	92%	95%
Pentane	20	75%	73%	65%	69%
	2	76%	72%	65%	70%

3.3 Optimisation of the SIM–GC-MS

As the amount of these esters in wine was suspected to be low, a stable isotope dilution assay was chosen for the quantification. The measurement is based on the close physical properties of the deuterium-labelled and of the unlabelled molecules (especially the partition coefficients, boiling points and FID responses).[2]

The standard esters were deuterated on their alcohol moiety, as their mass spectra were known to include a strong molecular ion peak, which was chosen for the SIM–GC-MS study. The ions 178 and 181, 176 and 179, 151 and 154, and 165 and 168 were thus selected to quantify A, B, C and D using the standards A-d_3, B-d_3, C-d_3 and D-d_3 respectively.

The SIM–GC-MS detection limit of the esters was also determined from a model organic solution. With the concentration factor involved in the preparation of the wine extract, we could theoretically quantify those esters in wine from 50 ng l^{-1}.

3.4 Determination of the Concentrations of the Esters

All measurements (triplicates) on the 33 wines were made within three weeks, using standard curves renewed regularly, plotting ester–standard intensity ratio *versus* ester–standard amount ratio. As seen from Table 3, the calibration was fairly constant during this period.

Table 3 *Calibration Characteristics*

Ester code	A	B	C	D
Mean slope	0.49	0.90	0.80	0.86
Standard deviation of slope	2%	4%	4%	5%

Table 4 *Significance of the Differences between Wines – Analysis of Variance*

Ester code	A	B	C	D
F ratio[*]	35	53	145	453
Associated Probability, P (H$_0$)	10^{-30}	10^{-35}	10^{-49}	10^{-65}

[*] Critical value: F = 2.12 (P (H$_0$) <0.01).

The results of the quantitation are presented in Figure 1. The four esters were always found in the wine samples, but only at low concentrations. These concentrations vary between 0.8 and 3.2 ppb for A, 0.5 and 1.6 ppb for B, 0.06 and 0.6 ppb for C and 0.6 and 4.8 ppb for D. However, the variation of the quantification estimated from three replicates was always lower than 10%, which is excellent considering the average concentration of those compounds.

When the data were analysed (ANOVA), ethyl anthranilate was found to be the most discriminant compound between wines: this can also be seen from the histograms. It was also found to have the highest olfactory index among the four esters in a previous CHARM analysis of Pinot Noir wine extracts.[1]

Theoretically, the method developed allows the determination of ethyl cinnamate and methyl anthranilate at sub-threshold concentrations since their thresholds in water are 16 ppb[3] and 2 ppb[4] respectively The amounts of these two esters in the wines studied were lower than these values. Their contribution to the flavour of wine, suggested by the chemical analysis on an Echezeau wine,[1] is therefore not obvious since their detection thresholds are most probably higher in wine.[3] Concerning the two other esters, no data could be found concerning their olfactory thresholds. More work has therefore to be done in order to check and to measure their thresholds in this particular type of wine. A quantitative descriptive analysis of the flavour of 30 other Pinot noir wines is currently underway to observe if a correlation exists between the concentrations of these four esters and some of the wine sensory characteristics.

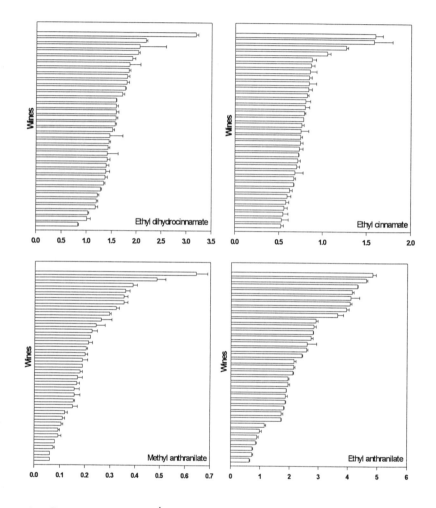

Figure 1   *Concentrations (μg l⁻¹) determined for the four esters in the 33 wines*

REFERENCES

1. L. Moio and P.X. Etiévant, *Am. J. Enol. Viti.*, 1995, **46**, 392.
2. J. Gilbert, 'Applications of Mass Spectrometry in Food Science', Elsevier, London, 1987, Chapter 2, p.89.
3. P.X. Etiévant, S.N. Issanchou and C.L. Bayonove, *J. Sci. Food Agr.*, 1983, **34**, 497.
4. T. Hirvi and E. Honkanen, *Z. Lebensm. Unters. Forsch.*, 1982, **175**, 113.

THE IMPACT OF DEALCOHOLIZATION ON THE FLAVOUR OF WINE – RELATING CONCENTRATION OF AROMA COMPOUNDS TO SENSORY DATA USING PLS ANALYSIS

U. Fischer,[*] R.G. Berger,[†] Å. Håkansson[§] and A.C. Noble[‡]

[*]Staatliche Lehr- und Forschungsanstalt für Landwirtschaft, Weinbau und Gartenbau, Breitenweg 71, 67435 Neustadt, Germany

[†]Institut für Lebensmittelchemie, Universität Hannover, Wunstorfer Str. 14, 30453, Hannover, Germany

[§]VIN & SPRIT AB, R&D, P.O. Box 47 313, 100 74 Stockholm, Sweden

[‡]Department of Viticulture and Enology, University of California Davis, Davis, CA 95616, U.S.A.

1 INTRODUCTION

Although the first patent for dealcoholization of wine was filed in 1908,[1] only in the last two decades has health consciousness and preference for 'light' food products stimulated interest in dealcoholized wines. Although many methods of dealcoholization, including thermal processes, membrane-based technologies and CO_2 extraction, have been utilized, none has yielded a wine with an acceptable flavour in comparison to that achieved in alcohol-free or alcohol-reduced beers. Most research in dealcoholization of wine has focused on the technical feasibility, and not on the effects of the procedures on wine flavour and volatile composition. In the present study, the effect of dealcoholization of wine by vacuum distillation (VD) and reverse osmosis (RO) on wine flavour was investigated by chemical and sensory analysis. Modelling of chemical and sensory data was done to elucidate the relationship between aroma stimuli and sensory perception, and to govern the optimization of dealcoholization technology towards a commercially acceptable product.

2 EXPERIMENTAL

A commercial Sauvignon blanc wine (11.3% ethanol [v/v]) from California, Chardonnay-Semillon wine (11.5% ethanol [v/v]) from Australia, and a Muskat Ottonel wine (12.2% ethanol [v/v]) from Austria were dealcoholized by a vacuum distillation pilot plant (with a capacity of $500 \, l \, h^{-1}$) and by a laboratory reverse osmosis unit at VIN&SPRIT in Stockholm, Sweden. Experimental wines included the base wine (BW), dealcoholized wines of 0.5% ethanol [v/v] generated by both technologies (VD05, RO05), two wines which were compensated for ethanol reduction by adding neutral spirit to the dealcoholized wines to 11% ethanol [v/v] (VD11, RO11) and a reconstituted wine (VD11A) prepared by recombining the alcoholized wine with the vacuum distillate. The aroma and taste of the wines were characterized by descriptive analysis, which was performed in triplicate by 21 trained judges. Volatiles were isolated by liquid–liquid extraction (pentane–dichloromethane 2:1), separated into three fractions on a silica gel column utilizing a pentane–diethylether gradient and quantified following separation by gas chromatography as reported elsewhere.[2] Volatiles were identified by gas chromatography–mass spectrometry (GC–MS), and the sensory significance of the volatiles was determined by two judges sniffing the GC effluent (GC–olfactometry) of aroma extracts, which were diluted in steps of 10 (AEDA).[3] Partition coefficients were measured by static headspace GC[4] at 20 °C in a model wine (10 g l^{-1}

glycerol, 1 g l^{-1} potassium, pH 3.2) at 0.5% and 12% ethanol [v/v], respectively. Sensory and instrumental data were analysed by analysis of variance and by principal component analysis using SAS 6.03 for the PC,[5] while instrumental and sensory data sets were modelled by partial least square regression analysis (PLS) using Unscrambler II.[6]

3 RESULTS AND DISCUSSION

3.1 Sensory Impact of Dealcoholization

Highly significant differences in intensity for each of the ten aroma and five taste attributes were observed. The effect of ethanol removal on wine aroma and taste is illustrated for Sauvignon blanc in Figure 1 which shows the factor scores and attribute loadings for the first two principal components (PC).

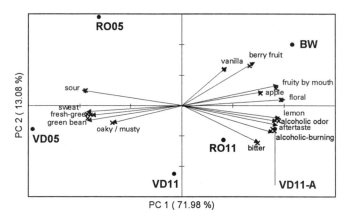

Figure 1 *Principal component analysis of sensory attributes (vectors) of Sauvignon blanc wine dealcoholized by vacuum distillation (VD) and reverse osmosis (RO)*

The first PC, which accounted for 72% of the variation, separated the wines on the basis of ethanol content. With the exception of VD11, wines with 11% ethanol (BW and the reconstituted wines) were higher in intensity in the fruity, spicy, floral attributes and ethanol-linked properties, while the two wines with 0.5% ethanol (RO05 and VD05) were lower in these terms and higher in sour taste and vegetative, musty and sweaty attributes. The second PC separates the BW, which was lower in the trigeminal alcoholic-burning attribute, from the reconstituted wines. Comparing both technologies, the wines made by reverse osmosis (RO05, RO11) show higher loadings on PC1 and PC2, reflecting higher intensities of fruity, spicy and floral attributes than their vacuum distilled counterparts (VD05, VD11).

3.2 Mass Balance of Aroma Compounds

The mass balance of aroma compounds revealed an overall reduction of 75% for vacuum distillation, which was mainly due to the transfer of esters and alcohols into the distillate. More polar alcohols like 2-phenylethanol, lactones, phenols and short chain fatty acids however, were not removed. In contrast, during reverse osmosis, permeation through and adsorption to the membrane occurred evenly across all the different compound classes and accounted for 40% loss of total aroma compounds. In addition to the transfer of aroma compounds into the distillate (VD) and permeate (RO), upon the removal of ethanol, the rate of hydrolysis of ethyl esters increased, accounting for substantial reductions in ethyl ester concentrations.

3.3 Relating Instrumental to Sensory Data

To model the sensory data by PLS, all GC peaks, which were detected by GC–O, were used initially. In a second step, the instrumental data set was reduced from the starting set of 107 GC peaks to 27 by retaining only those which had flavour dilution (FD) values of at least 100.[7] While the first model explained 60% of the chemical information and 81% of the sensory properties,[8] the latter yielded a model explaining 68% of chemical composition and 79% of sensory information.

Thus, the prediction of the sensory data was not diminished by including only those GC peaks which could be detected by GC–O after 100-fold dilution of the extract. The fit of the aroma composition was even improved due to the reduced variance in the instrumental data set. In a third model, the 27 peaks with FD ≥ 100 were expressed as the concentration in the headspace using the partition coefficients determined in model wine solutions of 0.5% and 12% vol. ethanol by headspace GC.[4] This model explained 70% of variation in the GC data and 78% of sensory space, and more effectively separated the different dealcoholization procedures in the PLS model.

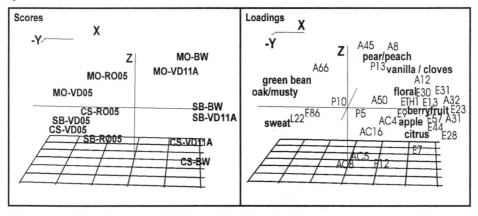

Figure 2 *Three-dimensional PLS model relating headspace concentration of odour active aroma compounds (FD ≥ 100) (X matrix) to sensory odour attributes (Y matrix). Left: Scores plot of experimental wines (CS: Chardonnay-Semillon blend, MO: Muscat Ottenal; SB: Sauvignon blanc). Right: loadings plot of instrumental and sensory variables*

In Figure 2, factor scores for each wine are plotted for the first three dimensions of the PLS model in the left-hand graph, while loadings for the sensory data and instrumental variables are shown in the right-hand plot. The PLS analysis across all three varieties showed a similar structure to that obtained by the PCA of sensory analysis of Sauvignon blanc (Figure 1) and of chemical composition (shown elsewhere[2]). The plot of variable loadings reveals that the fruity components are associated with esters (E) and some odour active alcohols (A) suggesting that the reduction in fruitiness in dealcoholized wines may be attributed to the removal of esters. Additionally, no compounds, with the exception of the 5-decanolide (L22) retained due to its high boiling point, were associated with the vegetative, sweaty and musty flavours. This suggests that the increase in intensity of these three negative terms in dealcoholized wines is a function of the removal of components such as esters, which initially masked vegetative, sweaty and musty flavours, rather than of production of new compounds during dealcoholization eliciting these aromas.

PLS permits further tentative interpretation of the relationship of the GC peaks to specific terms. For example, eugenol (P13) plots close to the vanilla–cloves attribute, and is highly associated with it. Similarly, 2-phenylethanol (A66) is associated with the pear–peach attribute. Even though two non-aromatic varieties were included in the trial, α-terpineol (A50) is very closely associated with the floral term, while linalool (A45) is less strongly related. The fruity attributes, berry fruit, apple and citrus are surrounded mostly by esters and the two 3-hexenol isomers (A31, A32) which exhibit a fresh green smell.

In the score plot of Figure 2, the two dealcoholized wines (VD05, RO05) are close together, which is also true for base wine (BW) and reconstituted wine (VD11A). Additionally, on the third dimension, the four Muscat Ottonel wines were separated from the other two varieties. Presumably, this is a function of stronger intensities of pear–peach, vanilla–clove and floral terms, and higher concentrations of the terpenes (A45, A50) in the Muskat Ottonel.

4 CONCLUSIONS

Both dealcoholization processes significantly reduced all fruity attributes. At the same time, unpleasant attributes were enhanced. The modification in taste and mouthfeel was primarily related to the removal of ethanol. The changes in aroma produced during vacuum distillation were due to loss of highly volatile esters and alcohols, while lactones, short chain fatty acids and volatile phenols remained in the dealcoholized wine. Reverse osmosis showed an equal reduction of all aroma compounds, regardless of polarity, volatility and molecular weight.

According to PLS analysis, the increase of unpleasant vegetative, sweaty and musty odours during dealcoholization was not caused by generation of new aroma compounds nor by increased levels of volatiles. Selection of 27 GC peaks with flavour dilution values greater than 100 explained the same variance as with use of 107 peaks. Using individual partition coefficients to calculate the headspace concentrations of these 27 GC variables improved the modelling of the sensory data and yielded a better separation of the alcoholization technologies.

ACKNOWLEDGEMENT

The authors thank VIN & SPRIT AB in Stockholm, Sweden for funding this research and technical assistance during dealcoholization.

REFERENCES

1. C. Jung, Swiss patent 44 090, 1908.
2. U. Fischer, R.G. Berger, Å. Håkansson, *J. Agric. Food Chem.*, 1996, submitted.
3. F. Ullrich and W. Grosch, *Z. Lebensm. Unters. Forsch.*, 1987, **184**, 277.
4. C. Fischer, U. Fischer and L. Jakob, *Am. J. Vitic. Enol.*, 1996, submitted.
5. SAS/STAT, version 6.03 for PC, SAS Institute Inc. Cary, NC 27513, USA, 1988.
6. Unscrambler II, version 5.0, CAMO A/S, Trondheim, Norway, 1993.
7. U. Fischer and R.G. Berger, *J. Agric. Food Chem.*, 1996, submitted.
8. U. Fischer, Ph.D. thesis, University of Hannover, 1995.

THE INFLUENCE OF ODOURANT BALANCE ON THE PERCEPTION OF CHEDDAR CHEESE FLAVOUR

C.M. Delahunty, F.Crowe and P.A. Morrissey

Department of Nutrition, University College, Cork, Ireland

1 ABSTRACT

Complex foods contain many distinct odourous compounds and a proportion of these will be released during consumption. Volatiles released together will mix, and the greater the number in a mixture the greater may be the chance of an overlapping of olfactory response patterns resulting in a 'blended' flavour perception. It was proposed that the easily identifiable character of a poor, or distinct, product may be induced by a compound or group of compounds dominating the volatile mixture. A more harmonious product may have a mix of volatiles nearly equal in odour intensity. To study this proposal the flavours of eight cheddar cheeses were characterised by a sensory panel of 18 assessors using 'Free-Choice Profiling'. The cheeses were also analysed using the buccal headspace of three cheese consumers coupled with GC–MS. The volatility profile of one cheese, 'cheesy' and 'clean' in character, was compared with those of the seven other cheeses.

2 INTRODUCTION

A proposed flavour component balance theory,[1] still widely accepted, suggests that perceived cheddar flavour is induced by the balance of volatile compounds released from a cheese during consumption. However, 44 years later, identification of the important compounds is still proving difficult. Reasons for this are outlined in a review paper.[2] Buccal headspace analysis (BHA)[3,4] was used to measure volatile compounds released from cheddar cheese during consumption. Using this method, developed to measure the volatiles responsible for flavour perception in their correct concentration ratios, cheese is consumed in a near normal manner and released volatiles are displaced through the nose and trapped. However, more than 30 measurable compounds are present in extracts, and it is likely that more cannot yet be measured. To determine which are important remains difficult. It may be better to eliminate compounds that contribute little to differences in flavour. In this regard, humans can only identify up to three or four odourants in a mixture.[5] If more are present the overall perception is 'balanced'. For example, humans identify the complex odour of chocolate as a single entity.[6] However, time during consumption should be considered, as the overall flavour perception will change constantly as the released volatile mixture changes.[2] This paper considers the 'time-averaged' sensory and instrumental data by studying the balance of some volatiles released from a selection of characterised cheddar cheeses.

3 EXPERIMENTAL

Eight cheddar cheeses were selected for analysis. Cheeses 1–5 were produced on the same day by the same manufacturer and were analysed at six months' maturity. Cheese 6 was also from the same plant, but was produced three months earlier. Cheese 7 was a mild cheese, Cheese 8 was a 12-month cheese and each was produced by a different manufacturer. All analyses were carried out in duplicate. For sensory evaluation, cheeses were assessed by a panel of 18 using 'Free-Choice Profiling'[7] and the data were analysed using analysis of variance (ANOVA) and generalised procrustes analysis (GPA).[8] BHA of three cheese consumers was used for instrumental evaluation and data were analysed by ANOVA and principal components analysis (PCA).[9] To illustrate quantitative differences in volatile profiles between cheeses, log transformed compound peak areas were compared with those of a 'balanced' cheese chosen by sensory evaluation (Table 1). Sensory–instrumental relationships were explored by partial least squares regression (PLS),[10] using the instrumental data for prediction.

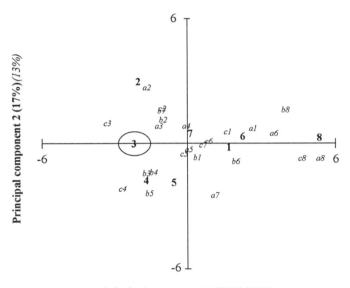

Figure 1   *PCA scores on PC1 and PC2 for eight cheeses (1–8) assessed in duplicate by buccal headspace analysis using three subjects (a–c). (See text for explanation.) The pooled SD for the analysis is represented by the ellipse on cheese 3*

4 RESULTS

Dimensions 1 and 2 of the GPA of free-choice data (which accounted for 42% and 14% of the variance respectively; not illustrated) separated cheeses by age (dim1, $p = 0.002$) and cheeses of similar age (dim2, $p = 0.000$). Mouldy and rancid character descriptions (Cheeses 6 and 8) were at the opposite end of a pole to clean (Cheeses 4 and 5), processed (Cheeses 6 and 7) was opposite to cheesy (Cheese 5) and creamy and milky (Cheeses 4 and 7) were opposite to pungent and intense (Cheese 8). Cheese 5, cheesy and clean in character, was

Table 1 Volatile Compound Concentration Differences between Cheese 5 and the other 7 Cheeses, represented as log transformed peak areas. Cheese 5 was characterised as 'cheesy' and 'balanced' by a sensory panel

Cheese	Compound (referred to by name or number)											
	1	2	3	DMDS	5	6	7	8	CDS	10	EB	TP
1	0.88*	-0.12	-0.51	-0.34	-0.40	-0.53	-0.36	0.60	0.45	-0.33	0.13	-0.02
2	1.98†	0.04	-0.01	0.34	-0.09	-0.10	-0.02	2.52†	0.79*	0.03	-1.06†	-2.17†
3	0.19	-0.48	0.21	0.62	0.16	0.01	0.20	1.78†	0.86*	0.22	-0.41	-0.15
4	-0.80*	0.11	-0.03	0.01	0.10	0.09	0.11	0.42	0.13	0.29	-0.32	-0.60
5	0.00	0.00	0.00	0.00	0.00	0.00	0.00	0.00	0.00	0.00	0.00	0.00
6	1.45†	-1.09†	-0.36	-0.13	-0.54	-0.52	-0.37	0.29	-0.15	-0.28	-0.08	-0.66
7	1.30†	-1.24†	-0.11	-1.14†	-0.11	-0.12	-0.11	0.58	0.09	-0.09	-1.14†	-0.72*
8	1.46†	-1.59†	-0.76*	0.18	-0.80*	-0.69	-0.71*	1.17†	-0.58	-1.29†	0.80*	0.66

* ± > 5× compound concentration in Cheese 5;

† ± > 10× compound concentration in Cheese 5.

selected as the most 'balanced'. Thirty-one compounds were isolated by BHA and 12 of these, which showed greatest variation between cheeses, were selected for further data analysis. The scores and loadings of these 12 compounds from the first two components of a PCA on log transformed data are shown in Figure 1 and Figure 2 respectively. Figure 1 depicts two PCAs. The first (in italics) was calculated using individual headspace data (duplicates averaged) and the second (bold type) using the overall cheese average. ANOVA showed significant differences between cheeses for both components (p = 0.000 and p = 0.006 for PC1 and PC2 respectively) but not between subjects (p = 0.298 and p = 0.523 for PC1 and PC2 respectively). Table 1 shows the quantitative differences between Cheese 5 and the other seven cheeses. PLS1 successfully predicted sensory dimension 1 (r = 0.98) but not dimension 2 (r = 0.67).

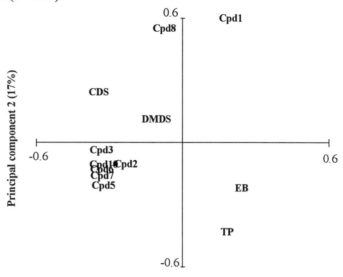

Figure 2 *PCA loadings for Figure 1 (bold type; overall average). DMDS = dimethyl disulfide; CDS = carbon disulfide; EB = ethyl butyrate; TP = thio-propene. The other compounds have not been identified*

5 DISCUSSION

Although there are many factors contributing to volatile release in the mouth[11] which may be influenced by physiological differences, it was found that individual profiles obtained by BHA were similar for each cheese (Figure 1). Figure 2 and Table 1 show which compounds differ most between one cheese and another. Cheese 8, a 12-month cheese, was most different from Cheese 5. Compounds 1, 2, 8 and 10 differ by a factor of 10 or greater. Compounds 3, 5, 7 and ethyl butyrate differ by 5 times or greater. Cheese 7, a mild cheese, was also considerably different, but the direction of change was opposite for dimethyl disulfide, ethyl butyrate and thio-propene. PLS successfully predicted the first sensory dimension from the volatile data and this dimension separated cheeses by age. Qualitative differences within cheeses of similar age were more difficult to predict. Cheese 3 was most rancid and mouldy of the 6-month cheeses. Increases in the concentration of Compound 8 and carbon disulfide may have been responsible. However, these two compounds also increased in Cheese 2 but it was not described as mouldy and rancid. In this case their impact may be lessened by the increase in concentration of Compound 1 and the decrease in that of ethyl butyrate and thio-propene. In conclusion, although some compounds show considerable variation between cheeses, others do not. It is most likely that those which do not change contribute little to changes in flavour character, although they may be important for a 'typical' cheddar flavour and for changes in flavour intensity. Also, the time taken to consume a piece of cheese means that 'time-averaged' product differences are insufficient for a complete understanding and that time-intensity studies are required.

REFERENCES

1. H. Mulder, *Neth. Milk Dairy J.*, 1952, **6**, 157.
2. C.M. Delahunty and J.R. Piggott, *Int. J. Fd. Sci. Tech.*, 1995, **30**, 555.
3. C.M. Delahunty, J.R. Piggott, J.M. Conner and A. Paterson, in 'Trends in Flavour Research', eds. H. Maarse and D.G. van der Heij, Elsevier, Amsterdam, 1994, p. 47.
4. C.M. Delahunty, J.R. Piggott, J.M. Conner and A. Paterson, *Ital. J. Fd. Sci.*, in press.
5. D.G. Laing, *Fd. Qual. Pref.*, 1994, **5**, 75.
6. D.G. Laing and B.A. Livermore, in 'Chemical Signals in Vertebrates VI' eds. R.L. Doty and D. Muller, Schwartz Plenum, New York, 1992, p. 587.
7. A.A. Williams and S.P. Langron, *J. Sci. Fd. Agr.*, 1984, **35**, p. 558.
8. D.C. Oreskovich, B.P. Klein and J.W. Sutherland, in 'Sensory Science Theory and Applications in Foods' eds. H. Lawless and B.P. Klein, Marcel Dekker, New York, 1991, p. 353.
9. J.R. Piggott and K. Sharman, in 'Statistical Procedures in Food Research' ed. J.R. Piggott, Elsevier Applied Science, London, 1986, p. 181.
10. M. Martens and H. Martens, in 'Statistical Procedures in Food Research. ed. J.R. Piggott, Elsevier Applied Science, London, 1986, p. 293.
11. P.Overbosch, W.G.M. Afterof and P.G.M. Haring, *Fd. Rev. Int.*, 1991, 7, 137.

SENSORY AND CHEMICAL CHANGES OF SALMON DURING FROZEN STORAGE

H.H.F. Refsgaard, P.M. Brockhoff[*] and B. Jensen

Danish Institute for Fisheries Research, Department of Seafood Research, Technical University of Denmark, Building 221, DK-2800 Lyngby, Denmark

[*]Department of Mathematics and Physics, Royal Veterinary and Agricultural University, Thorvaldsensvej 40, DK-1871 Frederiksberg C, Denmark

1 INTRODUCTION

During frozen storage, the sensory quality of salmon deteriorates due to hydrolysis and oxidation processes. Lipids in salmon are characterized by high levels of polyunsaturated fatty acids which are very susceptible to deterioration by oxidation. Lipid oxidation results in rancidity due to formation of a variety of different compounds, including a large number of volatile compounds, mainly carbonyls, hydrocarbons and alcohols. Some of these volatile compounds have very intense flavours and affect the sensory quality even in low concentration. In the present investigation, sensory and chemical changes during 15 weeks of frozen storage of salmon fillets were determined.

2 MATERIALS AND METHODS

2.1 Salmon

Salmo salar of mass 4–4.5 kg was purchased from Sekkingstad A/S (Skogsvåg, Norway). The salmon were stored as fillets at −10 °C or −20 °C protected by aluminium foil and polyethylene bags. Salmon used as reference standards were stored whole at −30 °C after having been frozen, glazed and wrapped in aluminium foil and polyethylene bags. After storage, the fillets were thawed in water at 0 °C for 16 hours and the third cutlet from each fillet was used for headspace sampling of volatiles while the others cutlets were assessed sensorially.

2.2 Descriptive Sensory Analysis

Nine odour attributes were selected for raw salmon and ten for cooked salmon. Salmon samples of 60 g were placed in porcelain bowls covered with aluminium foil, and heated for 16 min at 100 °C in an oven. For each sensory attribute, the scoresheet carried unstructured scales of 15 cm anchored 1 cm from the ends with terms which limited the attributes. The panel consisted of ten trained persons (three male and seven female) aged between 20 and 60 who were selected by use of screening tests.[1] At each assessment, a salmon stored at −30 °C was presented to the panellists and the intensities of the sensory attributes of the declared reference were marked on the scoresheet. The panellists evaluated four randomized cooked samples, including a hidden reference, and four raw samples, also including a hidden reference, after a declared raw reference sample. The panellists were not informed that the hidden reference had been included nor that the experiment was a storage experiment.

2.3 Headspace Sampling

The cooked and raw samples were cut into pieces and frozen in liquid nitrogen and powdered. To salmon powder (20 g) and a standard mixture (0.07 g) of heptane, decane, tridecane and hexadecane in oil were added 25 ml of water. The samples were purged with 340 ml min^{-1} nitrogen at 45 °C for 30 min and volatiles were collected on Tenax GR traps (225 mg in $^1/_4$" steel tubes; Perkin Elmer, Buckinghamshire, U.K.). Water was blown off with 50 ml min^{-1} nitrogen at ambient temperature for 10 min. All collections were run in triplicate.

2.3.1 Headspace Sampling of Standards. For quantification purposes, seventeen C_6–C_{10} aldehydes and the C_6–C_{10} alkanes dissolved in oil in four sets of concentrations were added to samples of fresh salmon with low concentration of volatiles. The standards were collected as described above.

2.4 GC–MS

Volatiles were desorbed from the Tenax traps by use of an automatic thermal desorber system ATD 400 from Perkin Elmer. A Hewlett Packard (Palo Alto, U.S.A.) 5890 IIA GC equipped with an HP 5972A mass selective detector was used for identification and quantification. A DB 1701 column (30 m × 0.25 mm × 1 µm; J&W Scientific, Folsom, U.S.A.) with a flow of 1.3 ml min^{-1} and the following temperature programme was used: 25 °C for 1 min, 25 °C to 175 °C at 4 °C min^{-1} and 175 °C to 240 °C at 20 °C min^{-1}. The mass selective detector used ionization at 70 eV and 50 µA emission and scans were performed in the range 30–350 amu at a rate of 2.2 scan s^{-1}.

2.4.1 Quantification. Results from the collections described in Section 2.3.1 were used to prepare calibration curves for each of the seventeen aldehydes by use of HP ChemStation software.

2.5 Statistical Analysis

For the ten odour attributes for cooked salmon and the nine for raw salmon, analysis of variance was performed to test for the significance of any kind of treatment or time effects. For these analyses, 240 observations (= 10 assessors × 4 times × 3 treatments × 2 salmon) were used except for missing values. The effects of time and temperature and their interaction were tested *versus* the assessor-interaction effect in accordance with a mixed model analysis of variance with effects due to assessors considered random. For the seventeen aldehydes, we also performed analysis of variance on the raw and the cooked salmon data separately. For this analysis 24 observations (= 4 times × 3 treatments × 2 salmon) were used, since the average was taken from the replicate measurements on the same salmon.

Correlations between the sensory data and aldehyde concentrations were investigated by partial least squares (PLS) regression and the Unscrambler software package (version 3.54) (CAMO A/S, Trondheim, Norway) was used for the analysis with full systematic cross-validation as the validation method.

3 RESULTS

3.1 Sensory Changes During Frozen Storage

Only minor sensory changes were detected after 15 weeks' storage of the salmon as fillets at −10 °C and −20 °C. Salmon stored for two weeks had a sweet and earthy odour and in addition a boiled potato odour was characteristic for the cooked samples and a cucumber odour for the raw ones. The intensities of green and fish oil odour increased significantly during storage independent of storage temperature. The intensity of earthy odour of raw salmon was higher for salmon stored at −30 °C than for salmon stored at higher temperatures. The train oil odour in raw samples was higher for salmon stored at −10 °C than for salmon stored at lower temperatures. Salmon stored at −30 °C had an odour less sour than that of salmon stored at higher temperatures.

3.2 Chemical Changes During Frozen Storage

The concentration of most of the seventeen aldehydes increased significantly during storage as seen in Table 1. For pentanal, *trans*-2-pentenal, hexanal, *trans, trans*-2,4-heptadienal, *trans*-2-octenal, *trans*-2-nonenal and decanal, significant time and temperature interaction effects were found (Table 1). The concentration of pentanal, *trans*-2-pentenal and hexanal increased faster and to a higher level in salmon stored at −10 °C than in salmon stored at the lower temperatures. *Cis*-3-hexenal is proposed as an indicator substance for the rancid or the fatty, fishy off-flavour of boiled trout.[2] This compound could not be detected in salmon stored at −10 °C for 15 weeks. Interestingly, the level of another oxidation product, *trans, trans*-2,4-heptadienal, was higher in salmon stored at −30 °C than in salmon stored at higher temperatures.

Table 1   *Concentration of Aldehydes ($\mu g\ kg^{-1}$ salmon) in Raw Salmon Stored at −10 °C and Results from the Analysis of Variance of the Aldehydes*

Aldehyde	RI	Storage Time (weeks)				Results from ANOVA		
		2	7	11	15	Time	Temp.	Time x Temp.
Pentanal	779	2.5	3.0	5.2	12	*r, ***c	*c	*c
t-2-Pentenal	862	13	2.3	2.6	11	***	***r	***r
Hexanal	884	14	10	18	54	***	***	***r
t-2-Hexenal	964	3.3	7.2	5.5	10	*r, ***c		
Heptanal	988	7.7	51	49	22	*c	*r	
c-4-Heptenal	992	2.0	7.2	6.4	8.9	***r	***r	
t,t-2,4,-Hexadienal	1043	3.7	2.3	3.6	5.8	**		
t-2-Heptenal	1073	0.93	3.9	3.4	1.8	***c		
Octanal	1093	5.0	9.2	6.3	14	***	**r	
t,t-2,4-Heptadienal	1148	2.2	1.8	2.8	6.5	***	***r, **c	***r, **c
t-2-Octenal	1179	2.0	2.4	3.2	4.5	***c, **r	*r	*r
Nonanal	1197	18	27	27	68	*r, **c		
t,c-2,6-Nonadienal	1285	0.00	0.0	0.0	0.0			
t-2-Nonenal	1285	2.5	9.6	7.3	8.9	**r		**r
Decanal	1302	7.6	29	20	15	***r, *c	**r	***r
t,t-2,4-Nonadienal	1363	0.15	0.0	0.55	0.0	*c		
t,t-2,4-Decadienal	1475	13	1.4	7.1	8.5	*c		

*, **,***: Significant effects at $p < 0.05$, $p < 0.01$, $p < 0.001$, respectively;
c: Cooked samples;
r: Raw samples.

3.3 Correlation between Sensory and Chemical Changes

Some relationships caused by the difference between raw and cooked salmon were observed. The intensity of sweet and fish oil odour was established to be higher in cooked reference samples than in the raw reference samples and *vice versa* for earthy and cucumber odour. This was corroborated by the chemical data, where pentanal, *trans*-2-pentenal, hexanal and *trans*, *trans*-2,4-heptadienal clearly had higher levels in cooked samples.

Due to the minor sensory changes, little of the variation could be explained and predicted by the chemical data. Of the significant treatment effects mentioned in Section 3.1, the sourish odour of boiled salmon had a correlation between observed and predicted values with two PLS-factors of 0.670.

ACKNOWLEDGEMENT

The authors are grateful to Kim Mathiasen for technical assistance and the panellists for participation in the sensory assessments. The support of the Danish Research and Development Programme for Food Technology through the LMC Centre for Advanced Food Studies is acknowledged.

REFERENCES

1. ISO 3972, 'Sensory Analysis – Methodology – Method of Investigating Sensitivity of Taste', International Standardization Organization, Genève, 1991.
2. C. Milo and W. Grosch, *J. Agric. Food Chem.*, 1993, **41**, 2076.

FLAVOUR CONTRIBUTION OF ETHANOL, A NEGLECTED AROMA COMPOUND

M. Rothe and R. Schrödter

Deutsches Institut für Ernährungsforschung Potsdam-Rehbrücke, Germany

1 INTRODUCTION

Ethanol has a special role within flavour perception because in yeast-fermented food products its concentration exceeds that of other aroma compounds by several orders of magnitude. It also contributes to flavour through trigeminal effects. There is, however, no clear knowledge of the relative amounts and proportions of the olfactory and trigeminal stimulation.

Another important characteristic of ethanol in the flavour field is its use as a positive or negative quality indicator. Elevated ethanol contents in beverages of fruit origin indicate fruit storage under anaerobic conditions, which are often accompanied by microbial formation of off-flavour compounds.[1] In fresh wheaten bread, high amounts of ethanol indicate both a prolonged flavour-improving fermentation time and a proposed trigeminal effect.[2] In this case, ethanol may well contribute to the preference for and popularity of freshly baked bread.

Special sensory testing procedures and a trained sensory panel have been used in order to elucidate some of these questions. These included:

1 Sensory tests using different ethanol concentrations in aqueous solution to determine flavour intensity with the samples introduced from squeeze bottles;
2 Determination of alcoholic irritation intensity in-mouth when the nose was closed with a clip.

Sensory profiling methods and a ranking procedure were also applied.

2 RESULTS

The results of the concentration–sensory response relationship in aqueous solution indicate a strong validity of Steven's Law for the olfactory and the trigeminal perception. Threshold values calculated from the functions in Figure 1 were found at a concentration of 3.5×10^{-6} for the olfactory and 5.3×10^{-5} for the trigeminal perception. Above the threshold, the higher slope of the trigeminal function indicates an increasing contribution of this part to the overall flavour perception. The effect of the trigeminal irritation finally overcomes that of the olfactory irritation near an ethanol concentration of 5%. Thus, the trigeminal effect dominates in alcoholic beverages which normally have an alcohol content near or above this level.

In grape juice, increasing amounts of ethanol can influence the intensity of the perceived gustatory attributes. Sensory profiling studies demonstrate that especially in the case of sweet taste, a small reduction of the perceived intensity parallel to an increasing ethanol content is registered. The contribution of the alcoholic note was found high in systems with 5% alcohol or more.

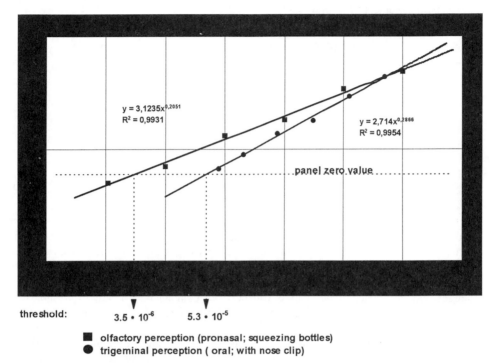

threshold: $3.5 \cdot 10^{-6}$ $5.3 \cdot 10^{-5}$

■ olfactory perception (pronasal; squeezing bottles)
● trigeminal perception (oral; with nose clip)

Figure 1   *Relationship between ethanol concentration and perceived irritation intensity*

Ethanol is a suitable indicator for intensity and acceptability of wheaten bread flavour. In dough systems with normal and shortened fermentation time, ethanol content was found to be positively correlated with the flavour quality of the bread crumb. This fact corresponds with the old observation that the alkanol amount found in baking oven gases correlates positively with the flavour quality of the bread.[3]

REFERENCES

1. D.G. Richardson and M. Kosittrakun, in 'Fruit Flavors – Biogenesis, Characterization and Authentication', eds. R.L. Rouseff and M.M. Leahy, A.C.S. Symposium Series 596, American Chemical Society, Washington, D C., 1995, p. 220.
2. M. Rothe, 'Aroma von Brot', Akademie-Verlag, Berlin, 1974, p. 101.
3. R. Coffman, D.F. Meisner and D.E. Terry, *Cereal Sci. Today*, 1964, **9**, 305.

ODOUR ANALYSIS OF STORED ORANGE JUICE BY GC–ODOUR PROFILING

D. Tønder, L. Poll and M.A. Petersen

Department of Dairy and Food Science, The Royal Veterinary and Agricultural University, Rolighedsvej 30, 1958 Frederiksberg C, Denmark

1 INTRODUCTION

In Denmark, reconstituted orange juice, aseptically packed in a Tetra Brik, is given a shelf life of eight months at room temperature. During storage at room temperature for *twelve* months, sensory evaluations showed that the juice changes by becoming progressively less orange in odour and taste, more oxidized in odour and taste and more bitter in taste. The aroma of juice stored at low temperature (5 °C or frozen) did not change (unpublished data).

To determine which aroma compounds are important contributors to the flavour changes, gas chromatographic analysis can be used combined with odour thresholds or different gas chromatographic–olfactometric methods, such as dilution analysis[1,2,4] and Osme.[3]

The purpose of the work reported here was to use GC–odour profiling with a sniffing panel – a technique that has been used in few studies.[4,5] The objective was to find the most important odours in the orange juice and to find odour differences between samples stored for one year at –18 °C and 20 °C.

2 EXPERIMENTAL

Commercially reconstituted orange juice aseptically packed in a Tetra Brik was used. The juice samples were stored for one year at –18 °C ('fresh' juice) and 20°C ('stored' juice) and then analysed.

The volatile compounds were isolated using direct extraction with diethyl ether–*n*-pentane (2:1). After stirring for 15 minutes, the sample was held at –18 °C until the water phase was frozen. The two phases were separated and the diethyl ether–*n*-pentane phase was concentrated by gently blowing N_2 over the surface. This extract was injected into the GC.

Five trained assessors were asked to judge the odour intensity at the GC's sniffing port (on an intensity scale from 1 to 5) and to describe the odours. No odour descriptions were given in advance. The odours were, when possible, identified using GC–MS and/or retention time of reference compounds.

Table 1 Intensity and description of the odours detected by three or more assessors during sniffing of the 'fresh' and the 'stored' orange juice. The odours are presented according to decreasing intensity. For odours with equal intensity the odours detected by most assessors are presented first

Ret. time	Description of Odours	Identification	Intensity for Assessors 1–5 and Average Intensity						Sign. Diff.[a]
			1	2	3	4	5	Average	
'Fresh' juice (F)									
43.8	Close, wood, plastic, sausage, smoked, hot meal	A sesquiterpene	3.0	3.0	0.0	0.0	5.0	2.2	F>S
23.4	Fruit, citrus, orange flower	1-Octanal	2.0	0.0	3.0	4.5	1.0	2.1	F>S
31.6	Nauseous, hot meal, acetic acid, boiled potato	Acetic acid	2.0	3.5	0.0	1.0	4.0	2.1	
8.0	Orange, fruit, flower, sweet	Ethyl butanoate	1.0	0.0	1.0	2.0	4.0	1.6	F>S
12.2	Tree-like, sweet drops, marigold-like, body odour	β-Pinene	2.0	1.0	0.0	2.0	2.0	1.4	F>S
36.7	Sweet, flower, orange	1-Octanol	0.0	2.0	2.0	3.0	0.0	1.4	
5.8	Fruit, sweet drops, sweet, caramel	2-Pentanone	2.0	0.0	0.0	2.0	2.0	1.2	
18.4	Orange, liquorice	Limonene	0.0	1.0	1.5	2.0	1.0	1.1	
29.7	Fruit, sourish, orange, flower	Limonene	2.0	0.0	0.0	1.0	1.0	0.8	
7.2	Resin, nutty, dry, flower	α-Pinene	1.0	1.0	0.0	0.5	0.0	0.5	
'Stored' juice (S)									
31.6	Close, boiled vegetable/potato, sourish	Acetic acid	3.0	2.5	2.0	0.5	5.0	2.6	
18.4	Flower, sweet, liquorice	Limonene	3.0	0.0	2.0	2.0	1.0	1.6	
34.8	Unpleasant, flower, rubber	Carvone	0.0	0.0	2.5	1.0	4.0	1.5	F<S
44.1	Unpleasant, smoked, hot meal	Butanoic acid	0.0	0.0	2.5	1.0	3.0	1.3	
39.6	Stuffy, close, burnt		3.0	0.0	2.0	0.0	1.0	1.2	
31.0	Nutty, burnt fruit, off		1.0	0.0	2.0	0.0	2.0	1.0	
36.7	Flower, boiled fruit,	1-Octanol	0.0	3.0	1.0	1.0	0.0	1.0	F<S
46.8	Fruit, sour, peach, lemon		2.0	1.0	0.0	1.0	0.0	0.8	
8.0	Fruit, orange, lemon	Ethyl butanoate	1.0	0.0	0.0	1.0	1.0	0.6	F>S

[a] Significant difference (p < 0.05) between the intensity of the 'fresh' and the 'stored' juice. F>S: Highest intensity for the 'fresh' juice. F<S: Highest for the 'stored'.

3 RESULTS AND DISCUSSION

During GC–sniffing the assessors detected 44 odours in the sample of the 'fresh' juice and 43 odours in the 'stored' juice. In total, 68 different odours were detected in the two juices. However, the degree of agreement between assessors varied greatly (Figure 1). Only once did all five assessors agree with an odour in the stored juice identified as acetic acid. The odours detected by four assessors were, for the 'fresh' juice, ethyl butanoate, β-pinene, limonene, octanal and acetic acid. For the 'stored' juice, limonene was found by four assessors. All assessors found approximately the same number of odours.

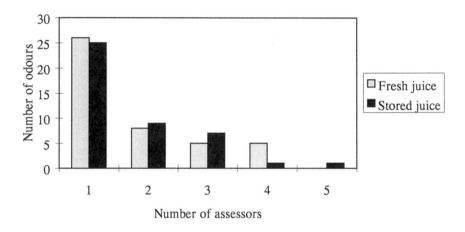

Figure 1   *Number of odours detected by respectively 1, 2, 3, 4 and 5 assessors during sniffing of 'fresh' and 'stored' orange juice*

Table 1 presents the odours detected by three or more assessors. The odours are presented in order of decreasing average intensity. It is seen that there is a great difference between the odour impressions of the 'fresh' and the 'stored' juice, which was also the case with sensory evaluations (unpublished data). The odour intensity of a sesquiterpene (retention time 43.8 min), 1-octanal, ethyl butanoate and β-pinene were significantly higher for the 'fresh' juice, while the intensity of the unidentified compounds at the retention times 34.8 min and 31.0 min were higher for the 'stored' juice.

The most odour-active compounds in the 'fresh' juice are, according to the sniffing, a sesquiterpene (43.8 min), 1-octanal, acetic acid, ethyl butanoate, β-pinene, 1-octanol, 2-pentanone, limonene, α-pinene and one unidentified compound. In the 'stored' juice the most odour active compounds are acetic acid, limonene, carvone, butanoic acid, 1-octanol, ethyl butanoate and three unidentified compounds.

Among the 15 most odour-active compounds, found in orange juice by dilution analysis, Marin et al.[6] identified citral, linalool, vanillin, ethyl 2-methylbutyrate, ethyl butyrate and limonene. For most of the odours this is different from what is found in the present study. Marin et al. found that limonene only had a trace odour activity. Here it was found that limonene plays a more important role especially in the 'stored' juice. In the 'fresh' juice, however, the activity of ethyl butanoate is similar to that found by Marin et al.

The reason for the differences between the results could be that the juice used by Marin *et al.* was not made from concentrate as were the ones presented here.

For some of the odours (*e.g.* 1-octanal, acetic acid), there are great differences between the intensities detected by each assessor. The variation between assessors is greater than the variation between samples and between individual odours. For other odours the variation is small (*e.g.* 2-pentanone, α-pinene). The variation could most likely be minimized by using repetition of the sniffings, more assessors and further training of the assessors.

4 CONCLUSIONS

It is possible to use a sniffing panel to identify important odour-active compounds in orange juice, and to find differences between 'fresh' and 'stored' orange juice. The results can also be handled statistically.

The most odour-active compounds in the 'fresh' juice were, according to the sniffing, a sesquiterpene, 1-octanal, acetic acid, ethyl butanoate, β-pinene, 1-octanol, 2-pentanone, limenene , α-pinene and one unidentified compound. In the 'stored' juice the most odour-active compounds are acetic acid, limonene, carvone, butanoic acid, 1-octanol, ethyl butanoate and three unidentified compounds.

The odour activity of a sesquiterpene, 1-octanal, ethyl butanoate and β-pinene was found to be significantly highest in the 'fresh' juice, while the odour intensity of two unidentified compounds were highest for the 'stored' juice.

ACKNOWLEDGEMENT

This work was done in collaboration with the Association of Danish Juice and Fruit Drink Manufacturers.

REFERENCES

1. T.E. Acree, J. Barnard and D.G. Cunningham, *Food Chem.*, 1984, **14**, 273.
2. P. Schieberle and W. Grosch, *Z. Lebensm. Unters. Forsch.*, 1984, **178**, 479.
3. M.R. McDaniel, R. Miranda-Lopez, B.T. Watson, N.J. Micheals and L.M. Libbey, in 'Flavors and Off-Flavors', ed. G. Charalambous, Elsevier Science Publishers B.V., Amsterdam, 1990, p. 23.
4. T.S. Chamblee, Jr. and B.C. Clark, in 'Bioactive volatile compounds from plants', eds. R. Teranishi, R.G. Buttery and H. Sugisawa, American Chemical Society, Washington, D.C., 1993, p. 88.
5. F. Cormier, Y. Raymond, C.P. Champagne and A. Morin, *J. Agric. Food Chem.*, 1991, **39**, 159.
6. A.B. Marin, T.E. Acree, J.H. Hotchkiss and S. Nagy, *J. Agric. Food Chem.*, 1992, **40**, 650.

PEATY CHARACTERISTIC OF SCOTCH MALT WHISKY

S.J. Withers, J.R. Piggott, J.M. Conner and A. Paterson

Centre for Food Quality, University of Strathclyde, 204 George Street, Glasgow, G1 1XW, U.K.

1 INTRODUCTION

The introduction of peat smoke during the malting of barley, known as 'peating', is an old and well-established process, which contributes to the characteristic peaty flavour of many Scotch malt whiskies. This peaty characteristic, described as *phenolic* in this study, is very distinctive and easily identified by both trained and untrained panellists. The level of peat smoke introduced to the malted barley can be varied according to the level of 'peatiness' desired by the distiller. Generally, the greater the concentration of peat smoke adsorbed by the barley, the greater the *phenolic* character of the finished product: a lightly peated malt contains 1–5 ppm total phenols; a medium peated malt contains 5–15 ppm total phenols; and a heavily peated malt contains 15–50 ppm total phenols.[1] The degree of peating is usually measured in terms of the concentration of phenolic compounds in the peated malt.[2] Previous studies[3,4] have detected a range of phenolic compounds including guiacol, phenol, cresols and eugenol in heavily peated malt whiskies and in the peated malt itself. It is these phenolic compounds which are thought to be associated with the peaty characteristic of heavily peated malt whiskies.[3] In this study, sensory and chemical analyses were used to determine the attributes of a range of Scotch malt whiskies, with the aim of producing a chemical and sensory profile of heavily peated malt whiskies.

Table 1 *Matured and New Distillates Samples with Abbreviations*

Lagavulin (Laga)	Bunnahabhain (Bunna)	Auchroisk (Auch)
Ardbeg (Ard)	Benrinnes (Benr)	Auchentoshan (Auchen)
Caol Ila (Caol)	Braeval (Braev)	Glenlossie (Glenlo)
Laphroaig (Laphr)	Glenfarclas (Glenfa)	Aultmore (Ault)
Bowmore (Bow)	Allt A'Bhainne (Allt)	Glengoyne (Glengo)
Highland Park (Highl)	Macallan (Maca)	Glen Elgin (Glenel)

2 EXPERIMENTAL

2.1 Materials

Eighteen cask samples of mature Scotch malt whisky and eighteen samples of new distillate (Table 1) were chosen to give a mixture of heavily peated, light to medium peated malt whiskies and one unpeated malt whisky – Glengoyne. All the matured samples were fifteen years old, except for the eighteen-year-old Ardbeg.

2.2 Methods

2.2.1 Sensory Analysis. A panel of 22 trained assessors described the aroma of the samples, using a vocabulary of 24 terms[5,6] on an intensity scale from 0 to 5. All samples were nosed at 23% v/v alcohol concentration in tulip shaped nosing glasses similar to standard wine-tasting glasses (BS5586:1978) but of approximately 150 ml capacity, covered with watch glasses and assessed under red light to minimise colour differences. Samples were assessed in duplicate, with sample order approximately balanced within sessions. Data were collected using the PSA-System.

2.2.2 Chemical Analysis. Solid phase microextraction (SPME)[7] was used to determine the phenolic compounds in the two sets of whisky samples. Analysis was performed on a Finnegan-MAT ITS-40 GC–MS with septum programmable injector (230°C). The column was a 30 m × 0.25 mm Carbowax BP20 (df = 0.25 μm) with helium at 1.8 ml min^{-1} as carrier gas.

2.2.3 Statistical Analysis. Chemical analytical data and descriptive sensory data were analysed using partial least squares regression analysis (PLS).[8] Partial least squares was used to model relationships between the compositional and sensory data, for both the new and matured samples using the Unscrambler.

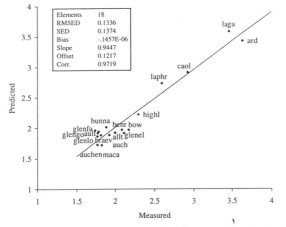

Figure 1 *Plot of predicted* versus *measured values of phenolic by PLS using the first principal component for mature Scotch malt whisky samples*

3 RESULTS AND DISCUSSION

For both sets of data, good predictions were achieved for *phenolic* and *woody* (Figures 1 and 2) on the first component, which accounted for 44% of the variance for the mature samples and 60% of the variance for the new distillates. The loadings plot for both the matured and the new distillate displayed a cluster of compounds cresol, 4-ethylguaiacol, 2-ethylphenol, 4-ethylphenol, ethylmethylphenol, dimethylphenol, isopropylphenol, dimethylphenol, ethyldimethylphenol and trimethylphenol, with sensory descriptors *phenolic, pungent* and *woody* (Figure 3). The scores plot for the mature and new distillate whiskies (Figure 4) displayed Lagavulin, Ardbeg, Laphroaig and Caol Ila with high positive values, which were associated with the chemical and sensory variables with high positive loadings. The PLS plots of the mature and new distillate samples were similar in their separation of heavily peated from lightly peated, therefore only the mature plots are shown in Figures 1–4.

It has been known for many years that heavily peated malt whiskies contain a range of phenolic compounds not detected in light to medium peated malt whiskies.[3] In this study, Lagavulin, Laphroaig, Caol Ila and Ardbeg displayed the sensory and chemical characteristics often present in heavily peated malt whiskies, for both the mature and new distillate samples. The PLS plots of the new distillates and mature samples did not display any obvious chemical or sensory differences. However, this experiment was carried out in terms of phenolic compounds; future experiments may involve pyridines and pyrazines, which are also thought to be important in the distinctive flavour of heavily peated malt whisky.[9,10]

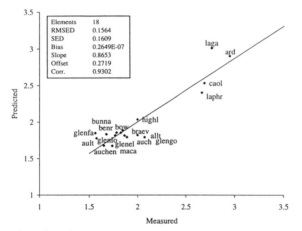

Figure 2   *Plot of predicted versus measured values of woody by PLS using the first principal component for mature Scotch malt whisky samples*

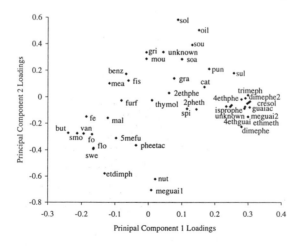

Figure 3   *PLS loadings plot of chemical and sensory variables for mature Scotch whisky samples on the first and second principal components*

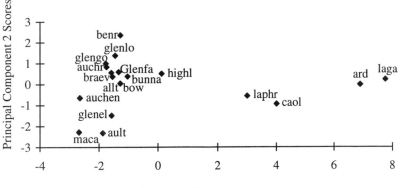

Figure 4 *PLS scores plot of mature Scotch whisky samples on the first and second principal components*

ACKNOWLEDGEMENTS

The U.K. Biotechnology and Biological Sciences Research Council, and the Chivas and Glenlivet Group provided financial support and technical assistance.

REFERENCES

1. G.N. Bathgate and R. Cook, in 'The Science and Technology of Whiskies', eds. J.R. Piggott, R. Sharp and R.E.B. Duncan, Wiley and Sons, New York, 1989.
2. G.N. Bathgate and A.G. Taylor, *J. Inst. Brew.*, 1977, **83**, 163.
3. J.S. Swan and D. Howie, in 'Current Developments in Malting, Brewing and Distilling', eds. F.G. Priest and I. Campbell, Institute of Brewing, London, 1983.
4. M. Lehtonen, *Chromatographia*, 1982, **16**, 201.
5. J.R. Piggott, in 'Sensory Science Theory and Applications in Food', eds. H.T. Lawless and B.P. Klein, Marcel Dekker, New York, 1991.
6. J.R. Piggott and P.R. Canaway, in 'Flavour '81', ed. P. Schreier, Walter de Gruyter, Berlin, 1981.
7. R.E. Shirley, *Reporter*, 1995, **14**, 6.
8. M. Martens and H. Martens, in 'Statistical Procedures in Food Research', J.R. Piggott, Elsevier Applied Science, London, 1983.
9. C.M. Delahunty, J.M. Conner, J.R. Piggott and A. Paterson, *J. Inst. Brew.*, 1993, **99**, 479.
10. J.R. Piggott, A. Paterson, J.M. Conner and G. Haack, in 'Recent Developments in Food Science and Nutrition: Proceedings of the 7th International Flavour Conference', ed. G. Charalambous, Elsevier, Amsterdam, 1993.

FLAVOUR BINDING AND FLAVOUR RELEASE

TIME COURSE PROFILING OF VOLATILE RELEASE FROM FOODS DURING THE EATING PROCESS

R.S.T. Linforth, K.E. Ingham and A.J. Taylor

Department of Applied Biochemistry and Food Science, University of Nottingham, Sutton Bonington Campus, Loughborough, Leicestershire, LE12 5RD, U.K.

1 INTRODUCTION

Many of the compounds that contribute to the aroma of foods have been identified and quantified in a wide range of foods. However, far less is known about the release of these aroma compounds from the food matrix during the eating process and their subsequent passage to the olfactory epithelium. It is likely that both the concentration and the rate of change of concentration of aroma compounds present at the olfactory epithelium will influence the perception of aroma during eating.[1] Consequently, we need to develop methods to determine the effects of the food matrix and human physiological factors on aroma release. There are three main approaches which can be used to investigate aroma release from foods: theoretical, model systems and direct measurement.

The theoretical approach results in mathematical models which can be used to predict the effects (and the relative significance) of specific parameters on aroma release. This may enable us to evaluate the relative significance of the diffusion of compounds through the food matrix, or the partitioning of aroma compounds between the food matrix and saliva, on aroma release.

Model systems allow us to breakdown whole systems into their component parts. Using these simpler systems it is possible to test experimentally the effect on aroma release of modification(s) to the food matrix. A typical application would be to study the effects of aroma binding and release to different biopolymers or lipids.

Direct measurement of the concentration of volatiles in the mouth ('mouthspace') or, breath expired from the nose ('nosespace') enables us to observe the influence of human physiology on aroma release from different food matrices. Consequently, if the chewing action or salivary flow rates change in response to textural or flavour differences, then direct measurement of aroma release would allow us to monitor corresponding changes in the volatile release profile.

Each of the three approaches has its advantages and disadvantages, but, by combining the advantages of all three approaches there is the potential to substantially advance our understanding of the processes affecting aroma release that take place during eating. Our main focus of research in this area has been to improve the methodology available for the direct measurement of volatiles present in breath. Initially we sampled nosespace onto Tenax traps over several minutes whilst eating different foods.[2,3] Similar techniques were used by Delahunty and co-workers in the analysis of cheese.[4] However, such sampling techniques do not yield any temporal information on the changes in intensity and composition of the aroma profile as food is eaten. As a result we have investigated other methods of volatile analysis, which can be used to study the dynamics of volatile release.

2 ANALYSIS OF THE DYNAMICS OF VOLATILE RELEASE

2.1 General Considerations

Methods for the analysis of volatile compounds present on the breath will involve the collection of gaseous samples from either the nose or the mouth. Samples from the mouth will directly follow changes in volatile release over the entire eating process. In contrast, samples from the nose will only contain aroma compounds when the subject exhales. These aroma compounds will be present as a result of chewing, which forces some of the gas phase present in the mouth into the respiratory tract and will be influenced by the partitioning of the volatiles between the gas phase and the surfaces of the respiratory tract.

Eating each mouthful of food is a relatively quick process. Therefore the method of volatile detection should be capable of responding quickly to changes in volatile concentration in order to follow these processes. Sensitivity is another major consideration, because the nose is extremely sensitive to many compounds (*e.g.* the odour threshold of (*Z*)-3-hexenal is 0.66 ppb,[5] by volume). Furthermore if whole breath is to be introduced into a detection system, it should be capable of monitoring several different compounds simultaneously. The most common detectors offering the required selectivity, sensitivity and response rates are mass spectrometers. However, the sensitivity of electron impact sources of mass spectrometers can be significantly reduced when moisture laden breath is introduced.[3]

Some workers have overcome this problem by measuring the amount of volatile compounds still present in the bolus over the eating time-course,[6] or the decline in the bolus mass,[7] in order to study the eating process. These results may not be directly comparable with the concentration of aroma compounds in the breath, due to changes in the saliva flow rate during eating. This will have varied the extent to which the aroma compounds released were diluted during the eating process and their subsequent concentration on the breath.

2.2 Membrane Separators

One of the methods that has allowed the introduction of breath into electron impact sources utilises a membrane to prevent water and oxygen entering the source whilst volatiles pass through the membrane where they can be ionised and detected. Using this method it was possible to determine the aroma release profile of 2-pentanone and 2-butanone from oil–water systems.[8,9] These profiles could then be compared with sensory time–intensity profiles of the perceived aroma.

However, the method has some disadvantages. Membranes exhibit selectivity which may exclude some compounds with a low affinity for the membrane. Other compounds may have long response times due to slow rates of diffusion through the membrane. In addition, the use of an electron impact ionisation mass spectrometer will lead to fragmentation of the compounds of interest, producing complex spectra if several compounds are introduced simultaneously (this is usually the case with real foods) which may make it difficult to interpret the data.

2.3 Tenax Trapping

An alternative method is to collect samples of breath onto Tenax traps which effectively separates the volatile compounds of interest from air and, to a great extent, water. The volatiles can subsequently be desorbed and chromatographed, thereby allowing the identification and quantification of the compounds present. Using this method, large samples of breath can be collected by sampling several times over from the same time window of the eating time course, which will improve sensitivity. The compounds are then chromatographically separated which allows breath containing complex mixtures of aroma compounds to be analysed.

This method does however have several disadvantages. To produce each time point of a time course, a chromatogram must be run. Consequently a large number of chromatograms will be needed to produce an aroma release profile, especially if data points are to be replicated and the differences between several foods investigated. In addition, chromatographic variation and changes in the eating pattern (*e.g.* saliva flow rate and breathing patterns) will increase the variability of the data which will decrease the capacity to observe minor differences in volatile release.

Legger and Roozen[10] used this methodology and were able to collect a series of breath samples from the mouths of people eating chocolate. They demonstrated that the release profile of 2-methyl butanal (which they found to be the chromatographic component with the aroma most characteristic of chocolate) was very similar to the time intensity profile of chocolate flavour perceived by a sensory panel. The number of data points of the volatile release profile was however limited (samples collected every 30 s) which probably obscured any minor changes in 2-methyl butanal concentration.

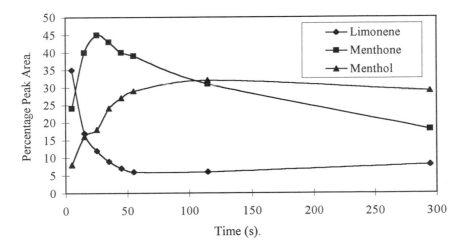

Figure 1 *Average volatile release curves for menthol, menthone and limonene released from mint flavoured sweets during eating*

Our preliminary time course experiments involved sampling nosespace from people eating mint-flavoured sweets[11] and demonstrated that menthone, menthol and limonene were present in different proportions over time (Figure 1). We found that, although the absolute amounts of each compound varied considerably between individuals (no attempt was made to standardise the eating technique), the relative proportions of each compound in the breath were very similar. The compounds that appeared the most persistent in the breath after the mints were swallowed, were the more polar compounds.

Attempts were made to improve our ability to monitor changes in the absolute amounts of volatiles present in breath during the eating process. For this work,[12] we chose a simple food system. Strawberries were used as the test food because they had a high water content, were low in fat and contained a range of compounds from the same chemical class (the esters). Samples of breath from five different individuals eating strawberries (to a specified eating pattern) were collected onto one Tenax trap. This minimised differences between individual fruit and the breathing patterns of people, whilst minimising sensitivity problems associated with short sampling times. The trends observed showed that the amounts of esters in the breath increased rapidly in the early stages of mastication, reached a maximum at the point of swallowing (10 s) and then fell rapidly after the bolus was swallowed (Figure 2). Using Tenax trapping to trap volatiles from several subjects over short time windows of the eating process, it was possible to produce profiles of changes in the concentration of volatiles in the breath with a reasonable number of data points, even if the technique was time consuming.

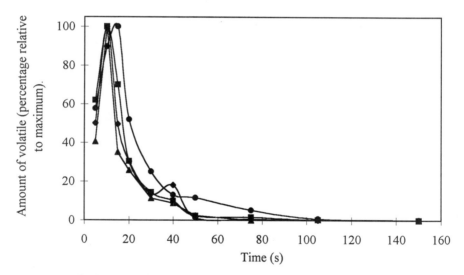

Figure 2   *Volatile profile of methyl acetate (●● methyl butanoate (◆), methyl 3-methylbutanoate (■) and methyl hexanoate (▲) released from strawberries during eating*

The improved Tenax trapping method was subsequently applied to the study of the effects of hydration on volatile release from rye crispbreads[13] and biscuits with different fat contents.[14] This showed that hydration was a major factor in the release of volatile compounds from these foods, which was consistent with the findings of Dalla Rosa and co-workers,[15] who studied the headspace above model and real food systems at different water activities. However, the volatile release studies using Tenax trapping and GC–MS were restricted by the time taken to produce the flavour release profile of a given food sample. Consequently a parallel research programme was initiated which pursued the objective of measuring aroma release on a real-time, breath-by-breath basis.

2.4 Breath-by-Breath Analysis of Volatile Release

Electron impact mass spectrometers are not ideally suited to the direct introduction of breath samples. However, other ionisation techniques such as atmospheric pressure chemical ionisation (APCI) where compounds can be ionised in air containing water vapour at atmospheric pressure (and the ions are subsequently sampled into the high vacuum region of the analyser), allow easier interfacing. In addition, APCI exhibits high sensitivity and rapid response times[16] which are essential for the detection of aroma compounds, breath-by-breath.

APCI sources are typically designed for the analysis of samples in aqueous solvents. In order to analyse breath samples, it was necessary to modify the APCI source of the mass spectrometer (Platform II, Micromass, Manchester, U.K.) to allow the introduction of gas phase samples. APCI is a soft ionisation technique, which adds a proton (in positive ionisation mode) to the compound of interest and does not normally induce fragmentation. Consequently compounds present in the breath were monitored in selected ion mode (MH^+ ion), further enhancing sensitivity. The peaks corresponding to compounds present in individual breaths were integrated to produce peak height data for each compound monitored over the duration of the aroma release experiment.

In our preliminary experiments, we used mint-flavoured sweets as the test food system. We were able to detect peaks corresponding to volatiles present in individual breaths and confirmed the greater persistence of menthol relative to menthone (Figure 3) observed in earlier experiments (Figure 1). These differences can clearly be seen in the combined volatile release curves from four replicates (Figure 4). The volatile release profiles were found to be reasonably consistent when mints were eaten by one individual (Table 1), a fact that had previously been difficult to demonstrate.

Table 1   *Mean Parameters (± SD) for the Release of Menthol and Menthone from Mint-flavoured Sweets. Each result is the mean of four replicates*

Compound	Parameter			
	$T_{50}{}^a$	$T_{max}{}^a$	$I_{max}{}^b \times 10^{-4}$	$T_{50'}{}^a$
Menthol	0.35 ± 0.04	0.72 ± 0.11	1603 ± 214	1.11 ± 0.05
Menthone	0.28 ± 0.03	0.59 ± 0.04	7710 ± 855	0.94 ± 0.06

[a]   Time (min) to maximum intensity (T_{max}), time to 50% of maximum intensity (T_{50}) and time to decrease to 50% of maximum intensity ($T_{50'}$);

[b]   The maximum intensity (I_{max}) in peak height integrator units.

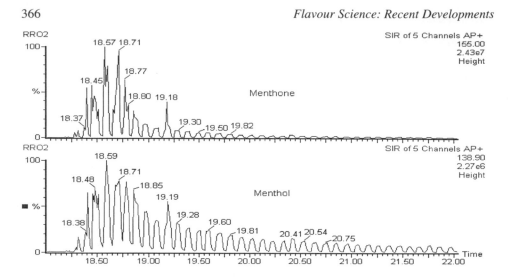

Figure 3   *Volatile release profiles of menthol and menthone from mint flavoured sweets: raw data*

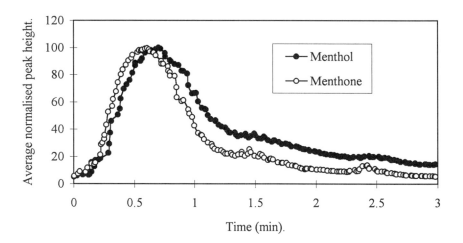

Figure 4   *Volatile release profiles of menthol and menthone from mint flavoured sweets: average of four replicates*

The APCI method monitors the volatile release pattern in real time, which has enabled us to run replicate analyses so that we can investigate differences between individuals or food matrices. The results from 12 replicate volatile release profiles for two individuals eating strawberry-flavoured sweets are shown in Table 2. These data demonstrated that although one individual was reasonably consistent (see also Table 1), there could be significant differences in both T_{max} ($P < 0.05$) and I_{max} ($P < 0.001$) between different people. This is probably associated with factors such as the saliva flow rate, chewing efficiency and breathing pattern of the individual.

Table 2 *I_{max} and T_{max} ($\pm$ SD) of Two Subjects eating Strawberry-flavoured Sweets. Each result is the mean of twelve replicates*

Subject	T_{max}	I_{max} x 10^{-4}
Rob	0.51 ± 0.11	213 ± 41
Kate	0.42 ± 0.07	392 ± 104

Tomatoes represent an interesting aroma release system, because some of the aroma compounds are present in the intact fruit, but the 'fresh notes' are generated upon homogenisation as a result of enzyme–substrate interaction. The rate of volatile generation in-mouth may be one of the major factors affecting the concentration of volatiles in the breath and therefore the perceived aroma. Aroma generation from tomatoes in-mouth is relatively easy to follow using APCI, which can monitor the dynamics of the whole process. Figure 5 shows that there are clear differences in the amounts of different volatiles over the eating time course. Isobutylthiazole is present in the intact fruit and is released quickly, hexenal (*E* and *Z* isomers unresolved) is generated in mouth during eating and shows a delay in its increase in the breath. Hexenol (*E* and *Z* isomers unresolved) is generated from hexenal through the action of alcohol dehydrogenase; this compound shows the greatest delay and subsequent persistence (presumably due to continued generation and greater polarity relative to hexenal).

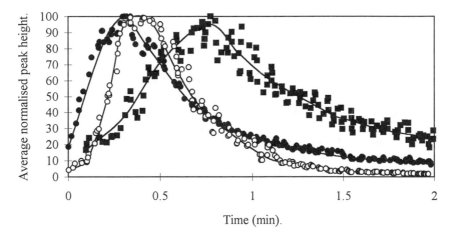

Figure 5 *Isobutylthiazole (●), hexenal (○) and hexenol (■) present in the breath as tomatoes are eaten; average of five replicates*

3 CONCLUSION

The progressive development of mass spectrometers, software systems and interfaces has enabled the production of improved methods of analysing volatile compounds present in the breath as foods are eaten. This allows us to monitor the pattern of volatile release in real time as food is eaten. This can be used to study the effects of aroma encapsulation or the effects of food reformulation (such as reducing the fat content) on aroma release.

Now that there are methods capable of following the time course of volatile release, there is the potential to link up with researchers who are studying the theoretical approach to aroma release and examine the correlation between the mathematical prediction and in-mouth aroma release.

ACKNOWLEDGEMENTS

We would like to thank BBSRC for funding our work with both the Tenax trapping and APCI systems. The APCI project is jointly funded by MAFF and BBSRC in a LINK scheme with our industrial partners who include Firmenich and Micromass (formerly VG Organic).

REFERENCES

1. B. Berglund and T. Lindvall, 'The nose, upper airway, physiology and the atmospheric environment', Elsevier, Amsterdam, 1982.
2. R.S.T. Linforth and A.J. Taylor, *Food Chem.*, 1993, **48**, 115.
3. A.J. Taylor and R.S.T. Linforth, 'Trends in Flavour Research', Elsevier, Amsterdam, 1994.
4. C.M. Delahunty, J.R. Piggott, J.M. Conner and A. Paterson, 'Trends in Flavour Research', Elsevier, Amsterdam, 1994.
5. R. Teranishi, R.G. Buttery and D.G. Guadagni, 'Standardized Human Olfactory Thresholds', IRL Press, Oxford, 1990.
6. K.B. de Roos and K. Wolswinkel, 'Trends in Flavour Research', Elsevier, Amsterdam, 1994.
7. B.P. Hills and M. Harrison, *Int. J. Food Sci. Tech.*, 1995, **30**, 425.
8. W.J. Soeting and J. Hidema, *Chem. Senses*, 1988, **13**, 607.
9. P.G.M. Haring, 'Flavour Science and Technology', Wiley, Chichester, 1990.
10. A. Legger and J.P. Roozen, 'Trends in Flavour Research', Elsevier, Amsterdam, 1994.
11. K.E. Ingham, R.S.T. Linforth and A.J. Taylor, *Lebensm. Wiss. u. Technol.*, 1995, **28**, 105.
12. K.E. Ingham, R.S.T. Linforth and A.J. Taylor, *Food Chem.*, 1995, **54**, 283.
13. K.E. Ingham, A.R. Clawson, R.S.T. Linforth and A.J. Taylor, *J. Food Sci.*, in press.
14. K.E. Ingham, A.J. Taylor, F.F.V. Chevance and L.J. Farmer, 'Flavour Science: Recent Developments', 1996.
15. M. Dalla Rosa, P. Pittia and M.C. Nicoli, *Ital. J. Food Sci.*, 1994, **6**, 421.
16. S.N. Ketkar, J.G. Dulak, W.L. Fite, J.D. Buchner, and S. Dheandhanoo, *Anal. Chem.*, 1989, **61**, 260.

EFFECT OF THE FOOD MATRIX ON FLAVOUR RELEASE AND PERCEPTION

Jokie Bakker, Wendy Brown, Brian Hills, Nathalie Boudaud, Claire Wilson and Marcus Harrison

Institute of Food Research, Reading Laboratory, Earley Gate, Reading, RG6 6BZ, U.K.

1 INTRODUCTION

There are many factors influencing consumer choice and acceptance of foods. However, it is considered that the sensory perception of flavour of a food forms an important aspect of the enjoyment people get from eating, and hence influences consumers' acceptability. The perceived flavour characteristics are due to a combination of the taste, imparted by the non-volatile components during eating, and the smell, imparted by the volatile components during eating. With the growing range of new foods available, many with lower fat or lower sugar formulations than the traditional foods, it is becoming increasingly important to understand the factors which affect the perception of flavour, including how flavour is released from food matrices, in order to deliver an acceptable flavour from these foods. Most detailed studies on flavour release have been made on simple liquid systems, and relatively little research has been done on the release from solid or semi-solid foods, having different structures. Hence the mechanisms of flavour release still need to be elucidated, and the formulation of flavour systems of new foods tends to be carried out by using extensive experience, rather than based on scientific knowledge regarding the physics of the flavour in the various food matrices.

An important parameter is the partition coefficient, defined as the equilibrium concentration of a flavour in two phases at rest. Numerous studies have been done, mainly on the determination of partitioning, and are reviewed elsewhere.[1] Since many flavours are hydrophobic, fat as a food ingredient is an excellent solvent of many food flavours and even the addition of a small amount of fat to a flavour solution has a considerable effect on the food–air partition coefficient. Binding of flavours to food ingredients (proteins and also carbohydrates) can also lessen the concentration of free flavour, and hence will affect partitioning of flavour.

The perceived quality and intensity of the flavour of a food is related to the concentrations of volatile components released into the air space of the mouth while eating. For simplicity we ignore the mechanism of perception, and assume that the concentration of a flavour released into the airspace is quantitatively and qualitatively related to the sensory perception. What are the factors affecting the rate of release of flavour during eating? When eating a food, the food structure is broken down by mastication, mixed with saliva, the food may be partially dissolved, depending on the structure, and subjected to an air flow, before being swallowed. All these factors can affect the rate with which the flavours are released, and hence can influence flavour perception. The composition and matrix of the food as well as how the food is broken down during eating are all expected to influence flavour perception.

There are very few studies looking at the rate of flavour release. In the case of emulsions the structure itself has been shown to affect the release rate of flavour.[2,3] Flavour release rates can be measured by sensory methods and instrumentally. In our research, we use sensory methods involving time–intensity measurements of flavour release to ascertain the temporal aspect of flavour perception as a function of the food matrix. In addition, mastication studies are done to determine the effect of the breakdown of the food matrix on flavour release. Robust instrumental methods of sufficient sensitivity to measure flavour release still need to be developed and this forms an important aspect of our work. Using mathematical modelling to describe the various aspects of flavour release gives a fundamental understanding of the mechanism of flavour release. In this paper we will discuss flavour release from liquid and solid matrices, which have been thickened with gels.

2 MODELLING

Mathematical models are used to describe in physical terms the events leading to flavour release, and allow predictions regarding the factors of importance for flavour release from defined food structures. Flavour release is essentially a three phase problem: the flavour is released from the food via a layer of saliva to the air phase. In recent years progress has been made in developing mathematical models for flavour release by exploiting the physics of interfacial mass transport. A two-film theory of flavour release from solids has been developed by Hills and Harrison.[4] The authors proposed the existence of a stagnant saliva layer adjacent to the interface of the food; the thickness of this stagnant layer determines the rate of release, since the actual volatile concentrations can be approximated by a linear concentration across this stagnant layer through which the volatiles diffuse into the bulk saliva, from which they are released into the mouth airspace. The rate of mass transfer across this boundary is determined by the time required for the volatiles to diffuse through the boundary layer and is enhanced by an increase in surface area and an increase in concentration gradient. The thickness of a stagnant layer cannot be determined directly experimentally. However, its thickness will be influenced by other factors such as the viscosity of the liquid and how fast the liquid is being stirred.

Numerous factors will influence the release of flavour from a food, but which are critical will depend on the type of food and what its fate is during eating. For example, if one assumes that one aspect of eating a particular food can be approximated by stirring, than diffusion from the bulk to the surface is not a limiting factor. Hence when we are eating a food, the release of flavour is expected to be greatly influenced by the way the food is eaten, for example if the food is sucked the release will be slower than when the food is crunched into numerous pieces, resulting in a rapid increase in surface area. Recently the thermal transport equations have been combined with those for interfacial mass transport to model flavour release from gelatine–sucrose gels.[5] This theoretical paper illustrates that the parameters to be considered in flavour release studies depend on the properties of the food. The driving force for flavour release depends on the bulk melting temperature of the gel. In gels with melting points lower than the mouth temperature, flavour release is determined by the rate heat can diffuse into the gel matrix and initiate melting. For harder gels with melting points above mouth temperature, the diffusion of sucrose from the surface of the gel into the adjacent saliva phase is predicted to be the rate limiting step for flavour release, since this lowers the melting temperature of the surface layer. Changes will take place continuously while eating, and these may affect flavour release, for example an increase in saliva flow while eating a gelatine sample will give changes in volume and dilution of the volatile. The viscosity of the saliva can also change during eating either due to physiological factors or because it is being mixed with food components. This is expected to have an effect on the mass transfer through the boundary layer, since the diffusion is slower with increasing viscosity. An important consideration is the magnitude of the effects that can be predicted with the theoretical models, in order to determine which factor has the most significant effect on perception. When studying flavour release from liquids, the two-film theory can still be applied. A stagnant boundary layer will be present at the interface between the liquid and the air.

3 CHEMICAL MEASUREMENTS

In order to systematically examine and quantify the release of a flavour from a food, an instrumental method is necessary such that the mechanisms of perception and the individual differences between people which form part of the sensory assessments of flavour can be eliminated. In order to represent perception, the instrumental method should be able to mimic aspects of eating and monitor the release of flavour in real time into an air flow which represents breathing, which, in turn, requires a flow rate. Aspects of the oral breakdown of the food, such as solids being chewed, and mixing with saliva should also be incorporated. This method should also match the sensitivity and response time of the human nose. A method of studying dynamic flavour release allowing real-time detection of a number of volatiles has been developed.[6] The method allows stirring of the liquid food at a defined stirring rate to mimic the shear forces for liquid foods in the mouth, while a flow of helium is maintained over the sample, in order to mimic the air flow while breathing.

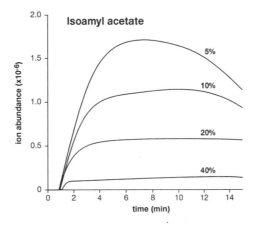

Figure 1 *Release of isoamyl acetate (10 mg l^{-1}) from four aqueous solutions containing 5%, 10%, 20% and 40% ethanol determined under dynamic conditions (20 ml sample, stirred at 300 rpm, helium flow rate 30 ml min^{-1}, 37 °C)*

Although both factors only approximate what happens during eating, this set-up can give useful information for predicting flavour perception. A mass spectrometer was used as a detector to monitor the concentration of a number of selected volatiles from the food. The volatiles need to be present in sufficiently high concentrations to be detected, and have unique mass weights in order to separate the signals.[6] The release curve for each volatile is akin to a time–intensity curve. The maximum intensity is reached when the release rate of a volatile from the food is in dynamic equilibrium with the rate at which the volatile is swept away from the headspace with the helium flow. Using this technique the release of a solution of isoamyl acetate (10 mg l^{-1}) was determined from four aqueous solutions containing 5%, 10%, 20% and 40% ethanol (Figure 1). The release curves are comparable with the sensory time–intensity curves, although the time scale is different. From this curve, the slope of the increase and the maximum intensity under dynamic flow conditions can be determined.

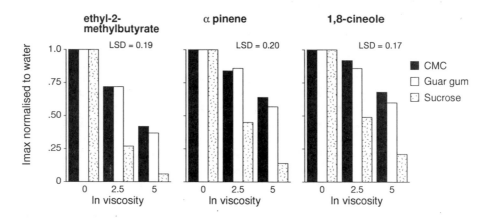

Figure 2   *Effect of viscosity on release of flavour compounds from solutions of sucrose, guar gum and carboxymethylcellulose (20 ml sample, stirred at shear rate of 100 s^{-1}, helium flow rate 30 ml min^{-1}, 37 °C)*

This experimental set-up was used to examine the effect of viscosity and thickening agent on volatile release from solution. Release of seven volatiles (α-pinene, methyl anthranilate, 1,8-cineole, ethyl 2-methylbutyrate, maltol, vanillin and 2-methoxy-3-methylpyrazine) from equiviscous solutions of three thickeners (carboxy methyl cellulose, guar gum and sucrose) was measured.[7] The dynamic release profiles of all volatiles were collected simultaneously while stirring the samples at a shear rate of 100 s^{-1}, at 37 °C, with an helium flow of 30 ml min^{-1} through a vessel of 125 ml volume containing 20 ml sample. The maximum intensity for each flavour was normalised to the appropriate aqueous solution. The aim was to test the effect of viscosity on aroma release and to determine whether decreases in aroma release were due to differences in mass transfer or to binding of the flavours by the thickeners. The results showed an increased viscosity resulted in a detectable decrease in release of the volatiles with high volatility (α-pinene, 1,8-cineole, ethyl 2-methylbutyrate, see Figure 2), although not all the differences were significant at a 5% level. For the less volatile flavours (methyl anthranilate, maltol and vanillin), no detectable effect was found. There were no significant differences between the two gums. If the reduction in flavour release with increasing viscosity was only due to changes in mass transfer resulting from lower diffusion rates in viscous liquids, the reduction in flavour release would be expected to be much less. Hence other mechanisms also affect flavour release. Sucrose gave a significantly lower release at both viscosities for α-pinene, 1,8-cineole and ethyl 2-methylbutyrate.

4. SENSORY MEASUREMENTS

Time–intensity assessment of perceived flavour provides useful information regarding the temporal aspects of flavour perception. An important parameter determining flavour release in the mouth is the change in surface area of the food, as discussed above. The surface area may be increased rapidly because the food is melting in the mouth (which results in a rapid increase in surface area) and would be expected to give a rapid release of flavour. The surface area can also be increased because the food is broken down into small pieces by chewing. We have done a study to investigate the breakdown in the mouth of gels of different mechanical and physical properties as determined by subjects' mastication patterns on the perception of flavour.[8] Three gelatin gels, containing 5%, 10% and 20%, with 10% sucrose and all flavoured with 1% banana flavour, were evaluated by eight trained subjects. Time–intensity data were recorded simultaneously with electromyographs of the activity of the masticatory muscles (see Brown[9]).

Higher gelatin concentration samples result in gels with higher melting points, breaking strength and elastic modulus. The panel-averaged time–intensity curves for flavour showed that an increase in gelatin concentration resulted in a lower maximum intensity and a later time to reach maximum intensity (Figure 3). Examination of the mastication data showed an increase in chewing time with increasing gelatin concentration, reflecting the fact that lower concentration gels with a melting point below mouth temperature melted rapidly in the mouth with rapid release of flavour.

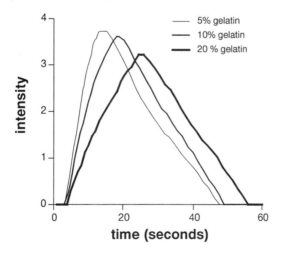

Figure 3 *Time–intensity curves (n = 24) for flavour perception for gelatin samples sweetened with 10% sucrose*

For firmer gels of higher concentration, oral breakdown occurred principally by chewing, resulting in a slower increase in surface area from which flavour was liberated and a slower rate of flavour perception. Although the mastication patterns for the samples differed between individuals the temporal pattern of flavour perception could be linked to the mastication patterns of the individual subjects. The perception of maximum intensity generally coincided with swallowing, hence subjects with larger chewing times recorded a later T_{max}. Such a result would be predicted from simple models of flavour release under dynamic conditions involving differences in the rate of increase of surface area with time. These models also explained differences in perception patterns (time–intensity curves) as a result of differences in their mastication patterns. A more detailed discussion of this work by Brown and Wilson is published later in this book.

5 CONCLUSIONS

Flavour release is essentially a three-phase problem: the flavour is released from the food via a layer of saliva to the air phase. A two-film theory of flavour release from solids has been applied to predict the release from different food systems, taking into consideration how the food would be eaten in the mouth, and allowed the interpretation of the results in terms of physical parameters. Instrumental measurements of flavour release in real time using a mass spectrometer to monitor the release of flavour of volatiles allowed us to measure simultaneously the release of seven volatiles from three different thickeners, prepared to be equiviscous solutions. The results showed release under dynamic conditions could in part be explained by mass transfer, while other mechanisms of flavour interactions also occurred. Sensory analysis of flavour perception showed that the temporal pattern of flavour perception could be linked to mastication patterns. Softer gels gave a higher maximum flavour intensity than harder gels, while the timing of the maximum was linked to the individuals chewing strategy. Further research is ongoing to verify the sensory data with chemical measurements and to ascertain the relative importance of other factors involved in the release and perception of flavour.

REFERENCES

1. J. Bakker, 'Ingredient Interactions – Effects of Food Quality', Marcel Dekker, Inc., New York, 1995.
2. P. Overbosch, W.G.M. Afterof and P.M.G. Haring, *Food Rev Intl.*, 1991, **7**, 137.
3. D. Salvador, J. Bakker, K.R. Langley, R. Potjewijd, A. Martin and J.S. Elmore, *J. Food Qua. Pref.*, 1994, **5**, 103.
4. B.P. Hills and M. Harrison, *Int. J. Food Sci. Technol.*, 1995, **30**, 425.
5. M. Harrison and B.P. Hills, *Int. J. Food Sci. Technol.*, in press.
6. J.S. Elmore and K.R. Langley, *J. Food Sci.*, in press.
7. D.D. Roberts, J.S. Elmore, K.R. Langley and J. Bakker, *J. Agric. Food Chem.*, 1996, **44**, 1321.
8. C.E. Wilson and W.E. Brown, *J. of Sensory Studies*, in press.
9. W.E. Brown, *J. Texture Studies*, 1994, **23**, 1.

FLAVOUR RELEASE AND FLAVOUR PERCEPTION IN OIL-IN-WATER EMULSIONS: IS THE LINK SO CLOSE?

C. Brossard, F. Rousseau and J.P. Dumont

Laboratoire d'Etude des Interactions des Molécules Alimentaires, Institut National de la Recherche Agronomique, BP 1627, 44316 Nantes Cedex 03, France

1 INTRODUCTION

Nowadays, it is commonplace to state that in viscous[1] or solid foods,[2] mass transfer of flavour molecules is impeded, with the immediate consequence of a slower rate of surface renewal at the air–food interface in the mouth.[3] An extreme situation occurs when the aqueous solvent (saliva) is unable to mix with the food, leaving flavour supply solely dependent on the molecular diffusion in the hydrophobic fraction. In such a situation, for instance, the half-life for flavour equilibration of butter in the mouth may exceed two hours.[4]

Recently, a non-equilibrium partition model[1] for flavour release has been developed to supercede earlier versions[5] unaffected by flavour volatility, resistance to mass transfer and hydrophobicity. In this paper, sensory properties of flavoured oil-in-water (o/w) emulsions were studied to examine the degree of the fit between intensity of perceived flavour and concentration of available flavour calculated from the models.

2 MATERIALS AND METHODS

2.1 Materials

Benzaldehyde, miglyol (oil made of a mixture of saturated triglycerides with fatty acids ranging from C_8 to C_{14}) and commercial β-lactoglobulin (purity > 90 %) were used respectively as the flavour, the fat phase and the emulsifying agent.

2.2 Sample Preparation

Benzaldehyde solutions in water (1.7, 5 and 15 ppm) were prepared daily from stock solution and presented as reference samples to the sensory panel. Usually, 750 ml of a 60/40 (miglyol–water) emulsion ($\Phi_i = 0.6$) was prepared in a single batch from water, containing the dispersed emulsifying agent (0.004 w/w), and miglyol containing benzaldehyde (650 ppm). A two-step procedure was introduced involving dispersion by means of an agitator (Polytron PT 45-80) and subsequent homogenisation for 15 minutes with a homogeniser (ALM 0 Guérin S.A.). Emulsions with 0.5 and 0.3 oil volume fractions ($\Phi_i = 0.5$ or $\Phi_i = 0.3$) were obtained by diluting the stock emulsion. Samples (20 ml) were poured in 100 ml odour-free brown glass bottles closed with screw caps and left overnight at room temperature (22 °C) before sensory evaluation.

2.3 Physicochemical Characterisation of Samples

Distribution of oil droplets in emulsion samples was determined in triplicate by means of a MasterSizer 3600 laser granulometer (Malvern). Flow curves of emulsion samples at 30 °C were obtained using a stress-controlled rheometer (Carrimed CSL 50) in a cone–plate geometry. Applied shear stress was increased linearly for two minutes from the lower to the upper value and then decreased at the same rate.

2.4 Sensory Analysis

The panel consisted of eight subjects (three male and five female) who had previous experience in the evaluation of odour intensity of pure odorants dissolved in either water or miglyol. In each session, two sets of three samples (first the series of aqueous samples and secondly the emulsions), arranged in random order, were presented to each panellist. Subjects were instructed to sniff samples and to record them on an axis by reference to the intensity of the bitter almond note (boundaries being 'None' and 'Very High'). A pause of 35 s was imposed before testing the next sample to minimize adaptation effects. After a while, subjects were asked to taste the samples and to rate the intensity of the perception according to three descriptors: sample fluidity (ranging from 'Slightly Liquid' to 'Very Liquid'), oiliness (from 'None' to 'Very High') and bitter almond flavour (from 'None' to 'Very High'). Four replications were carried out, of which the first one was considered as a rehearsal and corresponding data were disregarded.

3 RESULTS AND DISCUSSION

Data dealing with physicochemical characteristics of emulsions are shown in Table 1. It appeared that dilution of the stock emulsion with water has had practically no effect on the size distribution of oil droplets as very close values, in the range of 0.5 m^2 cm^{-3}, have been calculated for the specific area of droplets in all the investigated samples.

Table 1   *Physicochemical Characteristics of the Studied Emulsions*

Parameter	Emulsion 1	Emulsion 2	Emulsion 3
Oil volume fraction	0.3	0.5	0.6
Viscosity (mPa s^{-1})	12	34	550
Solute in water (ppm)[a]	15.4	15.8	16
Solute in oil (ppm)[a]	614	634	640
Fraction of solute contained by water	0.058	0.025	0.016

[a]   On the basis of a value of 40 for oil–water partition coefficient.[6]

According to the data shown in Table 1, it can be concluded that emulsions were similar in flavour but were very different in viscosity. It should also be noted that despite large differences in the water and oil contents of the samples, the proportion of flavour molecule in the aqueous phase is always low.

Results from the sensory panel were digitized and data dealing with odour and flavour notes of benzaldehyde solutions in water were used to construct models according to Stevens equation:

$$S = KC^n \tag{1}$$

where S is the intensity of sensory perception, C the concentration of the active molecule in air or in a solvent, n the constant related to the active molecule and K a constant.

Data dealing with odour and flavour parameters of benzaldehyde in the emulsions were plotted on a graph and the equivalent concentrations of solute in the water phase were estimated. The results are shown in Table 2.

Perceived odour intensities of the emulsions are similar as might have been expected from the calculated concentration in the aqueous phase. This suggests that, on average, the models used work satisfactorily. Benzaldehyde concentrations in the aqueous phase deduced from the Stevens plot, however, are a third lower than their calculated nominal values. A deviation of the same order has been shown previously. The odour intensity of flavour components dissolved in either pure miglyol or water was measured in order to determine 'operating' partition coefficients.[6]

Flavour intensities show a very different pattern. Samples ranked in an inverse order when considered according to Stevens estimate or to calculated values, that assumed instant even dilution of the sample by a fixed proportion of saliva. Furthermore, comparing Table 1 and Table 2, it is clear that sample ranking according to the Stevens estimate for flavour, reproduces the order of increasing viscosities. Obviously, in the present experiment, something has gone wrong with commonly accepted propositions like 'only the flavour in the aqueous phase can stimulate perception'[5] or 'flavor release may be reduced with increased viscosity of foods and beverages'.[2] Before trying to bridge the gap between conflicting evidence, let us consider Figure 1 in which, assuming sample dilution by an equal volume of saliva, data on the calculated, the actual and the eventual concentrations of solute in aqueous phase of the emulsions are matched.

Table 2 *Estimated Values for Benzaldehyde Concentration in the Aqueous Phase of the Studied Samples (ppm)*

Parameter	Aqueous Solution	Emulsion 1	Emulsion 2	Emulsion 3
Solute in water	15	15.4 [a]	15.8 [a]	16 [a]
Stevens estimation based on odour note	15	10.4	10	11
Stevens estimation based on flavour note	15	9.4	10.2	12.4
Calculated (dilution 1:1)	7.5	6.2 [a]	5.3 [a]	4.8 [a]
Sensory (dilution 1:1)	7.5	4.7	5.1	6.2

[a] On the basis of a value of 40 for oil–water partition coefficient.[6]

It is obvious that none of the considered samples meets the 50% equilibration criterion, introduced by McNulty and Karel,[7] which estimates the time food stays in the mouth. Data reported by these authors would suggest an equilibration time less than 10 seconds. This is certainly not true and casts doubt on the validity of some aspects of the proposed model. In most circumstances, calculated values relying on full equilibration of the system, are of little practical significance as this is unlikely to occur in the span of food ingestion.

Considering the filled and the open square symbols in Figure 1, there is a suggestion that flavour perception has been delayed in the case of the emulsion with $\Phi_i = 0.3$, has broken even in the intermediate sample and was boosted by solute transfer from oil in the sample with the higher oil content. Whilst the former proposition is consistent with most previous data[2,3] showing an adverse effect of viscosity or gel mechanical strength on flavour release, the latter two appear contradictory as mass transfer should be severely impeded by increasing viscosities. Moreover, the picture may look puzzling if, recalling Table 1, we conclude that perceived intensity of the flavour is negatively related to the quantity of benzaldehyde initially contained by the aqueous phase.

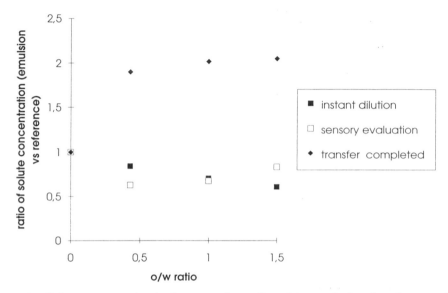

Figure 1   *Solute concentration in the water phase of emulsions calculated on the grounds of the different schemes considered for flavour release*

The flaw is likely be found in the assumption made by[5] that 'oil and water are extreme cases of o/w emulsions with values of Φ_i = 1.0 and 0 respectively' Overlooking the inversion step (transition from o/w to w/o), the proposition carries implicitly the belief that the bulk of possible emulsified systems is arranged linearly along a continuous scale. As a consequence, sample mixing with saliva is simply seen as the dilution of a stock emulsion with an aqueous solvent and the assumption that 'only the flavour in the aqueous phase can stimulate perception' looks common sense. We consider that flavour equilibration at the air–food interface occurs where the aqueous phase had equilibrated or not. Viscous o/w emulsions do not leave much water available for air–food exchanges and there are opportunities for other contributions to build up the mouthspace. As our data show, flavour concentration in the mouthspace is not entirely dependent on the pace of flavour equilibration in the aqueous phase. We assume that, for a transient period which can look a steady state if the time scale of aqueous phase equilibration outlasts ingestion, unsoaked parts of the food (especially oil droplets) contribute to mouthspace formation.

Whereas odour evaluation deals with systems containing solute at equal partial pressure in both solvents, flavour measurement on o/w viscous emulsions triggers a succession of transient non-equilibrated systems. We postulate that in the early stages of mouthspace formation, the concentration of volatiles in the air depends rather on oil than on water and the distribution of solute's partial pressure in the sample is not even at the difference of models[5] bound to estimate intensity of the sensorial stimulus under the provision that solute is evenly distributed in the water phase. We think that concept of non-equilibrium partition, proposed by De Roos and Wolswinkel[1] as an alternative to the release model[5] can be applied more extensively than one perhaps might have thought. The partition model is based upon the assumption that the time required for flavour renewal at the air–sample interface is short compared to the time scale of mass transfer in the sample. As a consequence, the concept of the non-equilibrium state applies to solute spatial distribution as well as unidirectional distribution currently involved in models. In living organisms or systems, volatile flavour components may locally show high concentration with a corresponding high potential for flavour release. Bursts of volatiles in the mouthspace may result which are totally unpredictable on the grounds of data dealing with mean concentrations.

4 CONCLUSION

Understanding flavour availability on eating food cannot be understood simply by a unique model for flavour release. Data calculated for release of flavour molecules from food, on the grounds that 'only the flavour in the aqueous phase can stimulate perception', and the measured sensory impact are closely related so long as the samples involved are liquid or reasonably fluid. When tasting highly viscous food systems, the link between flavour release and perception appears to be ruled by an intricate combination of factors, namely maximum solute concentration in the fat phase, velocity of the oil-to-water mass transfer, spatial distribution of pockets of intact sample and the possibility of solute exchanges between the fat phase and the mouthspace.

REFERENCES

1. K.B. De Roos and K. Wolswinkel, in 'Trends in Flavour Research', eds. H. Maarse and D.G. van der Heij, Elsevier Science B.V., Amsterdam, 1994, p. 15.
2. J.X. Guinard and C. Marty, *J. Food Sci.*, 1995, **60**, 727.
3. P. Overbosch, W.G.M. Agterof and P.G.M. Haring, *Food Rev. Int.*, 1991, 7, 137.
4. P.B. McNulty, in 'Food Structure and Behaviour', eds. J.M.V. Blanshard and P.J. Lillford, Academic Press, London, 1987, p. 245.
5. P.B. McNulty and M. Karel, *J. Food Technol.*, 1973, **8**, 309.
6. F. Rousseau, C. Castelain and J.P. Dumont, *Food Qual. & Pref.*, 1996, in press.
7. P.B. McNulty and M. Karel, *J. Food Technol.*, 1973, **8**, 415.

RELATIONSHIPS BETWEEN ODOROUS INTENSITY AND PARTITION COEFFICIENTS OF δ-DECALACTONE, DIACETYL AND BUTYRIC ACID IN MODEL EMULSIONS

C. Guyot,[*] C. Bonnafont,[*] I. Lesschaeve,[*] S. Issanchou,[*] A. Voilley[†] and H.E. Spinnler[§]

[*]Institut National de la Recherche Agronomique, Laboratoire de Recherches Sur les Arômes, 17, Rue Sully, B.V. 1540, 21034 Dijon Cedex, France

[†]Ecole Nationale Supérieure de Biologie Appliquée à la Nutrition et à l'Alimentation, Laboratoire de Génie des Procédés Alimentaires et Biologiques, Campus Universitaire, 21000 Dijon, France

[§]Institut National Agronomique Paris-Grignon, Chaire de Technologie Alimentaire, 78850 Thiverval Grignon, France

1 INTRODUCTION

Foods are complex systems containing mainly water, proteins, carbohydrates and lipids. The non-volatile molecules, and particularly lipids, interact with aroma compounds in modifying their volatility and eventually in changing the olfactory perception of the food. Some papers have reported that most flavour compounds have a lower vapour pressure (and higher odour thresholds) in oil than in aqueous solutions.[1] The physical properties of the solvent in which aroma compounds are dispersed affect their odour threshold, e.g., the more apolar is the solvent, the lower will be the volatility of an apolar molecule.[2] Up to now, very few studies have tried to understand and predict the changes in sensory response when food composition changes.[3] Reconstructing the interactions between the volatile and the non-volatile compounds requires the evaluation of the behaviour of aroma compounds in model systems similar to the original product. Moreover, while studies dealing with vapour–liquid partition phenomena may have reported the effects of medium composition on the headspace concentrations at equilibrium, they have not connected the physical properties with sensory scores[1,4] by model equations.

Several investigations have recognized the role of pH when volatile fatty acids are essential flavourants, and have showed that the flavour perception is affected by the state of dissociation of these acids.[5,6] However, the effect of pH on odour and vapour–liquid partition coefficient has received very little attention.

In the present study, we have examined the odour intensity in parallel with the liquid–liquid and the vapour–liquid partition coefficients of three aroma compounds in model emulsions, where the oil content and the pH were changed. The aroma compounds in this study were diacetyl, δ-decalactone and butyric acid and were chosen because of their different physical properties and their presence in numerous food products. We compare our results with the values estimated from the Buttery[7] model equations.

2 EXPERIMENTAL

2.1 Liquid–Liquid Partition Coefficients

Aqueous solution containing the aroma compound (diacetyl, butyric acid and δ-decalactone at concentrations of 100, 200 and 200 ppm respectively) was in contact with paraffin oil. The two phases were gently stirred at 25±1 °C until equilibrium was reached. For butyric acid, analyses were carried out on a Hewlett Packard HP5890 Series II gas chromatograph equipped with an FID and a 30 m × 0.32 mm cross-linked capillary FFAP® column. For diacetyl and δ-decalactone, analyses were carried out on a Chrompack CP9000 gas chromatograph equipped with an FID and a 3.0 m × 2.0 mm stainless steel Carbowax 20M® column packed with chromosorb W-AW (100-120 mesh). The equilibrium concentrations were determined with respect to a calibration scale and the liquid–liquid partition coefficient (P) was finally expressed as the ratio between the concentration of aroma compound in oil and the concentration of aroma compound in water (in $\mu l\ l^{-1}$).

2.2 Vapour–Liquid Partition Coefficients

The measurements were made in the infinite dilution domain where the aroma concentrations (100, 200 and 500 ppm for diacetyl, δ-decalactone and butyric acid respectively) do not affect the partition coefficient. The equilibrium concentrations of the aroma compounds between the liquid and vapour phases were measured at 25 °C for diacetyl and butyric acid and at 80 °C for δ-decalactone. Nitrogen passed through the liquid phase at a constant flow rate (20, 60 and 100 ml min⁻¹ for diacetyl, butyric acid and δ-decalactone respectively) and carried the volatile compounds into the headspace. A sample of the vapour phase was automatically injected at regular intervals into the gas chromatograph. For diacetyl and δ-decalactone the apparatus and the columns were the same as for determination of liquid–liquid partition coefficients. For butyric acid, the column was replaced by a 0.5 m × 2.0 mm stainless steel HayeSep Q® column (80-100 mesh). The vapour–liquid equilibrium was considered to have been reached when the concentration of aroma compounds in the gas phase remained constant. The vapour–liquid partition coefficient (K) was expressed as the ratio between the concentration of aroma compound in the gaseous phase to the concentration of aroma compound in the liquid phase (in $\mu l\ l^{-1}$) determined with respect to a calibration scale.

2.3 Sensory Evaluation

2.3.1 Samples. The concentrations of diacetyl, δ-decalactone and butyric acid in the emulsions were 10, 10 and 20 ppm respectively. Paraffin oil constituted the oily phase. The emulsifier, a sucroester (SP 50®, Sisterna, Netherlands), was used at a concentration of 1% (w/w). For each aroma compound, six emulsions were evaluated with water contents of 0%, 16%, 50%, 84%, 99% and 100% (w/w). This last medium was investigated in order to evaluate the effect of the emulsifier. For each water content, one pH on either side of butyric acid's pK was studied – pH 4.5 and 5.2. The emulsions were made up at 4 °C in order to minimize loss of the aroma compound.

2.3.2 Panellists. Twenty panellists, having previous sensory experience, constituted the jury. They were trained and selected especially for their capacity to recognize and memorize odours and to describe their perceptions when evaluating food products.

2.3.3 Sensory Test Procedure. 20 ml of each sample were presented at 21±1 °C in coded 60 ml brown capped flasks. The booths were lit with red light. A sample presentation was established for taking into account the serving order and carry-over effects of samples by the use of latin square.[8] A 10-point category scale effected with isoamyl acetate solutions (0.04 to 10 ppm with a four-fold geometrical ratio) was used for the intensity evaluation as a reference scale adapted from a previous technique.[9] Panellists were instructed to smell each sample, to rate its intensity on the reference scale and to describe the odour that they perceived. Thus all the samples were evaluated with respect to the same odour category scale. Two replicates were made for each odorous compound.

2.4 Statistical Analysis

All the statistical analyses were done with Statistical Analysis System, SAS (SAS Institute, Inc., Cary, NC, U.S.A., 1987). An analysis of variance (ANOVA) was used on the sensory and physical results to determine significant differences among samples. A Student-Newman-Keuls test was used in order to perform a multiple comparison of means.

3 RESULTS

3.1 δ-Decalactone

3.1.1 Sensory Analysis and Vapour–Liquid Partition Coefficients. At pH 4.5 as at pH 5.2, the media without oil present the highest odour intensity, but when the fat content increases, the odour intensity decreases (Figure 1A). In spite of the temperature (80 °C instead of 25 °C), vapour–liquid partition coefficients for water content 99% and 100% were the only ones that could be measured. A pH effect on odour intensity and vapour–liquid partition coefficient is observed in the emulsified medium without oil and in the non-emulsified medium without oil respectively. The emulsifier effect is only observed at pH 5.2. However, if the odour intensity is significantly more intense in the presence of emulsifier, the vapour–liquid partition coefficient decreases significantly and disagrees with the sensory results.

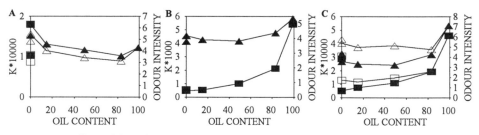

Figure 1   *Effect of the oil content on odorous intensity and vapour–liquid partition coefficient (K) of: (A) δ-decalactone; (B) diacetyl; and (C) butyric acid. Key: K at pH 4.5 (□) and pH 5.2 (■). Odorous intensity at pH 4.5 (Δ) and pH 5.2 (▲). Odour intensities and (K) values are respectively means of 2 and 3 replicates*

3.1.2 Liquid–Liquid Partition Coefficients. At pH 4.5 and 5.2, without emulsifier, the liquid–liquid partition coefficients were respectively 9.79±0.79 and 10.03±1.78 (mean ± SD) and with emulsifier were 2.88±0.13 and 3.77±0.20 respectively. These results show the great affinity of δ-decalactone for the oil and illustrate the hydrophobicity of this molecule, estimated with log P = 3.4.[10] The molecule is mainly dissolved in the organic phase. If pH does not affect the results, the emulsifier effect is significant and emulsifier tends to retain the molecule in the aqueous phase.

3.2 Diacetyl

3.2.1 Sensory Analysis and Vapour–Liquid Partition Coefficients. All the measurements were made at pH 5.2 because the diacetyl structure does not change with pH. The higher the oil content, the more intense is the odour. These data agree with vapour–liquid partition coefficients. The emulsifier had no effect on sensory or physical results (Figure 1B).

3.2.2 Liquid–Liquid Partition Coefficients. The liquid–liquid partition coefficient is 5.48×10^{-2} ($\pm 0.42 \times 10^{-2}$) and 3.20×10^{-2} ($\pm 0.43 \times 10^{-2}$) in the absence and in the presence of emulsifier respectively. These results show that diacetyl has a higher affinity for water than for oil. This is confirmed by the calculated log P equal to -2.[10] The emulsifier effect is significant and tends to retain diacetyl in the aqueous phase.

3.3 Butyric Acid

3.3.1 Sensory Analysis and Vapour–Liquid Partition Coefficients. Whatever the pH, the odour intensity and the vapour–liquid partition coefficient increase when the paraffin oil content increases (Figure 1C). At pH 4.5, the physical and the sensory values are significantly higher than those obtained at pH 5.2. The emulsifier causes a significant decrease of the vapour–liquid partition coefficients, but this effect was not perceived by the panellists.

3.3.2 Liquid–Liquid Partition Coefficients. Without emulsifier, the liquid–liquid partition coefficient is 12.7×10^{-2} ($\pm 0.58 \times 10^{-2}$) at pH 4.5 and 15.4×10^{-2} ($\pm 0.59 \times 10^{-2}$) at pH 5.2. In the presence of emulsifier, the partition coefficient is 2.77×10^{-2} ($\pm 0.29 \times 10^{-2}$) at pH 4.5 and 3.44×10^{-2} ($\pm 0.32 \times 10^{-2}$) at pH 5.2. Butyric acid has a high affinity for water, in agreement with the hydrophilic behaviour of the molecule which has a calculated log P equal to 0.8.[10] In the absence of emulsifier, a pH effect is seen – values at pH 5.2 are significantly higher than those at pH 4.5. Whatever the pH, an emulsifier effect is observed.

4 DISCUSSION

4.1 Behaviour of the Odorous Compounds

4.1.1 Hydrophobic Molecules. The hydrophobic behaviour of δ-decalactone is so strong that it was not possible to measure vapour–liquid partition coefficients in media containing paraffin oil. As δ-decalactone is highly retained by the oily phase, its concentration in the gaseous phase decreases and leads to weaker odour intensities.

4.1.2 Hydrophilic Molecules. Unlike δ-decalactone, diacetyl and butyric acid have a higher affinity for the *aqueous* phase. When the oil content increases, their concentration in the gaseous phase increases and leads to higher odour intensities and vapour–liquid partition coefficients.

4.2 pH Effect

In the pH range of this study, the structures of diacetyl and δ-decalactone probably remain the same. For δ-decalactone, the open form (δ-hydroxy decanoic acid) is unstable and tends to lactonize spontaneously.[11] However, sensory and physical results are always higher at pH 5.2 than at pH 4.5, even if there is not always a significant pH effect. For butyric acid, the pH effect is significant. With a pK equal to 4.8, 67% of protonated forms (which are odorous) are present at pH 4.5, and only 29% at pH 5.2. Thus, the concentration of the odorous form increases in the gaseous phase when the pH falls below the pK, leading to higher odour intensities and vapour–liquid partition coefficients. This pH effect has been observed previously for flavour.[5,6]

4.3 Emulsifier Effect

In the case of liquid–liquid partition coefficients, the emulsifier could act in forming a barrier between oil and water, preventing the aroma compound from diffusing in the oily phase and leading to a decrease of the liquid–liquid partition coefficient. In the case of the vapour–liquid partition coefficients and odorous intensity, the emulsifier seems to interact with the molecular forms of the aroma compound leading to lower sensory and physical results.

4.4 Comparison of Headspace Equilibrium Concentrations with those calculated with the Buttery Model

Equations have been developed[7] for estimating the vapour–liquid partition coefficient (K) in a three-phase system with the air–water partition coefficient (K_w), the air–oil partition coefficient (K_o), the volume fraction of water in mixture (F_w) and the volume fraction of oil in mixture (F_o):

$$K = \frac{1}{\dfrac{F_w}{K_w} + \dfrac{F_o}{K_o}}$$

(1)

As the liquid–liquid partition coefficient is $P = K_o/K_w$, equation (1) can be re-written as:

$$K = \frac{K_w}{F_w + P F_o}$$

(2)

4.4.1 δ-Decalactone. The high P value which characterizes the affinity of δ-decalactone for oil leads to an underestimation of the calculated values of K estimated with equation (2). Though these values have not been checked because our equipment has too low a sensitivity, the measurements have shown that the oil in the medium leads to a solubilisation of the molecule in the organic phase and to a decrease of the gaseous phase concentration and of the odour intensity (Figure 2A). This is in agreement with the flavour thresholds of lactones which are higher in fatty phase than in the aqueous phase.[11]

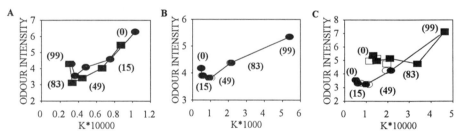

Figure 2   *Correlation between odour intensity and vapour–liquid partition coefficient (K) of (A) δ-decalactone; (B) diacetyl; and (C) butyric acid in emulsions with different oil contents (…). Key: pH 4.5 experimental (□); pH 4.5 theoretical (■); pH 5.2 experimental (○); pH 5.2 theoretical (●). Sensory and (K) values are means of 2 and 3 replicates respectively*

4.4.2 Diacetyl. The correlation between the measured values and those calculated from equation (1) is excellent (Figure 2B). The equation of the regression line established with 13 degrees of freedom is: (Experimental K) = 0.960 (±0.039) × (theoretical K), with a correlation coefficient of 0.990. When oil is added to the emulsion, the odour seems less intense. The diacetyl behaviour seems to be influenced by the homogeneity of the medium as reported previously.[12] Thus diacetyl release was shown to be greater from an oil-in-water emulsion than from a water-in-oil emulsion.

4.4.3 Butyric Acid. At pH 5.2, there is strong correlation between values calculated from equation (1) and the experimental results (Figure 2C). The equation of the regression line established with 13 degrees of freedom has a correlation coefficient equal to 0.981 and is: (Experimental K) = 0.970 (±0.053) × (theoretical K). However, the residues are not randomly distributed around the straight line of the model. At pH 4.5, the estimated values are superior to experimental results. The linear regression established with 13 degrees of freedom is: (Experimental K) = 0.960 (±0.146) × (theoretical K), but the correlation coefficient is equal to 0.893 because of the non-random data distribution. As for diacetyl, the structure of the medium influences the molecule behaviour.

4 CONCLUSION

Three aroma compounds with different physico-chemical properties have been shown to have various behaviours according to the medium composition. In most cases, physical results fit with the sensory data. Hydrophobicity of the molecule is a key factor in predicting the equilibrium between the emulsion and the gaseous phase. The highest vapour–liquid partition coefficients and odour intensities are observed if hydrophilic molecules (diacetyl and butyric acid) are in hydrophobic media. The reverse is found for δ-decalactone. The pH effect was also considered to explain vapour and odour levels successfully. For diacetyl and butyric acid at pH 5.2, the Buttery model allows the estimation of the vapour–liquid partition coefficients from the measurement of two partition coefficients regardless of the oil content.

REFERENCES

1. M.A.J.S. Van Boekel and R.C. Lindsay, *Neth. Milk Dairy J.*, 1992, **46**, 197.
2. D.A. Forss, *J. Agric. Food Chem.*, 1969, **17**, 681.
3. K.B. de Roos and K. Wolswinkel, in 'Trends in Flavour Research', eds. H. Maarse and D.G. van der Heij, Elsevier Science, 1994, p. 15.
4. D.G. Land, in 'Progress in Flavor Research', eds. D.G. Land and H.E. Nurstens, Applied Science Publishers, Barking, U.K., 1979, p. 53.
5. R.E. Baldwin, M.R. Cloninger and R.C. Lindsay, *J. Food Sci.*, 1973, **38**, 528.
6. P. Hartwig and M.R. McDaniel, *J. Food Sci.*, 1995, **60**, 384.
7. R.G. Buttery, D.G. Guadagni and L.C. Ling, *J. Agric. Food Chem.*, 1973, **21**, 198.
8. J.H. Macfie and N. Bratchell, *J. Sens. Studies*, 1989, **4**, 129.
9. P.H. Punter, N.D. Verhelst and A. Verbeek, 'Odour Intensity Measurement using a Reference Scale', symposium: 'Characterization and Control of Odoriferous Pollutants in Process Industries', 1984, Louvain-la-Neuve, Belgium.
10. R.F. Rekker, in 'Pharmacochemistry Library', eds. W. Nauta and R. Rekker, Elsevier Scientific Publishing Compagny, Amsterdam, 1977.
11. J.E. Kinsella, *Food Technol.*, 1975, **5**, 82.
12. S. Desamparados, J. Bakker, K.R. Langley, R. Potjewijd, A. Martin and J. Elmore, *Food Qual. Pref.*, 1994, **5**, 103.

EFFECT OF FAT CONTENT ON VOLATILE RELEASE FROM FOODS

K.E. Ingham, A.J. Taylor, F.F.V. Chevance[*] and L.J. Farmer[*]

Department of Applied Biochemistry and Food Science, University of Nottingham, Sutton Bonington Campus, Loughborough, Leicestershire, LE12 5RD, U.K.

[*]Department of Food Science, The Queen's University of Belfast, Newforge Lane, Belfast, BT9 5PX, U.K.

1 INTRODUCTION

Until recently, the measurement of aroma release in the mouth has been limited to analysis by sensory or instrumental means using time–intensity methods and headspace techniques respectively. Modern advances include the development of headspace model mouth systems[1,2,3] which attempt to mimic mouth conditions and the development of in-mouth systems which sample volatiles directly from the mouth. At the University of Nottingham, a method has been developed which collects some of the air from the mouths or noses of people eating a food, concentrates the volatiles onto traps containing Tenax GC and analyses them by gas chromatography–mass spectrometry (GC–MS). This type of analysis has been termed mouthspace or nosespace analysis and has recently been used in a number of laboratories to study volatile release from mint-flavoured sweets,[4,5,6] Ryvita Crispbreads,[7] cheese[8] and various dehydrated vegetables.[9]

Lipids are among the major components of food products and, therefore, affect the food quality in several ways. They act as a solvent for fat-soluble volatiles and are a precursor of many volatile compounds. Triglycerides also play an important part in the mouthfeel or texture of a food during eating. Removal or reduction of these lipids leads to an imbalanced flavour, often with a much higher intensity than the original full fat food. This is because the non-polar volatiles are no longer dissolved in the lipid phase and are released from the food as soon as eating begins. Bennett[10] used an expert sensory panel to determine time–intensity curves for a range of fatty and reduced fat foods. It was concluded that the pattern of flavour release from non-fat foods was very different to flavour release from fatty foods.

The aim of this paper was to measure the effect of triglyceride content on volatile release from two foods during eating. The release of flavour volatiles in sweet biscuits and Frankfurter sausages, each prepared with two concentrations of triglyceride, was studied by monitoring their concentrations in the mouth and nose respectively. Volatile concentration in the mouth is generally higher than in-nose, so in-mouth analysis is preferable where volatile concentrations are low. A model mouth system was also used to examine the effect of hydration on the release of aroma compounds from the sweet biscuits.

2 MATERIALS AND METHODS

2.1 Materials

Sweet biscuits containing 4.6% and 16.5% triglyceride in the freshly baked biscuits ('4.6% fat' and '16.5% fat') were prepared using a recipe in which the fat content could be varied easily. Almond essence was added to the biscuits as a source of easily measurable volatiles. Frankfurter sausages with 5% and 30% triglyceride in the cooked sausages ('5% fat' and '30% fat') were prepared in conjunction with the National Food Centre, Dublin.

2.2 Methods

2.2.1 Model Mouth System. Crushed 16.5% fat biscuits were hydrated by the addition of water on a weight basis over the following range, 0% (*i.e.* no water added to biscuit), 10%, 25% and 50%, and the 4.6% fat biscuits were hydrated to 0%, 10%, 25%, 50% and 75% levels. Volatiles were sampled from the headspace onto traps containing Tenax GC and analysed by GC–MS. The peak areas relative to the internal standard (valeronitrile) of benzaldehyde were measured. This was carried out twice for each level of hydration.

2.2.2 In-Mouth System. A biscuit was placed in the mouth and chewed. Volatiles were sampled from the mouth by suction through traps containing Tenax at 0, 5, 10, 30, 40 and 60 s after beginning chewing and were analysed by GC–MS. Swallowing occurred at 35 s. The peak areas relative to the external standard (valeronitrile) of benzaldehyde were calculated as described previously.[7] Portions of Frankfurter sausages were chewed and volatiles were sampled from the nose at 0–10, 10–20, 20–30, 30–40, 40–50, 50–60 and 110–120 s. Swallowing occurred at 35 s. The peak areas of a number of terpenes were measured. For both foods, each analysis was conducted in triplicate and mean peak areas calculated.

3 RESULTS AND DISCUSSION

Figure 1 shows the release of benzaldehyde from 4.6% and 16.5% fat biscuits in the model mouth system. The model mouth system attempted to simulate hydration by saliva in the mouth. There was clearly a difference in the pattern of aroma release between the 4.6% and 16.5% fat biscuits. Benzaldehyde release from the 16.5% fat biscuit increased with increasing hydration to a maximum of approximately 25% hydration, after which the volatile level decreased. Volatile release from the 4.6% fat biscuit increased with hydration to a maximum at 75% hydration, which was the highest level of hydration measured. The amounts of benzaldehyde released from the model mouth were similar between 0% and 25% hydration for both the 4.6% and 16.5% fat biscuits, but at 50% hydration, larger amounts were released from the 4.6% fat biscuit.

Figure 2 shows the release of benzaldehyde in the mouth from 4.6% and 16.5% fat biscuits. Although the standard deviations were high for some points there was a significant difference between the release of benzaldehyde from the 4.6% and 16.5% fat biscuits. Higher amounts of volatiles were released from the 4.6% fat biscuit than the 16.5% fat biscuit. The pattern of release was also different between the two biscuits. Release from the 16.5% fat biscuit reached a maximum at 5 s and then remained constant whereas release from the 4.6% fat biscuit increased to a maximum at approximately 20 s and then decreased to a lower level.

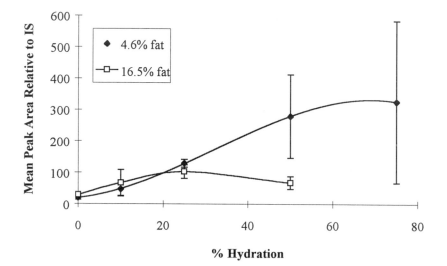

Figure 1   *Release of benzaldehyde from 4.6% and 16.5% fat biscuits in the model mouth.*
(Error bars represent ± one standard deviation)

For each biscuit, volatile release in the model mouth system was different to volatile release measured in-mouth which indicates that the model mouth headspace system cannot fully simulate the conditions in the mouth. Nevertheless, in both the model system and the in-mouth system there were clear differences in volatile release between the 4.6% and 16.5% fat biscuits. One possible explanation is the extent of dissolution of benzaldehyde in the fat phase; a greater proportion of benzaldehyde would have been dissolved in the fat in the 16.5% fat biscuit than the 4.6% fat biscuit leading to a lower level of volatile release into the mouth during eating or into the headspace in the model mouth system.

Figure 3 shows the release of a terpene into the nose from 5% and 30% fat Frankfurter sausages. Other terpenes showed similar patterns of release. Although variation between replicates was high for some points, there was a difference between the release of terpenes from the 5% and 30% fat Frankfurters. Higher amounts of terpenes were released from the 5% fat Frankfurter than the 30% fat Frankfurter. The pattern of release was also different between the two sausages. Release from the 5% fat Frankfurter reached a maximum at 15 s and then gradually decreased, whereas release from the 30% fat sausage gradually increased to a maximum at 35 s (possibly due to act of swallowing) and then decreased quickly after this point.

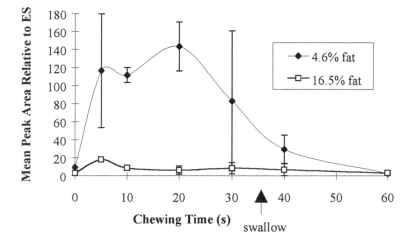

Figure 2 *Release of benzaldehyde from 4.6% and 16.5% fat biscuits in the mouth. (Error bars represent ± one standard deviation)*

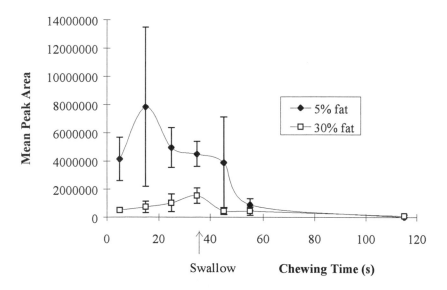

Figure 3 *Release of terpene (RT 10.6) from 5% and 30% fat Frankfurters in the mouth. (Error bars represent ± one standard deviation)*

Frankfurter sausages may be regarded as a complex emulsion in which globules of fat are stabilised in an aqueous–protein matrix.[11,12] While this description is simplistic, especially for the cooked sausage, it is interesting to interpret the flavour release in terms of models proposed to explain the release of flavour from emulsions.

A model described by McNulty,[13] represents appropriate foods as oil-in-water emulsions which are diluted by the addition of saliva during chewing. It is assumed that flavour compounds are transferred from the oil phase to the aqueous phase on dilution with saliva and that the perceived flavour is proportional to the concentration of flavour compound in the aqueous phase, C_w. If equilibrium were attained, the concentration of flavour compound in the aqueous phase (C_{we}) would be given by:

$$C_{we} = \frac{C_i}{\Phi_i \left(K_p - 1 \right) + D}$$

where C_i is the initial concentration of flavour compound, Φ_i the initial oil volume fraction, K_p the equilibrium partition coefficient and D the dilution factor. However, while equilibrium may be approached near the end of chewing, a non-equilibrium situation will exist immediately after dilution with saliva. The aqueous concentration of flavour compound (C_{wd}) will then be given by:

$$C_{wd} = \frac{C_{we} \left(1 - \Phi_i \right)}{D - \Phi_i}$$

Using this model, it is possible to calculate values for the relative aqueous concentrations of a terpene in Frankfurters containing 5% and 30% fat. For this purpose, C_i is taken as 100, Φ_i can be calculated as 0.06 and 0.37 respectively, D has been determined experimentally to be approximately 1.4, while K_p is *ca.* 200 for terpenes. Using these values, the initial and equilibrium concentrations have been calculated (Table 1).

Table 1   *Calculation of Initial Aqueous Concentrations (C_{wd}) and Equilibrium Aqueous Concentrations (C_{we}) of Terpenes on Chewing*

D^a	C_{wd}		R^b	C_{we}		R
	30%	5%		30%	5%	
1.4	0.81	5.4	6.7	1.3	7.7	5.8
2.0	0.51	3.8	7.3	1.3	7.4	5.6

[a]   The value (D=1.4) determined by chewing and spitting out portions of Frankfurter may be an underestimate of the true value as some saliva may be swallowed; however, a higher value (D=2) gives similar trends;

[b] $R = C_{5\%}/C_{30\%}$.

These data suggest that reducing the fat content from 30% to 5% increases both the initial and equilibrium aqueous terpene concentration by 5–7 times. Assuming that the concentration in the vapour phase is proportional to that in the aqueous phase, this model would therefore predict a similar difference between the concentrations measured in the nose from 30% and 5% fat Frankfurters; this is in agreement with the measurements illustrated in Figure 4. In addition, the model would predict that the difference between the two fat levels would be greater on initial dilution than at equilibrium; this is also observed.

As might be expected for a food not possessing an emulsion structure, the application of this model to sweet biscuits does not agree well with the observed results.

4 CONCLUSION

In vivo measurements of the release of flavour compounds from sweet biscuits and Frankfurter sausages have demonstrated that the low-fat versions release flavour compounds more rapidly than do the same products containing higher fat concentrations.

In Frankfurter sausages the release of volatiles can be explained reasonably well by the release model proposed by McNulty for emulsion systems. In the biscuits however, this model is inappropriate and a different release model which incorporates the rate of hydration would need to be developed. However, the measurement of volatile release in the mouth does show differences in release between high and low fat foods and could be used to optimise flavour delivery when reformulating food flavours in these products.

ACKNOWLEDGEMENTS

Two of us (F.F.V.C. and L.J.F.) gratefully acknowledge the funding and collaborative support received from the National Food Centre, Dublin as part of E.U. programme, AIR2-CT93-1691.

REFERENCES

1. W.E. Lee, *J. Food Sci.*, 1986, **51**, 249.
2. S.M. Van Ruth, J.P. Roozen and J.L. Cozijnsen, 'Trends in Flavour Research', eds. H. Maarse and D.G. Van der Heij, Elsevier, Amsterdam, 1994, p. 59.
3. D.D. Roberts and T.E. Acree, *J. Agric. Food. Chem.*, 1995, **43**, 2179.
4. R.S.T. Linforth and A.J. Taylor, *Food Chem.*, 1993, **48**, 115.
5. K.E. Ingham, R.S.T. Linforth and A.J. Taylor, *Flav and Fragr. J.*, 1995a, **10**, 15.
6. K.E. Ingham, R.S.T. Linforth and A.J. Taylor, *Lebensm.-Wiss. u.-Technol.*, 1995b, **28**, 105.
7. K.E. Ingham, A.R. Clawson, R.S.T. Linforth and A.J. Taylor, *J. Food Sci.*, 1996, in press.
8. C.M. Delahunty, J.R. Piggott, J.M. Connor and A. Paterson, 'Trends in Flavour Research', eds. H. Maarse and D.G. Van der Heij, Elsevier, Amsterdam, 1994, p. 47.
9. S.M. Van Ruth, J.P. Roozen and J.L. Cozijnsen, *Food Chem.*, 1995, **53**, 15.
10. C.J. Bennett, *Cereal Foods World*, 1992, 429.
11. L.L. Borchert, M.L. Greaser, J.C. Bard, R.G. Cassens and E.J. Briskey, *J. Food Sci.*, 1967, **32**, 419.
12. K.W. Jones and R.W. Mandigo, *J. Food Sci.*, 1982, **49**, 1930.
13. P.B. McNulty, 'Food Structure and Behaviour', eds. J.M.V. Blanshard and P.J. Lillford, Academic Press, London, 1987, p. 245.

THE RETENTION OF FLAVOUR IN ICE CREAM DURING STORAGE AND ITS RELEASE DURING EATING

B.M. King and N. Moreau

Quest International, 28 Huizerstraatweg, NL-1411 GP Naarden

1 INTRODUCTION

This paper considers flavour release from two different time perspectives: short-term in the mouth and long-term during storage. It is important to understand the storage properties of flavours used in ice cream because this product is hardly ever consumed 'fresh'. Although ice cream is usually stored under what might be considered as optimal conditions for flavour compounds (in cold, dark, sealed containers), there is ample opportunity for interactions between flavour compounds and the complex mixture of fat, protein, sugar, emulsifiers, stabilizers, water and air comprising an ice cream base. Most ice cream mixes are over-flavoured at production to counteract losses of flavour. Hansen[1] quantified losses for vanillin in refrigerated mix as 6%–8% in 24 hours, whereas up to 40% vanillin could be lost (bound to protein) during the pasteurization step.

Flavour release in the mouth has been reviewed.[2] More recently, Roberts and Acree[3] reviewed several methods for studying release by headspace analysis and proposed a new retronasal aroma simulator. Their study included several effects that are directly relevant to the evaluation of ice cream, in which the flavour perception is retronasal: saliva, temperature, shearing and oil. Recent studies of the volatility of flavour compounds from model food emulsions have looked at the type of emulsion as well as the roles of emulsifiers and the specific surface area present.[4,5] Because ice cream is a complex emulsion, many factors will be responsible for flavour release.

The work described in this paper was undertaken as a preliminary study of (a) the effect of a flavour molecule's aqueous solubility on its perception over time; and (b) the influence of fat dispersion in the emulsion on short-term flavour release in the mouth. Flavour perception was measured by a trained panel using profiling and time–intensity techniques. Sensory textural properties of stabilized ice cream have been studied previously by time–intensity,[6] but to our knowledge time–intensity measurements have not been made on ice cream with respect to flavour release.

2 EXPERIMENTAL

Ice cream samples evaluated in these experiments are listed in Table 1. The ice cream base for all samples contained butter oil (125 g kg^{-1}), skim milk powder (85 g kg^{-1}), sucrose (170 g kg^{-1}) and was stabilized with a mixture of locust bean gum (2.2 g kg^{-1}) and carrageenan (0.18 g kg^{-1}). Different levels of destabilized fat were obtained by varying the nature and amount of the emulsifier. The percent destabilized fat was determined by diethyl ether–

hexane extraction of the ice cream samples. All samples were flavoured with a vanilla extract (0.5 g kg^{-1}) in addition to either aubepine (*p*-methoxybenzaldehyde) or cyclotene (methylcyclopentenolone) in the dosages indicated in Table 1. Details of sample production as well as sample presentation for sensory evaluation by profiling have been given elsewhere.[7] For time–intensity measurements, samples of ice cream were presented on plastic spoons so that the entire portion could be taken into the mouth at one time. Samples A1, A2, A3, C1, C2 and C3, which were used for the storage tests, were maintained at −18 °C for three months and subsequently re-evaluated.

All samples were evaluated by flavour profiling. Only data for three flavour descriptors were considered in this paper: *anise*, *caramel* and *nutty*. Time–intensity measurements were made on samples A2, A3, A7, A9, C2, C3, C4 and C6. For aubepine-flavoured samples, the descriptor used for time–intensity was *anise*, for cyclotene-flavoured samples the descriptor *caramel* was chosen. Details regarding the sensory panel and all sensory methodology, including data processing, have been published elsewhere.[7,8]

Oil–water partition coefficients for aubepine and cyclotene were determined by dissolving 100 mg of the flavour compound in 20 g melted butter, adding 20 g water (pH 6.6) and shaking at 5 °C for about 50 hours. The concentration of flavour compound in the aqueous phase, after centrifugation and filtration, was determined by gas chromatography of an aliquot taken up in methanol.

Table 1   *Ice Cream Samples Evaluated*

Code	Aubepine (g kg^{-1})	Overrun %	Destabilized Fat %	Code	Cyclotene (g kg^{-1})	Overrun %	Destabilized Fat %
A1	0.025	90	15.5	C1	0.025	90	23
A2	0.1	90	15.5	C2	0.1	90	23
A3	0.4	90	15.5	C3	0.4	90	23
A4	0.8	65	7	C4	0.4	65	6
A5	0.8	65	17	C5	0.4	65	19
A6	0.8	65	19	C6	0.4	65	23
A7	0.8	100	7	C7	0.4	100	7
A8	0.8	100	19	C8	0.4	100	19
A9	0.8	100	31	C9	0.4	100	28

3 RESULTS AND DISCUSSION

Aubepine, which can be characterized by the sensory descriptor *anise*, is much less soluble in water than cyclotene which has a *nutty*, *caramel* character: the oil–water partition coefficients for these two flavour compounds were 19.2 and 0.23, respectively. Figure 1 gives the panellists' mean profiling scores for the descriptor *anise* plotted against the percent destabilized fat in each ice cream sample. The only significant differences are between samples flavoured with aubepine and those flavoured with cyclotene. The aubepine samples having 100% overrun do show a very slight but systematic increase in *anise* intensity (605-615-626) as the percent destabilized fat increases.

Figures 2 and 3 show that the cyclotene-flavoured samples scored significantly higher than the aubepine samples for the descriptors *nutty* and *caramel*, as could be expected. For the descriptor *nutty*, cyclotene-flavoured samples decreased slightly (not significantly) in intensity as the percent destabilized fat increased both at 65% overrun (491-477-473) and at 100% overrun (512-486-463). No trend was evident for the descriptor *caramel* despite the fact that cyclotene-flavoured samples were perceived as more *caramel* than *nutty*.

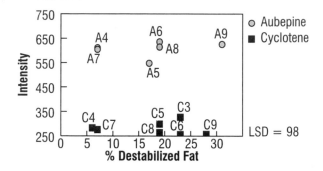

Figure 1   *Panellists' mean profiling scores for the descriptor ANISE in ice cream flavoured with aubepine (0.8 g kg⁻¹) or cyclotene (0.4 g kg⁻¹)*

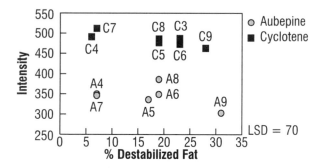

Figure 2   *Panellists' mean profiling scores for the descriptor NUTTY in ice cream flavoured with aubepine (0.8 g kg⁻¹) or cyclotene (0.4 g kg⁻¹)*

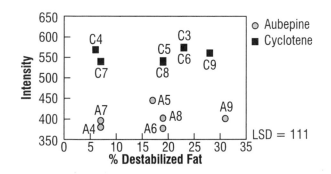

Figure 3   *Panellists' mean profiling scores for the descriptor CARAMEL in ice cream flavoured with aubepine (0.8 g kg⁻¹) or cyclotene (0.4 g kg⁻¹)*

It is evident from the data in Figure 4 that the differences in intensities measured by profiling for the three concentrations of each flavouring are much larger than the differences measured for the three levels of destabilized fat. For aubepine-flavoured samples, A3 was significantly more *anise* than A2 and A1: P = 0.0011 ($F_{2,36}$ > F). There was no significant difference between the two lower concentrations for this descriptor. The other two descriptors showed no significant differences for aubepine-flavoured samples.

Samples flavoured with cyclotene showed significant differences between the concentrations for the descriptors *caramel* and *nutty* but not for the descriptor *anise*. Sample C3 was significantly more *nutty* than C2 and C1: P = 0.0122 ($F_{2,34}$ > F). There was no significant difference between the two lower concentrations for this descriptor. For evaluations of the descriptor *caramel*, C3 was significantly more intense than C1 (P = 0.0407) whereas the differences between C1 and C2, or C2 and C3 were not significant.

Figure 4 also puts into perspective the changes observed in these samples over three months of storage. Aubepine-flavoured samples showed no significant decrease in *anise* although A1 decreased slightly in *caramel*: P = 0.0986, ($F_{1,17}$ > F). On the other hand, the *nutty* character of cyclotene decreased significantly for C3: P = 0.0047 ($F_{1,17}$ > F). Cyclotene-flavoured samples showed no significant decrease in *caramel* character during storage. Samples of unflavoured ice cream base showed no storage effects on these three descriptors.

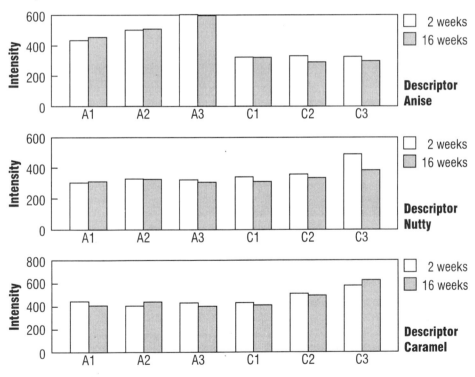

Figure 4 *Panellists' mean profiling scores showing storage effects on three concentrations of aubepine or cyclotene in ice cream*

Table 2   *Panel Means for Time–Intensity Parameters and P-Values from Two-Way*
ANOVA (Panellist, Product) for Aubepine-Flavoured Ice Cream

Parameter	Samples				Concentration	ANOVA Product Effects[a] Destabilized Fat
	A2	A3	A7	A9		
I_{max}	85	137	121	121	0.0208	0.8674
T_{max}	10.3	13.1	14.3	12.6	0.1210	0.2959
I_{swal}	36	58	50	49	0.1049	0.7829
T_{swal}	8.3	9.0	7.7	8.3	0.2035	0.0913
Release[b]	25	32	23	18	0.2006	0.1794
Area	1678	2825	2836	2960	0.0243	0.6323
$T_{50\%}$	25.2	28.9	28.4	32.2	0.0995	0.1477
$T_{30\%}$	32.7	39.7	41.8	42.4	0.0036	0.8301
$T_{20\%}$	38.3	46.9	49.3	48.7	0.0003	0.7827
$T_{10\%}$	46.3	54.7	57.1	55.2	0.0004	0.2324
$T_{5\%}$	54.5	59.3	60.8	60.7	0.0077	0.8896

[a] Prob $(F_{1,19} > F)$;
[b] I_{max}/T_{max}.

Panel means for the time–intensity parameters are given in Tables 2 and 3. Analysis of variance of these parameters showed significant concentration effects for both flavourings: maximum intensity, area under the curve, and the time to reach intensities equal to 50%, 30%, 20%, 10% and 5% of the maximum intensity. There were no significant effects on any time–intensity parameter for either flavouring due to differences in destabilized fat at the same overrun. These results corroborate the lack of significance shown by profiling tests although the trends discussed earlier are evident in the parameters Area for both flavours and the parameter I_{max} for cyclotene. The parameter Release showed no statistically significantly differences related to destabilized fat. Release was greater, however, for both flavours when the percent destabilized fat was lower, *i.e.*, the situation in which diffusion from oil phase to water phase is faster.

Time–intensity measurements have been used by many authors to study flavour release in foods of varying fat content. Usually the aspects of flavour investigated were basic tastes or mouthfeel.[9–11] These flavours were generally caused by water-soluble non-volatile molecules. Perception of these molecules from emulsions depends on their concentration in the aqueous phase. Shamil *et al.*[9] explained their data in terms of the 'effective concentration' of the hydrophobic flavour compounds: the rate of flavour release for sharpness, bitterness and astringency tended to be greater in the reduced-fat cheeses, although the difference was usually not significant. Saltiness, on the other hand, had a significantly lower flavour release in reduced-fat cheese despite its hydrophilic nature. This effect was explained in terms of the higher moisture content of reduced-fat foods which would decrease the effective concentration of NaCl.

Table 3 *Panel Means for Time–Intensity Parameters and P-Values from Two-Way ANOVA (Panellist, Product) for Cyclotene-Flavoured Ice Cream*

Parameter	Samples				Concentration	ANOVA Product Effects[a] Destabilized Fat
	C2	C3	C4	C6		
I_{max}	75	103	107	89	0.0436	0.1358
T_{max}	13.9	16.2	14.5	17.7	0.2139	0.1699
I_{swal}	32	47	47	39	0.1758	0.0816
T_{swal}	8.9	8.3	8.1	8.6	0.4265	0.1285
Release[b]	14	13	25	1.1	0.7599	0.1758
Area	1589	2430	2420	2262	0.0062	0.3755
$T_{50\%}$	28.0	33.1	30.3	34.8	0.0416	0.1451
$T_{30\%}$	35.6	42.2	41.9	45.1	0.0128	0.2580
$T_{20\%}$	41.7	49.2	48.0	52.3	0.0126	0.0892
$T_{10\%}$	48.4	55.6	54.8	58.8	0.0089	0.1262
$T_{5\%}$	52.6	59.1	58.7	61.0	0.0143	0.0935

[a] Prob $(F_{1,19} > F)$;
[b] I_{max}/T_{max}.

Prediction of retronasal perception for volatile flavours from foods or model emulsions is difficult due to the number of factors involved. Garlic was perceived as equally intense in fat-free and 10% fat mashed potatoes according to time–intensity studies reported by Rosin and Tuorila.[12] The present paper showed no significant difference in perception for either of the two ice cream flavours, independent of aqueous solubility, when the amount of destabilized fat in the emulsion was varied. These results support the model proposed by Overbosch *et al.*[2] in which flavour release was claimed to be independent of emulsion type, given the same flavour, oil and volume fraction of oil. Neither were significant differences observed between the two levels of overrun. Long-term flavour release after storage, on the other hand, was evident only for the aqueous-soluble flavour.

ACKNOWLEDGEMENTS

Appreciation is expressed to Susanne Schroff and Seeta Sockhal for their help in carrying out the sensory tests. Rien van Beek made all the ice cream samples and measured the destabilized fat content. The partition coefficients were determined by Louis Doorn.

REFERENCES

1. A.P. Hansen, *Modern Dairy*, 1989, **68** (6), 18.
2. P. Overbosch, W.G.M. Afterof and P.G.M. Haring, *Food Rev. Int.*, 1991, 7 (2), 137.
3. D.D. Roberts and T.E. Acree, *J. Agric. Food Chem.*, 1995, **43**, 2179.
4. D. Salvador, J. Bakker, K.R. Langley, R. Potjewijd, A. Martin and J.S. Elmore, *Food Qual. & Pref.*, 1994, **5**, 103.
5. P. Landy, J.-L. Courthaudon, C. Dubois and A. Voilley, *J. Agric. Food Chem.*, 1996, **44**, 526.

6. L.J. Moore and C.F. Shoemaker, *J. Food Sci.*, 1981, **46**, 399.
7. B.M. King, *Lebensm.-Wiss. u. -Technol.*, 1994, **27**, 450.
8. B.M. King and N. Moreau, in 'Consumer Preference and Sensory Analysis, ed. S. Porretta, Miller Freeman Technical Ltd., Maarssen, 1996, p. 98.
9. S. Shamil, L.J. Wyeth and D. Kilcast, *Food Qual. & Pref.*, 1991/92, **3**, 51.
10. C.R. Stampanoni and A.C. Noble, *J. Text. Studies*, 1991, **22**, 381.
11. H. Tuorila, C. Sommardahl and L. Hyvönen, *Food Qual. & Pref.*, 1995, **6**, 55.
12. S. Rosin and H. Tuorila, *Lebensm.-Wiss. u.-Technol.*, 1992, **25**, 139.

MODEL DEVELOPMENT FOR FLAVOUR RELEASE FROM HOMOGENEOUS PHASES

Deborah D. Roberts[*] and Terry E. Acree

Department of Food Science and Technology, Cornell University, New York State
Agricultural Experiment Station, Geneva, NY 14456, U.S.A.

1 INTRODUCTION

Flavour release devices simulating the release of aroma compounds in the mouth were used
to develop models of flavour release. The retronasal aroma simulator (RAS) was used to
investigate the effects of oil and temperature on flavour release. A device directly connected
to a mass spectrometer was used to determine the effect of viscosity. Addition of oil and
thickeners to water reduced aroma release, especially of highly volatile compounds.
Compounds that were non-polar were naturally more affected by oil than polar compounds.
Based on the results, a model for flavour release in homogeneous foods was developed from
simple foods with homogeneous phases. This model seeks to extend the predictive power of
Henry's law, which was developed for aqueous solutions. It is based on oil–water–air
partitioning theory and empirical relationships of viscosity and temperature. It was tested
with data from this work and found to exhibit a correlation of $R^2 = 0.90$ for all compounds
tested.

Models aim to predict a property of a system that is based on other properties in order
to give insight into governing factors. In food, knowledge of what factors influence aroma
release into the air will allow us to utilize aroma effectively.

One of the most basic theories describing volatile compounds in dilute solutions is
Henry's law. It states that the partial pressure of a solute above the solution is proportional
to its molar concentration in the liquid. This theory has been tested for dilute solutions
containing an homologous series of volatiles and found to be true up to the saturation point
of the flavour compound.[1,2] However, foods are usually not simple solutions and contain
many components such as fats, carbohydrates, proteins, salts and other components which
greatly affect the values. A revised 'food version' of Henry's law would be an asset to those
who are involved in flavouring food products.

Several reviews of flavour release studies[3,4] emphasized the need for a better
understanding of food-flavour interactions and under more complex food consumption
conditions. A recent model of flavour release utilized partitioning theory (air–oil, air–water,
air–oil) and incorporated the variable of percentage oil.[5] This model was developed for
equilibrium as well as dynamic conditions and was able to predict the released flavour in
milk under static conditions and a cake under dynamic conditions. Another model, proposed

---

[*]Present address: Nestle Research Center, Vers-Chez-les-Blanc, Case Postale 44, CH-1000
Lausanne 26, Switzerland

for solid foods and tested with boiled sweets, was based on a two layer stagnant film theory where both surface-area effects and mass and heat transfer were important.[6]

While much research remains to be done to understand basic mechanisms, a need also exists for predictions of aroma release in complex foods involving more than one phase and differing textures. In order to meet these needs, a model has been developed based on experimental results from homogeneous phases. This model attempts to predict aroma released from the primary properties of the food and the aroma compound.

2 EXPERIMENTAL

2.1 Retronasal Aroma Simulator

One of the devices used to produce data on flavour release was the retronasal aroma simulator,[7] a purge-and-trap device made from a blender. It simulated mouth conditions by regulating temperature to 37 °C, adding synthetic saliva and blending at a shear rate reported to occur during eating. Volatiles were collected on a silica trap and quantified by GC–FID and GC–MS to determine the volatilty rate constant.

Results from two experiments using the retronasal aroma simulator were used in model development.[7] In one study, which determined the effect of mouth temperature on flavour release, the release of five volatile compounds was compared at room temperature (23 °C) and body temperature (37 °C). Another study compared the release of six compounds in a water and soybean oil–water (83:17 v/v) mixture.

2.2 Direct MS Measurement of Flavour Release

This device simulated the shearing and temperature of the mouth and allowed a measurement in real time of the released aroma compounds. The protonated molecular ions for each compound were simultaneously monitored under chemical ionization conditions and the maximum ion abundance (I_{max}) was determined.[8] Carboxymethylcellulose, guar gum, and sucrose solutions of different viscosities (12 and 160 mPa s) were compared with water for their release of seven compounds.[9]

3 RESULTS AND DISCUSSION

3.1 Model Development

Model development was based on partition theory where partitioning of compounds between air and a solution of oil and water, defined as release (R), is shown by the following mass balance:

$$R_{air:oil-water} = \frac{X_{air}}{X_{oil-water}} = \frac{C_a V_a}{C_o V_o + C_w V_w} \tag{1}$$

where X is the amount in air or food (g), C the concentration of the compound (g l^{-1}), V the volume (l), and the subscripts a, o and w are air, oil and water respectively.

By definition, $P_{aw} = C_a/C_w$ and $P_{ow} = C_o/C_w$. Substitutions were made for C_a and C_o to give the following equation:

$$R_{air:oil-water} = \frac{P_{aw} V_a}{P_{ow} V_o + V_w} \tag{2}$$

In order to extrapolate this equation to phases other than just oil and water, a few assumptions were made. The effects of carbohydrates are mainly due to a change in viscosity. Although proteins exhibit specific binding with certain molecules, the general effect is hydrophobic in nature and is usually outweighed by the effect of oil in the system. Of all food components, oil has the largest influence on flavour compound volatility. Therefore, food components were broken down into two categories, oil and non-oil, where non-oil includes water, carbohydrates and proteins. P_{ow}, the oil–water partition coefficient wasn't readily reported for compounds; consequently, the octanol–water partition coefficient ($P_{oct-water}$) was substituted.

Equation (2) was extended to include parameters which had a significant influence on flavour release by the addition of empirically derived components. The proposed revision of this equation is

$$R_{air:food} = \left(\frac{P_{aw} V_a}{P_{oct-water} V_o + V_w^*} \right) \left(\frac{\eta_0}{\eta} \right)^{0.1} \exp\left(\frac{T - T_0}{21.5} \right) \tag{3}$$

where T is the temperature in K, η the viscosity of the food in cps, V_0 and V_w fractions which sum to unity, V_w the volume of non-oil components and T_0 and η_0 are standard conditions of 298 K and 1 cps (water).

Variables were included to account for the effects of temperature and viscosity. The data used in this modelling study were from diverse experiments which used different conditions, devices and analysis methods. While this was not ideal for model development, it may provide a realistic basis for other diverse methods. In all of the data, $P_{oct-water}$ was substituted for $P_{oil-water}$ and the volatility rate constants for the compounds determined with the retronasal aroma simulator were substituted for P_{aw}.[7] These substitutions were also not ideal and actual values for P_{ow} and P_{aw} would probably have given a better correlation. In order to put the data in a format that could be compared, the release data were transformed to values of percentage change from water. When comparing analytical data, in general, a percentage change of at least 30% is necessary for the human olfactory system to perceive a difference.[10]

Table 1 *Effect of Viscosity on Released Aroma. Values are % Decrease in I_{max} from Water and are Average Values for the Three Thickeners*

Compounds	Low Viscosity	High Viscosity
High Volatility		
α-Pinene	28	55
Ethyl-2-methylbutyrate	43	72
1,8-Cineole	24	50
Low Volatility		
2-Methoxy-3-methylpyrazine	4	15
Methyl anthranilate	8	10
Maltol	13	8
Vanillin	11	8
Predicted Value	22	40

Increasing viscosity (Table 1) reduced I_{max} for the highly volatile compounds but not for compounds with low volatility. For the three most volatile compounds, this decrease was 32% and 59%, respectively for low and high viscosity samples. The most volatile compounds fit best with $(\eta_0/\eta)^{0.15}$, with a 32% and 54% respective decrease in volatility.

However, the average values for all compounds fit best with $(\eta_0/\eta)^{0.1}$, giving a 22% and 40% decrease for the low and high viscosity samples. This difference in the viscosity effect, based on how volatile the compound is in water, may be due to a different rate determining step in the release mechanisms. For example, the highly volatile compounds were more affected by an increase in viscosity because their rate-limiting step to volatilization was the movement to the surface. However, the rate-limiting step for the low volatility compounds was the actual volatilization at the surface so there was little effect of changes in the viscosity.

Table 2   *Effect of Temperature Regulation to 37 °C on the Volatility Rate Constant*

Compounds	% increase from 23 °C water
1,8-cineole	110
2-methoxy-3-methylpyrazine	130
methyl anthranilate	120
2-acetyl pyridine	90
o-aminoacetophenone	90

A doubling of the volatility rate constant occurred with about a 15 °C increase in temperature (Table 2). Data with Henry's law constants[11] confirm this observation. The temperature component of equation (3), $\exp\{(T-T_0)/21.5\}$, incorporated the temperature effect.

Table 3   *Prediction Ability of Equation 3 for Compound Release in Oil–Water Model Systems*

Compound	Predicted % decrease from water	Actual % decrease from water
α-Pinene	99.9	99.99
Ethyl-2-methylbutyrate	64	99.2
1,8-Cineole	74	99
Butyric acid	16	0
2-Methoxy-3-methylpyrazine	71	85
Methyl anthranilate	94	38
Propanol (1% fat)	2	0 [a]
Propanol (5% fat)	7	8 [a]

[a]   See Schirle-Keller *et al.*[13]

Table 3 shows the accuracy prediction using equation (3). The more non-polar compounds would show a greater percentage decrease in release from water with added oil than the polar compounds. The $P_{oct-water}$ value for propanol $(2.51)^{12}$ and the equation were used to closely predict the percentage change in release with small amounts of added oil.[13] The degree to which the octanol–water partition coefficient predicts the oil—water partition coefficient influences the accuracy of the results.

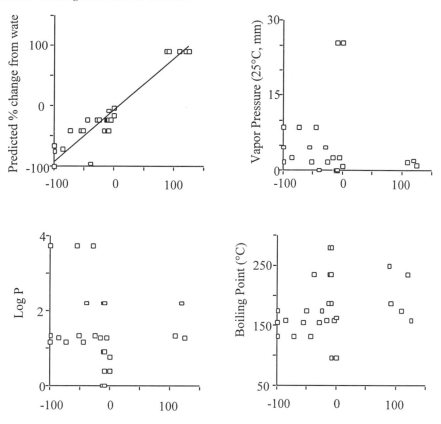

Measured % change in volatility from water

Figure 1 *Comparison of R from equation (3) with other compound properties for their ability to predict the measured percentage change in release from water*

Figure 1 shows the prediction ability of equation (3) and compares it to the prediction ability of other compound parameters. It is evident that the octanol–water partition coefficient, the boiling point and the vapour pressure do not predict release in foods with homogeneous phases. R from equation (3), however, showed a clear relationship between predicted and actual release (R^2 = 0.9). Equation (3)'s strength is its ability to predict and compare the release of aroma compounds in more complex food systems.

4 CONCLUSIONS

Analysis methods such as headspace analysis and gas chromatography–mass spectrometry do require equipment and time. Utilization of models based on experimental research may allow predictions of aroma release prior to product formulations and reduce trial and error product development. This model is a basic attempt to use research on homogeneous phases to produce a model that can be applied to actual foods. It takes into account factors of the food (oil content, viscosity and temperature) and factors of the aroma compound (air–water and oil–water partition coefficients) to give a basis for comparison between food prototypes. However, lacking in the literature are reports of the partition coefficients for aroma compounds. Generation of these basic data would be a key step forward for model development.

REFERENCES

1. C. Weurman, *Food Technol.*, 1961, **15**, 531.
2. R.G. Buttery, L.C. Ling and D.G. Guadiagni, *J. Agric. Food Chem.*, 1969, **17**, 385.
3. P. Overbosch, W.M. Afterof and P.G.M. Haring, *Food Rev. Int.*, 1991, **7**, 137.
4. J. Bakker, 'Flavor Interactions with the Food Matrix and their Effects on Perception', ed. J. Bakker, Marcel Dekker, New York, p. 411.
5. K.B. de Roos and K. Wolswinkel, 'Non-equilibrium Partition Model for Predicting Flavour Release in the Mouth', eds. K.B. de Roos and K. Wolswinkel, Elsevier, Amsterdam, 1994, p. 15.
6. B.P. Hills and M. Harrison, *Int. J. Food Sci. and Technol.*, 1996, **30**, 425.
7. D.D. Roberts and T.E. Acree, *J. Agric. Food Chem.*, 1995, **43**, 2179.
8. J.S. Elmore and K.R. Langley, *J. Agric. Food Chem.*, submitted.
9. D.D. Roberts, J.S. Elmore, K.R. Langley and J. Bakker, *J. Agric. Food Chem.*, 1996, **44**, 1321.
10. J.C. Walker and R.A. Jennings, 'Comparison of Odor Perception in Humans and Animals', eds. J.C. Walker and R.A. Jennings, Springer-Verlag, Berlin, 1991, p. 261.
11. D. Mackey, W.Y. Shiu, K.T. Valsaraj and L.J. Thibodeaux, 'Air–Water Transfer – The Role of Partitioning', eds. D. Mackey, W.Y. Shiu, K.T. Valsaraj and L.J. Thibodeaux, ASCE, New York, 1992, p. 34.
12. A. Leo, C. Hansch and D. Elkins, *Chem. Rev.*, 1971, **71**, 525.
13. J.P Schirle-Keller, G.A. Reineccius and L.C. Hatchwell, *J. Food Sci.*, 1994, **59**, 813, 875.

MEASUREMENT OF THE INFLUENCE OF FOOD INGREDIENTS ON FLAVOUR RELEASE BY HEADSPACE GAS CHROMATOGRAPHY–OLFACTOMETRY

S. Widder[*] and N. Fischer[†]

[*]Corporate Research Division and [†]Flavor Division, DRAGOCO Gerberding & Co. AG, D-37601 Holzminden, Germany

1 INTRODUCTION

The acceptability of a food depends mainly on its sensory qualities and in particular on its flavour. The flavour sensation is caused by flavour molecules released into the vapour phase during eating and subsequently transported to the olfactory epithelium. To perceive an aroma, the flavour compounds need to achieve a sufficiently high concentration in the vapour phase to stimulate the olfactory receptors.[1]

Concentration of aroma compounds, and therefore aroma perception during eating, depends on the nature and concentration of the volatiles present in the food as well as on their availability to perception. Availability is influenced in part by the process of eating, such as mastication, temperature and the effect of saliva, but mainly by interactions between aroma compounds and non-volatile food constituents, such as fats, proteins and carbohydrates.[2] The types of interactions vary with the nature of the food component and the volatile compounds. Examples are solution, entrapment, formation of covalent bonds, hydrogen bonds and physical adsorption via hydrophobic bonds.[3-5] Thus the composition of a food product greatly influences the performance of a flavouring and therefore the sensory quality. Any change in a food matrix requires a change or a modification of a given flavouring in order to optimize its performance. Consequently, with regard to the necessity to develop flavours that are optimally adjusted to different food systems, the knowledge of the nature and intensity of flavour–food interactions is of great interest for a flavourist.

The binding of flavour and flavour release can be studied by different methods. On the one hand sensory methods such as descriptive sensory analysis are used to describe and quantify the influence of the food composition on specific flavour attributes leading to flavour profiles. On the other hand flavour release can be investigated by analysing the volatiles in the gaseous headspace above the food sample.

Many studies concerning the interactions between volatile substances and non-volatile compounds have been carried out in simple systems composed of water,[6] proteins,[7,8] fat[6] and carbohydrates.[9] Since most of the headspace techniques applied had low sensitivity, especially in regard to strong flavour compounds occurring in trace amounts, the concentration of the flavour material used was relatively high. Recently, a headspace-gas chromatography–olfactometry (HS–GC–O) method with increased sensitivity was described.[10] Trace compounds of sensory relevance not detectable by means of HS–GC–FID or HS–GC–MS can be detected using the human nose for detection. Using this method, common flavouring levels can be applied.

The object of this study was (a) to investigate the influence of fat content and protein content on the flavour release from simple model systems as well as from more complex systems such as milk by an optimized HS–GC–O method; (b) to establish the overall flavour profile of the different samples by descriptive sensory analysis; and (c) to correlate the results of the technical and the sensory analysis in order to obtain information required for optimally adjusting flavours to a given matrix.

In our study, the process of flavour release during eating was simulated in a very simplified way by investigating the flavour composition in the headspace above samples which were continuously stirred at physiological temperature (37 °C).

2 EXPERIMENTAL

2.1 Sample Preparation

2.1.1 Model Emulsions. Water (50 °C) and a mixture of oil and emulsifier (70 °C) were homogenized by means of an Ultraturrax high-speed mixer (4000 min^{-1}, 30 sec). After cooling to room temperature a mayonnaise flavour cocktail (0.5%) was added and the mixture equilibrated overnight. Ingredients of the flavour cocktail were diacetyl, (Z)-3-hexenol, butyric acid, allyl isothiocyanate, allyl thiocyanate and (E,Z)-2,6-nonadienol.

2.1.2 Protein Models. Casein (3%, 6%, 9% and 12%, according to Hammersten) was dissolved in a buffer system (41.3% (v/v) KH$_2$PO$_4$, 0.07 mol l^{-1}; 58.7% (v/v) Na$_2$HPO$_4$ 0.2 mol l^{-1}). After adjusting the pH to 7, a melon flavoured cocktail (0.01%) was added and the mixture equilibrated overnight. Ingredients of the flavour cocktail were ethyl butyrate, ethyl-2-methyl butanoate, ethyl hexanoate, heptanal, (E,Z)-2,6-nonadienal, (E)-2-nonenal, damascenone, β-damascone and β-ionone.

2.1.3 Flavoured Milk. Milk containing different amounts of fat (0.3%, 1.5% and 3.5%) was spiked with the melon flavoured cocktail (0.01%). Samples were equilibrated overnight.

2.2 Headspace Analysis

2.2.1 Volatile Sampling. For sampling the headspace volatiles, the following experimental design was used. The system comprised two gas-tight syringes, coupled by a valve. Both were placed in a thermostated environment (37 °C). The first syringe represents the flavour release vessel and contains the sample (20 g), a stirring bar and a defined headspace volume above the sample. After reaching equilibrium, the headspace was transferred into the second syringe by simultaneously moving both plungers of the syringe without changing the headspace volume. After closing the valve, the second syringe was removed and the headspace sample was injected into the GC.[10]

2.2.2 Analysis of Volatiles (HS–GC and HS–GC–O). The headspace volatiles were analysed by means of a gas chromatograph (Siemens Sichromat 1-4) using a thin film capillary DB-1 (30 m × 0.32 mm). For detection, an FID (HS–GC) and a sniffing device (HS–GC–O) was used. Furthermore, the gas chromatograph was equipped with a PTV (programmable temperature vaporizer), filled with 2 mg Tenax TA 80-100 mesh. After injection of the headspace sample into the cooled PTV (2 °C), the trapped volatiles were thermally desorbed onto the capillary by raising the temperature (800 °C min^{-1}) to 170 °C. Simultaneously, the oven temperature of 60 °C was raised by 4 °C min^{-1} to 250 °C.

The intensity of the flavour compounds perceived during HS–GC–O by sniffing the GC effluent was scored on a scale from 1 ('Weak Aroma') to 4 ('Strong Aroma').

2.3 Sensory Analysis

First, specific flavour attributes for the description of the sensory profile were chosen both by GC–O analysis and by discussion within the sensory panel consisting of five experienced persons. The attributes for the different samples were scored on a scale from 1 ('Weak') to 4 ('Strong').

3 RESULTS AND DISCUSSION

3.1 Influence of Fat Content on Flavour Release

The fat content is an important variable in a food matrix. It is often reduced in order to reduce calorie intake to make food healthier. Indeed, fat cannot simply be removed as it makes a significant contribution to the sensory properties of foods in several ways.[11] One important point is that fat is a good solvent of flavour compounds and influences the vapour pressure of the volatiles, thereby affecting the perceivable aroma profile. Hence good fat based flavourings tend to become unbalanced or even off-flavoured in aqueous or reduced fat systems.[12, 13]

In order to investigate the influence of fat content on flavour release and aroma profile, we first analysed mayonnaise flavoured model emulsions containing 1%, 5% or 20% fat by means of HS–GC and HS–GC–O. The concentration of the flavouring was equal to the concentration normally used for the production of 'real' mayonnaise.

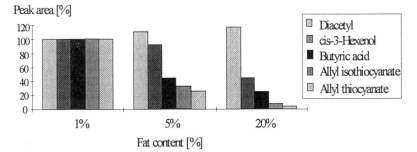

Figure 1 *Influence of fat content on flavour release of individual flavour components of model emulsions containing a mayonnaise flavour cocktail*

The results of the HS–GC analysis shown in Figure 1 imply that increasing fat content leads to a decrease in the concentration of all flavour compounds with the exception of diacetyl. This is reasonable since these compounds are more soluble in the fat phase than in water. Therefore, an increasing fat content results in a dilution of these compounds in the fat phase. Hence, their vapour pressure and consequently their concentration in the headspace is decreased. Due to their somewhat higher polarity, (Z)-3-hexenol and butyric acid were less influenced than allyl thiocyanate and allyl isothiocyanate. On the other hand, the influence of fat concentration on the vapour pressure of diacetyl – a water soluble flavour compound – was minimal. A slight increase could be observed.

Table 1   *Influence of Fat Content on Flavour Release of Individual Flavour Component (Headspace GC–O Analysis)[a]*

Compound	Flavour Quality	Intensity[b]		
		1%	5%	20%
Diacetyl	Buttery	3	4	>4
(Z)-3-Hexenol	Green	2	1	—
(E,Z)-2,6-Nonadienol	Green, fatty	4	<1	—
Allyl isothiocyanate	Pungent, mustard-like	4	3	<1
Allyl thiocyanate	Pungent, mustard-like	4	3	<1

[a]   See experimental;

[b]   Sensory intensity of the compound during Headspace–GC–olfactometry scored on a scale from 1 ('Weak') to 4 ('Strong'). Values are means of three replicates.

The effect on (E,Z)-2,6-nonadienol, a further ingredient of the mayonnaise flavour cocktail could not be detected by using the FID. Obviously, its concentration in the headspace was below the detection limits of the instrument.

For such compounds, especially those of sensory importance, the HS–GC–O method is much more suitable as it uses the human nose for detection.

Table 1 summarizes the intensities of flavour compounds perceived during HS–GC–O analysis of the different model emulsions. With this method, (E,Z)-2,6-nonadienol could be detected and it particularly exhibited a high intensity in the low fat model. Increasing the fat content results in a decrease in its intensity. In the sample containing 20% fat, (E,Z)-2,6-nonadienol was of no sensory importance. Concerning the other flavour compounds, good agreement with the results in Figure 1 could be observed. Table 1 also demonstrates the capability of the HS–GC–O method in discriminating between volatiles and flavour compounds of sensory relevance: butyric acid positively detected by the FID was not perceivable during GC–O.

Finally the comparison of the different models suggested that, except for diacetyl, the intensities of all flavour compounds were low in the high fat model, whereas in the low fat models the intensity of all components was high.

As shown in Figure 2, this change in balance of flavour compounds directly affected the sensory profile. The three models exhibited totally different aroma profiles.

In the model emulsion containing 20% fat – the fat level for which the flavour was developed – the overall aroma was described as creamy, mayonnaise-like and balanced. The profile analysis gave high scores for the attributes buttery and fatty which can be related to the high intensity of diacetyl observed during HS–GC–O, and the high fat content. Decreasing the amount of fat changed the profile to more pungent and green notes, whereas the intensity of the buttery attribute was lowered.

These results demonstrate a good correlation with the results of the HS–GC–O method since the attribute 'pungent' can be related to allyl thiocyanate and allyl isothiocyanate and the attribute green to (Z)-3-Hexenol and (E,Z)-2,6-nonadienol. Thus, using these two methods, information for the necessary corrections for the concentration of individual components of the flavour cocktail leading to an optimised flavour performance in a given matrix can be obtained.

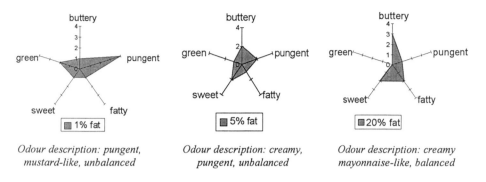

Odour description: pungent,
mustard-like, unbalanced

Odour description: creamy,
pungent, unbalanced

Odour description: creamy
mayonnaise-like, balanced

Figure 2 *Influence of fat content on the sensory profile of model emulsions (intensities of the flavour attributes were scored on a scale from 1 to 4)*

3.2 Influence of Protein Content on Flavour Release

In addition to fats, proteins belong to another important class of components in food systems capable of influencing flavour release. We used casein in order to investigate flavour–protein interactions. Model solutions were prepared by dissolving different concentrations of casein (0% (reference), 3%, 6% and 12%) in a buffer system and finally by spiking with a melon flavour cocktail (see Section 2 for composition).

During HS–GC analysis, ethyl butyrate, ethyl-2-methylbutyrate, ethyl hexanoate and heptanal could be detected by the FID. Comparison of the peak areas (data not shown)[14] implied a slight but continuous decrease of the esters with increasing protein content up to 6%. Further increase of protein lowered the ester concentrations much more markedly. The weak influence of protein observed at the beginning can be attributed to hydrophobic interactions. The much stronger effect in the more concentrated protein solutions is probably caused by the high viscosity produced by the formation of a gel network. A decrease in flavour release with increasing viscosity has already been reported.[15] Unlike the esters, the level of the aldehyde, heptanal, was reduced greatly even by *minor* protein concentrations. This behaviour can be explained by the possible reactions of aldehydes with amino groups of the proteins leading to covalent bonds. Again, higher protein concentration results in a further strong reduction of the headspace concentration of heptanal.

The results of the HS–GC–O analysis for the other flavour compounds not detectable by FID are given in Table 2. For the esters and heptanal, the results were in agreement with those obtained by HS–GC–FID analysis. Furthermore a decrease of the aldehydes (E,Z)-2,6-nonadienal and (E)-2-nonenal, having cucumber and fatty-like flavour qualities, and of the ketones β-damascone and β-ionone, having sweet and floral odour qualities, could be observed. β-Damascenone was scarcely influenced. Obviously, its concentration was still too high to perceive differences in perception.

Comparison of these data with the results of the descriptive sensory analysis (data not shown)[14] again showed a good correlation. Cucumber and fatty-like aspects attributed to the aldehydes were perceived only in the pure buffer system. The fruity attributes related to the esters first showed a slight decrease, then a further reduction in the more concentrated protein solutions.

Table 2   *Influence of Casein Content on Flavour Release of Individual Flavour*
Components (Headspace–GC–O)[1]

Compound	Flavour Quality	Intensity[2]			
		0%[3]	3%	6%	12%
Ethyl butanoate	Fruity	4	3	3	3
Ethyl 2-methyl butanoate	Fruity	3	2	2	1
Ethyl hexanoate	Fruity	4	4	4	3
Heptanal	Fatty	3	2	1-2	—
(E,Z)-2,6-Nonadienal	Cucumber	1–2	—	—	—
(E)-2-Nonenal	Fatty	2	—	—	—
Damascenone	Sweet, floral	4	4	4	3
β-Damascone	Sweet, floral	2	1	1	<1
β-Ionone	Sweet, floral	3	—	—	—

[1]   See experimental;
[2]   Sensory intensity of the compound during Headspace–GC–olfactometry scored on a
       scale from 1 ('Weak') to 4 ('Strong'). Mean values of three replicates;
[3]   Buffer system.

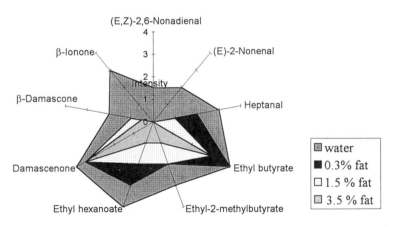

Figure 3   *Headspace gas chromatography–olfactometry analysis of flavoured milk:*
influence of fat content on the intensity of different flavour components

3.3 Influence of Fat Content on the Flavour Release from Milk

In order to investigate flavour release, not only in simple models but also in real
complex foods, we spiked milk of different fat content with the melon flavour cocktail and
analysed the headspace composition by HS–GC–O. Water was used as a reference.
Descriptive sensory analysis was performed and the results are given in Figures 3 and 4.

The overall flavour of the water reference was described as green, cucumber- and melon-like. Whereas the overall flavour of the milk samples was described as sweet, fruity and raspberry-like, already indicating an influence of the food matrix on flavour release. Comparison of the sensory profiles (Figure 4) shows a lack of the fatty and cucumber attributes in the milk samples explaining the shift to a more fruity and sweeter overall flavour.

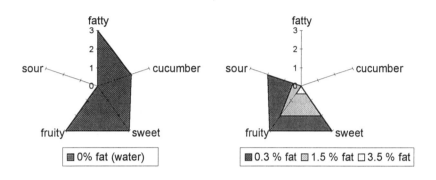

Figure 4 *Influence on fat content on the sensory profile of milk containing a melon flavour cocktail*

The HS–GC–O analysis (Figure 3) evaluating the single flavour compounds showed the presence of all components of the flavour in the headspace of the reference. In milk containing 0.3 % fat, β-ionone and the aldehydes (E,Z)-2,6-nonadienal and (E)-2-nonenal were no longer perceptible, thus explaining the lack of the fatty and cucumber odour qualities and the shift in the sensory profile. With the exception of ethyl butyrate the intensities of the other flavour compounds were lower than in the reference. Increasing fat content in milk results in a further decrease of the intensities of all the components. These results showed a good correlation with the decrease of the overall flavour intensity observed during sensory profile analysis.

The differences between water and milk containing 0.3% fat were caused by interactions with the ingredients of the milk, especially proteins, and to a lesser extent carbohydrates, salts and the small fat portion. As shown by the sensory profile analysis, the interactions with the complex matrix of milk resulted in a complete change in the overall flavour impression, whereas increasing fat content only led to a decrease in intensity.

4 CONCLUSIONS

For measurement of flavour release, a headspace gas chromatography–olfactometry method was used which allows the application of authentic flavouring levels. This method makes it possible to measure the influence of food ingredients on the release behaviour of individual flavour components. If the results of HS–GC–O, concerning the perceived odour qualities, are used to determine descriptors for sensory analysis, the correlations between individual flavour compounds and the overall flavour properties can be used to adjust the performance of a flavour compound in a given matrix.

Furthermore, the effect of food ingredients on the performance of a flavour compound in a simple as well as in a complex matrix could be studied by this method.

ACKNOWLEDGEMENT

We like to thank Mrs. E. Gruber for her valuable technical assistance and Mr. S. Färber for the development of the flavour cocktail.

REFERENCES

1. L. Buck and R. Axel, *Cell*, 1991, **65**, 175.
2. P. Overbosch, W.G.M. Agterot and P. Haring, *Food Rev. Int.*, 1991, 7, 137.
3. H.G. Maier, *Angew. Chem.*, 1970, **9**, 917.
4. J. Solms, F. Osman-Ismail and M. Beyeler, *Can. Inst. Food Sci. Tech. J.*, 1973, **6A**, 10.
5. A. Voilley, C. Lamer, P. Dubois and M. Feuillat, *J. Agric. Food Chem.*, 1990, **38**, 248.
6. R.G. Buttery, D.G. Guadagni and L.C. Ling, *J. Agric. Food Chem.*, 1973, **21**, 198.
7. M. Le Tanh, P. Thibeaudeau, M.A. Thibaut and A. Voilley, *Food Chem.*, 1992, **43**, 129.
8. P. Landy, C. Druaux and A. Voilley, *Food Chemistry*, 1995, **54**, 387.
9. W.W. Nawar, *J. Agric. Food Chem.*, 1971, **19**, 1057.
10. N. Fischer and T. van Eijk, A.C.S. Symposium Series, 1996, in press.
11. D.A. Forss, *J. Agric. Food Chem.*, 1969, **17**, 681.
12. L.C. Hatchwell, *Food Technol.*, 1994, **2**, 98.
13. H. Plug and P. Haring, *Trends Food Sci. Technol.*, 1993, **4**, 150.
14. N. Fischer and S. Widder, lecture presented at the I.F.T. Symposium, 1996.
15. D.D. Roberts, J.S. Elmore, K.R. Langley and J. Bakker, 'Bioflavor '95', eds. P. Etievant and P. Schreier, INRA Editions Paris, 1995, p. 127.

CHANGES IN THIOL AND DISULFIDE FLAVOUR COMPOUNDS RESULTING FROM THE INTERACTION WITH PROTEINS

Donald S. Mottram, Ian C. Nobrega, Andrew T. Dodson and J. Stephen Elmore

The University of Reading, Department of Food Science and Technology, Whiteknights, Reading RG6 6AP, U.K.

1 INTRODUCTION

Sulfur-containing volatile compounds generally have low odour threshold values and many are important in determining the aroma characteristics of foods. Thiol-substituted furans, such as 2-methyl-3-furanthiol (1) and 2-furanmethanethiol (3), and the corresponding disulfides, 2 and 4, have been shown to have meat-like or roasted, coffee-like aromas at low concentrations. During thermal processing, such compounds may be formed in the Maillard reaction[1-3] or from the degradation of thiamin.[4,5] These thiols and disulfides are widely used as components of flavourings for soups, savoury products and meat substitutes, where they are either added to the flavourings as nature-identical chemicals or as components of reaction-product flavourings.

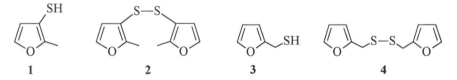

It was recognised in the early 1970s that sulfur-substituted furans, such as 1 and 2, had meat-like aroma characteristics and exceptionally low odour threshold values.[6,7] However, it was not until quite recently that compounds of this type were isolated from cooked meat.[8-10]

In aqueous solution, thiols are readily oxidised to the corresponding disulfides, and mixtures of different thiols readily form mixed disulfides.[11,12] It was also shown that, in boiling aqueous solution, disulfides were hydrolysed to thiols, which were detected using 4-vinylpyridine as a thiol trapping agent.[13] In the absence of such a reagent, free thiols were not detected because they re-oxidise to the disulfide, but in a mixture of deuterated and unlabelled disulfides mixed disulfides were obtained. Such changes could result in modification of the sensory properties of foods containing thiols and disulfides. In proteins, reduction–oxidation (redox) reactions involving interchange of sulfhydryl and disulfide groups, within the protein or with external thiol groups, are well known.[14] This raises the possibility of thiol and disulfide flavour compounds interacting with proteins in foods and causing changes in their relative concentrations. Recently, we demonstrated that when disulfides 2 and 4 were heated in aqueous solution with a protein, such as egg albumin, some of these disulfides were lost but a large proportion of each was converted to the corresponding thiol.[15] Interaction with sulfhydryl groups in the protein appeared to be responsible.

Meat contains a significant number of free sulfhydryl groups and disulfide bonds within the proteins and interactions with thiol and disulfide aroma compounds during cooking or post-cooking treatments, such as volatile extraction during aroma analysis, could influence the relative proportions of thiols and disulfides found in the meat.

This paper reports marked differences which have been found in the quantities of volatile thiols and disulfides in cooked meat isolated by two different aroma extraction methods, and provides evidence that the differences arose from interaction with protein.

2 EXPERIMENTAL

2.1 Preparation of Cooked Beef

Beef fillet (*M. Psoas major*) was purchased from a local butcher; any remaining fat was trimmed and 100 g portions were vacuum packed and stored at −20 °C until required. The meat was chopped in small pieces, minced in a blender, transferred to 500 ml bottles, closed with PTFE-lined lids and cooked in an autoclave at 140 °C for 30 min.

2.2 Addition of Disulfides to Meat

Bis (2-methyl-3-furanyl) disulfide (**2**) was obtained as a gift from a flavour company, and bis (2-furanylmethyl) disulfide (**4**) was purchased from Aldrich Chemical Co. Solutions of the disulfides were prepared in ethanol (1 mg ml^{-1}) and 0.5 ml portions of each were added to 100 g cooked minced beef before volatile extraction using SDE.

2.3 Simultaneous Distillation-Extraction (SDE)

Each sample (cooked beef, cooked beef + disulfides, or disulfides alone) was blended with 750 ml of distilled water and volatiles extracted using SDE in a modified Likens-Nickerson apparatus. Extractions were carried out for two hours with a mixture of re-distilled n-pentane (27 ml) and diethyl ether (3 ml) as the solvent. After extraction, an internal standard (1,2-dichlorobenzene, 130 μg in 0.1 ml diethyl ether) was added, and the solvent was dried and concentrated to about 0.5 ml by distillation in a Kuderna-Danish apparatus.

2.4 Headspace Concentration on Tenax TA

Cooked beef was chopped and transferred to 250 ml conical flask and volatiles were collected in a glass-lined stainless steel trap (105 mm × 3 mm internal diameter) containing 85 mg Tenax TA (SGE Ltd.), as described by Madruga and Mottram.[10] During the collection, the conical flask was maintained at 60 °C and the volatiles were swept onto the trap using a flow of oxygen-free nitrogen (40 ml min^{-1}); the collection was continued for 1.5 hours. At the end of this time, the flask was removed, an internal standard (1,2-dichlorobenzene, 100 ng in 1 μl of diethyl ether) was added to the front end of the trap which was connected directly to the nitrogen supply for 5 min (40 ml min^{-1}) to remove moisture prior to GC–MS analysis.

2.5 Gas Chromatography–Mass Spectrometry (GC–MS)

A Hewlett-Packard HP5890 gas-chromatograph coupled to a HP5972 mass selective detector was used. A split/splitless injection was used to introduce 1 μl of each SDE extract onto the GC column. The headspace traps were thermally desorbed, at 250 °C, directly on to the cooled (0 °C) GC column using a CHIS injection port (SGE Ltd.). A BPX5 fused silica capillary column (50 m × 0.32 mm; 0.5 μm film thickness; SGE Ltd.) was used for all analyses, with helium at 1.6 ml min^{-1} (35 cm s^{-1}) as carrier gas. The oven was initially held at 50 °C for 2 min, and then programmed at 4 °C min^{-1} to 250 °C. The interface of the GC to the MS was maintained at 280 °C, and the MS was operated in the electron impact mode with an ionisation energy of 70 eV and a scan rate of 1.9 scans s^{-1} over the mass range 33–400 amu.

Table 1 *Comparison of the Quantities[a] of some Volatile Sulfur Compounds Isolated from Meat by SDE and Headspace Concentration*

Compound	SDE	Headspace
3-Mercapto-2-butanone	20 (1)	–
3-Mercapto-2-pentanone	13 (3)	–
2-Mercapto-3-pentanone	8 (1)	–
2-Methyl-3-furanthiol	4 (1)	–
2-Furanmethanethiol	62 (17)	tr
Dimethyl disulfide	–	356 (201)
Dimethyl trisulfide	9 (2)	253 (82)
Dimethyl tetrasulfide	–	29 (14)
Dipropyl disulfide	–	7 (1)
2-Furanylmethyl methyl disulfide	2 (1)	23 (7)
2-Methyl-3-furanyl methyl disulfide	–	7 (2)
2-Furanylmethyl methyl trisulfide	–	6 (3)
2-Methyl-3-furanyl methyl trisulfide	–	tr
bis(2-Furanylmethyl) disulfide	–	tr
2-Furanylmethyl 2-methyl-3-furanyl disulfide	–	tr
4-Methylthiazole	–	11 (2)
4,5-Dimethylthiazole		5 (1)
2-Acetylthiazole	10 (3)	26 (7)
3-Methylthiophene	–	66 (32)
2-Thiophenecarboxaldehyde	7 (1)	21 (7)
3-Methyl-2-thiophenecarboxaldehyde	9 (3)	5 (3)
5-Methyl-2-thiophenecarboxaldehyde	2 (1)	7 (2)
2,3-Dihydro-6-methylthieno[2,3c]furan	5 (2)	12 (0)

[a] Quantities expressed as relative total chromatogram (GC–MS) peak area × 10,000. Values are the mean of triplicate analyses and standard deviations are shown in parentheses;
– Not detected;
tr Trace.

A solution containing C$_6$–C$_{22}$ n-alkanes was also analysed to allow calculation of linear retention indices (LRI) for each sample component. Identifications were made by comparison of mass spectra and LRI with those obtained from GC–MS analysis of authentic compounds, or by comparison of mass spectra with those in the NIST/EPA/MSDC spectral collection or in published literature. Quantities of compounds **1–4** in SDE extracts were determined by comparison of the peak areas in the total ion chromatogram with areas of the compounds in analysis of standard solutions, using 1,2-dichlorobenzene as internal standard. Other compounds were quantified by comparison of TIC peak areas with those of the internal standard, using a response factor of 1.

3 RESULTS AND DISCUSSION

The method used for volatile extraction influenced the relative proportions of sulfur-containing compounds isolated from the boiled meat (Table 1). The effect was particularly pronounced with thiols and disulfides. Extracts prepared by simultaneous steam distillation solvent extraction (SDE) contained very few disulfides and no furan-substituted disulfides, although a number of thiols, including **1** and **2**, were found. When volatiles were examined using headspace concentration on Tenax TA, followed by thermal desorption, disulfides were found but with much lower quantities of thiols than in the SDE extracts.

Table 2   *Recovery of Thiols and Disulfides from Cooked Meat Systems containing added bis(2-Furanylmethyl) Disulfide and bis(2-Methyl-3-furanyl) Disulfide*

Compound	Quantity added (μg)	Quantity recovered (μg)	
		Water + Disulfides	Cooked Beef + Disulfides[a]
bis(2-Furanylmethyl) disulfide	500	437	104 (27)
bis(2-Methyl-3-furanyl) disulfide	500	532	126 (27)
2-Furanmethanethiol	1[b]	< 1	294 (41)
2-Methyl-3-furanthiol	1[b]	< 1	302 (35)
2-Furanylmethyl 2-methyl-3-furanyl disulfide	—	16	39 (6)
2-Methyl-3-furanyl methyl disulfide	—	—	7 (1)
2-Furanylmethyl methyl disulfide	—	—	3 (0)

[a]   Mean of triplicate analyses, standard deviations shown in parentheses;
[b]   The disulfides contained very small amounts of the corresponding thiols as impurities;
—   Not detected.

These results suggested that redox reactions occurred during extraction and/or analysis by GC–MS. In order to determine how the extraction conditions affected such reactions in meat, the disulfides **2** and **4** were added to meat, after it had been cooked, and the meat was subjected to SDE. Recoveries of disulfides and thiols were compared with those obtained in extractions from aqueous solution. In the presence of meat, marked decreases in the concentration of the disulfides were observed with increases in the corresponding thiols (Table 2). A small quantity of the mixed disulfide, 2-furanylmethyl 2-methyl-3-furanyl disulfide, was also found, together with the methyldithio-derivatives 2-methyl-3-furanyl methyl disulfide and 2-furanylmethyl methyl disulfide. However, quantitatively all of the lost disulfides were not recovered as thiols or mixed disulfides, indicating that binding to the protein may have occurred. No thiols were found in SDE of the aqueous solutions containing the disulfides without meat although a small amount of the mixed disulfide, 2-furanylmethyl 2-methyl-3-furanyl disulfide, was found.

The most likely explanation for these observations is interaction between the thiol and disulfide flavour compounds with sulfhydryl groups and disulfide bridges, from cysteine and cystine amino acid units in the protein. This appears to involve conversion of the furan disulfides to the corresponding thiols and the formation of new disulfide links between protein and thiol with the associated loss of flavour compound from the volatile extract.

Similar observations have been made in our laboratory when the individual disulfides were added to egg albumin and the mixture extracted by SDE.[15] With casein there was only a relatively small change in the disulfides and there was no evidence of significant interaction of the disulfides with the carbohydrate substrate, maltodextrin (Table 3).

Table 3   *Quantities of Disulfides and Thiols Recovered from Aqueous Systems containing Protein or Carbohydrate Substrates. Adapted from Mottram et al.[15]*

Compound	Quantity added (µg)	Quantity recovered (µg) [a]		
		Egg Albumin	Casein	Malto-dextrin
Experiment 1				
bis(2-Furanylmethyl) disulfide	500	3 (1)	377 (63)	492 (10)
2-Furanmethanethiol	1 [b]	127 (14)	12 (4)	1 (1)
Experiment 2				
bis(2-Furanylmethyl) disulfide	500	11 (6)	393 (23)	566 (17)
2-Furanmethanethiol	40	158 (14)	16 (1)	1 (0)
Experiment 3				
bis(2-Methyl-3-furanyl) disulfide	500	6 (2)	272 (21)	378 (24)
2-Methyl-3-furanthiol	1 [b]	226 (54)	67 (8)	1 (1)

Footnotes – see Table 2.

Albumin and meat protein contain much greater proportions of –SH groups than casein, which explains the different behaviours of the two proteins towards the flavour compounds. Such sulfhydryl–disulfide interchange reactions are well known in protein chemistry.[14,16] An example of such interchanges is seen in bread-making where sulfhydryl–disulfide interchanges between glutathione and the flour proteins are important in relation to dough rheology.[17,18]

The odour threshold values of the thiols may differ from those of the corresponding disulfides, *e.g.*, the odour threshold value for the disulfide **2** is reported as 2×10^{-5} µg kg^{-1} while that of the corresponding thiol **1** is 5×10^{-3} µg kg^{-1}. The aroma characteristics of these compounds may also change with concentration. In general, at low concentrations approaching the odour threshold values, the compounds have pleasant savoury or roasted aromas but at higher concentration they become more sulfurous and unpleasant.[19–21] Redox-induced changes in the relative concentrations of these thiols and disulfides due to protein–sulfhydryl interchanges could result in significant changes in the sensory properties when the compounds are used in flavourings for food products. The thiol-disulfide exchanges may also contribute to the different aroma characteristics of meat cooked under different conditions. The results demonstrate that chemical, as well as physical, interactions between aroma compounds and food components are important determinants of food flavour.

REFERENCES

1. L.J. Farmer, D.S. Mottram and F.B. Whitfield, *J. Sci. Food Agric.*, 1989, **49**, 347.
2. L.J. Farmer and D.S. Mottram, in 'Flavour Science and Technology', eds. Y. Bessière and A.F. Thomas, Wiley, Chichester, 1990, p. 113.
3. T. Hofmann and P. Schieberle, *J. Agric. Food Chem.*, 1995, **43**, 2187.
4. L.M. van der Linde, J.M. van Dort, P. de Valois, B. Boelens and D. de Rijke, in 'Progress in Flavour Research', eds. D.G. Land and H.E. Nursten, Applied Science, London, 1979, p. 219.
5. P. Werkhoff, J. Brüning, R. Emberger, M. Güntert, M. Köpsel, W. Kuhn and H. Surburg, *J. Agric. Food Chem.*, 1990, **38**, 777.
6. W.J. Evers, H.H. Heinsohn, B.J. Mayers and A. Sanderson, in 'Phenolic, Sulfur and Nitrogen Compounds in Food Flavors', eds. G. Charalambous and I. Katz, American Chemical Society, Washington, D.C., 1976, p. 184.
7. G.A.M. van den Ouweland and H.G. Peer, *J. Agric. Food Chem.*, 1975, **23**, 501.
8. U. Gasser and W. Grosch, *Z. Lebensm. Unters. Forsch.*, 1988, **186**, 489.
9. P. Werkhoff, J. Brüning, R. Emberger, M. Güntert and R. Hopp, in 'Recent Developments in Flavor and Fragrance Chemistry', eds. R. Hopp and K. Mori, VCH, Weinheim, 1993, p. 183.
10. M.S. Madruga and D.S. Mottram, *J. Sci. Food Agric.*, 1995, **68**, 305.
11. D.S. Mottram, M.S. Madruga and F.B. Whitfield, *J. Agric. Food Chem.*, 1995, **43**, 189.
12. T. Hofmann, P. Schieberle and W. Grosch, *J. Agric. Food Chem.*, 1996, **44**, 251.
13. H. Guth, T. Hofmann, P. Schieberle and W. Grosch, *J. Agric. Food Chem.*, 1995, **43**, 2199.
14. P.C. Jocelyn, 'Biochemistry of the SH group', Academic Press, London, 1972.
15. D.S. Mottram, C. Szauman-Szumski and A. Dodson, *J. Agric. Food Chem.*, 1996, **44**, 2349.
16. G.M. Whitesides, J. Houk and M.A.K. Patterson, *J. Org. Chem.*, 1983, **48**, 112.
17. X. Chen and J.D. Schofield, *J. Agric. Food Chem.*, 1995, **43**, 2362.
18. W. Grosch, in 'Chemistry and Physics of Baking', eds. J.M.V. Blanshard, P.J. Frazier, and T. Galliard, Royal Society of Chemistry, London, 1986, p. 602.
19. S. Arctander, 'Perfume and Flavor Chemicals', published by the author, Monclair, NJ, 1969.
20. R. Tressl and R. Silwar, *J. Agric. Food Chem.*, 1981, **29**, 1078.
21. S. Fors, in 'The Maillard Reaction in Foods and Nutrition', eds. G.R. Waller and M.S. Feather, American Chemical Society, Washington, D.C., 1983, p. 185.

INTERACTIONS BETWEEN WOOD AND DISTILLATE COMPONENTS IN MATURED SCOTCH WHISKY

J.M. Conner, J.R. Piggott, A. Paterson and S. Withers

Centre for Food Quality, University of Strathclyde, Department of Bioscience and Biotechnology, Glasgow, G1 1XW, U.K.

1 INTRODUCTION

During the maturation of Scotch whisky, the dissolution of wood components is of prime importance and has been used to predict the development of mature characteristics in the spirit.[1] Concurrently, less pleasant characteristics of new distillate, described as oily, soapy and grassy, are lost or masked. The chemical composition of the distillate however remains remarkably unchanged during maturation; indeed the concentrations of most components increase due to the evaporation of ethanol and water. The presence of wood components has also been shown to affect the solubility of medium- to long-chain esters when the spirit is diluted for sensory analysis.[2] On dilution with water, a supersaturated solution of the alcohol soluble ester is formed and the excess ester forms agglomerates. These agglomerates also incorporate shorter chain esters, alcohols and aldehydes from the solution, decreasing their free solution concentration, which results in lower headspace concentrations.[3] The presence of wood components, extracted during maturation, greatly increases the incorporation of hydrophobic compounds into agglomerates, further reducing their free solution concentration and consequently the headspace concentration. There is, however, a minimum free solution concentration that may be achieved by the addition of either wood extractives and/or agglomerate forming esters.

The headspace effects of wood components are least obvious in heavy-bodied whiskies with high concentrations of long chain esters. Wood extracts have been shown to affect the size and stability of the ester agglomerates formed on dilution.[4] It is therefore possible that the stable agglomerates formed in the presence of wood-extract release lower concentrations of distillate components in the mouth and so affect the perceived flavour of the matured spirit. The two major changes that occur when the spirit is taken into the mouth are an increase in temperature and dilution by saliva. This presentation details an investigation of the effect of temperature on the size and stability of agglomerates and on the release of ethyl esters into the headspace.

2 RESULTS AND DISCUSSION

2.1 Size and Stability of Agglomerates

Particle sizes were determined using photon correlation spectroscopy (PCS). Correlator functions were analysed by the methods of cumulants[5] and the hydrodynamic diameter of the particles was calculated by the Stokes–Einstein relation.[6] Recorded sizes were found to be time dependent, particularly for new distillate and ester models. Consequently, samples were diluted, immediately filled into the sample cell and the size recorded at 5 minute intervals for at least two hours. Solutions were characterised by a mean rate of size change and an initial size (by extrapolation to 0 min).

Samples of a malt distillate aged for 5 years in glass and in four types of casks were characterised (Table 1). Significant differences were found in the rate of change for the new and used cask- and glass-matured samples. No significant differences were observed for initial size. The difference could be reproduced using solutions containing the major medium- and long-chain esters of the distillate either with wood extract added ('mature' model) or with no added wood extract ('immature' model).

Table 1   *PCS Measurements at 25 °C for the Same Malt Distillate Matured for Five Years in Glass, New Charred and Three Ages of Used Casks (increasing age from 1 to 3)*

Sample	Initial size (nm)	Rate of Change (nm min^{-1})	Size after 4 hours (nm)
Glass	107	0.477	199
New char	99	0.278	136
Used 1	96	0.455	176
Used 2	93	0.484	187
Used 3	103	0.574	194
Immature model	108	0.539	217
Mature model	86	0.236	136

Experiments were then repeated for the model systems at 20, 25, 30, 35 and 40 °C but valid results were only obtained up to 30 °C. Above this temperature, size determination became unreliable due to a loss of signal. Signal strength was maintained by increasing the concentration of ethyl hexadecanoate to 50 mg l^{-1}. For the higher concentration, initial sizes of both models showed no significant change with increasing temperature (mean of 113.6 nm for 'immature' model; 80.5 nm for 'mature' model). For the 'immature' model a significant increase in the rate of change was observed only between 25 and 30 °C (0.288 nm min^{-1} for 20 and 25 °C; 0.512 nm min^{-1} for 30, 35 and 40 °C). For the 'mature' model the rate increased steadily from 0.035 nm min^{-1} at 20 °C to 0.208 nm min^{-1} at 40 °C. Differences in the initial sizes and rates were observed between models containing low and high concentrations of ethyl hexadecanoate. Longer chain esters have been found to form smaller agglomerates and also decrease the size of agglomerates formed by shorter chain esters in mixed solutions. Temperature appears to have a very limited effect on the size or rate of change of ester agglomerates. The largest difference in rates observed in the 'immature' models (between 25 and 30 °C) is most probably related to the melting point of ethyl hexadecanoate (24 °C).

2.2 Headspace Effects

Ethyl decanoate was used in the experiments as a typical hydrophobic aroma compound. Two model solutions were used to measure the effect of temperature on the headspace concentration of ethyl decanoate:

1. With long chain ester concentrations typical of bottled malts;

2. With long chain ester concentrations typical of cask strength spirits (before chill filtration).

Samples were analysed with esters only ('immature' models) and with added wood extract ('mature' models).

Headspace concentrations of ethyl decanoate solutions were determined using previously described methods.[2] Headspace concentrations of ethyl hexadecanoate were determined from solutions prepared in the published manner, using a 7 μm polymethylsiloxane coated micro-extraction fibre. The fibre was exposed in the analyte headspace for 5 minutes, then desorbed in the SPI injector of a Finnegan MAT ITS40 GC–MS. Peak areas were determined by the sum of responses to *m/z* 88, 101 and 284. The solubility of both esters was followed using solute activity. Solute activity has been shown to approach 1 as the limit of solubility is reached.[2]

Results from the model spirits are shown in Figures 1 and 2. 'Mature' models at both higher ester concentrations gave linear increases in headspace concentration with temperature. For 'immature' models headspace concentrations were higher. Changes in the rate of release occurred between 30 and 35 °C for the bottled malt model and between 35 and 40 °C for the cask strength model. The differences between the mature and immature models suggest that there are two mechanisms of volatile suppression. The suppression caused by wood extractives appears to be more stable to raised temperatures. However the results of the particle size analysis do not indicate a direct link with the size or stability of agglomerates.

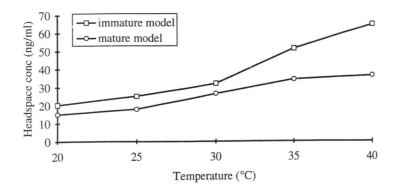

Figure 1 *Ethyl decanoate headspace concentrations of bottled malt models at different temperatures*

For 'immature' models, suppression is partly due to the displacement of decanoate from the solution by other esters. Increasing temperature reduced the solute activity of ethyl decanoate indicating an increase in the solubility of the ester (Figure 3). At 20 and 25 °C the ethyl decanoate concentration is close to its limit of solubility (activity is close to one).

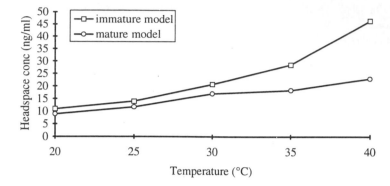

Figure 2 *Ethyl decanoate headspace concentrations of cask strength malt models at different temperatures*

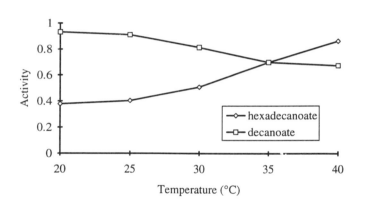

Figure 3 *Activities of 25 mg l⁻¹ solutions of ethyl esters at different temperatures*

Higher temperatures increase the limit of solubility and consequently, the concentration of unit activity. Thus for a solution with a constant concentration, as the temperature increases, the activity decreases. However ethyl hexadecanoate activity increased with temperature. At the lower temperatures the limit of solubility of ethyl hexadecanoate is < 1 mg l⁻¹ so most hexadecanoate forms agglomerates. The increased solubility at higher temperatures caused more hexadecanoate to dissolve, reducing the excess concentration. Theoretically, the concentration of hexadecanoate free in solution should remain at its limit of solubility until the limit is greater than the solution concentration. This should result in a measured solute activity close to one until this limit is passed, after which the activity should start to decrease. The low activities at the lower temperatures suggest that the presence of high concentrations of ester agglomerates can displace ester from solution, reducing solute activity and consequently headspace concentration.

Figure 4 shows the results from simplified whisky models containing only two esters. It showed an increased release of ethyl decanoate at 40 °C for 25 mg l^{-1} ethyl hexadecanoate but no such change at the higher concentration. It thus seems likely that the presence of high concentrations of ester agglomerates can suppress the release of volatiles from spirits. The earlier results indicate that while the higher ester concentrations found in whisky, particularly bottled malts, are sufficient to suppress the release of volatiles at room temperature, they are not sufficient to prevent greater release in the mouth. The results also show that the addition of wood components can reduce the release of volatiles without ester agglomerates. When these are present however, the effect is greatly enhanced and remains so even at higher temperatures.

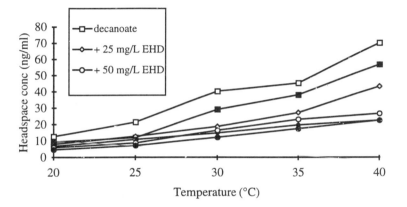

Figure 4 *Ethyl decanoate headspace concentrations of models with different concentrations of ethyl hexadecanoate (EHD). Filled symbols are for models with added wood extract*

3 CONCLUSIONS

The results suggest different mechanisms of volatile suppression in mature and immature spirits. In immature spirits the presence of ester agglomerates results in suppression of volatile release at low temperatures. Increasing the temperature of the solution causes more ester to dissolve. This reduces the excess concentration that forms agglomerates and the resultant volatile suppression. This results in increased headspace concentrations of hydrophobic aroma compounds. In mature spirits wood components can suppress the release of hydrophobic aroma compounds even when there are no agglomerates present. However the presence of ester agglomerates enhances the suppression of volatile release even at higher temperatures.

These results indicate that interactions between wood and distillate components may be of particular importance for the reduced perception of new distillate and immature characters of a mature whisky in the mouth.

ACKNOWLEDGEMENTS

This research is funded by the Biotechnology and Biological Sciences Research Council of the U.K. and the Chivas and Glenlivet Group, Paisley, Scotland. Dr. T. Whateley and Dr. N. Shankland, Department of Pharmaceutical Sciences provided practical and theoretical assistance for the particle size determinations.

REFERENCES

1. J. Piggott, J. Conner, A. Paterson and J. Clyne, *Int. J. Food Sci. Technol.*, 1993, **28**, 303.
2. J. Conner, A. Paterson and J. Piggott, *J. Sci. Food Agric.*, 1994, **66**, 45.
3. J. Conner, A. Paterson and J. Piggott, *J. Agric. Food Chem.*, 1994, **42**, 2231.
4. A. Paterson, J. Conner, D. Horne and J. Piggott, in 'Proceedings of the Fourth Aviemore Conference on Malting, Brewing and Distilling', eds. I. Campbell and F. Priest, Institute of Brewing, London, 1994, p. 222.
5. D. Koppel, *J. Chem. Phys.*, 1972, **57**, 4814.
6. D. Brooksbank, C. Davidson, D. Horne and J. Leaver, *J. Chem. Soc. Faraday Trans.*, 1993, **98**, 3419.

INTERACTION OF AROMA COMPOUNDS WITH β-LACTOGLOBULIN

E. Jouenne and J. Crouzet

Laboratoire de Génie Biologique et Sciences des Aliments, Unité de Microbiologie et Biochimie Industrielle Associée à l'INRA, Université de Montpellier II, 34095 Montpellier II 34095 Montpellier Cedex 05, France

1 INTRODUCTION

Whilst the interaction between aroma compounds and proteins has been known about empirically for a long time, it has been investigated scientifically only recently. The interest in such research has been prompted by the difficulties encountered in removing off-flavours from proteinaceous products such as soy or fish meals,[1] and by the interest shown in the flavouring of food products such as light foods.[2]

To date, several more or less purified protein systems, soy protein isolates, actomyosin, bovine serum albumin, caseinates, β-lactoglobulin and a small number of volatile compounds, have been investigated, generally using equilibrium methods developed for the study of non volatile ligands: liquid–liquid partition, equilibrium dialysis, gel filtration and ultrafiltration.[3]

Since these methods are not particularly suited to the study of interactions involving aroma compounds, headspace methods based on the volatility of these molecules have been developed. Both static and dynamic approaches have been used.[3]

In the present work, headspace techniques have been used to study interactions between several aroma compounds (including an homologous series of methyl ketones and ethyl esters) and β-lactoglobulin, a globular protein possessing a well-defined tridimensional structure. This protein is known to bind retinol and vitamin A as well as related compounds such as β -ionone and other volatile compounds.[4,5]

2 HEADSPACE MEASUREMENT

A preliminary study was carried out using static headspace measurement. A decrease in the concentration of methyl ketone and ethyl ester in equilibrium, at 25 °C and pH 3.4, with aqueous solutions (3%, w/w) of commercial β-lactoglobulin (> 90 % purity) containing these compounds, relative to pure water is shown in Figure 1. This indicated the retention of these compounds by the protein. The decrease of aroma concentration in the headspace was determined by gas chromatography and expressed as the peak decrease. This varied from 5% for 2-hexanone to 60% for 2-undecanone in the homologous series of methyl ketones. Similar results were obtained for ethyl esters, the decrease of aroma concentration in the atmosphere in equilibrium with the solution varied from 39% for ethyl hexanoate to 46% for ethyl octanoate. A discrepancy was, however, noticed for ethyl nonanoate, where the decrease was only 40%. The increase in interactions resulting from an increase in chain length indicated the hydrophobic nature of the interactions. This result is in good agreement with previously reported data obtained using equilibrium dialysis to determine the interaction of β-lactoglobulin and aromatic hydrophobic compounds[6] or alkanones.[5]

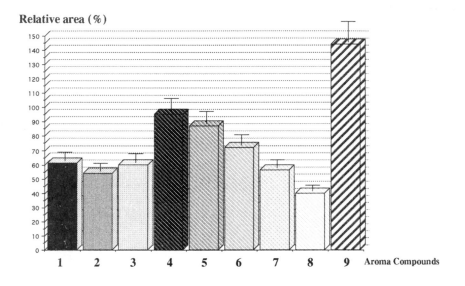

Figure 1   *Relative area of volatile compounds present in the headspace in the presence of β-lactoglobulin (3%, w/w) – 1: ethyl hexanoate; 2: ethyl octanoate; 3: ethyl nonanoate; 4: 2-hexanone; 5: 2-heptanone; 6: 2-octanone; 7: nonanone; 8: 2-undecanone; 9: limonene*

Retention of 15%–20% was found for several aroma molecules: acetophenone, benzaldehyde, linalool and 2,5-dimethyl pyrazine. Surprisingly the retention obtained for β-ionone was only 30%, with poor reproducibility of measurements. When limonene was used as the ligand, on the other hand, there was a salting-out effect, as indicated by a large increase of volatility (144%).

3 EXPONENTIAL DILUTION

When a dilute, aqueous solution of a volatile compound contained in an equilibrium cell is stripped by an inert gas, the concentration of the solute in the gas phase decreases according to an exponential law. The concentration decrease is related to the infinite dilution activity coefficient γ^∞ by the expression:

$$\ln S \;=\; \ln S_0 + \frac{D}{RT}\frac{p_i^s}{N}\gamma^\infty t$$

where S is the GLC peak area, S_0 the GLC peak area extrapolated to zero time, D the gas carrier flow rate, N the number of solvent moles, R the gas constant, p_i^s the vapour pressure of the pure solute, T the temperature and t the time. The infinite dilution activity coefficient can be calculated from the slope of the straight line obtained by plotting ln S *versus* time.

Anti-foaming agent was used to prevent β-lactoglobulin foaming. Interaction was evaluated according to changes in the reduced infinite dilution coefficient measured at 25 °C.[7] This coefficient was defined as the ratio of the value of the volatile compound activity coefficient measured for a β -lactoglobulin water solution (2%, w/w) in the presence of the anti-foaming agent and the value obtained for water in the presence of the anti-foaming agent. A decrease of this ratio indicates fixation by the solute whereas an increase implies a salting-out effect.

Table 1 *Log P Values, Relative Infinite Dilution Activity Coefficients and Retention of Volatile Compounds in the Presence of β-Lactoglobulin (2% w/w, pH 3.0)*

Compound	$\gamma^{\infty}(a)$	$\gamma^{\infty}(b)$	$\gamma^{\infty}(b/a)$	Retention (%)	log P
2-Heptanone	3955	3658	0.925	+7.5	1.82
2-Octanone	6869	5640	0.82	+18	2.35
2-Nonanone	19026	11177	0.59	+41	2.88
Ethyl hexanoate	18934	12454	0.66	+34	3.62
Ethyl octanoate	151850	77546	0.51	+49	4.68
Ethyl nonanoate	230468	183041	0.79	+21	5.21
Limonene	22704	24009	1.06	−6	4.40
Myrcene	20805	26586	1.28	−28	4.75

a γ^{∞} determined in the presence of anti-foaming;

b γ^{∞} determined in the presence of anti-foaming and β-lactoglobulin.

3.1 Determinations at pH 3.0

The hypothesis that hydrophobic interactions exist between methyl ketones, ethyl esters and β-lactoglobulin is supported by the results obtained for the relative infinite dilution activity coefficient determined at pH 3.0 (Table 1). Furthermore, the salting out effect observed when using the static headspace technique for limonene was confirmed and also detected for another terpene, myrcene.

Surprisingly, the behaviour of the terpene compounds having log P values (calculated according to Rekker[8]) higher than those calculated for methyl ketones and ethyl esters cannot be explained hydrophobically. Other factors might be involved in the retention of these compounds, one possibility being that steric hindrances limit the access of these compounds to hydrophobic sites, which would also explain the unusual behaviour of ethyl nonanoate in the ethyl ester series.

3.2 Determinations at Different pH

Interactions between β-lactoglobulin and aroma compounds were determined at several pH values from pH 2.0 to pH 11.0, and expressed as retention percent (Figure 2).

The retention of methyl ketones and ethyl esters was approximately the same at pH 2.0 as at pH 3.0, due to the presence of the protein at these pH in a stable monomeric form, without denaturation. An important increase in retention occurred at pH 6.0 and pH 9.0, due to changes in the monomer–dimer equilibrium at room temperature. When the pH was increased from 3.0 to 6.0 there was an increase in the dimeric form, whereas between pH 6.0 and pH 9.0 a disaggregation process accompanied by changes in conformation occurs. This process was followed by a partial alkaline denaturation.[9] The determination carried out at pH 11.0 revealed a large decrease in retention for 2-octanone, 2-nonanone and ethyl octanoate whereas an inversion of the phenomenon, is observed for less hydrophobic compounds 2-heptanone and ethyl hexanoate or for ethyl nonanoate. Previously reported data indicate irreversible conformational changes associated with ionisation phenomena and protein defolding in alkaline media.[9]

Retention (%)

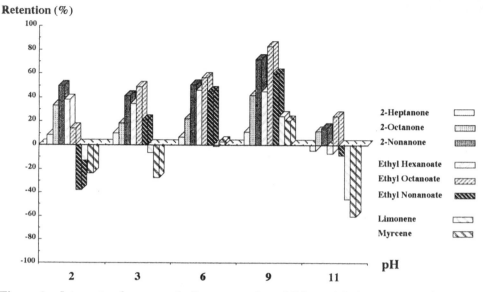

Figure 2   *Interaction between volatile compounds and β-lactoglobulin in water solution (2% w/w) at several pH from 2.0 to 11.0*

The hypothesis formulated from the results obtained for methyl ketones and ethyl esters concerning the existence of hydrophobic interactions between these compounds and monomeric and dimeric form of the protein is supported by the results obtained for terpene compounds. An increase of salting-out at pH 2.0 relative to pH 3.0, together with a low retention increasing from pH 6.0 to pH 9.0, was found for limonene and myrcene. These facts and the considerable salting-out effect, 50%–60% at pH 11.0 for these compounds possessing high log P values (Table 1), clearly indicated changes in the nature of hydrophobic sites available on β-lactoglobulin according to its conformational status. The occurrence of ionisation phenomenon in addition to conformational changes, induced a decrease in apparent hydrophobicity.

Otherwise, no competition was detected using static headspace measurement when solutions containing ethyl hexanoate, β-ionone (considered as bound to the hydrophobic pocket of the protein), and β-lactoglobulin[10] were used. This result supports our contention that surface hydrophobicity occurs.

Further works concerning the effect of ionic strength, of dissociating agents like urea, and thermal denaturation, are presently in progress in order to elucidate the effect that the conformation and the structure of the β-lactoglobulin have on the retention of aroma compounds.

REFERENCES

1. J.E. Kinsella and S. Damodaran, 'The Analysis and Control of Less Desirable Flavors in Foods and Beverages', Academic Press, New York, 1980, p. 95.
2. H. Plug and P. Haring, *Trends Food Sci. Technol.*, 1993, **4**, 150.
3. S. Langourieux and J. Crouzet, 'Food Macromolecules and Colloids', Royal Society of Chemistry, Cambridge, 1995, p. 124.
4. R. Fugate and P.S. Song, *Biochem. Biophys. Acta*, 1980, **625**, 28.
5. T. O'Neil and J.E. Kinsella, *J. Agric. Food Chem.*, 1987, **35**, 770.
6. H.M. Farrel, Jr., M.J. Behe and J.A. Enyeart, *J. Dairy Sci.*, 1987, **70**, 252.
7. A. Sadafian and J. Crouzet, 'Progress in Terpene Chemistry', Editions Frontières, Gif sur Yvette, 1986, p. 165.
8. R.F. Rekker, 'The Hydrophobic Fragmental Constant' Elsevier Scientific Pub. Co., Amsterdam, 1977, p. 110.
9. H.L. Casal, U. Köhler and H.H. Mantsch, *Biochem. Biophys. Acta*, 1988, **957**, 11.
10. E. Dufour and T. Haertlé, *J. Agric. Food Chem.*, 1990, **38**, 1691.

UNUSUAL FLAVOUR INSTABILITY IN FUNCTIONAL DRINKS

Willi Grab, Klaus Gassenmeier and Hanspeter Schenk

Givaudan-Roure, CH8600 Dübendorf, Switzerland

1 INTRODUCTION

The stability of food flavours is mainly influenced by oxygen, heat, pH and enzyme activity. The oxidative degradation of food and beverages during storage is the main concern of the producers, especially in the case of citrus flavours. Antioxidants like ascorbic acid and tocopherol are widely used to protect the products against oxidation.

During the development of a functional drink containing citrus flavours, vitamin C, vitamin E, carotene and minerals, we observed a specific, unusual reduction of the aldehydes octanal, decanal, dodecanal and citral to the corresponding alcohols octanol, decanol, dodecanol, nerol and geraniol, leading to an unbalanced and unacceptable soapy flavour profile.

Investigation of the phenomenon shows, that the reaction is not just a simple reduction of the aldehydes, but that it may involve a cascade of reactants and reaction steps.

It is well known that flavours are sensitive products and are subject to physical and chemical changes by oxygen, heat, pH and enzyme activity.[1-7] We often observe oxidation (*e.g.* limonene to limonene oxide), hydrolysis of esters (*e.g.* ethyl butyrate to butyric acid), *trans* esterification (*e.g.* triacetin to benzyl acetate in benzyl alcohol), acetalisation of aldehydes (*e.g.* hexanal to hexanal diethylacetal in alcoholic beverages) and fat oxidation (to yield aldehydes and ketones). The reduction of aldehydes to the corresponding alcohols is known to occur in biological systems and in a reducing chemical environment. So far we have not observed a reduction of flavouring components during storage in normal, pasteurized food.

2 OBSERVATIONS

During the development of functional drinks containing orange, lemon and lime flavours, vitamin C, vitamin E, carotene, citric acid, vegetable oil, gum Arabic, xanthan gum, maltodextrin and minerals (Ca, Mg, Na, Cl, phosphate), we observed in some samples the development of an unbalanced and unacceptable soapy flavour profile. The off-note appeared irregularly in about 10%–30% of the finished drink samples only. The phenomenon showed up in stored cans (filled by a bottling company) and in laboratory filled bottles but not in soft packs. The off-note developed in one case during four hours from a diluted powder drink.

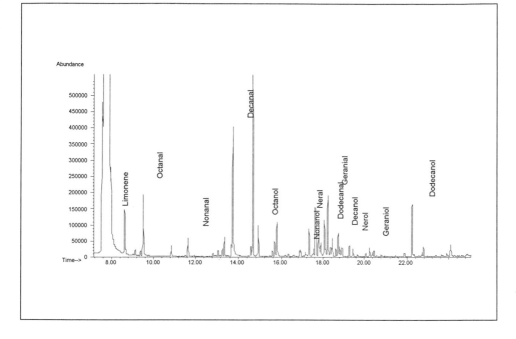

Figure 1 *Orange flavour with off-note*

2.1 Analysis

The GC–MS- and GC–sniffing analysis of the contaminated samples showed that octanol, decanol, nerol and geraniol were responsible for the unusual off-note (Figure 1). The concentrations were about the same as normally used in flavours. Further analysis showed, that the increase of the alcohol content went in parallel with the decrease of the aldehyde content.

A summary of the problem is given in Table 1.

2.2 Search for Contamination

In a first set of analyses we tried to find a source of contamination. We checked all used raw materials, bases, ingredients, water and the cleaning system of the bottler using a well documented batch history. No contamination could be identified. A contamination by micro-organisms was eliminated by growing experiments: no biological activity could be identified.

2.3 Search for Catalysts

Harel[8] describes the influence of NaCl on the metal redox cycle in an ascorbic acid–iron–copper system leading, in a cascade of reactions, to the oxidizing hydroxyl radical. We therefore checked for any type of catalyst such as Fe, Fe^{2+}, Fe^{3+}, Zn, Al, Cu^{2+}, Ag^+. It was not possible to reproduce the effect with the addition of these catalysts.

Table 1   *Summary of the Problem*

Composition	Speculation	Hypothesis
Vitamin C	Cause of contamination / reaction:	Complex catalyzed chain reaction
Vitamin E	– used raw materials	
Carotene	– during bottling	
Citric acid	– cleaning agents	
Vegetable oil	– microbiological	
Gum Arabic	– degradation of raw materials	
Xanthan gum	– reduction by ingredients	
Glucose	– catalytic reduction	
Sugar		
Maltodextrin		
Minerals		
(Ca, Mg, K, Na		
Cl, phosphate)		

3 CONCLUSIONS

We could not identify the real reason for this unusual reduction process. The effect was never really reproducible and did not show up a year later. We assume that a reaction cascade induced by ascorbic acid and the presence of unknown catalysts could occur in the complex matrix environment of an emulsion of minerals, vitamins, natural gums and vegetable oils. We cannot completely refute biological or enzymatic contamination, although our results didn't give any indication of this.

REFERENCES

1. E.J. Freeburg, B.S. Mistry, G.A. Reineccius and J. Scire, *Perfumer & Flavorist*, 1994, **19**, 23.
2. P. Schieberle and W. Grosch, *Z. Lebensm. Unters. Forsch.*, 1989, **189**, 26.
3. P. Schieberle, H. Ehrmeier and W. Grosch, *Z. Lebensm. Unters. Forsch.*, 1988, **187**, 35.
4. P. Schieberle and W. Grosch, *Lebensm.-Wiss. u. -Technol.*, 1988, **21**, 158.
5. W. Grab, in 'Trends in Flavor Research', eds. H. Maarse and D.G. van der Heij, Elsevier, Amsterdam, 1994, p. 381.
6. W. Grab in 'European Brewery Convention, Proceedings of the 22nd Congress, Zurich', 1989, p. 93.
7. W. Grab in 'Sensory Quality in Foods and Beverages', eds. A.A. Williams and R.K. Atkins, Ellis Horwood Ltd., Chichester, 1983, p. 412.
8. S. Harel, *J. Agric. Food Chem.*, 1994, **42**, 2402.

INTERACTIONS OF β-LACTOGLOBULIN WITH FLAVOUR COMPOUNDS

M. Charles, B. Bernal and E. Guichard

INRA, Laboratoire de recherches sur les Arômes, 17 rue Sully, 21034 Dijon Cedex, France

1 INTRODUCTION

Proteins often cause a decrease in the volatility of flavour compounds.[1] In general, the volatiles are retained by proteins by reversible hydrophobic interactions, adsorption or absorption, and by chemical bonds of various strengths. The strength of the interactions depends on the ability of flavour volatiles to induce the unfolding of the protein. The interactions depend also on the specific nature of the volatile compounds and of the macromolecules. Aldehydes, for instance, are particularly reactive towards proteins. They are more strongly bound to soya protein or to bovine serum albumin than alcohols or ketones.[2] Carboxylic acids have a much lower affinity for these proteins.[3]

In the present study, we worked with a well characterized protein, β-lactoglobulin (β-lac). High odour impact volatile compounds were studied, for which no previous data existed: benzaldehyde, linalool, isoamyl acetate, ethyl hexanoate, 2-nonanone and acetophenone. The concentrations used were close to those found in real food systems.

2 MATERIAL AND METHODS

2.1 Headspace Analysis

Static headspace analyses were carried out in triplicate, in amber flasks (40 ml) closed with mininert valves (Supelco). The purity of the flavour compounds was evaluated by GC–MS (> 95%). All solutions were prepared in pure water containing NaCl (50 mM) adjusted to pH 3 with HCl. The purity of β-lac (Besnier 735) was evaluated at 90%. The solutions of protein were prepared in NaCl (50 mM) solution adjusted to pH 3 with the addition of sodium azide (0.02% w/w) and stored for less than ten days at 4 °C. Analysed solutions, with and without protein, were equilibrated at 30 °C. Injections of 1 ml, 100 μl, 50 μl or 25 μl of vapour phase were made on a Carlo Erba 8000 gas chromatograph with a DB-Wax column (J&W Sci., 0.3 mm, 30 m). The temperature of injector and detector were 250 °C; the rate of flow of the carrier gas (H_2) was 1.9 ml min^{-1}.

2.2 Determination of the Thermodynamic Constants

Concentrations of free ligand (L) were determined using the following equation:[4]

$$L = \frac{R}{T}I \qquad (1)$$

where R (mol l^{-1}) is the measured concentration of ligand in the headspace with protein; T (mol l^{-1}) is the measured concentration of ligand in the headspace without protein and I (mol l^{-1}) the initial concentration of ligand in the solution.

The maximum number of binding sites (n) can be calculated using the saturation curve:[5]

$$v = \frac{nL}{\left(\dfrac{1}{K_a} + L\right)} \tag{2}$$

For a protein having a number of equivalent and independent binding sites (n), the reversible interactions between flavour molecules and protein can be represented thermodynamically by the Scatchard equation:[6]

$$\frac{v}{L} = K_a n - K_a v \tag{3}$$

or the double reciprocal equation:

$$\frac{1}{v} = \frac{1}{n} + \frac{1}{nK_a L} \tag{4}$$

where v is the number of moles of ligand bound per mole of protein, K_a the intrinsic affinity constant and L the concentration of free ligand calculated from equation (1).

However the protein can have some non-equivalent and dependant binding sites. Thus the binding of flavour compounds on a site facilitates the binding of other molecules on other sites. A co-operation between the sites exists and the interactions can be represented thermodynamically by the Hill equation[7] (5), its double reciprocal form[8] (6) or the linearized Scatchard plot[9] (7):

$$v = \frac{n}{\left(\dfrac{1}{(K_a L)^h} + 1\right)} \tag{5}$$

$$\frac{1}{v} = \frac{1}{n} + \frac{1}{n(K_a L)^h} \tag{6}$$

$$\frac{v}{L^h} = nK_a^h - vK_a^h \tag{7}$$

where h is the Hill coefficient reflecting the co-operation between the sites.

Table 1   *Binding Parameters Calculated with the Different Equations*

Volatile Compounds	Correlation Coefficient, R^2	n	K_a	Global Affinity nK_a
2-Nonanone *0 to 40 ppm*	0.99 (4)	0.2	6250	1250
45 to 200 ppm	0.92 (4)	0.5	1667	833
Ethyl hexanoate	0.97 (3)	0.8	1053	842
0 to 300 ppm	0.99 (4)	0.9	983	885
Isoamyl acetate	0.94 (2)	0.08	6173	494
0 to 250 ppm	0.98 (6)	0.08	7867	630
	0.97 (7)	0.09	8424	758

3 RESULTS AND DISCUSSION

Among the volatile compounds studied (Figure 1), acetophenone and linalool were not significantly bound to β-lac, benzaldehyde showed weak interactions whereas isoamyl acetate, ethyl hexanoate and 2-nonanone were more significantly bound to β-lac.

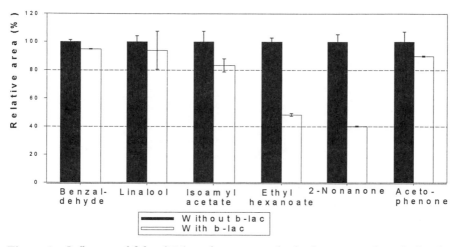

Figure 1 *Influence of β-lac (3%) on the amount of volatile compounds in the headspace (initial concentration in water: 100 ppm)*

The binding parameters were calculated with the different equations mentioned above for isoamyl acetate, ethyl hexanoate and 2-nonanone (Table 1). Saturation of the binding sites was only reached for isoamyl acetate (Figure 2a), due to the low solubility of 2-nonanone (0.4 g l^{-1})[10] and ethyl hexanoate (0.52 g l^{-1}).[11] The linear Scatchard equation (3) or its double reciprocal form (4) could only be applied to the data obtained with 2-nonanone and ethyl hexanoate. The Hill equation was used for isoamyl acetate to determine a Hill coefficient of 1.65, showing a positive co-operation between the binding sites (Figure 2b).

The global affinity of β-lac for 2-nonanone is higher than for the esters. For this methyl ketone, two phenomena occur depending on the concentration. We can thus postulate that two types of site exist. Increasing the concentration of 2-nonanone could induce a conformational change of the protein, creating a new type of binding site. Our results differ from those reported by Jasinski and Kilara[12] and O'Neil and Kinsella[5] who worked at pH between 6 and 7 (dimeric form of β-lac) with a lyophilised sample, which can affect the amount of bound ligand. Concerning the esters, the global affinity is lowest for isoamyl acetate. As the number of binding sites differs for these two esters, the mechanism of interaction could be different. We can postulate that the linear hydrophobic chain of ethyl hexanoate and 2-nonanone could bind to the site constituted by the hydrophobic pocket of β-lac.[5]

Further research is under way on different flavour compounds and on mixtures of compounds, in order to study competition between ligands for the same binding site.

a   b

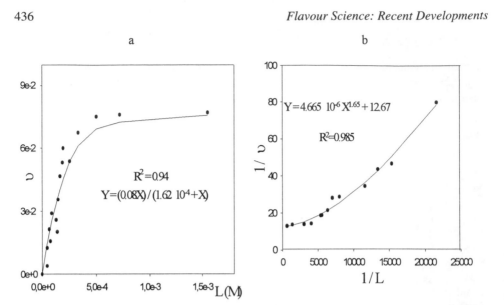

Figure 2   *Interactions between β-lac (3%) and isoamyl acetate: (a) Saturation curve (2); (b) Double reciprocal form of Hill equation (6).*

ACKNOWLEDGEMENTS

This programme was part of a collaborative study with ENSBANA (Dijon), ENSIA (Massy), GBSA (Montpellier), and INRA (Dijon, Nantes), and was partly financed by the French Ministry of Agriculture and Fisheries.

The authors thank D. Aravantinos (IFF-Dijon) for providing the flavour compounds.

REFERENCES

1. K. Franzen and J. Kinsella, *J. Agric. Food Chem.*, 1974, **22**, 675.
2. S. Damodaran and J.E. Kinsella, *J. Agric. Food Chem.*, 1980, **28**, 567.
3. S. Damodaran and J.E. Kinsella, *J. Agric. Food Chem.*, 1981, **29**, 1249.
4. S.F. O'Keefe, L.A. Wilson, A.P. Resurreccion and P.A. Murphy, *J. Agric. Food Chem.*, 1991, **39**, 1022.
5. T. O'Neil and J. Kinsella, *J. Agric. Food Chem.*, 1987, **35**, 770.
6. G. Scatchard, *Ann. N.Y. Acad. Sci.*, 1949, **51**, 660.
7. A.V. Hill, *J. Physiol.*, 1910, **40**, 190.
8. P. Landy, C. Druaux and A. Voilley, *Food Chem.*, 1995, **54**, 387.
9. M.A. Ruschmann, J. Heiniger, V. Pliska and J. Solms, *Lebensm.-Wiss. u. -Technol.*, 1989, **22**, 240.
10. M. Espinoza-Diaz and A. Voilley, personal communication.
11. M. Le Thanh, T. Lamer, A. Voilley and J. Jose, *J. de Chimie Phys.*, 1993, **90**, 545.
12. E. Jasinski and A. Kilara, *Milchwissenschaft*, 1985, **40**, 596.

MODELLING THE UPTAKE OF DIACETYL IN EXTRUDED STARCH

Miranda Y.M. Hau, Stephen Hibberd[*] and Andrew J. Taylor

Department of Applied Biochemistry and Food Science, University of Nottingham, Sutton Bonington Campus, Loughborough, Leicestershire, LE12 5RD, U.K.

[*]Department of Theoretical Mechanics, University of Nottingham, University Park, Nottingham, NG7 2RD, U.K.

1 INTRODUCTION

The migration of flavour molecules during storage determines, to a great extent, the keeping quality of some foods. An understanding of the factors affecting the migration rates of flavour molecules, such as matrix composition and water content, is therefore important. In these cases, there is a need for quantitative information about the migration processes involved (*e.g.* diffusion, adsorption).

This paper discusses various approaches to understanding the mechanisms involved in the binding of diacetyl to extruded starch by linking several possible theoretical approaches to experimental findings. Experimentally, a known concentration of the volatile in air was brought into contact with a cylinder of extruded starch in a sealed system and the rate of diacetyl uptake was determined by headspace analysis. In an attempt to model real food systems, relatively large samples of starch were used rather than the thin samples which are often favoured in this type of investigation.[1-3] The amount of volatile used was also substantially less than in other sorption studies,[1,4-6] the concentration being closer to real foods. Data showing the loss of diacetyl from the headspace were obtained using the method reported previously[7] at 25 °C and over a time range of 2 to 50 hours. The basic binding curves of uptake against time were then analysed.

2 BINDING MODELS

In attempting to quantify the migration process, several mathematical treatments of the data were investigated which gave an insight into the mechanisms that occur during binding. Within the experimental approaches, the headspace was kept well stirred and consequently the binding mechanisms could be considered as uniform over the starch surfaces so that a one-dimensional analysis for volatile uptake into the starch was valid. The uptake of volatiles within the starch was initially considered to be a diffusion process controlled by a constant diffusion coefficient, D. Diffusion was assumed to be Fickian *i.e.* satisfying Fick's second equation:

$$\frac{\partial c}{\partial t} = D\frac{\partial^2 c}{\partial x^2} \tag{1}$$

where c is the concentration of diffusing substance per unit surface area, D the diffusion coefficient, x the space co-ordinate measured normal to the section and t time. The equation has to be supplemented by appropriate boundary conditions at the starch surfaces. It is implicitly assumed that no adsorption of volatiles occurs within the starch. Further considerations are that the volatile may bind to the starch surface prior to diffusional effects into the starch and that partition effects will need to be considered.

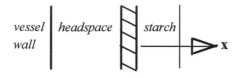

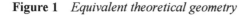

starch/volatile   starch lower surface
interface

Figure 1 *Equivalent theoretical geometry*

Three theoretical models are presented that will serve to evaluate the relative importance of possible physical processes active in the binding of volatiles. These are:
1. Diffusion dominated with a constant (saturated) interface concentration C_s;
2. Diffusion limited by starch extrudate thickness;
3. Effect of surface adsorption.

2.1 Diffusion Model 1

In this model, the volatile within the headspace is assumed to rapidly establish a saturation concentration of volatile (C_s) at the interface. Diffusion into the starch is then driven by this surface concentration which is continually maintained by volatiles within the headspace. This mechanism will be modified after a period of time t_s, after which the headspace concentration is sufficiently depleted to support only an interface concentration lower than C_s. In this latter stage, the diffusion is now driven by a progressively lower concentration. It is convenient to assume that the diffusion of volatile elements will not fully penetrate the extrudate, in which case, the extrudate may mathematically be taken to be of semi-infinite extent. The assumptions of the model are as follows:
1. A maximum (saturation) concentration exists within the extrudate;
2. The time for the surface to reach saturation level is negligible;
3. The amount of volatile stored on the surface is negligible;
4. The extrudate is sufficiently thick so as not to unduly restrict the diffusion process.

For time $t < t_s$, an exact mathematical solution to the diffusion problem is given by Crank[8] together with an expression for the rate of volatile uptake into the starch. As the volatile entering the starch arises from the headspace, a straightforward conservation relation enables an expression for the volatile concentration within the headspace gas phase to be established as:

$$\frac{C_g(t)}{C_g(0)} = 1 - \frac{2}{\pi} \frac{C_s}{C_g(0)} \frac{S}{V} (Dt)^{1/2} \qquad (2)$$

where $C_g(t)$ is the concentration of volatile in the gas phase at time t and $C_g(0)$ the initial concentration, S the surface area of the exposed starch extrudate and V the volume of the headspace.

The relationship (2) is valid for $t < t_s$, after which time the solution must be modified – a procedure that requires a numerical computation for the diffusion equation. A significant prediction is that the relative concentration $C_g(t)/C_g(0)$ decays as $t^{1/2}$ or, consequently, the initial volatile uptake is proportional to $t^{1/2}$. Figure 2 shows the percentage of diacetyl taken up by the extrudate when plotted against $t^{1/2}$ and demonstrates the expected initial linear relationship; the slope of which can be used to determine an associated diffusion coefficient. Diffusion coefficient values of 0.492×10^{-4} to 8.86×10^{-4} m^2 s^{-1} have been calculated which are in stark contrast to values of 10^{-10} m^2 s^{-1} obtained by Menting[1] for a maltodextrin system suggesting that additional factors may need to be considered to describe binding in the system.

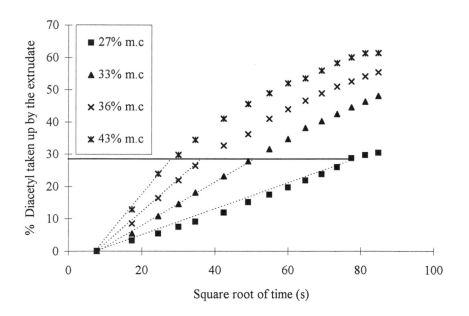

Figure 2 *Comparison of diacetyl binding to extruded starch of various moisture content*

2.2 Diffusion Model 2

The second model studied was that used by Menting[1] for measuring the diffusion coefficient of organic volatiles into maltodextrin sheets by the sorption and desorption of vapour to a layer of maltodextrin solution. However, in these experiments, very thin samples and saturating concentrations of volatiles were used. The solution for the diffusion equations is given in Crank[8] for diffusion from a limited volume system into a cylinder.

The assumptions are:

1. The amount of volatile stored on the surface is negligible;
2. Samples are thin enough for a steady state concentration to be established.

The problem can be solved in terms of an infinite series or approximately as:

$$\frac{C_g(t)}{C_g(0)} = \exp\left(\frac{T}{\alpha^2}\right) erfc\left(\frac{T}{\alpha^2}\right)^{\frac{1}{2}} \tag{3}$$

$$T = \frac{Dt}{\ell^2}, \qquad \frac{C_\infty}{C_g(0)} = \frac{\alpha}{1+\alpha}, \qquad \alpha = \frac{a}{\ell}$$

$$a = \text{representative length in gas} = \frac{V}{s}$$

However, rather than solving equations of this type, solutions in the form of a standard graph for different percentage uptake were used. Table 1 illustrates the typical diffusion coefficient values calculated. The solutions take into account the final equilibrium uptake of the volatile.

This method requires a reliable final equilibrium value which was only achieved after 12 hours. When these equilibrium experiments were carried out, the curves showed evidence of multicomponent behaviour. It was felt that a more physically reasonable method of analysis which involved surface adsorption, in addition to diffusion, would be more appropriate.

Table 1   *Calculated Values for the Diffusion Coefficient of Diacetyl in Starch at Various Moisture Contents using Model 2*

Moisture content (%) w/w	Diffusion coefficient ($m^2\ s^{-1}$)
27	1.98×10^{-11}
33	1.25×10^{-11}
36	4.31×10^{-11}
43	9.87×10^{-11}

2.3 Surface Adsorption and Diffusion Model 3

It is clear from the disparity in results from Model 1 that the observed decay in volatile concentration is occurring at a rate much faster than can be attributed to diffusion alone. A third, composite, model is considered by taking:

$$\frac{C_g(t)}{C_g(0)} = 1 - \frac{C_d(t)}{C_g(0)} - \frac{C_s(t)}{C_g(0)} \tag{4}$$

where $C_d(t)$ is of the form given in (2) and $C_s(t)$ the uptake arising from binding to the surface of the extrudate. Surface adsorption can be described by the relationship:

$$C_s(t) = C_s(\infty) + \left(C_g(0) - C_s(\infty)\right)\exp\left(-\gamma t\right) \tag{5}$$

where γ is a rate of uptake parameter and $C_s(\infty)$ a final saturation limit (for a given system) in the absence of diffusion. The rapid decay due to the exponential function in (5) means that surface adsorption will be substantially complete in the time scale of order γ^{-1} whilst diffusion is expected to contribute over a much larger time scale (of the order ℓ^2/D). Hence for $\gamma^{-1} \ll \ell^2/D$ then $C_d(t)$ is very much less than $C_s(t)$ and these measurements can be used to compute the parameter $C_s(\infty)$ and γ.

Table 2 indicates the typical values calculated. These values are still relatively high compared to the corresponding diffusion coefficients calculated by Menting but Model 3 is really calculating the permeability coefficient rather than the diffusion coefficient. The relationship between the two is P = DS, where P is the permeability, D is the diffusion and S is the solubility coefficient. The partition coefficient calculated for diacetyl in this system is approximately 662. This was calculated from residual concentration, indicating that solubility of the volatile masks the calculation of the true diffusion coefficient. From Table 3, the true diffusion coefficient can be seen and is comparable to that calculated from Model 2.

Table 2 *Calculated Values for the Diffusion Coefficient of Diacetyl in Starch at Various Moisture Contents using Model 3*

Moisture content (%) w/w	Diffusion coefficient ($m^2 s^{-1}$) – Model 3	Corrected diffusion coefficient ($m^2 s^{-1}$) – Model 3
27	2.29×10^{-10}	3.46×10^{-13}
33	6.03×10^{-10}	9.11×10^{-13}
36	1.47×10^{-9}	2.21×10^{-12}
43	1.87×10^{-9}	2.82×10^{-12}

3 CONCLUSION

Various mathematical models have been proposed to explain the binding process of diacetyl to starch resulting in the conclusion that both surface and diffusive mechanisms are responsible for the overall binding process. The apparent high values for the diffusion coefficient from some models can be attributed to a partition factor, a function of the solubility of the diacetyl in the system, which, when corrected for, gives a true diffusion coefficient value in the region of 10^{-12} m^2 s^{-1}.

REFERENCES

1. L.C. Menting, B. Hoogsteg and H.A.C. Thijssen, *J. Food Technol.*, 1970A, **5**, 111.
2. A.M. Roland and J.H. Hotchkiss, in 'Food and Packaging Interactions II', eds. S.J. Risch and J.H. Hotchkiss, A.C.S. Symposium Series 473, American Chemical Society, Washington, D.C., 1991.
3. R.C. Mason, P.T. DeLassus, G. Strandburg and B.A. Howell, *Tappi Journal*, June 1992, 163.
4. H. Rasmussen and H.G. Maier, *Chem. Mikrobiol. Technol. Lebensm.*, 1974, **4**, 119.
5. M. Le Thanh, P. Thibeaudeau, M.A. Thibaut and A. Voilley, *Food Chemistry*, 1992, **43**, 129.
6. G.W. Halek and J.F. Luttmann, in 'Food and Packaging Interactions II', eds. S.J. Risch and J.H. Hotchkiss, A.C.S. Symposium Series 473, American Chemical Society, Washington, D.C., 1991.
7. M.Y.M. Hau, D.A. Gray and A.J. Taylor, Interaction of Flavor with Non-Flavor Components, American Chemical Society Symposium Series, 1996, in press.
8. J. Crank, 'The Mathematics of Diffusion', Oxford University Press, 2nd edition, Oxford, 1975.

INFLUENCE OF PHYSICO-CHEMICAL PROPERTIES OF AROMA COMPOUNDS ON THEIR INTERACTIONS WITH OTHER CONSTITUENTS

M. Espinosa-Díaz,* A.M. Seuvre† and A. Voilley*

*ENSBANA and †IUT Biologie Appliquée, Université de Bourgogne, 21000 Dijon, France

1 INTRODUCTION

Flavour is one of the major organoleptic characteristics of food; it depends mainly on the nature and the quantity of aroma compounds involved. Foods are biological systems with complex interactions, particularly between volatile and non-volatile components.[1] Some workers[2,3] have shown the decrease of the sensory perception of aroma compounds in the presence of lipids or proteins. Hansen and Heinis,[4] for example, have observed a decrease of benzaldehyde and d-limonene flavour intensity in the presence of whey proteins or casein. Aroma compounds are generally retained by proteins, so that the concentration of volatiles decreases in the gas phase above and around the food.[3]

The intensity of the interactions depends on the nature of the aroma compounds. Some chemical functions are more reactive with proteins than others. The aldehydes are more strongly bound to soy proteins or bovine serum albumin than ketones and alcohols, while carboxylic acids and aniline have little or no affinity for these proteins.[5] Dufour and Haertlé[6] propose that the aroma compound structure is also important: they have shown that α-ionone does not bind to β-lactoglobulin while β-ionone does.

In this paper, the behaviour of aroma compounds is followed in various media. The influence of physico-chemical properties of aroma compounds on their interactions with other constituents is also studied.

2 MATERIAL AND METHODS

2.1 Reagents

The chosen volatile compounds belong to different chemical classes such as aldehydes, esters, pyrazines, alcohols and ketones. The purity of the flavour compounds is greater than 98% except for the 2,5-dimethylpyrazine (85%).

The protein is β-lactoglobulin with a monomer molecular weight of 18300 Da and an isoelectric pH of 5.2. The protein was purified before utilization. The lipid (miglyol) is a triacylglyceride of caprylic and capric acids with a molecular weight of 484.4 g mol^{-1}.

2.2 Vapour–Liquid Partition

2.2.1 Headspace Analysis. This method measures the equilibrium concentration of the aroma between the liquid and vapour phases at 25 °C. An inert gas (N$_2$) passes through the liquid phase (solution containing the aroma compound at infinite dilution, *i.e.* 100 g solution containing 10 µl of pure volatile compound) at a constant flow rate (30–60 ml min^{-1}) and carries the volatile compound into the headspace.

A sample (1 ml) of the vapour phase was automatically injected into the gas chromatograph at regular intervals. The vapour–liquid equilibrium was considered to be reached when the concentration of aroma compounds in the gas phase remained constant.[7] The vapour–liquid partition coefficient, K_i^{∞}, is the ratio of the molar fractions of the component, i, in the vapour and the liquid phases respectively at equilibrium.

2.2.2 Exponential Dilution. The exponential dilution consists of exhausting the aroma compound from the liquid phase in equilibrium with the vapour phase.[8]

The determination of the retention percentage, r, of the aroma compound by the protein is estimated from the calculation of the relative difference of the vapour–liquid partition coefficient of the aroma compound in the presence of the protein with that measured in water.

2.3 Liquid–Liquid Partition

Liquid–liquid partition consists of measuring the equilibrium concentration of a solute between aqueous and organic phases. The liquid–liquid partition coefficient, P_i, is the ratio of the mass fractions of the component, i, in the organic and the aqueous phases respectively.

2.4 Gas Chromatography Conditions

A Chrompack CP 9000 gas chromatograph with a flame ionization detector (FID) and a Hewlett Packard 3380A integrator were used. The stainless steel column (3 m × 2.2 mm internal diameter) was packed with 100/120 mesh Carbowax 20M-10%. The injector and detector temperatures were maintained at 190 and 200 °C respectively. The column was operated at 80–160 °C (depending on the compound). The nitrogen flow rate was 16 ml min^{-1}; the hydrogen flow rate was 25 ml min^{-1} and the air flow rate was 250 ml min^{-1}.

3 RESULTS AND DISCUSSION

The physico-chemical properties of the aroma compounds are given in Table 1: the water solubility was experimentally determined at 25 °C. The values range from 2.4 g l^{-1} for isoamylacetate to infinity for acetaldehyde and 2,5-dimethylpyrazine.

The hydrophobicity constants, log P, were estimated from Rekker's method.[9] The log P value represents the hydrophobicity of the volatile compound: a negative value implies the compound is hydrophilic.

Generally, water solubility and log P values are linked: components that are infinitely soluble in water have negative log P values, while less water soluble components present the highest log P values except for linalool which has a high log P value although this molecule has a relatively good water solubility.

Saturated vapour pressure is directly linked to the volatility of the pure compound, while K_i^{∞} value indicates the aroma compound volatility in solution. Lee-Kesler's method[10] was used to estimate the vapour pressure for which values vary from 0.2 for linalool to 848 mm Hg for acetaldehyde.

Physico-chemical characteristics of aroma compounds may allow a better understanding of their behaviour under different conditions. We have chosen aroma compounds divided into three classes according to their hydrophobicity: acetaldehyde and

2,5-dimethylpyrazine are the more hydrophilic molecules; isoamylacetate and linalool are hydrophobic and benzaldehyde has an intermediate behaviour.

Table 1 *Physico-chemical and Thermodynamic Characteristics of Aroma Compounds (25 °C)*

Aroma Compound	Water Solubility $(g\ l^{-1})$	Hydrophobicity $(log\ P)$ estimated[9]	Saturated Vapour Pressure $(mm\ Hg)$ estimated[10]	Vapour–liquid Partition Coefficient K_i^{∞}	P_i (miglyol–water)
Acetaldehyde	∞	−0.8	848	3.9[11]	0.2 (3)
2,5-Dimethyl pyrazine	∞	−1.6	2.3	0.1[12]	1.4 (4)
Benzaldehyde	7.1 (9)	1.5	0.7	1.7 (4)	44 (2)
Isoamylacetate	2.4 (5)	2.2	6.6	38.7 (1)	186 (1)
Linalool	2.6 (6)	4.0	0.2	2.3 (6)	302 (3)

Numbers in brackets: coefficient of variation %.

Concerning the vapour–liquid partition experiments, we remained with only benzaldehyde and linalool because they belong to different hydrophobic groups. The studied media were water, a β-lactoglobulin solution (3%) and a lipid medium (miglyol). For linalool, no significant difference in the vapour–liquid partition coefficient, K_i^{∞}, between aqueous media with and without β-lactoglobulin was observed. On the other hand, the presence of this protein at 3% concentration halves the benzaldehyde K_i^{∞}. In the lipid medium, benzaldehyde and linalool volatilities decreased strongly by 1000-fold. Both components have a great affinity for miglyol.

In order to study the behaviour of aroma compounds in biphasic media (miglyol–aqueous solution), we measured their transfer kinetics at the interface. The measurement of the transfer rate of benzaldehyde at the lipid (miglyol)–water interface indicated that the flux depended on the flavoured phase, but it did not depend on the presence of β-lactoglobulin in the aqueous phase (Table 2). However, Harvey et al.[13] have shown a resistance to transfer at the interface of ethylbutanoate and 2,5-dimethylpyrazine in the presence of sodium caseinate.

Table 2 *Kinetics of Benzaldehyde Partitioning between Miglyol and Aqueous Solution*

Flavoured Phase	β-Lactoglobulin (%)	Aroma Flux at the Interface at 25 °C $(mg\ h^{-1}\ cm^{-2})$
Aqueous	0	0.2
	3	0.2
Lipidic	0	0.008
	3	0.009

Table 1 presents the liquid–liquid partition coefficients of aroma compounds between miglyol and water. Both benzaldehyde and linalool had a very noticeable affinity for miglyol: their lipid–water partition coefficients (P) were 44 and 302 respectively, and the coefficients obtained with and without protein are not significantly different.

The retention of volatile compounds varies with the media composition: in the literature, some investigators[4,5] have shown a decrease of benzaldehyde volatility in the presence of proteins. Accordingly, we observed a retention of around 40% for the aqueous solution of β-lactoglobulin and of around 100% for miglyol. It seemed that there was a retention of benzaldehyde by β-lactoglobulin in the presence of miglyol, but the value was not significant. Consequently, the retention of benzaldehyde by β-lactoglobulin may be masked by the presence of miglyol.

4 CONCLUSION

Thermodynamic measurements have determined the volatility of aroma compounds and their retention by the constituents of the medium. The aroma flux at the interface has been calculated from kinetic measurements. The behaviour of the volatile compounds in different media (water, aqueous solution of β-lactoglobulin, miglyol) depended on their physico-chemical characteristics and particularly on their hydrophobicity.

ACKNOWLEDGEMENTS

This programme was part of a collaborative study with INRA (Dijon, Nantes), ENSIA (Massy) and USTL (Montpellier), partly financed by the French Ministery of Agriculture and Fisheries.

The authors wish to thank D. Aravantinos (IFF-Dijon) for providing the flavour compounds.

REFERENCES

1. J. Bakker, in 'Ingredient Interactions – Effects on Food Quality', ed. A.G. Gaonkar, M. Dekker Inc., New York, 1995, p. 411.
2. J.E. Kinsella, *Int. News on Fats Oils and Related Materials*, 1990, **1**, 215.
3. J. Solms, in 'Interactions of Food Components', eds. G.G. Birch and M.G. Lindley, Elsevier Applied Science Publishers, London, 1986, p. 189.
4. A.P. Hansen and J.J. Heinis, *J. Dairy Sci.*, 1992, **75**, 1211.
5. M. Beyeler and J. Solms, *Lebensm-Wiss. u. -Technol.*, 1974, **7**, 217.
6. E. Dufour and T. Haertlé, *J. Agric. Food Chem.*, 1990, **38**, 1691.
7. A. Voilley, D. Simatos and M. Loncin, *Lebensm.-Wiss. u. -Technol.*, 1977, **10**, 285.
8. F. Sorrentino and A. Voilley, *AIChE J.*, 1986, **32**, 1988.
9. R.F. Rekker, in 'The Hydrophobic Fragmental Constant', eds. W.Th. Nauta and R.F. Rekker, Elsevier Scientific Publishing Company, Amsterdam, 1977, Volume 1, p. 389.
10. R.C. Reid, J.M. Prausnitz and B.E. Poling, 'The Properties of Gases and Liquids', McGraw-Hill Book Company, New York, 4th edition, 1987, p. 205.
11. S. Langourieux, Thesis, USTL Montpellier II, 1993.
12. M. Le Thanh, T. Lamer, A. Voilley and J. José, *J. Chim. Phys.*, 1993, **90**, 545.
13. B.A. Harvey, C. Druaux and A. Voilley, in 'Food Macromolecules and Colloids', eds. E. Dickinson and D. Lorient, Royal Society of Chemistry, 1995, p. 154.

THE USE OF TIME–INTENSITY ANALYSIS FOR THE DEVELOPMENT OF FAT-FREE FOODS

E.M. Vroom, J. Mojet, J. Heidema, W. den Hoed and P.G.M. Haring

Unilever Research Laboratories Vlaardingen, P.O. Box 114, 3130 AC Vlaardingen, The Netherlands

1 INTRODUCTION

In home-prepared food, both the ratio of flavour compounds and the release of flavour are perceived as natural. In contrast, fabricated food products are often produced in a non-traditional way – the preparation of the food is different, other ingredients are used, the fat content is diminished or even totally removed, *etc.* This has an impact on the release of flavour from the food matrix, and consequently an influence on perceived flavour quality. Therefore, understanding the processes that influence the release of taste and aroma from the food matrix is of major importance in understanding flavour quality.[1] Using this knowledge, we will be able to control and manipulate the release of flavour.

A sensory measurement to describe flavour release is time–intensity analysis. This method is an extension of the classical scaling method providing temporal information about perceived flavour sensation. By having judges continuously monitor their perceived sensations, one is able to quantify the continuous perceptual changes.

For some 40 years, time–intensity quantification has undergone an evaluation as food scientists and psychophysicists alike have attempted to record the human response. Cliff and Heymann[2] give a good overview of the development of this technology upto late-1993. Most of these investigations have evaluated sweetness and bitterness intensity in both solution and food products. Less effort has been given to the evaluation of perceived aroma intensity. In contrast to taste, which is studied by assessment of the separate taste compounds, aroma is mostly assessed as a total flavour. One notable exception to the latter has been reported by Overbosch *et al.*[3] They studied the perception of single aroma compounds by time–intensity analysis.

The procedure for evaluating the samples by time–intensity has been well established, but the analysis of time–intensity curves themselves is still under discussion. Although it is common knowledge that inter-panellist variability can be large,[4] most researchers still average the raw data over the panellists. Due to this method, differences in perception will vanish if an averaged panel curve is used. It is noted that it cannot be verified on the basis of the data whether differences occur due to scale use, perception or other factors. Therefore, it is difficult or impossible to produce meaningful averaged panel curves.

This paper discusses a time–intensity analysis, that will be an improvement with respect to data organisation and analysis of curve parameters in order to overcome the problems due to inter-panellist variability. Sensory assessment of food products with various fat levels will show the applicability of this improved time–intensity method.

2 TIME–INTENSITY METHODOLOGY

Four different products, High Fat, Low Fat 1, Low Fat 2 and Fat-Free, flavoured with three different flavour components, A (hydrophilic), and B and C (both lipophilic) each in three different concentrations, were evaluated by 14 panellists.

The panellists selected were screened for normal smell and taste ability and submitted to training before the actual measurements took place. The tasting sessions were carried out in panel booths. Each panellist assessed the different samples three times. The samples were randomized over the tasting sessions. The samples were presented on plastic spoons for assessment. Ordinary tap water and cream crackers were used as palette cleansers between tastings.

The measurements were carried out by means of an automated computer system, with a computer in every panel booth. On the screen of this computer, a slide of an 'intensity meter' is shown. Up and-down movements of the mouse correspond to up and down movements of this slide. The slide was marked 'Very weak' at the lower end of the scale and 'Very strong' at the upper end. The panellists started recording the perceived intensity immediately after taking the sample into their mouth. The sample was ejected from the mouth after 30 s and recording stopped after 3 min.

The data analysis of the (t, I) curves obtained consisted of the following steps:
1. Extraction of 29 parameters (Figure 1) from each time–intensity curve;
2. Standardisation of the curve parameter values between panellists;
3. Factor analysis of the standardised data set to investigate the data structure and to select a representative attribute for each relevant factor, thus reducing the number of parameters;
4. Analysis of variance of the selected attributes;
5. A Student-Newman-Keuls analysis to investigate which treatments differ from each other.

3 RESULTS AND DISCUSSION

Due to inter-panellist variability, (t, I) curves differ widely from panellist to panellist (both in length and shape). A standardization step is therefore useful before any inter-panellist analysis (such as averaging) is performed. All parameter data of each panellist were standardized to a mean of 0 and a standard deviation of 1.

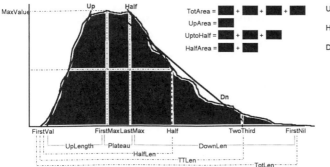

Figure 1 *Parameters used to describe time–intensity curves*

Table 1   *Results of Factor Analysis of the Parameters*

Factor 1	Factor 2	Factor 3	Factor 4	Factor 5	Factor 6	Factor 7	
FirstNil TotLen TwoThird TTLen	HalfArea	BetaHalf AngleHalf	FirstMax		Plateau	R2Dn	$|r| > 0.9$
DownLen	MaxValue UptoHalf	BetaDown AlfaHalf	UpLength R2Up	AlfaUp		R2Half	$0.8 < |r| < 0.9$
	TotArea UpArea	AlfaDn AngleDn	TotArea BetaUp AngleUp	FirstVal			$0.6 < |r| < 0.8$

Many of the 29 parameters obtained are intercorrelated and thus contain redundant information. To investigate this correlation we applied a factor analysis to the standardised data. The factor analysis applied was a principal component analysis, followed by varimax rotation to obtain optimum correlation of the variables with the main axes of the factor space. The factor analysis yielded seven factors with eigenvalues > 0.9 and which together accounted for about 90% of the total variance. The factors are correlated with the parameters as given in Table 1. Based on these results the following parameters were selected as representatives of the factors and can be described as:

- Factor 1 – FirstNil, relates to the time over which flavour is perceived;
- Factor 2 – HalfArea and MaxValue, relates to the total of perceived flavour;
- Factor 3 – BetaHalf, relates to the decline of the perceived flavour;
- Factor 4 – FirstMax, relates to the time until maximum intensity;
- Factor 5 – FirstVal, relates to the time until first impression;
- Factor 6 – Plateau, relates to the duration of maximum perception;
- Factor 7 – R2Dn, relates to the steadiness of the decline of perceived flavour intensity.

Factor analysis of data sets from other time–intensity panels assessing different products yielded the same factor structure.

From the Tables 2 to 4, it is clear that in many cases the factors yielded significant differences ($p < 0.05$) between the products tested on the selected parameters. Only for R2Dn and Plateau were no statistically significant differences between the products investigated found.

Table 2   *Student-Newman-Keuls Multiple Comparison Test for the Fat Levels*

Factor	Parameter	Full Fat	Low Fat 1	Low Fat 2	Fat-Free
1	FirstNil	bc	ab	a	c
2	HalfArea	a	a	a	a
	MaxValue	b	a	ab	ab
3	BetaHalf	a	a	a	a
4	FirstMax	ab	a	a	b
5	FirstVal	a	a	a	a
6	Plateau	a	a	a	a
7	R2Half	a	a	a	a

Products having a letter in common do not differ according to the SNK-test, a > b > c.

Table 2 shows that for the factor 'fat level' significant differences for the parameters, FirstNil, MaxValue and FirstMax could be found. This indicates that, independent of the type or concentration of the flavour, changing the amount of fat in this set of products has an influence on the length of flavour perception and on both the time and intensity of the maximum of flavour perception. However, the data do not show a *straightforward* influence of fat level on these three parameters.

Table 3 indicates that most significant differences between flavour component A on one hand and flavour component B and C on the other hand could be perceived. These differences were expected because B and C are both lipophilic whereas A is hydrophilic.

Table 3 *Student-Newman-Keuls Multiple Comparison Test for the Flavours*

Factor	Parameter	Flavour Component		
		A	B	C
1	FirstNil	a	b	b
2	HalfArea	a	b	b
	MaxValue	b	b	b
3	BetaHalf	a	a	b
4	FirstMax	a	b	b
5	FirstVal	a	ab	a
6	Plateau	a	a	a
7	R2Half	a	a	a

Products having a letter in common do not differ according to the SNK-test, a > b > c.

From Table 4 it is clear that only a difference between high flavour concentration on the one hand and low and medium concentration on the other hand could be perceived. The lack of significant difference between low and medium could be due to the difference in concentration being too small. The differences are to be expected: the higher the concentration then the higher the intensity of flavour perception, MaxValue, the higher the total perceived flavour, HalfArea, and the longer the length of perception, FirstNil.

Table 4 *Student-Newman-Keuls Multiple Comparison Test for the Concentrations*

Factor	Parameter	Low	Medium	High
1	FirstNil	b	b	a
2	HalfArea	b	b	a
	MaxValue	b	b	a
3	BetaHalf	a	a	a
4	FirstMax	a	a	a
5	FirstVal	a	a	a
6	Plateau	a	a	a
7	R2Half	a	a	a

Products having a letter in common do not differ according to the SNK-test, a > b > c.

5 CONCLUSIONS

The method of time–intensity panelling, data handling, analysis and interpretation presented here, enables an evaluation of differences in flavour perception expected from the set of samples assessed, without the blurring effect of inter-panellist variance. However, the large inter-panellist variance and relative small intra-panellist variance indicate that consumers might be segmented in their perception over time. The underlying factors influencing the in-mouth perception over time have still to be investigated.

REFERENCES

1. H. Plug and P.G.M. Haring, *Trends Food Sci. Techn.*, 1993, **4**, 150.
2. M. Cliff and H. Heymann, *Food Research Int.*, 1993, **26**, 375.
3. P. Overbosch, J.C. van der Ende and B.M. Keur, *Chem. Senses*, 1986, **11**, 331.
4. G.B. Dijksterhuis, M. Flipsen and P. Punter, *Food Quality Pref.*, 1994, **5**, 121.

INDIVIDUALITY OF FLAVOUR PERCEPTIONS – THE INFLUENCE OF MASTICATION PATTERNS

Wendy E. Brown and Claire E. Wilson

Institute of Food Research, Reading Laboratory, Earley Gate, Reading, RG6 6BZ, U.K.

1 INTRODUCTION

The volatile components of foods are released into the air from exposed surfaces of the food. They may be detected by the olfactory receptors as aromas in air that is drawn across the olfactory epithelium during breathing. When food is placed in the mouth, volatile components may be released into the saliva as the food sample melts or dissolves, or they may partition into the saliva from exposed surfaces of the food during mastication. Partitioning of the volatile components from the saliva into the airspace in the mouth, and transfer to the olfactory epithelium in the respiratory air flow, may result in perception. Perception of a volatile compound depends on the concentration of the compound at the receptors, our sensitivity to the particular compound and the duration of exposure. However perception or recognition or identification of a flavour depends on a system of pattern recognition based on the relative concentrations and temporal ordering of a variety of volatile components at the olfactory receptors. Consequently individuals may have different concepts of what constitutes a particular flavour by virtue of their semantic classification of sensory perceptions, the amounts and rates of release of volatiles from the food matrix during mastication and their sensitivities to the specific volatile components of the food. This work addresses the influence of oral breakdown of the food matrix on the temporal release and perception of volatiles in foods. Using a simple food matrix of gelatin, which at different concentrations is broken down in different ways in the mouth, we examined flavour perception using continuous recording of flavour intensity together with mastication patterns for a number of subjects.

2 METHODS

2.1 Samples

Gelatin gels were prepared containing either 5%, 10% or 25% w/w gelatin and 10% w/w sucrose. The gelatin and sucrose were dissolved in water at 78 °C for 30 min, and left to cool to 40 °C at which point 1% w/w of banana flavour (Langdales Ltd., Bury St. Edmunds, Suffolk, U.K.) was added and the solution poured into stoppered glass tubes of 16 mm internal diameter and allowed to solidify at 4 °C. Prior to use, the cylindrical gels were removed from the glass tubes and cut into 15 mm lengths.

2.2 Subjects

Eight members of a sensory panel, trained extensively in continuously recording their flavour perceptions using the Time Intensity (TI) technique, took part in the study which was approved by the Institute Ethics committee, and for which they gave their informed consent.

2.3 Mastication Patterns

These were obtained from electromyograph (EMG) records using surface electrodes placed on the skin overlying the masseter and temporalis muscles bilaterally, as described previously.[1] The number of chews and chewing time were determined from each record.

2.4 Sensory Assessment

The subjects were asked to record their flavour perception continuously over time by moving a pointer on a sliding potentiometer along a line scale labelled 'Not at all' to 'Extremely' (left to right) for flavour perception. The voltage output from the potentiometer was input into the same data acquisition system (1401 plus, Cambridge Electronic Design, U.K.) as the EMG data which allowed synchronisation of the two types of records. Each subject attended for three recording sessions on separate days. TI curves were averaged using the method of Liu and MacFie.[2]

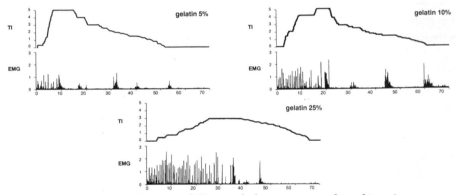

Figure 1   *Mastication patterns and TI records for one session for subject 4*

3 RESULTS AND DISCUSSION

Figure 1 shows the mastication pattern for a single subject for the soft, medium and hard gelatin samples. Each peak on the EMG trace indicates a single chew and the difference between the samples is reflected in the mastication patterns for these samples. The shortest chewing time occurs with the softest sample. This has a low melting point and the sample melts rapidly in the mouth through heat transfer within the matrix as it is deformed and comminuted. Conversely the hardest sample is comminuted progressively during a prolonged mastication phase. Since the melting point of this sample is above mouth temperature, complete melting does not occur, although as sucrose is leached from the exposed surfaces of the gel fragments, some melting of these surfaces occurs providing a smooth viscous feel to the particles. The associated TI curves for this subject demonstrate for the gels, which all contained the same flavour concentration, an intense and early perception of flavour for the softest sample and a less intense and prolonged perception for the hardest gel.

Figure 2 shows the averaged TI curves for the softest and hardest sample for each of the eight subjects. Clearly perceptual differences are evident among the subjects concerning the flavour of these two samples. This is reflected in the relative positions of I_{max} (the maximum intensity of flavour perceived) and T_{max} (the time of maximum perceived flavour) of their TI curves.

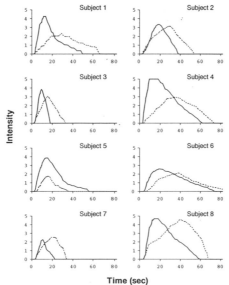

Figure 2 *Time intensity records for 5% gelatin (solid line) and 25% gelatin (dashed line) averaged across three repeats for each subject*

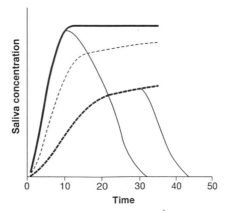

Figure 3 *Representation of flavour release into saliva for sample which melts rapidly (solid thick line) and which is comminuted slowly (dashed thick line) and rapidly (dashed thin line). Solid thin lines indicate hypothetical decay of flavour concentration after swallowing*

The mechanism for flavour release differs for the two extreme samples. For the soft sample, melting of the gelatin matrix quickly liberates the flavour components, resulting in a rapid increase in flavour concentration in the saliva. For the hard sample, flavour is released more slowly from the exposed surfaces. However as the sample is comminuted, the surface area for release is increased progressively, and the rate at which flavour is released increases accordingly. Thus the flavour concentration in the saliva with respect to time for the two samples could be represented by the thick lines in Figure 3. If we assume that flavour is released into the air phase in the mouth and transferred to the olfactory receptors in amounts proportional to the saliva concentration, we can begin to interpret some of the individual differences in perception. For these aqueous samples, the maximum salivary concentration of flavour occurs at the point at which the sample is swallowed since this action effectively removes the reservoir of flavour from the mouth. For each subject, the times of swallowing each of the samples were averaged for the three repetitions and are given in Table 1. In general the time of maximum perceived flavour intensity occurred slightly later than the time of swallowing (on average 4 s later for the soft sample and 1 s later for the hard sample).

Table 1   *Mean Chewing Times for each Subject (n=3)*

Sample	Subject							
	1	*2*	*3*	*4*	*5*	*6*	*7*	*8*
5% gelatin	5	7	8	10	4	8	6	24
25% gelatin	20	22	28	30	10	22	23	50

In Figure 3, an arbitrary scale has been marked on the time axis to represent the approximate time scale for the sensory responses in this experiment. By marking the time of swallowing of the soft sample on the thick continuous line in Figure 3, and that of the hard sample on the thick dashed line, for each subject, and assuming a gradual decline in salivary concentration thereafter, it is possible to obtain a good match between the hypothetical salivary concentration and the recorded TI patterns for the same subject. This has been indicated on the figure for subject 4 by the thin continuous lines.

The similarity between the projected salivary concentration and perceived flavour intensity is better for some subjects than for others. However the situation so far described assumes similar rates of oral breakdown of the samples by all subjects. Although this may be appropriate for the soft sample which relies more on oral temperature than mechanical breakdown, this is less likely for the harder sample. Individuals differ considerably in the efficiency with which they comminute foods.[3-7] A second dashed line on Figure 3 is shown to represent the salivary concentration of flavour for a more rapid rate of increase of surface area of the sample, representing greater efficiency in comminution. For some subjects, particularly subject 3 and 7, a closer representation of their TI curves is provided by substituting this curve for the thick dashed line.

In a study on cheeses, Jack *et al.*[8] reasoned that all subjects masticated samples to a common textural endpoint and normalised TI data in the time dimension only. Applying a similar method to the data in this study for the soft and hard samples separately (since the endpoint these two samples is clearly different) would not eliminate the individual differences in the intensity of flavour perception recorded.

It is widely reported that individuals differ in their sensory perceptions of food flavour, and in particular in the temporal aspects of these perceptions. These studies are providing information to aid our interpretation of these differences and our understanding of the mechanisms underlying such differences. The flavour of a food does not rely simply on the concentrations of the various volatile components within it, but is greatly affected by the variety of ways consumers interact with different types of food matrices during mastication.

REFERENCES

1. W.E. Brown, *J. Texture Studies*, 1994, **25**, 1.
2. Y.-H. Liu and H.J.H. MacFie, *Food Technology*, 1992, **46**, 92.
3. R.S. Manly and L.C. Braley, *J. Dent.*, 1950, **29**, 448.
4. E. Helkimo, G.E. Carlsson and M. Helkimo, *Acta Odont. Scand.*, 1977, **36**, 33.
5. J. Eldund and C.J. Lamn, *J. Oral Rehab.*, 1980, **7**, 123.
6. P.W. Lucas and D.A. Luke, *Br. Dent. J.*, 1985, **159**, 401.
7. W.E. Brown, C. Dauchel and I. Wakeling, *J. Texture Studies*, 1996, in press.
8. F.R. Jack, J.R. Piggott and A. Paterson, *J. Food Sci.*, 1994, **59**, 539.

SECTION 8

WORKSHOPS

WORKSHOP ON CURRENT AND FUTURE PROBLEMS IN FLAVOUR RESEARCH

Harry Nursten[*] and Gary Reineccius[†]

[*]Department of Food Science and Technology, University of Reading, Whiteknights, Reading, RG6 6AP, U.K.

[†]Department of Food Science and Nutrition, University of Minnesota, St. Paul, MN 55108, U.S.A.

1 INTRODUCTION

This workshop was conducted on two afternoons with about 75 participants in total. Four topics were discussed:
1. The problem of flavour deterioration during storage;
2. Criteria for choosing the key components of a food aroma;
3. International coordination of flavour meetings; and
4. The vocabulary and standards for aroma description.

2 FLAVOUR DETERIORATION DURING STORAGE

The discussion on flavour deterioration during storage focused on the mechanisms responsible, including aroma loss through evaporation, reaction of aroma constituents with the food base or other aroma constituents and 'masking' due to the appearance of undesirable and other aromas. The issue of aroma reaction became the focus in this discussion. Data were presented by Reineccius, demonstrating the loss of aroma constituents when a model system was stored at elevated temperatures in the presence of microcrystalline cellulose. A large but variable decrease in headspace concentration of all aroma compounds tried was observed in the first 3–4 hours of storage. After some discussion, it was agreed that the rapid rate of headspace loss was likely to have been due to sorption of aroma compounds by the cellulose. This conclusion was based on the loss of all aroma constituents, including those one would consider inert (e.g., nonane), and the very rapid initial rate of change, followed by a slower rate of aroma loss from the headspace, coupled with the selective nature of the continued loss. Rizzi pointed out that interactions should be slow, since aroma components are rarely present > 100 ppm. However, the losses of furfurylthiol, 1-methylpyrrole and pyrrole continued during subsequent storage and were dramatic. Furfurylthiol was lost in less than two days at 37 °C. This was considered as due to radical mechanisms. The discussion continued with speculations about possible mechanisms of aroma compound loss, the instability of furfurylthiol, 2-acetyl-thiazoline and furaneol being specially mentioned, and resulted in recommendations for study in this area to be continued.

3 KEY VOLATILES

The second topic centred on how best to focus on the key aroma compounds in foods for further study. While there is little disagreement that the use of odour activity values (OAV) is a part of this process, it is not clear what steps should be taken subsequently. Data were presented by Reineccius on odour thresholds and intensity *vs.* concentration plots for vanillin, furaneol, skatole and furfurylthiol. It was noted that, at equal sensory intensities, vanillin was present in solution at an OAV of 1585, furaneol at 318, skatole at 63 and furfurylthiol at 63. These data demonstrate that OAVs do not realistically rank aroma compounds by intensity (and therefore odour importance) in a food. We are then left with the question of how to establish priorities among odorants for further study.

Grosch discussed the procedure now adopted in his laboratory. OAVs are used to select odour-active compounds for further study, but not to assign any general ranking of importance. First, sensory analysis is used to determine the attributes/descriptors applicable to the given food. Then the analyst goes through the list of odour-active components and chooses compounds which display the sensory attributes selected by the sensory panel. OAVs are used to choose which compound is most likely to be the key odorant for a given attribute. For example, if 'nutty' notes were found in the product by the sensory panel, one would look through the odour-active compounds for nutty components. The compound with the highest OAV for nutty character would be chosen to represent this note. One would subsequently do the same for all the other sensory notes characteristic of the food until representative compounds for all the sensory notes have been selected. These compounds would then be used to formulate a model mixture for sensory evaluation. The model mixture would be compared to the real food product for similarity (or difference). If the model system differed from the food product, additional sensory work would be done to determine how the model system differed and the analyst would go back to adjust compound concentrations, remove unwanted notes, and/or add missing ones. One may even go back and do additional identifications and OAV determinations. In the end, one should have an aroma very similar to the real food system.

Additional discussion followed, pointing out that we have little information on power–function relationships characterising sensory intensity for aroma compounds in foods or knowledge of interactions between aroma compounds in foods. The problem of adequately evaluating the contribution of an odorant as an individual compound eluting from a gas chromatograph as opposed to its existence in the food matrix as part of a complex aroma profile is well recognised. Attention was also drawn to compounds, whose odour changes with concentration, such as damascenone.

4 INTERNATIONAL CO-ORDINATION OF FLAVOUR MEETINGS

The workshop organisers voiced concern that we have several international meetings which focus on the flavour of foods with little global co-ordination of timing or topics to be discussed. For example, in 1997, there will be flavour symposia in Florida (American Chemical Society meeting), in Greece (Charalambous Symposium), and in Germany (Wartburg Symposium), not to mention the (partly aroma) Maillard Symposium in London. The occurrence of three or more flavour-related symposia in one year taxes the ability to present new information as well as travel funds. Thus it was suggested that an effort be made to develop a means of informing each other of times and topics for flavour-related meetings. It was noted that the Flavor Subdivision of the American Chemical Society is forming a World Wide Web page, which could be used as a location for links to other international meetings. This was left to be investigated at the 1996 A.C.S. Meeting in Orlando, Florida.

5 VOCABULARY FOR SENSORY DESCRIPTORS AND STANDARD STIMULI

The historical development of a standardised vocabulary for sensory analysis of foods was presented as background. It was noted that Harper *et al.*[1] developed a list of 44 universally applicable descriptors, which was later expanded to 146 by Dravnieks.[2] Nursten expressed concern that several of the reference compounds listed to represent particular sensory descriptors were not readily available (2,2-dimethylpropanethiol, geranial), raised questions of toxicology (musk xylol), and/or were not necessarily the best, nor the most stable, representative of a particular sensory descriptor (herbal: acetaldehyde; musty: skatole). He felt that the descriptors, along with the corresponding odorants, are not being applied as widely as perhaps they should be. They had been developed in the hope that their use would standardise and thus simplify the interpretation of sensory work within and between laboratories. The discussion which followed the presentation was certainly not unanimous, but it was generally agreed that the use of such a list does facilitate the co-ordination of sensory work from different laboratories. However, concern was expressed regarding several areas. It would clearly be contradictory to restrict the choice of vocabulary of a sensory panel involved in free-choice profiling, but, if desired, one could readily incorporate the profiles of standard stimuli in the subsequent multidimensional analysis, thus facilitating between-study and between-laboratory communication. Another concern was the international use of a vocabulary based only on one language (English). Often a word does not have an equivalent in a different language and thus is not readily understandable in that language. Conversely, some languages have more extensive vocabularies relating to specific qualities. The last problem expressed was the difficulty of requiring a panellist, particularly one doing gas chromatography–odour assessment, to quickly choose a word from the list rather than permit selection of the word that most readily comes to mind at the moment of sniffing. It was generally agreed that a standardised vocabulary should be used judiciously, with the awareness that its application is not appropriate in several situations. A final comment was made by Acree, drawing attention to the fact that the American Society for Testing and Materials is about to publish a new lexicon of about 800 sensory descriptors.

REFERENCES

1. Harper *et al.*, *Br. J. Psychol.*, 1968, **59**, 231.
2. A. Dravnieks, *Science*, 1982, **218**, 799.

ELECTRONIC NOSE WORKSHOP

J. Mlotkiewicz[*] and J.S. Elmore[†]

[*]Dalgety Food Technology Centre, Station Road, Cambridge, CB1 2JN, U.K.

[†]Department of Food Science and Technology, University of Reading, Whiteknights, Reading, RG6 6AP, U.K.

1 INTRODUCTION

The electronic nose has been commercially available for approximately five years. However, very few publications containing data acquired using an electronic nose have appeared during this time. The aim of the workshop was to discover whether scientists were acquiring meaningful data using the electronic nose or whether there were problems with the equipment that needed to be addressed by the manufacturers. To this end, a 'devil's advocate' approach was used, in order to stimulate discussion among the workshop participants.

Two presentations were given. Jurek Mlotkiewicz suggested that there was a need for an aroma measurement device that was reliable, specific, stable, rugged, sensitive and cheap. He displayed claims made by the manufacturers, that the electronic nose was fast, reproducible, reliable and easy to use. He also presented data from work that had been carried out at Dalgety, which showed some element of discrimination in dry samples but highlighted problems with humidity interference in wet ones. There were also limitations in the statistical packages provided by the manufacturer, which meant that additional statistical methods were warranted, such as stepwise discriminant analysis, in order to fully exploit the results obtained. Other workers had shown that there were carry-over effects from sample to sample and the sensitivity of the sensors was not adequate. The conclusion was that there were discrepancies between the proposed ideal machine and what was available.

Thierry Talou, from the National Polytechnic Institute of Toulouse, gave an overview of the evolution of the electronic nose from early tin oxide gas sensors to surface acoustic waves and quartz piezo-electric sensors. He described how each of the different types of sensors worked and compared the different systems available, showing how the sample was introduced into each instrument. He reported on how the electronic nose had been successfully used for measuring migration of taints from packaging. In concluding, he suggested that the electronic nose was still evolving as an aroma analysis technique.

2 DISCUSSION

The discussion from the floor was illuminating. There was uncertainty about how the machine should be used and it was felt that this was a case of 'technology push' rather than 'market pull', i.e., the machine had been developed without an existing demand for it. There were worries about deterioration in the performance of the sensors with time and whether or not replacement sensors would be as reproducible as the original sensor. The sensors themselves were regarded as not being appropriate for aroma work and sensors were not specific enough.

Major improvements would be the development of sensors (a) which were not affected by humidity and (b) which were more specific. The machine could only be considered useful when sensors that were not affected by humidity had been developed. Reproducible sensors that could be bought off the shelf and plugged in when needed seemed an obvious requirement.

Another technique was available that could be used instead of the electronic nose. HS–MS, where the sample headspace is directly introduced into the mass spectrometer, gives a specific fingerprint at a similar cost to an electronic nose. As most flavour laboratories already possess GC–MS, this would appear to be a more practical solution at present.

It was concluded that there was a need for an instrument that used a sensor or sensor array to measure the aroma of a food. However, many concluded that what was needed was a rapid portable aroma discrimination technique, which would be used as a quality control device, rather than a research tool. This instrument could be fitted with different sensors at different times, in order to be used for different applications, for example, to determine if a fruit was ripe or if a beer contained an off-note. If such a machine could evolve from sensor technology, it would have many applications in the food industry.

THE CHEMICAL NATURE OF NATURALNESS

B.D. Baigrie

Reading Scientific Services Limited, Lord Zuckerman Research Centre, The University of Reading, Whiteknights, Reading, RG6 6LA, U.K.

1 INTRODUCTION

When asked what a customer would expect of a natural fruit flavour, the consensus view of delegates was that the consumer would expect it to have been derived from the named fruit. It was also agreed that, given a choice, a consumer would prefer a safe (added) flavour to a flavour defined as natural and that safety was the over-riding consideration.

2 EXISTING LEGISLATION

The initial discussion focused on the legal definitions of natural flavours. The definition under European law highlighted specific *physical* processes *e.g.* distillation and solvent extraction, by which flavours and flavouring preparations could be obtained. The latter included essential oils, concentrated essential oils, absolutes, tinctures and juices. Processes such as enzymolysis and hydrolysis were deemed to be physical processes, and therefore allowed under E.C. law. If the latter was not allowed, it was pointed out, then lime oil, for example, could not be natural. A separate definition of process flavours was required for materials such as Maillard reaction flavours.

The U.S. F.D.A. definition of natural flavours was examined and it was pointed out that these regulations did not exclude the use of non-food materials as a source of natural flavours.

3 EXAMPLES OF INCONSISTENCIES

Two examples were used to highlight the inconsistencies and anomalies that can arise from these legislative definitions. The first paradigm was the production of ethyl butanoate from natural ethanol and natural butanoic acid. Would the ethyl butanoate be deemed natural under current legislation? If catalysts of plant or natural origin *e.g.* enzymes, natural mineral acidic clays, *etc.*, were used, then probably the ester was natural. If strong mineral acid were used to catalyse the esterification reaction than the ethyl butanoate would not be deemed natural. Analysis of ethyl butanoate by SNIFF NMR or isotope ratio mass-spectrometry might possibly allow a decision to be made. The consensus of the group was that it would be difficult to prove naturalness unequivocally in this instance.

The second example used to highlight the difficulties in assaying naturalness was the production of 'natural' α-ionone from readily available natural citral and natural acetone via condensation and acid-catalysed cyclisation of the intermediate pseudo-ionone. The product from this reaction, deemed to be natural since it is produced from natural starting materials, actually exists as a racemic mixture. Yet it is known that α-ionone exists exclusively in nature as the (*R*-) enantiomer. Chiral analysis of the α-ionone produced from natural starting materials would conclude it was synthetic or, at least, not authentic. Other topics such as the use of synthetic substrates to feed micro-organisms, the listing of chemicals found in nature, *e.g.* allyl heptanoate, as an artificial substance, the use of genetically modified enzymes to produce natural flavours were discussed at some length.

4 CONCLUSION

The general conclusion drawn from the two workshop sessions was that because current legislation was so broad and non-specific it was often difficult to judge unequivocally whether a substance was natural, there being no case law in the U.K. or U.S.A. to rely on for precedent. Because of its ambiguity, it was agreed that the use of the term 'natural' to describe flavouring should be discontinued.

Author Index

Note: Names have been indexed under the initial letter of the main part of the surname. Thus, ter Burg is found under "B", de Jong under "J", van der Schaft under "S", *etc.*

Subject Index